SOLAR VARIABILITY AND PLANETARY CLIMATES

Springer

INTERNATIONAL SPACE SCIENCE INSTITUTE

Space Sciences Series of ISSI

Space Sciences Series of ISSI

Volume 23

The International Space Science Institute is organized as a foundation under Swiss law. It is funded through recurrent contributions from the European Space Agency, the Swiss Confederation, the Swiss National Science Foundation, and the University of Bern. For more information, see the homepage at http://www.issibern.ch

# SOLAR VARIABILITY AND PLANETARY CLIMATES

*Edited by*

Y. CALISESI
*International Space Science Institute (ISSI),*
*Bern, Switzerland*

R.-M. BONNET
*International Space Science Institute (ISSI),*
*Bern, Switzerland*

L. GRAY
*Reading University,*
*Reading, UK*

J. LANGEN
*European Space Agency – European Space*
*Research and Technology Centre (ESA-ESTEC)*
*Noordwijk, The Netherlands*

M. LOCKWOOD
*CCLRC Rutherford Appleton Laboratory,*
*Didcot, UK*
*and*
*University of Southhampton,*
*Southhampton, UK*

Reprinted from *Space Science Reviews*, Volume 125, Nos. 1–4, 2006

Springer

INTERNATIONAL SPACE SCIENCE INSTITUTE

Space Sciences Series of ISSI

A.C.I.P. Catalogue record for this book is available from the Library of Congress

ISBN: 978-0-387-48339-9

Published by Springer
P.O. Box 990, 3300 AZ Dordrecht, The Netherlands

Sold and distributed in North, Central and South America
by Springer,
101 Philip Drive, Norwell, MA 02061, U.S.A.

In all other countries, sold and distributed
by Springer,
P.O. Box 322, 3300 AH Dordrecht, The Netherlands

www.springer.com

*Printed on acid-free paper*

To the memory of

Yoram J. Kaufman (1948–2006)

## TABLE OF CONTENTS


PAUL J. CRUTZEN / Foreword 1–3

J. D. HAIGH / Introduction 5–15

**SECTION I: SOLAR OUTPUT VARIABILITY**

R.-M. BONNET / Assessing Solar Variability: *Introductory Paper* 17–24

S. K. SOLANKI and N. A. KRIVOVA / Solar Variability of Possible Relevance for Planetary Climates 25–37

G. ROTTMAN / Measurement of Total and Spectral Solar Irradiance 39–51

C. FRÖHLICH / Solar Irradiance Variability Since 1978: *Revision of the PMOD Composite during Solar Cycle 21* 53–65

J. BEER, M. VONMOOS and R. MUSCHELER / Solar Variability Over the Past Several Millennia 67–79

B. HEBER, H. FICHTNER and K. SCHERER / Solar and Heliospheric Modulation of Galactic Cosmic Rays 81–93

M. LOCKWOOD / What do Cosmogenic Isotopes Tell us About Past Solar Forcing of Climate? 95–109

**SECTION II: TROPOSPHERIC AEROSOLS, RADIATION BUDGET, AND CHANGES**

A. OHMURA / Observed Long-Term Variations of Solar Irradiance at the Earth's Surface 111–128

U. LOHMANN / Aerosol Effects on Clouds and Climate 129–137

Y. J. KAUFMAN / Satellite Observations of Natural and Anthropogenic Aerosol Effects on Clouds and Climate 139–147

D. ROSENFELD / Aerosol-Cloud Interactions Control of Earth Radiation and Latent Heat Release Budgets 149–157

J. CURTIUS, E. R. LOVEJOY and K. D. FROYD / Atmospheric Ion-Induced Aerosol Nucleation 159–167

F. ARNOLD / Atmospheric Aerosol and Cloud Condensation Nuclei Formation: A Possible Influence of Cosmic Rays? 169–186

## SECTION III: CLIMATE CHANGE: DETECTION AND ATTRIBUTION


L. BENGTSSON / On the Response of the Climate System to Solar Forcing: *Introductory Paper* 187–197

W. J. INGRAM / Detection and Attribution of Climate Change, and Understanding Solar Influence on Climate 199–211

M. CRUCIFIX, M. F. LOUTRE and A. BERGER / The Climate Response to the Astronomical Forcing 213–226

H. GOOSSE and H. RENSSEN / Regional Response of the Climate System to Solar Forcing: The Role of the Ocean 227–235


## SECTION IV: MIDDLE ATMOSPHERIC RESPONSE TO SOLAR VARIABILITY


M. A. GELLER / Discussion of the Solar UV/Planetary Wave Mechanism: *Introductory Paper* 237–246

K. LABITZKE / Solar Variation and Stratospheric Response 247–260

M.-L. CHANIN / Signature of the 11-Year Cycle in the Upper Atmosphere 261–272

Y. CALISESI and K. MATTHES / The Middle Atmospheric Ozone Response to the 11-Year Solar Cycle 273–286

M. L. SALBY and P. F. CALLAGHAN / Influence of the Solar Cycle on the General Circulation of the Stratosphere and Upper Troposphere 287–303

S. BRÖNNIMANN, T. EWEN, T. GRIESSER and R. JENNE / Multidecadal Signal of Solar Variability in the Upper Troposphere During the 20th Century 305–317

K. KODERA / The Role of Dynamics in Solar Forcing 319–330

J. D. HAIGH and M. BLACKBURN / Solar Influences on Dynamical Coupling Between the Stratosphere and Troposphere 331–344

H. SCHMIDT and G. P. BRASSEUR / The Response of the Middle Atmosphere to Solar Cycle Forcing in the Hamburg Model of the Neutral and Ionized Atmosphere 345–356

L. J. GRAY, S. A. CROOKS, M. A. PALMER, C. L. PASCOE and S. SPARROW / A Possible Transfer Mechanism for the 11-Year Solar Cycle to the Lower Stratosphere 357–370


## SECTION V: RECENT SPACE DATA


J. LANGEN / Recent Space Data: *Introductory Paper* 371–379

C. H. JACKMAN, M. T. DELAND, G. J. LABOW, E. L. FLEMING and M. LÓPEZ-PUERTAS / Satellite Measurements of Middle Atmospheric Impacts by Solar Proton Events in Solar Cycle 23 381–391

A. HAUCHECORNE, J.-L. BERTAUX and R. LALLEMENT / Impact of Solar Activity on Stratospheric Ozone and $NO_2$ Observed by GOMOS/ENVISAT 393–402

M. LÓPEZ-PUERTAS, B. FUNKE, T. VON CLARMANN, H. FISCHER and G. P. STILLER / The Stratospheric and Mesospheric $NO_y$ in the 2002–2004 Polar Winters as Measured by MIPAS/ENVISAT 403–416

E. HILSENRATH, M. R. SCHOEBERL, A. R. DOUGLASS, P. K. BHARTIA, J. BARNETT, R. BEER, J. WATERS, M. GUNSON, L. FROIDEVAUX, J. GILLE and P. F. LEVELT / Early Data from Aura and Continuity from Uars and Toms 417–430

**SECTION VI: PLANETARY SCIENCE**

R.-M. BONNET / What Do We Know About the Climate of Terrestrial Planets? *Introductory Paper* 431–434

J.-L. BERTAUX / Solar Variability and Climate Impact on Terrestrial Planets 435–444

F. W. TAYLOR / Climate Variability on Venus and Titan 445–455

F. MONTMESSIN / The Orbital Forcing of Climate Changes on Mars 457–472

1
2
3
4
5
6
7
8
9
10
11
12
13
14
15
16
17
18
19
20
21
22
23
24
25
26
27
28
29
30
31
32
33
34
35
36
37
38
39
40
41
42
43
44
45
46
47
48

GROUP PICTURE

## SOLAR VARIABILITY AND PLANETARY CLIMATES ISSI WORKSHOP, JUNE 6–10, 2005, BERN, SWITZERLAND

1. Veerabhadran Ramanathan
2. Stefan Brönnimann
3. Karin Labitzke
4. Paul Crutzen
5. Manuel López-Puertas
6. Donal Murtagh
7. Jean-Loup Bertaux
8. Franck Lefèvre
9. Daniel Rosenfeld
10. Yasmine Calisesi
11. Fred Taylor
12. Gerald Keating
13. Vittorio Manno
14. Atsumu Ohmura
15. Gary Rottman
16. Roger-Maurice Bonnet
17. Heinz Wanner
18. Joerg Langen
19. Charles Jackman
20. William Ingram
21. Alain Hauchecorne
22. Hartmut Graßl
23. Michael Mann
24. Hugues Goosse
25. Oleg Korablev
26. Grant Raisbeck
27. Robert Hudson
28. Marie-Lise Chanin
29. Lesley Gray
30. Yoram Kaufman
31. Ernest Hilsenrath
32. John Burrows
33. Mark Baldwin
34. Joanna Haigh
35. Murry Salby
36. Marv Geller
37. Hauke Schmidt
38. Michel Crucifix
39. Phil Jones
40. Lennart Bengtsson
41. Juerg Beer
42. Mike Lockwood
43. Claus Fröhlich
44. Bernd Heber
45. Kunihiko Kodera
46. Franck Montmessin
47. Franck Arnold
48. Ulrike Lohmann

Not on this picture: Christian Beck, Martyn Chipperfield, Joachim Curtius, Rosine Lallement, Sami Solanki.

Picture taken in front of Bern's Opera House by Saliba Saliba.

# FOREWORD

With a global average irradiance of 342 W/m$^2$, the Sun is by far the largest source of energy for planet Earth. In comparison, the internal energy produced by Earth itself is only about 0.087 W/m$^2$ (Pollack *et al.*, 1993), which in turn is 3.5 times larger than the 0.025 W/m$^2$ of heat produced by the burning of fossil fuels.

About 31% (31 units) of the solar energy which arrives at the top of the atmosphere is reflected back to space by scattering from clouds, aerosols, and the Earth's surface. Almost 20 units of solar radiation are absorbed in the atmosphere. The remaining 49 units are absorbed at the surface. Evaporation of water at the Earth's surface consumes 23 units, and 7 units are transferred to the atmosphere by heat conduction. On balance 19 units are lost from the Earth's surface as infrared radiation, however consisting of 114 going upward and 95 returning from the atmosphere to the Earth's surface (see also Rosenfeld, 2006). There is thus a sixfold recycling of energy. This is the greenhouse effect, established by the presence of water vapor and of other greenhouse gases, $CO_2$, $CH_4$, $N_2O$, and CFCs, in the atmosphere. Due to human activities the latter have been increasing in the atmosphere, causing climate warming through an energy imbalance of 2.5–3 W/m$^2$, more than 100 times larger than the heat released into the atmosphere by the burning of fossil fuels.

The greenhouse warming of the Earth is a well-known physical process, which is quantitatively relatively well understood except for the role of clouds (coverage, height distribution). At short wavelengths, the main uncertainty regarding quantitative understanding of Earth's climate is caused by clouds and particulate matter, which scatter an uncertain fraction of sunlight back to space. Particles affect Earth's radiation budget both directly by backscattering on the particles and indirectly, as they serve as condensation nuclei, thereby affecting the hydrological cycle, cloud distribution and physico-chemical properties, and Earth's albedo. Small shifts in the global distribution of cloud properties can have a major influence on Earth's radiation budget at short and long wavelengths. Assuming constant relative humidity, Crutzen and Ramanathan (2003) estimated that as much as half of the anthropogenic infrared greenhouse forcing could be annihilated by an increase in global albedo by enhanced backscattering of solar radiation by aerosol and clouds. This creates a dilemma for international environmental policy makers. For ecological and health reasons, it is clearly desired to reduce air pollution from the lower atmosphere in the form of particulate matter, especially sulfates. However, this cleaning enhances climate warming by reduction of scattered sunlight to space. This process has already been observed. At the workshop, A. Ohmura presented data showing for many stations a remarkably strong dimming of global short-wave radiation

Space Science Reviews (2006) 125: 1–3
DOI: 10.1007/s11214-006-9041-x

in Europe and Asia by 0.7 W/m$^2$ per year from 1960 until 1980, followed by a brightening of similar magnitude for 75% of the stations, causing climate warming (Ohmura, 2006). The latter period coincides with strong reductions in the emissions of $SO_2$ to the atmosphere by 2.7% per year in the industrial world (Stern, 2005). To counteract the policy makers' dilemma, one proposal, which has been made, is to produce small amounts of sulfate particles (1–2% of the amount that is emitted in the troposphere), in the stratosphere, where it would reflect some of the incoming solar radiation to cool the planet (Crutzen, 2006). This idea is much debated and is in need of much research to make sure that no major environmental side effects exist.

Much of the discussion of the ISSI workshop in Bern during the week of June 6–10, 2005 was devoted to issues like the variability and cause of the total solar irradiance, and its wavelength dependence. The former exerts mainly an influence on dynamic and thermodynamic processes, the latter on chemistry, decreasing in relative importance from the top to the bottom of the atmosphere (thermosphere, mesosphere, stratosphere, troposphere). The solar radiation output over the past 11-year solar cycles has been varying by a mere 0.1%, too small to have an easily observable direct influence on climate, due to the buffering by internal processes, which are particularly strong in the troposphere. Nevertheless, in the stratosphere there have been observable solar cycle influences (e.g., Labitzke, 2006; Chanin, 2006) and it is justified to look for a coupling to the troposphere, using certified data and proper statistical analysis. A promising mechanism for coupling the dynamics of stratosphere and troposphere was presented by Haigh and Blackburn (2006).

Several papers presented at the workshop considered the effects of solar and galactic cosmic rays (GCR) on the upper atmosphere. For the former, there exist plenty of examples of chemical effects on ozone due to the stratospheric and mesospheric production of NO and OH during solar proton events, which have a direct influence on ozone. Whether galactic cosmic rays may play a role in cloud formation is an unresolved issue and a topic for further research. However, there is a long way from initial ion formation effects by GCR and nucleation to an actual influence of GCR on cloud properties. Compared to the energy supplied by the Sun, the GCR fraction is only $2 \times 10^{-6}$, requiring highly efficient feedbacks to reach any global or regional significance.

As mentioned by S. Brönnimann at the workshop a very interesting point is why global average temperature did rise during the first half of the past century (Brönnimann *et al.*, 2006). Total solar irradiance increased by 0.8–1.5 W/m$^2$, or globally averaged by 0.2–0.375 W/m$^2$, while mean global temperatures went up by about 0.5 K. If the temperatures increase was mainly due to variations in solar radiation, it would indicate a high climate sensitivity of 1.3–2.5 K m$^2$/W. It is also possible that internal climate variability contributed to the temperature rise.

The influence of particles on the energy budget, climate and hydrological cycle of the atmosphere was discussed by several authors. The uncertainties are still major, in need especially of observations, *in-situ* and by satellites. Several satellite projects have been conducted during the past decade by European and U.S. groups. One of the undisputed leaders from the U.S. side was Yoram Kaufman, the project scientist of the Terra satellite from 1997–2001, and one of the participants and authors in the ISSI workshop. Yoram is no longer with us. He died tragically on May 31, 2006 after having been hit by a car while riding his bike on the campus of NASA Goddard Space Flight Center. The scientific community lost a brilliant and highly respected member, who was also highly appreciated as a human being. This book is devoted to the memory of Yoram. In his most recent paper with Ilan Koren (2006), published as a report in Science Express, using ground based data from the AERONET network, they showed an increase in cloud cover with an increase in the aerosol column concentration, and an inverse dependence on the aerosol absorption of sunlight (Kaufman and Koren, 2006). Clearly Yoram had so much more to discover. It wouldn't be.

Paul J. Crutzen
Max Planck Institute for Chemistry, Mainz, Germany
Scripps Institution of Oceanography
University of California, San Diego
La Jolla, California
U.S.A.

## References

Brönnimann, S., Ewen, T., Griesser, T., and R. Jenne: 2006, 'Multidecadal signal of solar variability in the upper troposphere during the 20th century', *Space Sci. Rev.*, this volume, doi: 10.1007/s11214-006-9065-2.

Chanin, M.-L.: 2006, 'Signature of the 11-year cycle in the upper atmosphere', *Space Sci. Rev.*, this volume, doi: 10.1007/s11214-006-9062-5.

Crutzen, P. J.: 2006, 'Albedo enhancement by stratospheric sulfur injections: A contribution to resolve a policy dilemma?', *Climatic Change*, in press.

Crutzen, P. J. and Ramanathan, V.: 2003, 'The parasol effect on climate', *Science* **302**, 1679–1680.

Haigh, J. D. and Blackburn, M.: 2006, 'Solar influences on dynamical coupling between the stratosphere and troposphere', *Space Sci. Rev.*, this volume, doi: 10.1007/s11214-006-9067-0.

Kaufman, Y. J. and Koren, I.: 2006, 'Smoke and pollution aerosol effect on cloud cover', *Science Express*, doi:10.1126/science.1126232.

Labitzke, K.: 2006, 'Solar variation and stratospheric response', *Space Sci. Rev.*, this volume, doi: 10.1007/s11214-006-9061-6.

Ohmura, A.: 2006, 'Observed long-term variations of solar irradiance at the earth's surface', *Space Sci. Rev.*, this volume, doi: 10.1007/s11214-006-9050-9.

Pollack, H. N., Hurter, S. J. and Johnson, J. R.: 1993, 'Heat flow from the earth's interior: Analysis of the global data set', *Rev. Geophys.* **31**, 267–280.

Rosenfeld, D.: 2006, 'Aerosol-cloud interactions control of earth radiation and latent heat release budgets', *Space Sci. Rev.*, this volume, doi: 10.1007/s11214-006-9053-6.

Stern, D. I.: 2005, 'Global sulfur emissions from 1850 to 2000', *Chemosphere* **58**, 163–175.

# INTRODUCTION

J. D. HAIGH
*Blackett Laboratory, Imperial College of Science, Technology and Medicine, London SW7 2BW, UK*
*(E-mail: j.haigh@imperial.ac.uk)*

(Received 21 July 2006; Accepted in final form 16 August 2006)

**Abstract.** This Introduction attempts to place the 2005 ISSI Workshop on "Solar Variability and Planetary Climates" in the context of the recent history of the subject and to identify key areas where progress has been, and is being, made in understanding the many complex issues involved.

## 1. Background

Variations in solar activity, at least as observed in numbers of sunspots, have been apparent since ancient times but to what extent solar variability may affect global climate has been far more controversial. The subject had been in and out of fashion, with connections between the Sun and climate intermittently proposed and dismissed, for at least two centuries until the early 1990s when a number of factors combined to again bring it to the forefront of scientific research. The main driving force was the international push to understand, and attribute causes to, apparent global warming. The need to distinguish between natural and anthropogenic causes of climate change led meteorologists, even those sceptical of solar-climate links, to accept that an objective analysis of the Sun's role was required.

In addition to this essentially socio-economical factor advances were being made in key scientific areas. The absolute radiometers launched on satellites in the late 1970s produced indisputable evidence that the total solar irradiance (TSI) reaching the Earth varied systematically through the 11-year sunspot cycle, thus relegating to history the term "solar constant". Space-borne ultraviolet spectrometers confirmed larger variability in that spectral region, and understanding of how the compensating effects of (dark) sunspots and (bright) faculae on the Sun's surface contribute to total irradiance allowed estimates of variability across the whole solar spectrum (Lean, 1991).

Based on the resulting quasi-empirical relationship between sunspot numbers and solar irradiance, calibrated by the satellite measurements, it became possible to "reconstruct" TSI back in time. Such reconstructions were produced for at least the three centuries back to the Maunder Minimum in sunspot activity (Hoyt and Schatten, 1993). Using proxy indicators for temperature, such as the width and density of tree rings or the accumulation rate of $^{18}O$ in ice cores, reconstructions of surface temperature were also being produced (e.g. 500 years of Northern


DOI: 10.1007/s11214-006-9042-9

Hemisphere summer temperatures by Bradley and Jones, 1993) allowing direct comparisons between measures of climate and solar activity.

Some such studies (e.g. Friis-Christensen and Lassen, 1991; Reid, 1991) appeared to show an astonishingly strong influence by the Sun on surface temperatures, although theoretical estimates (e.g. Schlesinger and Ramankutty, 1992; Kelly and Wigley, 1992) suggested that the magnitude of the variations in TSI were too small to produce such an effect and that increasing concentrations of greenhouse gases were more likely responsible for global warming.

Recognising these controversies Elizabeth Nesme-Ribes of the Paris Observatory initiated and organised a NATO Advanced Research Workshop in Paris in October 1993 (Nesme-Ribes, 1994). This was a ground-breaking event in bringing together solar physicists, paleoclimatologists, atmospheric physicists and climate modellers to discuss the state of the science and what was needed to advance understanding of the complex issues involved. In his summary of the workshop Eugene Parker wrote

> *"... the physics of the solar activity, with which irradiance changes appear to be associated, is too vague to provide much guidance at the present time. Like the global climate models of the terrestrial atmosphere, the solar models provide a framework for discussion, and for understanding the contribution of individual effects. But in neither case do the models allow assertion of what is important... Only precise observation over the coming decades, and careful analysis of the information already in hand, can establish the full nature of the variation in solar irradiance, the terrestrial climate and what connection there exists between them."* (Parker, 1994)

The next few years saw an expansion of activity in many of the relevant research areas, and in 1999 the International Space Science Institute (ISSI) hosted a workshop on Solar Variability and Climate so that scientists from across the range of disciplines could update each other on developments over the period 1993–1999. Many of the new studies posed as many questions as they answered, so that the controversy surrounding the subject appeared to expand rather than reduce. One problem area was identified in the production of a composite of TSI data from the individual satellite instrument records: two studies suggested very different secular trends (Willson, 1997; Fröhlich and Lean, 1998). Regarding longer timescales, understanding of how cosmogenic isotopes are deposited in biogeophysical reservoirs was being improved, allowing more confidence in the interpretation of these as proxy indicators of solar activity (Masarik and Beer, 1999). However, questions remained as to linking these measures of solar magnetic activity to TSI, and thus to climate forcing (Beer *et al.*, 1998).

One widely-referenced study was that of Svensmark and Friis-Christensen (1997) which presented a striking correlation between global cloud cover and galactic cosmic rays. The physical mechanisms producing such a relationship were not clear and this revived an interest in the role of atmospheric ionization in the production of cloud condensation nuclei (Turco *et al.*, 1998) and also in any relationship between the global electric circuit and cloud microphysics (Tinsley, 1996).

Simulations of the impact on climate of variations in TSI using general circulation models (GCMs) (e.g. Cubasch *et al.*, 1997), and statistical optimisation techniques (Stevens and North, 1996; Crowley and Kim, 1996, 1999; North and Stevens, 1998), revealed detectable signals of solar activity in surface temperature on decadal to century timescales, but a relatively minor role for the Sun in 20th century global warming.

Meanwhile the signal of the solar sunspot cycle in stratospheric temperatures, first found by Labitzke (1987), was found to persist over another 11-year period (van Loon and Labitzke, 1999) and the potential for changes in the stratosphere to influence the troposphere was being realized (Kodera, 1995). Atmospheric modellers, picking up from the earlier innovative study of Geller and Alpert (1980), investigated the potential for changes in middle atmosphere temperature and ozone, produced by variations in solar ultraviolet radiation, to affect the atmosphere lower down (Haigh, 1994, 1996, 1999; Rind and Balachandran, 1995; Shindell *et al.*, 1999).

The 1999 ISSI workshop included representations of all these topics and promoted extremely useful debates and an interchange of ideas. The volume of proceedings (Friis-Christensen *et al.*, 2000) not only presents the individual contributions to the meeting but also reports some of the discussions that took place and provides summaries of key areas.

The ISSI workshop on Solar Variability and Planetary Climates held in 2005, of which this volume presents the proceedings, provided a forum for an update on the scientific advances that had been achieved over the following six years. Some of the key developments are briefly outlined below.

## 2. Solar Irradiance

One major advance was the launch, in January 2003, of the SORCE satellite on which are mounted instruments for measuring TSI and solar spectral irradiance from the near-infrared through to soft X-rays (Rottman *et al.*, 2005). Spectral variability in the visible and near-infrared has not previously been measured from space so SORCE data are providing new information on the portion of the solar spectrum that is most strongly absorbed in the lower atmosphere. Temporal variations in TSI measured by the SORCE TSI instrument show close agreement with those from contemporaneous measurements made by VIRGO and ACRIM III (Fröhlich, 2006). However, inter-calibration of the instruments reveals large differences so that the absolute value of TSI appears to be less well constrained than it did before. Thus the construction of a TSI composite, and the existence of any long-term trends, remain controversial.

The spectral coverage of SORCE data, however, has afforded unprecedented characterization of irradiance variability on solar rotation time-scales, and this has provided a check on the models which are used to predict 11-year cycle variations in TSI based on the contributions of sunspot-darkening and facular brightening. On centennial time-scales, variations in irradiance have previously been deduced from solar activity proxies (such as geomagnetic indices or cosmogenic isotopes) or from analogies with "Sun-like" stars. More recently, however, TSI reconstructions have been derived from historic sunspot observations. Such derivations are based either on the position and magnetic field strength of active solar regions, combined with the distribution of magnetic flux in these regions calculated from the Michelson Doppler Imager images (Foster, 2004), or obtained from theoretical modelling of open and closed surface magnetic flux related to sunspots (Wang *et al.*, 2005; Solanki and Krivova, 2006). These are suggesting a rather smaller difference in solar output between the Maunder Minimum and the present (0.5–2.2 $Wm^{-2}$) than found in previous studies. While this suggests a smaller climate forcing by the Sun since the 17th century, it also implies that detection/attribution studies may have underestimated factors acting to amplify the effects of solar irradiance on surface temperatures.

On millennial timescales knowledge of solar variability is based on cosmogenic radionuclide records. Careful analysis of $^{10}Be$ in ice cores, including removal of the geomagnetic signal, has provided a record of solar activity dating back 10,000 years and comparison with paleoclimate data provides evidence for the influence of solar variability on climate (Beer *et al.*, 2006). However, a physical understanding of the relationship between the heliospheric magnetic fields which modulate the cosmic ray flux and solar irradiance has yet to be developed.

## 3. Observations of Solar Influence on the Atmosphere

Due to the relatively short period over which data have been available, and to problems with the intercalibration of different instruments, some uncertainties still remain in the response of stratospheric ozone to solar UV variability (Hood, 2004; Calisesi and Matthes, 2006). The 11-year solar cycle signal in ozone column amount has been established at 2–3%, but the vertical structure of the ozone response is less well known. Satellite measurements suggest a structure with symmetric upper stratospheric maxima in mid-latitudes and a minimum, possibly even a negative response, in the tropical middle stratosphere. This profile has not been entirely reproduced by models, but longer observational records are needed to establish whether the apparent observed structure is real or an artefact of the data analysis.

Multiple regression analysis (Haigh, 2003; Crooks and Gray, 2005) of zonal mean temperature records from the forty-year reanalysis datasets produced by the U.S. National Centers for Environmental Prediction (NCEP) and the European Centre for Medium-Range Weather Forecasts (ECMWF) has revealed a detectable

11-year solar-cycle influence in the troposphere. These show bands of warming in mid-latitudes when the Sun is more active suggesting a latitudinal widening of the Hadley cells. Careful analysis of upper air data dating back eighty years has broadly confirmed these findings (Brönnimann *et al.*, 2006). Furthermore, the patterns identified are similar to those predicted by atmospheric GCMs for the influence of enhanced solar UV, thus tending to confirm the existence of such an influence of the Sun on the climate. Several recent studies (e.g. Kodera, 2002; Boberg and Lundstedt, 2002; Thejll *et al.*, 2003; Haigh and Roscoe, 2006) have found evidence of a solar influence on polar modes of variability (the North Atlantic Oscillation, or Northern Annular Mode, and Southern Annular Mode) which may be important for two reasons: firstly it suggests that a significant climate response to solar variability may be found preferentially in certain geographical locations, and secondly it provides a potential route for the transfer of a solar stratospheric influence down to the troposphere (e.g. Black, 2002, see also Section 4).

Evidence for a solar influence on cloud has been more difficult to establish. The relationship proposed by Svensmark and Friis-Christensen (1997) between galactic cosmic rays and global total cloud cover was revised by Marsh and Svensmark (2000) in favour of an influence on low marine cloud. The reliability of the International Satellite Cloud Climatology Project (ISCCP) infrared low cloud dataset has, however, been questioned by Sun and Bradley (2002) and Kristjánsson *et al.* (2002) and, given the high natural variability of cloud and the difficulties involved in performing reliable measurements, it may require several decades of further data acquisition before a relationship between cloud and solar variability can be reliably established or refuted.

## 4. Atmospheric Dynamics and Modelling

Another major advance over the past few years has been the development of coupled chemistry climate models (CCMs) of the lower and middle atmosphere (e.g. Beres *et al.*, 2005; Schmidt *et al.*, 2006). These models do not, to date, include the ocean circulation, and so cannot be said to simulate the whole climate system, but they do provide a means whereby the interactions between radiative, chemical and dynamical processes which determine the structure and composition of the atmosphere may be investigated. This is particularly important in assessing the influence of solar variability on ozone, radiative heating rates and temperatures in the upper/middle atmosphere and how these may influence the climate of the lower atmosphere through dynamical coupling. Schmidt and Brasseur (2006), with the example of the HAMMONIA model, summarise the achievements of CCMs to date in broadly simulating the observed zonal mean temperature and ozone response to solar variability. They also indicate areas requiring further attention. One of these is the fully-interactive simulation of the quasi-biennial oscillation (QBO), which is technically feasible but requires high vertical resolution and thus considerable computational effort. Gray *et al.* (2006) review recent studies that have shown that perturbations to the equatorial upper stratospheric zonal wind field can be subsequently transmitted to the winter polar lower stratosphere. This provides a route for QBO and/or solar modulation of the tropical middle atmosphere to influence the lower stratosphere, but a fully-coupled model simulation of these effects has yet to be achieved. Another point stressed by Schmidt and Brasseur (2006) is the importance of considering the full 4D structure of the response to forcings because of the potential importance of wave effects. This, however, produces further challenges in terms of validating results against short and incomplete data records.

In the troposphere one of the key debates, as nicely outlined by Brönnimann *et al.* (2006), concerns whether the effect of solar variability is primarily "top-down", i.e., driven from the stratosphere, or "bottom-up", i.e., forced by solar heating of the surface. Brönnimann *et al.* (2006) suggest that these two mechanisms would have different impacts on the tropical Hadley circulation and should therefore be testable using observations of the tropical upper troposphere. Unfortunately the data are not yet of sufficient quality/quantity to test this intriguing idea.

In an analysis of NCEP Reanalysis temperature data, Salby and Callaghan (2006) find higher correlations with solar activity when the data are filtered at periods shorter than 5 years than those found using multiple linear regression analysis (see the discussion in Section 3). These fluctuations are associated with the QBO, and Salby and Callaghan (2006) suggest that solar modulation of the QBO may account for much of the solar signal in the lower atmosphere through a mechanism involving changes to the stratospheric Brewer-Dobson (BD) circulation. They show correlations between the BD circulation and the Hadley circulation and thus imply a relationship between the latter and the Sun. Kodera (2006) summarises earlier work on the impact of solar variability on the BD circulation, and suggests means whereby this might impact both convective activity in the tropical troposphere and planetary wave propagation in the winter sub-tropics.

To investigate the mechanisms whereby a perturbation to the stratosphere, due to whatever cause, might influence the dynamics of the troposphere, Haigh and Blackburn (2006) use a simplified atmospheric general circulation model and observe the time development of its fields to an imposed stratospheric heating. The results show that the initial effect of a change in static stability at the tropopause is to reduce eddy momentum flux convergence there, and that this is followed by a transfer of the momentum anomaly to the ground. This suggests that stratospheric influence on synoptic-scale wave propagation is key.

## 5. Clouds and Aerosols

The suggestion that atmospheric ionisation by galactic cosmic rays (GCRs) might influence cloud cover, and that through this mechanism cloud might respond to solar variability, was a key factor in promoting the 1999 ISSI workshop. Although it seems to have proved difficult since that date to accrue additional observational evidence of any GCR-cloud link (see the discussion in Section 3 above), there have been several theoretical, laboratory and observational studies designed to investigate the processes potentially involved. Important factors include whether or not ionised aerosol particles act preferentially as condensation nuclei (CN) and, if they do, whether the resulting CN can grow fast enough to have significant impact on the number of cloud droplets, given the large number of other candidate aerosols in the atmosphere (see, e.g., the review by Harrison and Carslaw, 2003).

In parallel with this activity there has been a growing concern with the role of aerosols, produced either by human activity or by natural factors unrelated to solar variability, in the radiative forcing of climate change. While the direct effect of such particles on the Earth's radiation budget, mainly in reflecting solar radiation back to space, are relatively well known, the possible indirect effects, through modification of cloud, are complex and uncertain. Proposed indirect effects include an increase in the number density and a decrease in the mean size of cloud droplets as well as an increase in cloud lifetime, in response to higher aerosol concentrations. Estimates of the magnitude of these effects suggest they could significantly increase the global albedo, and thus act to decrease the global warming due to greenhouse gases.

There are clear overlaps between the scientific understanding required by the "GCR-cloud" and "aerosol-climate" communities, and thus the 2005 ISSI workshop welcomed a number of leading authorities in cloud and aerosol physics to share their expertise. Satellite data are being used to provide information on atmospheric aerosol loading and properties, including their provenance, while studies of correlations between cloud albedo and aerosol loading provide empirical observational evidence of the aerosol effect on cloud (Kaufman, 2006). The picture emerging from modelling studies (Lohmann, 2006) is that since pre-industrial times, the effect of aerosols on cloud could have significantly offset global warming, but there are large uncertainties in all the input parameters. Based on examination of pyranometer measurements from about 400 ground stations over more than 50 years, Ohmura (2006) offers the intriguing possibility of the quantitative simultaneous evaluation of the effects of solar irradiance, aerosol loading, cloud and temperature on solar radiation at the ground.

Ion-induced nucleation, and subsequent particle growth, have been studied using laboratory measurements of the thermodynamics of the clustering sulphuric acid and water molecules on sulphate ions together with theoretical models and atmospheric observations (Arnold, 2006; Curtius *et al.*, 2006). These studies suggest that ion-induced nucleation is probably important in the cold upper troposphere but that subsequent particle growth is constrained by ambient concentrations of gaseous sulphur dioxide.

## 6. Conclusions

The influence of solar variability on climate is an exciting and flourishing area of research: it involves scientists from a breadth of disciplines, uses measurements from a wide range of observational campaigns and addresses an issue of significant environmental concern. The articles in this volume describe the advancements in knowledge and understanding that have been made over the past few years in key scientific fields but also identify several areas where questions remain to be answered.

One important issue is to establish the magnitude of any secular trends in total solar irradiance (TSI). This may be achieved by careful analysis and understanding of the satellite instruments involved in collecting data over the past two-and-a-half solar cycles, and must be continued through analysis of data from current and new satellites. For longer periods it requires a more fundamental understanding of how solar magnetic activity relates to TSI. This would not only facilitate more reliable centennial-scale reconstructions of TSI, from e.g., sunspot records, but also advance understanding of how cosmogenic isotope records may be interpreted as historical TSI.

With regard to the climate, further data-mining and analysis are required to firmly establish the magnitude, geographical distribution and seasonality of its response to various forms of solar activity. Understanding the mechanisms involved in the response then becomes the overriding objective. Current ideas suggest three main avenues where further research is needed. Firstly, the means whereby solar radiative heating of the upper and middle atmosphere may influence the lower atmosphere through dynamical coupling needs to be better understood. Secondly, it needs to be established whether or not variations in direct solar heating of the tropical oceans can be of sufficient magnitude to produce apparently observed effects. Thirdly more work is needed on the microphysical processes involved in ion-induced nucleation, and, probably more importantly, the growth rates of the condensation nuclei produced.

Perhaps when these questions are answered we will be confident that we really understand how changes in the Sun affect the climate on Earth.

## References

Arnold, F.: 2006, 'Atmospheric aerosol and cloud condensation nuclei formation: A possible influence of cosmic rays?', *Space Sci. Rev.*, this volume, doi: 10.1007/s11214-006-9055-4.

Beer, J., Tobias, S., and Weiss, N.: 1998, 'An active sun throughout the Maunder Minimum', *Solar Phys.* **181**, 237–249.

Beer, J., Vonmoos, M., and Muscheler, R.: 2006, 'Solar variability over the past several millennia', *Space Sci. Rev.*, this volume, doi: 10.1007/s11214-006-9047-4.

Beres, J. H., Garcia, R. R., Boville, B. A., and Sassi, F.: 2005, 'Implementation of a gravity wave source spectrum parameterization dependent on the properties of convection in the Whole Atmosphere Community Climate Model (WACCM)', *J. Geophys. Res.* **110**, doi: 10.1029/2004JD005504.

Black, R. X.: 2002, 'Stratospheric forcing of surface climate in the Arctic Oscillation', *J. Clim.* **15**, 268–277.

Boberg, F. and Lundstedt, H.: 2002, 'Solar wind variations related to fluctuations of the North Atlantic Oscillation', *Geophys. Res. Lett.* **29**, doi: 10.1029/2002GL014903.

Bradley, R. S. and Jones, P. D.: 1993, 'Little Ice Age summer temperature variations: Their nature and relevance to recent global warming trends', *The Holocene* **3**, 367–376.

Brönnimann, S., Ewen, T., Griesser, T., and Jenne, R.: 2006, 'Multidecadal signal of solar variability in the upper troposphere during the 20th century', *Space Sci. Rev.*, this volume.

Calisesi, Y. and Matthes, K.: 2006, 'The middle atmosphere ozone response to the 11-year solar cycle', *Space Sci. Rev.*, this volume, doi: 10.1007/s11214-006-9063-4.

Crooks, S. and Gray, L. J.: 2005, 'Characterisation of the 11-year solar cycle using a multiple regression analysis of the ERA-40 dataset', *J. Clim.* **18**, 996–1015.

Crowley, T. J. and Kim, K. Y.: 1996, 'Comparison of proxy records of climate change and solar forcing', *Geophys. Res. Lett.* **23**, 359–362.

Crowley, T. J. and Kim, K. Y.: 1999, 'Modeling the temperature response to forced climate change over the last six centuries', *Geophys. Res. Lett.* **26**, 1901–1904.

Cubasch, U., Voss, R., Hegerl, G. C., Waszkewitz, J., and Crowley, T. J.: 1997, 'Simulation of the influence of solar radiation variations on the global climate with an ocean-atmosphere general circulation model', *Clim. Dyn.* **13**, 757–767.

Curtius, J., Lovejoy, E. R., and Froyd, K. D.: 2006, 'Atmospheric ion-induced aerosol nucleation', *Space Sci. Rev.*, this volume, doi: 10.1007/s11214-006-9054-5.

Foster, S. S.: 2004, 'Reconstruction of solar irradiance variations for use in studies of global climate change: Application of recent SoHO observations with historic data from the Greenwich observatory', PhD. thesis, School of Physics and Astronomy, University of Southampton, UK.

Friis-Christensen, E. and Lassen, K.: 1991: 'Length of the solar cycle: an indicator of solar activity closely associated with climate', *Science* **254**, 698–700.

Friis-Christensen, E., Fröhlich, C., Haigh, J. D., Schüssler, M., and von Steiger, R. (eds.): 2000, *Solar Variability and Climate*, ISSI Space Science Series, Vol. 11, Kluwer Academic Publishers.

Fröhlich, C.: 2006, 'Solar irradiance variability since 1978', *Space Sci. Rev.*, this volume, doi: 10.1007/s11214-006-9046-5.

Fröhlich, C. and Lean, J.: 1998, 'The Sun's total irradiance: Cycles, trends and related climate change uncertainties since 1976', *Geophys. Res. Lett.* **25**, 4377–4380.

Geller, M. A. and Alpert, J. V.: 1980, 'Planetary wave coupling between the troposphere and middle atmosphere as a possible sun-weather mechanism', *J. Atmos. Sci.* **37**, 1197–1215.

Gray, L. J., Crooks, S. A., Palmer, M. A., Pascoe, C. L., and Sparrow, S.: 2006, 'A possible transfer mechanism for the 11-year solar cycle to the lower stratosphere', *Space Sci. Rev.*, this volume.

Haigh, J. D.: 1994, 'The role of stratospheric ozone in modulating the solar radiative forcing of climate', *Nature* **370**, 544–546.

Haigh, J. D.: 1996, 'The impact of solar variability on climate', *Science* **272**, 981–984.

Haigh, J. D.: 1999, 'A GCM study of climate change in response to the 11-year solar cycle', *Quart. J. Roy. Meteorol. Soc.* **125**, 871–892.

Haigh, J. D.: 2003, 'The effects of solar variability on the Earth's climate', *Phil. Trans. Roy. Soc. A* **361**, 95–111.

Haigh, J. D. and Blackburn, M.: 2006, 'Solar influences on dynamical coupling between the stratosphere and troposphere', *Space Sci. Rev.*, this volume, doi: 10.1007/s11214-006-9067-0.

Haigh, J. D. and Roscoe, H. K.: 2006, 'Solar influences on polar modes of variability', *Meteorol. Zeit* **15**, 371–378.

Harrison, R. G. and Carslaw, K. S.: 2003, 'Ion-aerosol-cloud processes in the lower atmosphere', *Rev. Geophys.* **41**, doi: 10.1029/2002RG000114.

Hood, L. L.: 2004, 'Effects of solar UV variability on the stratosphere', in: J. M. Pap and P. Fox (eds.), *Solar Variability and its Effects on the Earth's Atmospheric and Climate System*, Geophysical Monograph **141**, 283–303.

Hoyt, D. V. and Schatten, K. H.: 1993, 'A discussion of plausible solar irradiance variations', *J. Geophys. Res.* **11**, 18,895–18,906.

Kaufman, Y. J.: 2006, 'Satellite observations of natural and anthropogenic aerosol effects on clouds and climate', *Space Sci. Rev.*, this volume, doi: 10.1007/s11214-006-9052-7.

Kelly, P. M. and Wigley, T. M. L.: 1992, 'Solar-cycle length, greenhouse forcing and global climate', *Nature* **360**, 328–330.

Kodera, K.: 1995, 'On the origin and nature of the interannual variability of the winter stratospheric circulation in the northern-hemisphere', *J. Geophys. Res.* **100**, 14,077–14,087.

Kodera, K.: 2002, 'Solar cycle modulation of the North Atlantic Oscillation: Implication in the spatial structure of the NAO', *Geophys. Res. Lett.* **29**, doi: 10.1029/2001GL014557.

Kodera, K.: 2006, 'The role of dynamics in solar forcing', *Space Sci. Rev.*, this volume, doi: 10.1007/s11214-006-9066-1.

Kristjánsson, J. E., Staple, A., Kristiansen, J., and Sassi, F.: 2002, 'A new look at possible connections between solar activity, clouds and climate', *Geophys. Res. Lett.* **29**, doi: 10.1029/2002GL015646.

Labitzke, K.: 1987, 'Sunspots, the QBO, and the stratospheric temperature in the north polar region', *Geophys. Res. Lett.* **14**, 535–537.

Lean, J.: 1991, 'Variations in the Sun's radiative output', *Rev. Geophys.* **29**, 505–535.

Lohmann, U.: 2006, 'Aerosol effects on clouds and climate', *Space Sci. Rev.*, this volume, doi: 10.1007/s11214-006-9051-8.

Marsh, N. D. and Svensmark, H.: 2000, 'Low cloud properties influenced by cosmic rays', *Phys. Rev. Lett.* **85**, 5004–5007.

Masarik, J. and Beer, J.: 1999, 'Simulation of particle fluxes and cosmogenic nuclide production in the Earth's atmosphere', *J. Geophys. Res.* **104**, 12,099–12,111.

Nesme-Ribes, E. (ed.): 1994, *The Solar Engine and its Influence on Terrestrial Atmosphere and Climate*, NATO ASI Series 1, vol. 25, Springer-Verlag.

North, G. R. and Stevens, M. J.: 1998, 'Detecting climate signals in the surface temperature record', *J. Clim.* **11**, 563–577.

Ohmura, A.: 2006, 'Observed long-term variations of solar irradiance at the Earth's surface', *Space Sci. Rev.*, this volume, doi: 10.1007/s11214-006-9050-9.

Parker, E. N.: 1994, 'Summary comments', in: Elizabeth Nesme-Ribes (ed.), *The Solar Engine and its Influence on Terrestrial Atmosphere and Climate*, Proceedings of the NATO Advanced Research Workshop held 25–29 October, 1993 in Paris, France. Springer Verlag, Berlin Heidelberg, 527 pp.

Reid, G.: 1991, 'Solar total irradiance variations and the global sea surface temperature record', *J. Geophys. Res.* **96**, 2835–2844.

Rind, D. and Balachandran, N. K.: 1995, 'Modeling the effects of UV variability and the QBO on the troposphere-stratosphere system. Part II: The troposphere', *J. Clim.* **8**, 2080–2095.

Rottman, G., Woods, T., and George, V. (eds.): 2005, *The Solar Radiation and Climate Experiment (SORCE)*. Springer-Verlag, Dordrecht, The Netherlands.

Salby, M. L. and Callaghan, P. F.: 2006, 'Influence of the solar cycle on the general circulation of the stratosphere and upper troposphere', *Space Sci. Rev.*, this volume, doi: 10.1007/s11214-006-9064-3.

Schlesinger, M. E. and Ramankutty, N.: 1992, 'Implications for global warming of intercycle solar irradiance variations', *Nature* **360**, 330–333.

Schmidt, H. and Brasseur, G. P.: 2006, 'Response of the middle atmosphere to solar cycle forcing in the Hamburg model of the neutral and ionized atmosphere', *Space Sci. Rev.*, this volume, doi: 10.1007/s11214-006-9068-z.

Schmidt, H., Brasseur, G. P., Charron, M., Manzini, E., Giorgetta, M. A., Diehl, T., Fomichev, V. I., Kinnison, D., Marsh, D., and Walters, S.: 2006, 'The HAMMONIA chemistry climate model: Sensitivity of the mesopause region to the 11-year solar cycle and CO2 doubling', *J. Clim.*, in press.

Shindell, D., Rind, D., Balachandran, N., Lean, J., and Lonergan, P.: 1999, 'Solar cycle variability, ozone, and climate', *Science* **284**, 305–308.

Solanki, S. K. and Krivova, N. A.: 2006, 'Solar variability of possible relevance for planetary climates', *Space Sci. Rev.*, this volume, doi: 10.1007/s11214-006-9044-7.

Stevens, M. J. and North, G. R.: 1996, 'Detection of the climate response to the solar cycle', *J. Atmos. Sci.* **53**, 2594–2608.

Sun, B. M. and Bradley, R. S.: 2002, 'Solar influences on cosmic rays and cloud formation: A reassessment', *J. Geophys. Res.* **107**, doi: 10.1029/2001JD000560.

Svensmark, H. and Friis-Christensen, E.: 1997, 'Variation of cosmic ray flux and global cloud coverage – A missing link in solar-climate relationships', *J. Atmos. Solar-Terrestr. Phys.* **59**, 1225–1232.

Thejll, P., Christiansen, B., and Gleisner, H.: 2003, 'On correlations between the North Atlantic Oscillation, geopotential heights, and geomagnetic activity', *Geophys. Res. Lett.* **30**, doi: 10.1029/2002GL016598.

Tinsley, B. A.: 1996, 'Solar wind modulation of the global electric circuit and apparent effects on cloud microphysics, latent heat release, and tropospheric dynamics', *J. Geomag. Geoelect.* **48**, 165–175.

Turco, R. P., Zhao, J. X., and Yu, F. Q. A.: 1998, 'A new source of tropospheric aerosols: Ion-ion recombination', *Geophys. Res. Lett.* **25**, 635–638.

van Loon, H. and Labitzke, K.: 1999, 'The signal of the 11-year solar cycle in the global stratosphere', *J. Atmos. Solar-Terrestr. Phys.* **61**, 53–61.

Wang, Y.-M., Lean, J. L., and Sheeley, N. R. Jr.: 2005, 'Modeling the Sun's magnetic field and irradiance since 1713', *Astrophys. J.* **625**, 522–538.

Willson, R. C.: 1997, 'Total solar irradiance trend during solar cycles 21 and 22', *Science* **277**, 1963–1965.

# ASSESSING SOLAR VARIABILITY
*Introductory Paper*

R.-M. BONNET
*International Space Science Institute, Hallerstrasse 6, CH-3012 Bern, Switzerland*
*(E-mail: rmbonnet@issibern.ch)*

(Received 14 June 2006; Accepted in final form 15 June 2006)

## 1. Introduction

The Sun provides the main energy input to the Solar System, which is tied up to its star through gravitation, and is the first receptacle of the Sun's radiation, of its coronal mass ejections and of its solar wind. Located at roughly 150 million km from the Sun, right in the middle of the habitable zone and inhabited, the Earth is sensitive to these solar inputs. They, as well as the way the Earth reacts to them, determine the mean temperature of our planet. It is therefore essential to properly understand their effects and how their variations might influence the Earth's climate and that of similar bodies such as Venus and Mars.

This theme has been addressed in the past in many instances and meetings or books including the ISSI book "Solar Variability and Climate" published in 2000 by Kluwer. Naturally, if we were sitting on the surface of a rocky asteroid, our "climate" would be simply determined by the total solar radiation and its variations and by the distance to the Sun. No atmosphere thereon, no ocean, no El Niño, no Gulf Stream would blur the effect of any direct influence of the Sun's variations on the surface temperature of that asteroid. Fortunately, the Earth is not an asteroid! It has oceans, an atmosphere, geologic and anthropogenic activities, which however seriously complicate the issue. The famous "faint Sun paradox" illustrates the difficulty of assessing the effects of solar variability on the climate: even when the young Sun was 30% fainter than it is today, the natural atmospheric greenhouse effect at that time was powerful enough to maintain the Earth's temperature sufficiently warm for life to develop. This illustrates the delicateness of the issue, showing that understanding the Sun's influence is not easy and is bound to be open to controversies, especially when the competing anthropogenic influences on the climate become so highly sensitive issues as they touch on the industrial and political management of the planet and of its resources!

Therefore, it should not be surprising that understanding the effect of solar variability on our climate is a recurrent topic. The issue is indeed complex and has to be re-visited as we witness continuous progress in the acquisition of new and more accurate data, in the modeling of the Earth's atmosphere, and in a better understanding of how its different layers couple to each other down to the surface. This


DOI: 10.1007/s11214-006-9043-8

ISSI workshop had no other objective than re-assessing the most recent progress in that crucial area of science and in refining our understanding of the natural causes of climate variability.

## 2. A Controversial Issue

The Sun is variable on all time scales, ranging from seconds to billions of years. S. Solanki, in this volume, discusses the solar variables of possible relevance for influencing planetary climates. Five different solar or solar-modulated inputs might have an effect on our climate: Total Solar Irradiance (TSI), which measures the radiation integrated over all wavelengths, the Solar Spectral Irradiance or SSI, which measures the distribution of this radiation throughout the electromagnetic spectrum, galactic cosmic rays (GCRs), modulated by the Sun's open magnetic field, and particles ejected by the Sun, predominantly through the Solar wind and the Coronal Mass Ejections (CME) and the Sun's open and total magnetic flux. The variability of these inputs is strongly modulated by the 11-year solar cycle, and also over longer timescales that unfortunately extend back to periods of inexistent observations or of poor data quality. Direct measurements of the manifestations of solar variability are indeed very recent, essentially made by artificial satellites and covering less than three decades, with the exception of sunspot records which are unique for estimating the level of solar activity and go back 400 years before present, providing an essential tool for the reconstruction of the Sun's influences in the past. Reconstructing past solar variability through proxy data is therefore an essential part of the exercise and another issue open to controversy.

The TSI, representing an average energy input at the top of the Earth's atmosphere of 1365 W/m$^2$, is by far the dominant input for all planets of the Solar System. It has enough power to be able to influence the Earth's and the Sea Surface Temperature (SST). At the first ISSI workshop, Reid (2000) provided correlative evidence between the annual mean globally averaged SST and the 11-year cycle sunspot number. He did not reject the possibility of an effect of the 11-year cycle. He also suggested that the climate might well be affected by the effect of spectral irradiance on tropospheric circulation, because each layer of the atmosphere is differently affected by spectral variability. Five years later, Kodera (2006) suggests that the ocean is not radiatively heated, but that temperature variations are mainly due to the variation of the heat flux between the atmosphere and the ocean. He also proposes two possible processes in support of his conclusion. One involves changes in the vertical propagation of planetary waves and the resultant tropospheric circulation changes in the extra-tropics of the winter hemisphere. The other involves changes in the global meridional circulation in the stratosphere and associated convective activity changes in the tropics.

The effect of SSI variability on the thermosphere, the stratosphere and the troposphere appears non-controversial, as shown in the present volume by Labitzke,

Chanin and Rottman. In addition, solar proton events do have an effect on the chemical composition of the upper atmosphere, especially on ozone and $NO_2$ as reported here by Hauchecorne, Jackman, and López-Puertas. More and still controversial are the effects of cosmic rays, in particular on the formation of clouds, as discussed by Lockwood in one of the following book chapters.

This interplay between solar inputs and the mechanisms coupling the different layers of the Earth's atmosphere is at the heart of the main controversies which presently concern the community. This is put in striking contrast if we realize the smallness of the amplitude of solar irradiance variability: a few tenths of a percent for the TSI and, depending upon the wavelength range where it is considered, from a few percents in the near ultraviolet to an order of magnitude in the extreme ultraviolet for the SSI, which is still very small realizing that the total energy emitted in the ultraviolet and extreme ultraviolet is only a few percents of the total. For more than thirty years, solar physicists and climatologists have been arguing about the effects of these tiny variations. Five years after the first ISSI workshop on that topic, the issue is still present, but the situation has improved as more precise data and more refined models are introduced in the analysis.

## 3. Measuring Solar Variability

Measuring solar variability is a recent undertaking directly associated with the space era. Both G. Rottman for the measurement of the SSI and C. Fröhlich for that of the TSI, insist here that these two quantities can only be measured properly from space. Only from above the Earth's atmosphere can we have access to the totality of the electromagnetic spectrum of the Sun and are we able to measure solar variability independently of atmospheric perturbations such as turbulence, diffusion, cloud coverage and emission of light by the atmosphere itself. In orbit, the main remaining source of error and of uncertainty lies in the accurate knowledge of the photometric properties of the instruments and of the detectors, properties which cannot generally be directly accessed in space, which vary and degrade with time, and for which precise calibration is essential to the value of the data.

Indeed, these two authors appropriately discuss the key importance of absolute calibrations, an issue which is also directly relevant to a proper inter-comparison of data acquired by different instruments placed under the responsibility of different investigators, and to the continuity of the measurements. Fröhlich strikingly demonstrates the difficulties of placing a single set of measurements obtained with the same instrument on the same relative scale due to variations of sensitivity in orbit, and the even greater difficulty to reconcile measurements made with different instruments over long time periods. Hence, the need for proxy models as illustrated in Fröhlich's Figure 12 (Fröhlich, 2006). The methods of inter-calibration are complex, delicate and could even be subjective and prone to parochialism! In this context, we should consider the PMOD composite as described by Fröhlich, as

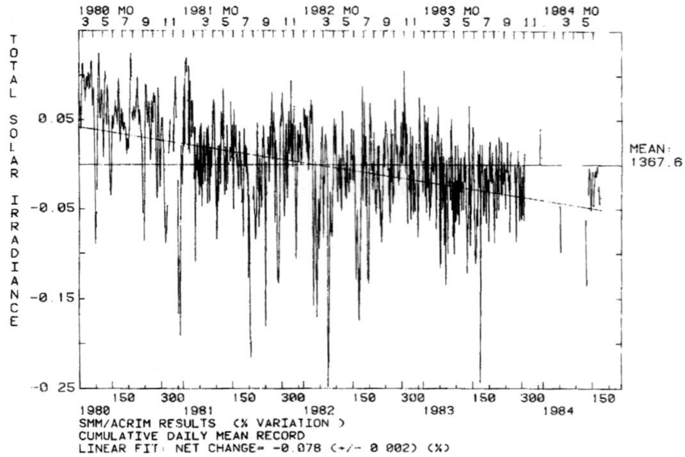


*Figure 1.* TSI measurements as of 1984, derived from measurements by the SMM/ACRIM experiment (Willson, 1985).

being the main observational reference for future work on the long-term variation of the TSI and the SSI.

We reproduce above on Figure 1 what we knew of the variability of the TSI in the mid 1980s. The comparison with Fröhlich's Figure 12 evidences the tremendous progress accomplished over a fifth of a century in the radiometric measurements of the Sun's radiation, what was still called at that time the *"Solar Constant"*. Despite such progress, even with the use of the most sophisticated and accurate techniques available in the space era, complemented with the most modern laboratory and ground based measurements, inaccuracies are still larger than the amplitude of the phenomena themselves. We may therefore wonder whether it is still of importance to continue monitoring solar variability with the accuracies reported in this volume, when the imprints of anthropogenic forcing is generally admitted to be much stronger. We discuss this point in the conclusion below.

Even though we do not have a similar composite as the PMOD composite for SSI variations, the excellent cooperation among the observers of solar radiation in the UV and EXUV leads to a trustworthy database on solar variability in that spectral range. Here also, the most critical issues are the need for continuity and inter-calibration. The longer the considered time span, the more reliable can we consider the reconstructed composites and the established proxy-models. Nevertheless, even the best SSI data sets are difficult to reconcile on an absolute scale. We should realize that the time constants that characterize the variability of the Earth climate are measured in tens of years if not in centuries. The characteristic time for relevant solar variability measurements can be set equal to one solar cycle, i.e., 11 years. The longer the time base, the more accurate will be the reconstruction of the variability. Requiring the best accuracy will therefore span more than one generation of scientists. This is a very serious problem! How can we ensure the permanence of these measurements, from space, over generations and even centuries?

This sends a strong message to the space agencies but also to the scientific community. It is not at all secured that there will always be a satellite in orbit with the proper instrumentation to provide the required uninterrupted data sets with the proper level of accuracy. Rottman gives the example of the 1980–1997 gap when no measurements were available in the XUV/EUV range, making it more difficult to correlate solar variability and its possible effects on Earth or to reconcile discontinuous measurements. Less secure even is the guarantee of permanence of the required scientific and experimental expertise. Will the scientific instances be able to attract, generate and form the talented experimenters that are needed to properly conduct these delicate measurements, bearing in mind that there is an intrinsic risk associated with the use of satellites and space techniques which might act as a dissuasive effect on the new generations? The need for continuous and routine observations is not what such instances consider as convincing arguments when comes the time of establishing priorities among the various disciplines of space science. Because the needs are clearly associated to the evolution of anthropogenic activities, these tasks should be considered of strategic relevance for the future management of our planet.

An alternative solution might be to reconstruct solar variability in the longer term through the use of more simple proxies, such as the solar magnetic field itself, which can be measured from the ground, and the use of models such as SATIRE as described by Solanki. This illustrates the crucial importance of such proxies and of reconstructions of the Sun's behavior, in view of ultimately forecasting the evolution of the Earth's climate.

## 4. Reconstructing Solar Variability

Of crucial importance in relation with the assessment of the effect of solar variability on the climate, is the tracking of the variations of past climate through historical records, together with the fluctuations of the Sun's influences as they can be reconstructed further back in time. Even though the accuracy of such reconstructions worsens with time the further we go in the past, this exercise provides the most solid basis on which to establish a clear relationship between climate and solar variability, especially over periods of time when anthropogenic influences were not dominant. The reconstruction of past climates and of past solar variability is addressed in this volume by S. Solanki and J. Beer, while M. Lockwood and B. Heber discuss the limits of the proxies and methods used in such processes.

According to Solanki, we are unable today to accurately track and forecast solar variability, be it on the short time scale (less than a solar cycle) or on the longer time scale. On the billions of years scale, we must rely on astrophysical models of the Sun's interior and on the comparison of the Sun with Sun-like stars. The level of accuracy of such modelizations is difficult to evaluate and bound to be uncertain. This is not so important, however, because such time scales span far longer than the few millions of years that characterize the evolution of human life on Earth and far longer also than what can be considered as a reasonable time (100,000 years?) for the survival of human life thereon.

On scales of millions of years, and certainly on shorter scales, intrinsic solar variability through all its aspects can be associated with the variability of the solar magnetic field and also, at the top of the Earth's atmosphere, to the variations of the Earth's orbital parameter as evidenced by the historical work of M. Milankovitch. The sunspot number as a proxy for evaluating the emergence of fresh magnetic field is therefore of crucial importance. In spite of tremendous progress accomplished in the recent past, in particular thanks to the ESA/NASA SOHO satellite, the detailed way in which the field is generated by the solar dynamo inside the convection zone, and how its energy is dissipated in the photosphere, the chromosphere and the corona, remain still a subject of active research. For example, the observed increase by more than a factor 2 of the open magnetic flux of the Sun as reported by Lockwood *et al.* (1999) remains unexplained by solar physics. In a very simplified model, only the upper layers of the Sun do contribute to the variability of relevance to climate: the magnetic field is generated in the convection zone, the TSI is emitted by the photospheric layers, the SSI and the energetic particles originate in the chromosphere and the corona. Recently, Letellier and Maquet (2006) discuss the generation of sunspots and the variation of solar activity from a purely mathematical point of view and conclude that they are of a chaotic nature. In order to better forecast solar magnetic activity, it is therefore necessary to study the convective zone where the field is generated, through the use of more precise helioseismology data, in particular along the magnetic axis of the Sun, and of more refined models of the dynamo. A real progress is reported here by Solanki, who is able to compute the Sun's magnetic field over time and to reproduce the measurements of the open magnetic flux since 1974, as well as the magnetic flux as reconstructed by Lockwood.

Establishment of a quantitative relationship between solar variability and solar forcing requires that we use extended records of solar variability much further back in time. Retracing solar variability of the past requires the use of different proxies such as $^{10}$Be, $^{14}$C and $^{36}$Cl cosmogenic isotopes which can be found in natural archives such as ice cores. These isotopes are produced in the Earth's atmosphere by the interaction of galactic cosmic rays with nitrogen, oxygen and argon. Galactic cosmic rays are modulated by the Sun's "open" magnetic field as they cross through the Heliosphere. This complex mechanism is well discussed in this volume by J. Beer and also by B. Heber in a well documented paper based on data obtained by the Ulysses, the two Voyagers and the Pioneer 10 spacecraft. They allow the reconstruction of solar activity back to at least 10,000 years. Obviously, the accuracy of that procedure decreases with time before present. According to Beer, for the time being, it doesn't seem wise to assess solar variability beyond 12,000 years before present. This time scale is certainly highly relevant for potential climatic effects of solar variability, and in his Figure 7, Beer provides an impressive comparison between the reconstructed solar modulation and the relative amount of ice rafted debris found in the bottom of the North Atlantic Ocean, which is an indicator of prevailing climate conditions.

This connection is also thoroughly discussed here by M. Lockwood with some skepticism as to which physical processes might explain it, given in particular the very small proportion of the "open" solar magnetic flux with respect to the total photospheric flux. Nevertheless, he shows that the production rate of $^{10}$Be can be used as a quantitative indicator of the TSI. Lockwood also addresses the very controversial question of the direct influence of cosmic rays on the formation of clouds, which has been the subject of intense critical debates after such a modulation was first reported by a Danish group (Friis-Christensen and Svensmark, 1977). He discusses the problem and possible mechanisms of cloud formation, as well as the value of the data sets and their (too short) duration. He concludes that a direct effect of cosmic rays may not be excluded, but that caution is necessary before definitely confirming that connection. It is therefore important to continue the monitoring of galactic cosmic rays and of the cloud coverage over short and long time periods.

## 5. Conclusion

It is clear from the contributions made to this volume that significant progress has been made in the past years as more and better data and more refined models become available. In particular, the identification of solar variability parameters of relevance to the climate has improved, as well as the reconstruction of these in the past.

Nevertheless, there is still considerable work left to improve the present situation. The continuity in data gathering is essential in particular for the SSI and the TSI, at least until we have been able to demonstrate that we can develop a reliable model for reproducing precisely these parameters as they are presently observed and for reconstructing their evolution in the past.

It is becoming every day more obvious that, given our present knowledge, solar forcing alone is not able to explain the recent increase noted in the Earth's surface temperature, and that anthropogenic effects have an increasing importance in the evolution of the temperature and most probably of the climate. We therefore might question the necessity to continue monitoring solar variability.

It is important to realize that at some point in time, and in an optimistic scenario based on political wisdom and vision, the kind of anthropogenic forcing we are witnessing today might be curbed down sometime in the future. In order to be in a position to assess this recovery effect, we will therefore need to be in possession of the best possible evaluations of solar forcing. We might then be able to properly quantify this expected recovery trend and make sure that we have re-entered an era of natural forcing. This is the reason why the most precise measurements of solar variability over the longest possible time span will be needed. For that, the availability of the required instrumentation, in particular in space, must be secured as well as a qualified community of scientists and engineers responsible for their development, their calibration and the reduction of their data.

Ultimately, it would be good to be able to predict solar variability. For that, we need better models of the solar convective zone and of the dynamo in order to forecast in a quantitative manner the level of solar activity. This is not for tomorrow, but we must not relinquish our ambition that sometime in the future we might be able to do that.

This progress needs to be constantly evaluated, and this might be done at a future ISSI workshop five years from now.

## References

Friis-Christensen, E. and Svensmark, H.: 1977, 'What do we really know about the sun-climate connection?', *Adv. Space Res.* **20**, 913–920.

Fröhlich, C.: 2006, 'Solar irradiance variability since 1978', *Space Sci. Rev.*, this volume, doi: 10.1007/s11214-006-9046-5.

Kodera, K.: 2006, 'The role of dynamics in solar forcing', *Space Sci. Rev.*, this volume, doi: 10.1007/s11214-006-9066-1.

Letellier, C. and Maquet, J.: 2006, 'Analyse de dynamiques chaotiques: une nouvelle approche de l'activité solaire', *Bull. Société Française de Physique* **154**, 10–13.

Lockwood, M., Stamper, R., and Wild, M. N.: 1999, 'A doubling of the Sun's coronal magnetic field during the last 100 years', *Nature* **399**, 437–439.

Reid, G. C.: 2000, 'Solar variability and the Earth's climate: Introduction and Overview', *Space Sci. Rev.* **94**, 1–11.

Willson, R. C.: 1985, Private communication, In *SOHO – Phase-A Study report*. ESA SCI(85)7, 78 pp.

# SOLAR VARIABILITY OF POSSIBLE RELEVANCE FOR PLANETARY CLIMATES

S. K. SOLANKI* and N. A. KRIVOVA
*Max-Planck-Institut für Sonnensystemforschung, Max-Planck-Str. 2, D-37191 Katlenburg-Lindau, Germany*
*(*Author for correspondence: E-mail: solanki@mps.mpg.de)*

(Received 26 August 2005; Accepted in final form 21 December 2005)

**Abstract.** The global variability of the Sun of relevance for planetary climates has been directly measured for the past few decades. For longer stretches of time models are required. Semi-empirical models can now accurately reproduce the measured records of solar total and spectral irradiance, as well as of the magnetic flux. They can also provide reconstructions of these quantities on longer time scales. Here a summary is given of some of the modelling efforts and of the results achieved so far.

**Keywords:** Sun: activity, Sun: irradiance, Sun: magnetic fields, solar-terrestrial relations, Sun: UV radiation

## 1. Introduction

Solar variability takes on different guises. Many of these correspond to local changes on the Sun (e.g., the appearance of a sunspot group, or the brightening of a system of coronal loops). Planetary atmospheres feel only global changes of the Sun. The global solar variables of possible relevance for planetary climates include the variable solar total radiative flux or total irradiance (i.e., the wavelength integrated radiative flux measured above the Earth's atmosphere), which determines the energy input into planetary atmospheres, the solar spectral irradiance, which affects stratospheric chemistry, the Sun's open and total magnetic flux (the open flux affects cosmic ray flux and planetary magnetospheres) and solar particle flux, including energetic particles (e.g., strength and speed of the solar wind around the ecliptic plane, number and strength of coronal mass ejections). We can safely neglect the direct influence on climate of the Sun's particle flux, since it carries only $2 \times 10^{-6}$ of the energy transported by radiation. However, solar energetic particles may play an indirect role in that they may to a certain extent affect proxies such as cosmogenic isotope concentrations used to reconstruct past solar activity. Note that the Sun's global variability is often driven by local processes and structures (e.g., flares and other eruptive processes, convection cells, sunspots, faculae, chromospheric network, coronal loops, coronal holes and streamers).

The Sun is variable at all time scales on which we can carry out measurements or computations, ranging from seconds to billions of years. The variability at different time scales, however, has different, often quite diverse causes. Thus, convection

Space Science Reviews (2006) 125: 25–37
DOI: 10.1007/s11214-006-9044-7

near the solar surface typically affects solar output on time scales shorter than roughly a day, oscillations are known to influence solar irradiance on time scales of minutes, while solar rotation introduces a period of 27–30 days as seen from Earth. At the other extreme, solar evolution, driven by the gradual change in chemical composition in the solar core, produces large, even drastic changes in the Sun's luminosity and radius at time scales of $10^6$–$10^{10}$ years. Finally, the Sun's magnetic field itself changes on a broad range of time scales (the most prominent one being the 11-year cycle) and in addition plays an often dominant role in producing changes in other global quantities. These changes driven by the magnetic field range from seconds (structure seen in radio bursts) to billions of years (evolution of the strength of the Sun's magnetic field and thus of magnetic activity).

In this overview we will concentrate on solar total and spectral irradiance variations. Measurements of these parameters with an accuracy that is sufficient to reliably display their variations are restricted to the last few solar cycles (see Fröhlich, 2006), which, although differing in some important aspects are relatively similar to each other (compared to some earlier cycles that were often much weaker). In order to obtain an estimate of the variation of the Sun on a longer time scale proxies and models are needed, preferably in combination.

There are a number of proxies that have been used to trace solar irradiance or some aspect thereof. These include sunspot number (available since 1610), facular area (available between 1874 and 1976), Mg II core-to-wing ratio (available from 1978), the total magnetic flux (regularly available from 1974), the geomagnetic aa-index (available from 1868), $\Delta^{14}C$ and $^{10}Be$ concentration (partly available for the whole Holocene). Models and the reconstructions of solar irradiance made therefrom often rely on some such proxy, since it is not yet possible to compute variations of the solar irradiance from first principles.

In the following we first consider reconstructions of irradiance over the last few decades, i.e., the period over which irradiance measurements are available (Sections 2 and 3). Then we discuss longer term reconstructions of irradiance (Section 4) and solar activity (Section 5). Finally, in Section 6 a brief summary and outlook is given.

## 2. Short-Term Total Solar Irradiance Variations

The measured total irradiance record (as put together in a composite (e.g., Willson and Mordvinov, 2003; Fröhlich, 2003, 2004; DeWitte *et al.*, 2004) shows two features that are striking. The first is the dips happening on time scales of a week or two, the second is the solar cycle variation of roughly 0.1% (i.e., roughly 1.3 $Wm^{-2}$, which converts into 0.24 $Wm^{-2}$ averaged over the whole Earth).

The dips, some of which can be as deep as 3 $Wm^{-2}$, are due to the passage of sunspot groups across the solar surface due to solar rotation. A group is visible for a maximum of roughly 14 days, but due to foreshortening and limb darkening (the limb of the Sun is significantly darker than the centre of its disk) the dips typically have a length of 7–10 days. As shown by Spruit (1982) the heat flux blocked by sunspots does not reappear at the solar surface immediately, but is distributed throughout the convection zone, reappearing on a time scale of around $10^5$ years. Therefore, a sunspot at the solar surface leads to a deficit in solar luminosity (i.e., the solar output integrated over all angles) for the period of time that the spot is present. In spite of this, the Sun is brighter at activity maximum, i.e., at a time when there are on average more sunspots on the solar surface. This seemingly inconsistent behaviour is due to the presence of increased amounts of faculae, i.e., bright structures. Sunspots and faculae are both formed together when active regions appear at the solar surface. In the early phases of an active region's development, the sunspots often dominate and lead to a darkening. However, the lifetime of sunspots is shorter than of the surrounding faculae and the sunspots decay into faculae. Hence active regions may be dark early in their life, but turn bright after some time (see Ortiz *et al.*, 2000 for an example).

This qualitative discussion can be tested using quantitative modelling. Different models that aim to reproduce solar total irradiance have been developed (Foukal and Lean, 1990; Chapman *et al.*, 1996; Fröhlich and Lean, 1997; Fontenla *et al.*, 1999, 2004; Fligge *et al.*, 2000; Preminger *et al.*, 2002; Ermolli *et al.*, 2003; Krivova *et al.*, 2003; Jain and Hasan, 2004; Wenzler *et al.*, 2004, 2005). Here we discuss one such set of models in greater detail, the so called SATIRE models (Spectral And Total Irradiance REconstruction). This model is based on the assumption that it is the magnetic field at the solar surface which is responsible for all irradiance variations on time-scales longer than roughly a day. The magnetic field lies at the heart of both the dark (e.g., sunspots) and bright features (e.g., faculae in active regions, and the network distributed over the whole Sun).

SATIRE models are semi-empirical. Spectra computed from model atmospheres are used to describe the radiative influence of different components of the solar atmosphere. These components are sunspot umbrae and penumbrae, faculae (including the network) and the quiet Sun. The model atmospheres are constructed using independent data (i.e., not the irradiance time series). Magnetograms and continuum images measured daily are used to separate the solar atmosphere into its components. Each pixel on the solar surface is then replaced by the corresponding spectrum. After summing over all pixels, the Sun's irradiance spectrum is obtained. After further integration over all wavelengths the total solar irradiance is found. The model has a single free parameter. A more detailed description of the model is given by Fligge *et al.* (2000), while the model atmospheres used are described by Unruh *et al.* (1999).

SATIRE gives a remarkably good correspondence (correlation coefficient of around 0.96) with the irradiance measured by the VIRGO instrument (Fröhlich *et al.*, 1995) flying on SOHO, irrespective of the source of the employed magnetograms (Krivova *et al.*, 2003; Wenzler *et al.*, 2004). On a longer time scale it still shows a good correspondence with the PMOD irradiance composite produced by Fröhlich, see Figures 1 and 2. In particular, there is no trend in the difference between the measured total solar irradiance and that reconstructed by the SATIRE model. This confirms firstly that the dominant part of the observed irradiance variations (on time scales of days to the solar cycle) are due to the surface magnetic field (Krivova *et al.*, 2003; Wenzler *et al.*, 2005, 2006). It also confirms that there is no trend of magnetic origin in the total solar irradiance since 1974 (Wenzler *et al.*, 2006). A putative trend, such as that proposed by Willson and Mordvinov (2003) and DeWitte *et al.* (2004), must therefore be caused by some effect that is independent of the magnetic field.

## 3. Short-Term Solar Spectral Irradiance Variations

Since SATIRE computes the spectrum at each point (pixel) on the solar surface, it is straightforward to reconstruct the Sun's irradiance spectrum. The results are reasonable at wavelengths longwards of 200 nm, since the assumption of LTE underlying the radiative transfer used for computing the model spectra breaks down at shorter wavelengths and more sophisticated techniques have to be used (Fontenla *et al.*, 1999, 2004; Haberreiter *et al.*, 2005). A comparison with SUSIM data in the

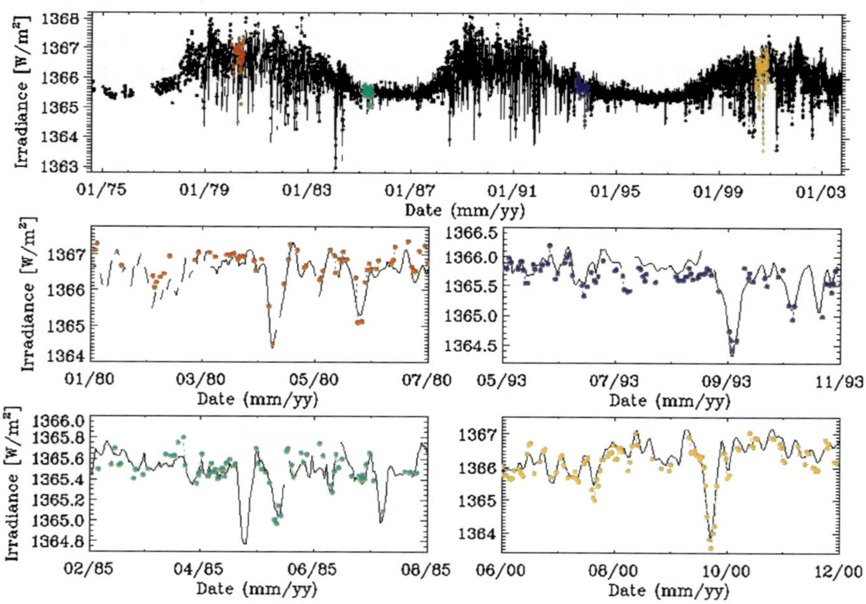


*Figure 1.* Total solar irradiance: composite of measurements of Fröhlich (solid lines) and reconstructed using SATIRE (dots). *Top panel:* Irradiance from 1974 to 2003. *Bottom 4 panels:* Shorter periods chosen at random, displaying more clearly the comparison between measured and modelled irradiance (from Wenzler *et al.*, 2006).

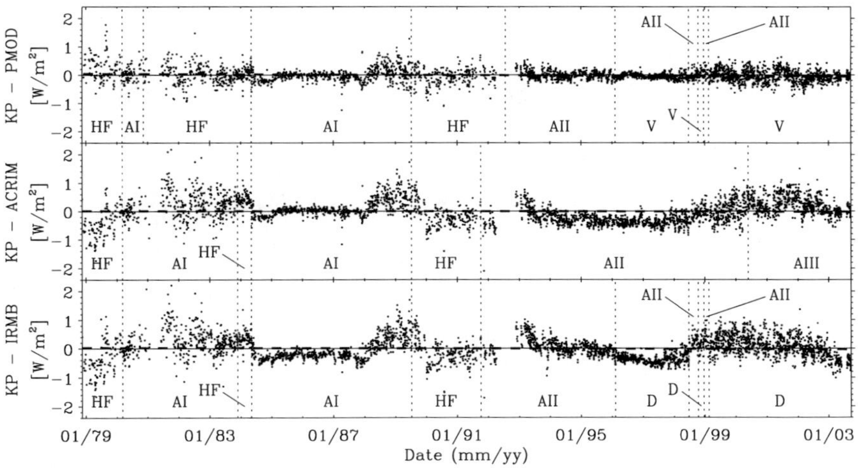


*Figure 2.* Difference between measured and modelled total solar irradiance. The three panels show the difference obtained for the three composites made by Fröhlich (2004) (top panel), by Willson and Mordvinov (2003) (middle panel), and by DeWitte *et al.* (2004) (bottom panel). The modelled irradiance is the output of the SATIRE models (from Wenzler *et al.*, 2006).

220–240 nm band is shown in Figure 3 for the rising phase of solar cycle 23. We stress that the same value of SATIRE's free parameter is used as for the reconstruction of the total irradiance for the same period of time. Although the agreement is not perfect (in particular, the reconstructed spectral irradiance is too low during the winter of 2000/2001), it is nevertheless quite satisfactory, giving a correlation coefficient $R_c = 0.97$. Krivova and Solanki (2005) have found a method to empirically extend the model spectrum to wavelengths as short as Ly-$\alpha$. This allows a more secure estimate of the contribution of the ultraviolet wavelength range ($\lambda < 400$ nm) to total irradiance variations to be made. The main uncertainty results from the wavelength range between 300 and 400 nm, where the SUSIM data are not sufficiently accurate to give reliable estimates of the small irradiance changes. The result is that around 60% of the total irradiance change between solar activity minimum and maximum is produced in the UV part of the spectrum, i.e., shortward of 400 nm, although only 8% of the radiation is emitted at these wavelengths. This result suggests that more attention should be paid to the influence of the Sun's varying UV radiation on the Earth's atmosphere (see Haigh and Blackburn, 2006; Schmidt and Brasseur, 2006).

## 4. Longer-Term Solar Variations

Once it has been shown that the surface magnetic field is responsible for solar irradiance variations via its manifestations such as sunspots and faculae, the reconstruction of solar irradiance on longer time scales requires the computation of the

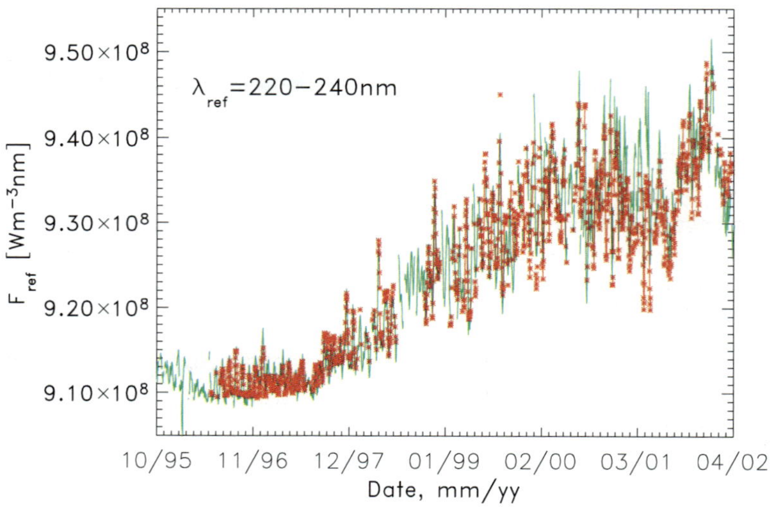


*Figure 3.* Comparison of solar spectral irradiance in the 220–240 nm band measured by the SUSIM instrument on UARS (solid line) with the SATIRE reconstruction (stars).

Sun's magnetic field for the same time scale. When doing this it can be useful to distinguish between cyclic variations, which can often be well described by proxies, and secular variations, which are more tricky to estimate.

The cyclic variations since 1700 have been computed with varying degrees of sophistication by Lean *et al.* (1995); Solanki and Fligge (1999), and others. The quality of the reconstructions is quite reasonable since 1874, but is increasingly lower at earlier times due to the lack of appropriate data (e.g., at earlier times the sunspot number records either are only available in monthly or yearly bins, or have large gaps in them). The change in cycle-averaged irradiance since the Maunder minimum due to the cyclic component is estimated to be 0.6 $Wm^{-2}$.

A potentially much larger effect can be produced by a secular trend in the Sun's irradiance. Evidence for such a trend originally came from Sun-like stars. Baliunas and Jastrow (1990) showed a double-peaked histogram of the number of field stars as a function of calcium emission (a commonly used measure of a star's magnetic activity). The peak at higher stellar activity was interpreted to be produced by cycling stars, while the lower peak was deemed to be populated by stars in a non-cycling, Maunder-minimum-like state. The present-day Sun displays Ca emission at levels corresponding to the higher-activity peak even during solar activity minimum. This was interpreted by Lean *et al.* (1992, 1995) to imply that the Sun was approximately 2.6 $Wm^{-2}$ less bright during the Maunder minimum than averaged over recent solar cycles.

This result has recently been questioned. Newer observations display a single-peaked distribution of activity, with non-cycling stars intermixed with stars exhibiting a cyclic behaviour of their magnetic activity (e.g., Wright, 2004; Giampapa, 2005; Hall and Lockwood, 2004). One clue to the possible cause of the discrepancy between these recent results and the older ones lies in the fact that observations of stars of the same age in the old open cluster M67 support the more recent observations of dwarf field stars. Thus, the older results of Baliunas and Jastrow (1990) may have been a result of an inappropriate inclusion of stars of too broad a range in ages into the diagram. In summary, stellar observations no longer provide any solid evidence for secularly changing solar activity and hence also not of a secular trend in irradiance.

New evidence for secular variations has emerged from another source, however. The interplanetary magnetic field, reconstructed by Lockwood *et al.* (1999) from the geomagnetic aa-index, exhibits a doubling over the last century, besides the normal 11-year cycle similar to that shown by sunspot number. Observations by the Ulysses spacecraft have revealed that the interplanetary field is very closely related to the Sun's open magnetic flux. The open flux is composed of magnetic field lines that are carried out by the solar wind into interplanetary space (closed field lines, by contrast, form loops with heights below a few solar radii). Since 1964 the reconstructed open flux agrees well with direct measurements made by spacecraft. The open magnetic flux of the Sun is responsible for the modulation of cosmic ray flux and hence of the production rate of cosmogenic isotopes. For the reconstruction of solar irradiance, the total magnetic flux is the basic quantity. The open flux directly contributes only a few percent to the total flux. Nonetheless, the open flux reconstruction by Lockwood *et al.* (1999) is an important time series also for irradiance reconstructions, since it runs considerably longer than the time series of total flux (which goes back only to the 1970s).

The mechanism proposed to explain the secular change in the Sun's magnetic field is based on the overlap of the magnetic flux between consecutive activity cycles (Solanki *et al.*, 2000, 2002). An overlap implies that the flux does not drop to zero at activity minimum (as is observed: even at activity minimum, the Sun is still covered by a network of magnetic field). An overlap can be produced either by the emergence of fresh flux belonging to the new cycle, while the old cycle is still in progress, or by the extended lifetime of some of the flux on the solar surface, so that flux that emerged during the old cycle is still present when the new cycle starts.

The basic recipe for computing the Sun's magnetic field over time is to use the sunspot number as a proxy for the emergence of fresh magnetic flux in active regions (since sunspots appear early in the life of an active region and decay relatively fast they are a reasonable proxy of freshly emerged flux). In addition to active regions, emergence of magnetic flux in smaller ephemeral regions is also considered. In a coarse model the overlap between consecutive cycles is achieved in two different ways. Firstly, the open flux is assigned a long lifetime. This applies mainly to the open flux that is built up at the poles of the Sun during solar activity minimum. Since the flux there is mainly unipolar, it decays slowly and is still present when the new cycle is already in full swing. Secondly, the ephemeral regions also introduce an overlap since they start to emerge earlier than the active regions, while the previous cycle is still running strong (ephemeral regions can be assigned to a particular cycle through the latitude of their emergence and, to a lesser extent, by Hale's polarity law). The first form of overlap mainly affects the Sun's open flux, but does not significantly influence the total flux, since the open flux is only a very minor constituent of the total magnetic flux.

Such a model not only reproduces the open magnetic flux reconstructed by Lockwood *et al.* (1999; see Figure 4), but also the measurements of total magnetic flux since 1974 (Harvey, 1994; c.f. Arge *et al.*, 2002; Figure 5), if the result of Krivova and Solanki (2004) is taken into account that in typical synoptic charts (such as those constructed from Kitt Peak magnetograms) around a factor of 2–3 of the magnetic flux in the quiet network is likely to be missed due to cancellation within a typical spatial resolution element.

Once the magnetic flux has been determined, then the time series of measured sunspot areas (or sunspot number) is employed to determine the magnetic flux in sunspots and hence darkening they produce. The remainder of the magnetic field is used to compute the brightening due to faculae. The reconstructed irradiance is compared with the PMOD composite of measurements and found to reproduce them simultaneously with the other data sets shown in Figures 4 and 5 (although the irradiance is reproduced with slightly lower accuracy than the reconstructions based on the magnetograms). The time series of the total solar irradiance reconstructed since the Maunder minimum is plotted in Figure 6. Various things can be seen from that figure. Firstly, the quality of the reconstruction is lower at earlier times. This has to do with the fact that the quality of the data decreases at earlier times (e.g., there are more data gaps). Secondly, a small, but significant secular increase of the irradiance since the Maunder minimum is found. Compared to the Maunder minimum the recent cycle-averaged total irradiance is 1.3 $\mathrm{Wm}^{-2}$ higher in this model (Balmaceda *et al.*, 2006, manuscript in preparation). Other models (Foster, 2004; Lockwood, 2005; Wang *et al.*, 2005) give irradiance rises ranging between 0.9 $\mathrm{Wm}^{-2}$ and 2.2 $\mathrm{Wm}^{-2}$.

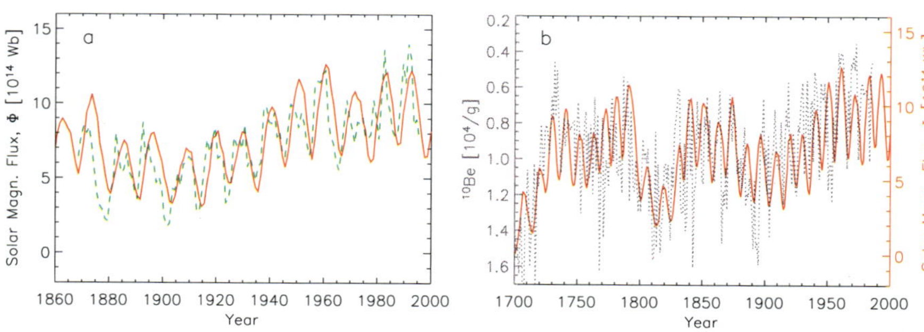


*Figure 4.* The open magnetic flux of the Sun computed by a coarse model (solid curve) compared to (a) the reconstruction by Lockwood *et al.* (1999) (dashed curve), and (b) the $^{10}$Be concentration in Greenland ice (Dye-3; Beer *et al.*, 1990) (dotted curve). Note the inverted scale for $^{10}$Be.

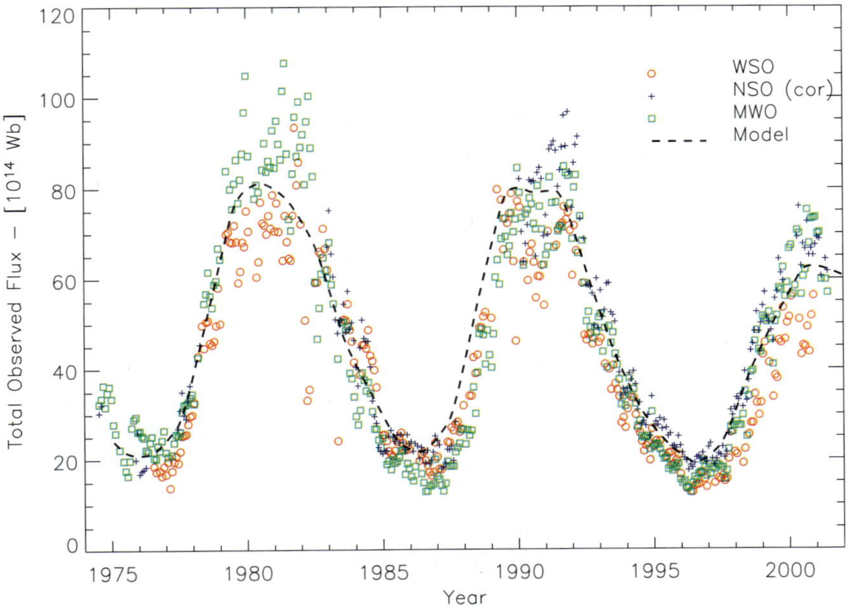


*Figure 5.* Comparison between measured (symbols) and computed (dashed line) total magnetic flux (each symbol is the magnetic flux in a synoptic chart).

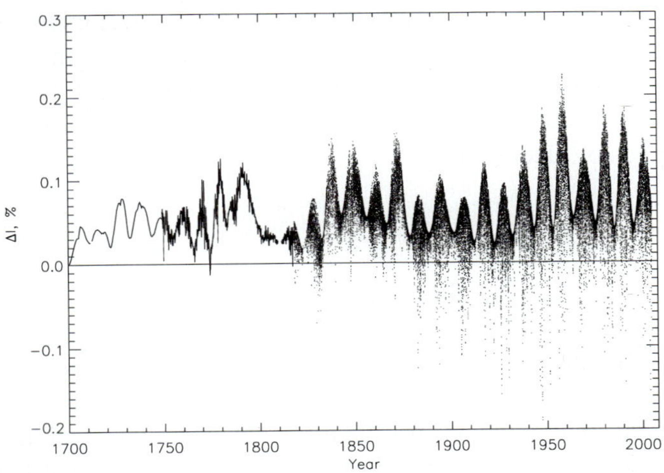


*Figure 6.* The reconstructed total solar irradiance since the Maunder minimum.

In addition to the processes described above, non-magnetic mechanisms are also conceivable for producing a secular change in solar irradiance. However, no concrete mechanism has so far been worked out in any detail and therefore it is not possible to judge how strong a secular change due to putative non-magnetic mechanisms might be.

## 5. Solar Variability Over Millenia

On an even longer time scale no direct measurements of solar variability are available (individual naked-eye observations of sunspots are far too incomplete to give a reliable picture of solar activity). Therefore, indirect proxies must be used, such as the cosmogenic isotopes $^{10}$Be and $^{14}$C, which are produced when high-energy cosmic rays enter the Earth's atmosphere and react with nitrogen and other atoms. Since the intensity of cosmic rays reaching Earth varies with solar activity (more particles make it through to 1 AU from the interstellar medium when the Sun's activity is lower, i.e., when the Sun's open flux is weaker). It is possible, from a measurement of the production rate of cosmogenic isotopes in terrestrial archives to reconstruct the solar modulation parameter, a parameter describing the influence of the Sun's magnetic activity on the cosmic ray flux and from that the strength of the Sun's open magnetic flux. This then allows the sunspot number to be reconstructed (Stuiver and Quay, 1980; Usoskin *et al.*, 2002). Only cycle averaged sunspot numbers can be reconstructed with any accuracy. The reconstructed sunspot numbers agree relatively well with the group sunspot numbers for the period that they overlap (Usoskin *et al.*, 2003, 2004; Solanki *et al.*, 2004).

The most surprising result obtained from these reconstructions is that the Sun is currently in a state of unusually high magnetic activity. The Sun spent only around 3% of the time in the last 11,400 years at a similar level of activity as in the last 60 odd years (see also the error bars; Figure 7). It appears likely that, given the statistics of previous periods of high activity seen in the reconstructed sunspot number, cycle averaged solar activity will decrease significantly within the next 50–100 years. More details on the Sun's behaviour on time scales of centuries to millenia are given by Beer (2006).

## 6. Summary and Outlook

The state of our quantitative knowledge and understanding of solar variability of importance for climate has made significant progress in past years. We can now reproduce observed solar total and spectral irradiance variations with high accuracy, have found a process that can explain a secular change in the solar magnetic field and thus in such quantities as cosmic ray flux and solar irradiance, and have improved methods of reconstructing specific parameters of solar activity throughout the Holocene.

In spite of this progress there is still considerable left work to be done. In the near term, reliable reconstructions of spectral irradiance need to be made for periods extending beyond the time for which direct measurements are available. Fligge and Solanki (2000) carried out a first such reconstruction, but it can be improved upon. A reconstruction of the solar (total and spectral) irradiance over the whole Holocene

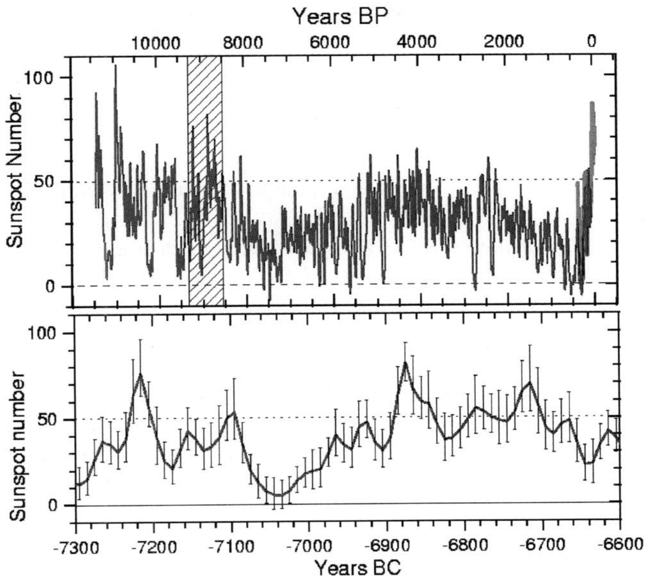


*Figure 7. Upper panel:* Cycle-averaged sunspot number reconstructed from $^{14}C$ over the last 11,400 years, combined with the Group Sunspot Number since 1610. *Lower panel:* Blow-up of the hatched region in the upper panel (adapted from Solanki *et al.*, 2004).

should also be possible with the help of the new sunspot number reconstructions. However, given the low temporal resolution of the data, this will be more tricky. The removal of the last remaining free parameter in the reconstructions of the observed irradiance record is needed and will hopefully soon be achievable. Finally, we do not have the capability to reliably predict solar irradiance. Here much work is needed and it is not likely that any quick successes will be achieved.

## Acknowledgements

We thank T. Wenzler and L. Balmaceda for providing material.

## References

Arge, C. N., Hildner, E., Pizzo, V. J., and Harvey, J. W.: 2002, 'Two solar cycles of nonincreasing magnetic flux', *J. Geophys. Res.* **107**, doi:10.1029/2001JA000503.

Baliunas, S. and Jastrow, R.: 1990, 'Evidence for long-term brightness changes of solar-type stars', *Nature* **348**, 520–523.

Beer, J., Blinov, A., Bonani, G., *et al.*: 1990, 'Use of $^{10}Be$ in polar ice to trace the 11-year cycle of solar activity', *Nature* **347**, 164–166.

Beer, J., Vonmoos, M., and Muscheler, R.: 2006, 'Solar variability over the past several millennia', *Space Sci. Rev.*, this volume, doi: 10.1007/s11214-006-9047-4.

Chapman, G. A., Cookson, A. M., and Dobias, J. J.: 1996, 'Variations in total solar irradiance during solar cycle 22', *J. Geophys. Res.* **101**, 13,541–13,548.

DeWitte, S., Crommelynck, D., Mekaoui, S., and Joukoff, A.: 2004, 'Measurement and uncertainty of the long term total solar irradiance trend', *Solar Phys.* **224**, 209–216.

Ermolli, I., Berrilli, F., and Florio, A.: 2003, 'A measure of the network radiative properties over the solar activity cycle', *Astron. Astrophys.* **412**, 857–864.

Fligge, M. and Solanki, S. K.: 2000, 'The solar spectral irradiance since 1700', *Geophys. Res. Lett.* **27**, 2157–2160.

Fligge, M., Solanki, S. K., and Unruh, Y. C.: 2000, 'Modelling irradiance variations from the surface distribution of the solar magnetic field', *Astron. Astrophys.* **353**, 380–388.

Fontenla, J., White, O. R., Fox, P. A., Avrett, E. H., and Kurucz, R. L.: 1999, 'Calculation of solar irradiances. I. Synthesis of the solar spectrum', *Astrophys. J.* **518**, 480–499.

Fontenla, J. M., Harder, J., Rottman, G., Woods, T. N., Lawrence, G. M., and Davis, S.: 2004, 'The signature of solar activity in the infrared spectral irradiance', *Astrophys. J. Lett.* **605**, L85–L88.

Foster, S.: 2004, 'Reconstruction of Solar Irradiance Variations for use in Studies of Global Climate Change: Application of Recent SOHO Observations with Historic Data from the Greenwich Observatory', Ph.D. thesis, University of Southhampton, Faculty of Science, School of Physics and Astronomy.

Foukal, P. and Lean, J.: 1990, 'An empirical model of total solar irradiance variation between 1874 and 1988', *Science* **247**, 556–558.

Fröhlich, C.: 2003, 'Solar irradiance variations', *ESA SP* **535**, 183–193.

Fröhlich, C.: 2004, 'Re-analysis of the long-term changes of the NIMBUS-7 radiometer and behaviour of total solar irradiance during solar cycle 21', *AGU Fall Meeting Abstracts*, p. A301.

Fröhlich, C.: 2006, 'Solar irradiance variability since 1978', *Space Sci. Rev.*, this volume, doi: 10.1007/s11214-006-9046-5.

Fröhlich, C. and Lean, J.: 1997, 'Total solar irradiance variations: The construction of a composite and its comparison with models', *ESA SP* **415**, 227–233.

Fröhlich, C., Romero, J., Roth, H., *et al.*: 1995, 'VIRGO: Experiment for helioseismology and solar irradiance monitoring', *Solar Phys.* **162**, 101–128.

Giampapa, M. S.: 2005, 'Stellar analogs of solar activity', in: *The Sun, Solar Analogs and the Climate, 34th 'Saas Fee' Advanced Course*, Berlin, pp. 307–415.

Haberreiter, M., Krivova, N. A., Schmutz, W., and Wenzler, T.: 2005, 'Reconstruction of the solar UV irradiance back to 1974', *Adv. Sp. Res.* **35**, 365–369.

Haigh, J. D. and Blackburn, M.: 2006, 'Solar influences on dynamical coupling between the stratopshere and troposphere', *Space Sci. Rev.*, this volume, doi: 10.1007/s11214-006-9067-0.

Hall, J. C. and Lockwood, G. W.: 2004, 'The chromospheric activity and variability of cycling and flat activity solar-analog stars', *Astrophys. J.* **614**, 942–946.

Harvey, K. L.: 1994, 'The solar magnetic cycle', in: R. J. Rutten and C. J. Schrijver (eds.), *Solar Surface Magnetism, Proceedings of the NATO Advanced Research Workshop*, Soesterberg, The Netherlands, 1993, Kluwer Academic Publishers, p. 347.

Jain, K. and Hasan, S. S.: 2004, 'Reconstruction of the past total solar irradiance on short timescales', *J. Geophys. Res.* **109**, doi:10.1029/2003JA010222.

Krivova, N. A. and Solanki, S. K.: 2004, 'Effect of spatial resolution on estimating the Sun's magnetic flux', *Astron. Astrophys.* **417**, 1125–1132.

Krivova, N. A. and Solanki, S. K.: 2005, 'Reconstruction of solar UV irradiance', *Adv. Sp. Res.* **35**, 361–364.

Krivova, N. A., Solanki, S. K., Fligge, M., and Unruh, Y. C.: 2003, 'Reconstruction of solar total and spectral irradiance variations in cycle 23: is solar surface magnetism the cause?', *Astron. Astrophys.* **399**, L1–L4.

Lean, J., Skumanich, A., and White, O.: 1992, 'Estimating the Sun's radiative output during the Maunder minimum', *Geophys. Res. Lett.* **19**, 1595–1598.

Lean, J., Beer, J., and Bradley, R.: 1995, 'Reconstruction of solar irradiance since 1610: Implications for climate change', *Geophys. Res. Lett.* **22**, 3195–3198.

Lockwood, M.: 2005, 'Solar outputs, their variations and their effects on earth', in: *The Sun, Solar Analogs and the Climate, 34th 'Saas Fee' Advanced Course*, Berlin, pp. 109–306.

Lockwood, M., Stamper, R., and Wild, M. N.: 1999, 'A doubling of the Sun's coronal magnetic field during the last 100 years', *Nature* **399**, 437–439.

Ortiz, A., Domingo, V., Sanahuja, B., and Sánchez, L.: 2000, 'An Example of Isolated Active Region Energy Evolution: NOAA AR 7978', in: *ESA SP-463: The Solar Cycle and Terrestrial Climate, Solar and Space weather*, pp. 395–398.

Preminger, D. G., Walton, S. R., and Chapman, G. A.: 2002, 'Photometric quantities for solar irradiance modeling', *J. Geophys. Res.* **107**, doi:10.1029/2001JA009169.

Schmidt, H. and Brasseur, G. P.: 2006, 'The response of the middla atmosphere to solar cycle forcing in the Hamburg model of the neutral and ionized atmosphere', *Space Sci. Rev.*, this volume, doi: 10.1007/s11214-006-9068-z.

Solanki, S. K. and Fligge, M.: 1999, 'A reconstruction of total solar irradiance since 1700', *Geophys. Res. Lett.* **26**, 2465–2468.

Solanki, S. K., Schüssler, M., and Fligge, M.: 2000, 'Evolution of the Sun's large-scale magnetic field since the Maunder minimum', *Nature* **408**, 445–447.

Solanki, S. K., Schüssler, M., and Fligge, M.: 2002, 'Secular variation of the Sun's magnetic flux', *Astron. Astrophys.* **383**, 706–712.

Solanki, S. K., Usoskin, I. G., Kromer, B., Schüssler, M., and Beer, J.: 2004, 'Unusual activity of the Sun during recent decades compared to the previous 11,000 years', *Nature* **431**, 1084–1087.

Spruit, H. C.: 1982, 'Effect of spots on a star's radius and luminosity', *Astron. Astrophys.* **108**, 348–355.

Stuiver, M. and Quay, P. D.: 1980, 'Changes in atmospheric Carbon-14 attributed to a variable sun', *Science* **207**, 11–19.

Unruh, Y. C., Solanki, S. K., and Fligge, M.: 1999, 'The spectral dependence of facular contrast and solar irradiance variations', *Astron. Astrophys.* **345**, 635–642.

Usoskin, I. G., Mursula, K., Solanki, S. K., Schüssler, M., and Kovaltsov, G. A.: 2002, 'A physical reconstruction of cosmic ray intensity since 1610', *J. Geophys. Res.* **107**, doi:10.1029/2002JA009343.

Usoskin, I. G., Solanki, S. K., Schüssler, M., Mursula, K., and Alanko, K.: 2003, 'Millennium-scale sunspot number reconstruction: Evidence for an unusually active Sun since the 1940s', *Phys. Rev. Lett.* **91**, 211,101–211,104.

Usoskin, I. G., Mursula, K., Solanki, S., Schüssler, M., and Alanko, K.: 2004, 'Reconstruction of solar activity for the last millennium using $^{10}$Be data', *Astron. Astrophys.* **413**, 745–751.

Wang, Y.-M., Lean, J. L., and Sheeley, N. R.: 2005, 'Modeling the Sun's magnetic field and irradiance since 1713', *Astrophys. J.* **625**, 522–538.

Wenzler, T., Solanki, S. K., Krivova, N. A., and Fluri, D. M.: 2004, 'Comparison between KPVT/SPM and SoHO/MDI magnetograms with an application to solar irradiance reconstructions', *Astron. Astrophys.* **427**, 1031–1043.

Wenzler, T., Solanki, S. K., and Krivova, N. A.: 2005, 'Can surface magnetic fields reproduce solar irradiance variations in cycles 22 and 23?', *Astron. Astrophys.* **432**, 1057–1061.

Wenzler, T., Solanki, S. K., Krivova, N. A., and Fröhlich, C.: 2006, 'Reconstruction of solar irradiance variations in cycles 21–23 based on surface magnetic fields', *Astron. Astrophys.*, in press.

Willson, R. C. and Mordvinov, A. V.: 2003, 'Secular total solar irradiance trend during solar cycles 21–23', *Geophys. Res. Lett.* **30**, 1199, doi:10.1029/2002GL016038.

Wright, J. T.: 2004, 'Do we know of any Maunder minimum stars?', *Astron. J.* **128**, 1273–1278.

# MEASUREMENT OF TOTAL AND SPECTRAL SOLAR IRRADIANCE

G. ROTTMAN
*Laboratory for Atmospheric and Space Physics, University of Colorado, Campus Box 590, Boulder, CO 80309-0590, USA*
*(E-mail: gary.rottman@lasp.colorado.edu)*

(Received 5 September 2005; Accepted in final form 10 March 2006)

**Abstract.** The Sun's electromagnetic radiation powers our solar system. In the case of the Earth it heats the lands and ocean, maintains our atmosphere, generates clouds, and cycles water. For other planets and minor bodies, similar and appropriate physical processes occur, also powered by the Sun. The Sun varies on all time scales and a precise knowledge of the Sun's irradiance and its variation is essential to our understanding of environments and physical conditions throughout our solar system. Measurements of solar irradiance and its variation can only be made from space, and almost thirty years of observation have now established that the total solar irradiance (TSI) varies by only 0.1 to 0.3%, while certain portions of the solar spectrum, the ultraviolet for example, vary by orders of magnitude more. This paper provides an overview of TSI observations and of spectral irradiance observations from the ultraviolet to the near infrared.

**Keywords:** solar irradiance, solar activity cycle, climate variability, atmospheric photochemistry, instruments and techniques

## 1. Introduction

Solar radiation is the dominant energy input to the Earth system. Globally averaged approximately 20% of this radiation is absorbed in the atmosphere, and establishes its temperature, composition, and structure (Kiehl and Trenberth, 1997). Another 30% is scattered and reflected back to space while the remaining 50% is absorbed at the surface where it warms the land and ocean and sustains life. A delicate balance is established between incoming solar radiation, the Earth's albedo (fraction of radiation reflected back to space), and outgoing long-wave infrared radiation emitted from the surface and warm atmosphere as altered by greenhouse gasses. Changes in solar irradiance will have both direct and indirect effects on the Earth climate system, and implications of a solar role are evident in most climate records (Lean *et al.*, 2005).

The integral of all electromagnetic radiation incident at the top of the Earth's atmosphere is referred to as the total solar irradiance or TSI and has a value of approximately 1365 W/m$^2$. The TSI is predominantly made up of visible and infrared radiation, but with small amounts of energetic ultraviolet and X-rays at short wavelengths, and microwaves and radio waves at long wavelengths. Sufficiently accurate observations of TSI only became possible after access to space, and the

Space Science Reviews (2006) 125: 39–51
DOI: 10.1007/s11214-006-9045-6

resulting thirty-year record shows a range of variability not exceeding a few tenths of one percent (Willson and Hudson, 1991). Based solely on radiative balance consideration the direct effect on the Earth's global temperature from such a small solar variation would be a change of only a fraction of a Kelvin.

Does the very small variation of TSI imply that the Sun has little influence on climate change? Likely not, for in parallel with early TSI observations, rockets and satellites were also making observations of the very energetic radiation from the Sun (Friedman, 1961). This ultraviolet and X-ray radiation exhibits a much larger range of variation – factors of two to ten, and even more. The small variation observed in TSI is easily reconciled with the highly variable short-wave radiation because the X-rays and ultraviolet (UV) make up less than 1% of the total. Because of the strong atmospheric absorption these more energetic photons do not have access to the Earth surface and lower atmosphere; and therefore, they do not have a direct effect on global surface temperature. However, they do have very important influence on the composition, temperature, and dynamics of the Earth's atmosphere, which in turn may couple to the lower atmosphere and have important indirect influence on local and global climate (Haigh, 2001; Rind, 2002). Such indirect effects are complicated and will eventually be understood through models that accurately portray the complex and interrelated processes occurring throughout the Earth's atmosphere.

## 2. Measurement of Solar Irradiance

Understanding the Sun's radiation and its possible variability has historically been deemed of great importance and has resulted in the prompt application of each new optical device and technique to study the Sun's irradiance. Until there was access to space this research endeavor was limited to ground observations. Figure 1 compares the solar spectral irradiance as observed at the ground, with the actual irradiance incident at the top of the atmosphere. The two spectra illustrate quite clearly that the atmosphere absorbs and scatters the radiation in a very wavelength dependent manner. The strong absorption bands in the infrared are due primarily to water vapor, although $O_2$, $CO_2$, ozone and other trace gases also play a role. The complete extinction of radiation with wavelengths shorter than 300 nm is caused by ozone, providing an effective shield against this biologically harmful radiation.

Figure 1 illustrates the difficulty of determining solar variability from the ground. Even in the visible where radiation is readily transmitted to the Earth's surface, the corrections for atmospheric scattering and absorption are large and introduce errors of 1 to 2% in extrapolated top of the atmosphere (TOA) irradiance values. The determination of true solar variability as deduced from such ground observations is greatly compromised and quite uncertain. In the ultraviolet, of course, no ground observations of solar irradiance are even possible.

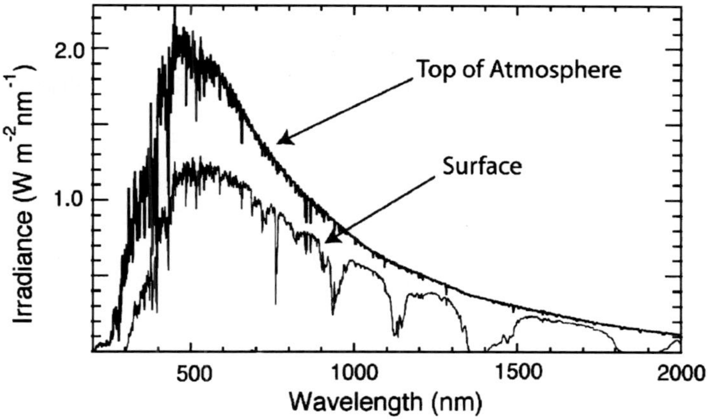


*Figure 1.* The solar spectral irradiance at the top of the atmosphere is compared with irradiance at the Earth's surface. The strong absorption bands in the infrared are primarily due to water vapor and the complete extinction at wavelengths below 300 nm is due to ozone.

## 2.1. Measurement of Total Solar Irradiance – TSI

Beginning in the late 1800s Langley, Abbot, and others (Menzel, 1949; Abbot, 1948) pursued a very active observational program to try to establish the TOA solar irradiance. They refined their techniques and deployed their instruments to remote and high altitude observatories. Their rigorous and careful programs continued well through the middle of the twentieth century, but, at best, could only conclude that the Sun varied less than their combined measurement uncertainties of a few percent. The amount of solar irradiance was then referred to as the "solar constant", a misleading description that was only retracted when space-based observations beginning in 1978 proved conclusively that the TSI did indeed vary (Hickey *et al.*, 1980; Willson *et al.*, 1981).

The measurement of TSI or radiant flux density requires an aperture to define the collection area followed by a sensor to measure the incident power. The instrument must be uniformly sensitive to all wavelengths from the very energetic and short wavelength X-rays to the very longest infrared wavelengths. One such device is a bolometer consisting of a black collector that absorbs radiation and converts it to internal heat. The temperature increase of the collector is then a measure of the incident radiation. A simple concept, but with sophisticated application, the bolometer has been the detector of choice for TSI observations during the past 100 years.

Since 1978 devices of this type have operated continuously on a number of different space missions including Nimbus-7 (Hickey *et al.*, 1988), SMM, UARS, and ACRIMSat (Willson, 1984, 1994), ERBE (Lee *et al.*, 1987), VIRGO (Fröhlich, 2004), and SORCE (Kopp and Lawrence, 2005; Kopp *et al.*, 2005). Figure 2 shows the combination of these TSI observations, where the offsets between the various

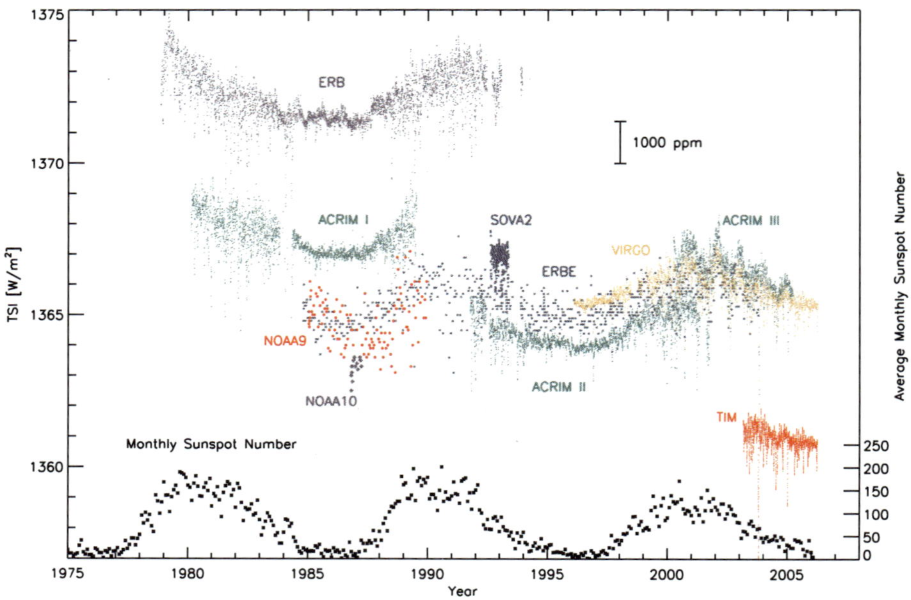


*Figure 2.* The complete record of modern space-based observations of the total solar irradiance (TSI). Data along the bottom are the monthly sunspot number to illustrate the 11-year solar cycle and periods of maximum and minimum solar activity. Differences between the various data sets are due to calibration differences (uncertainty).

individual data are a manifestation of slight differences in the calibration of each instrument – its traceability and correspondence to the SI units of Watt and meter squared – and to possible other instrumental effects, for example, scattered light. It is a great concern that the most recent TSI observations of the Total Irradiance Monitor, TIM, on the Solar Radiation and Climate Experiment (SORCE, Rottman, 2005) fall almost 5 W/m$^2$ below the concurrent observations of VIRGO and ACRIMSat. These differences are being closely examined, and hopefully a recommendation to the scientific community will soon be made as to a preferred and recommended value. The reader is directed to a review article by Fröhlich (2004) and an accompanying paper (Fröhlich, 2006) in these proceedings for an extended discussion of these TSI data and their interpretation. In combination these TSI instruments have now recorded almost three complete 11-year solar cycles, and for each cycle the levels of TSI are roughly 0.1% higher at the time of solar maximum (coinciding with the maximum levels of sunspots) than at solar minimum. These TSI observations show quite conclusively that the dominant solar variability over the 11-year cycle is due to magnetic activity in the photosphere with a positive contribution originating in the bright faculae and a negative contribution arising from the dark sunspots (Foukal and Lean, 1988). The best fit to the TSI data is achieved with a faculae contribution roughly twice the sunspot darkening, giving a net variation of about 0.1% (Fröhlich and Lean, 1998).

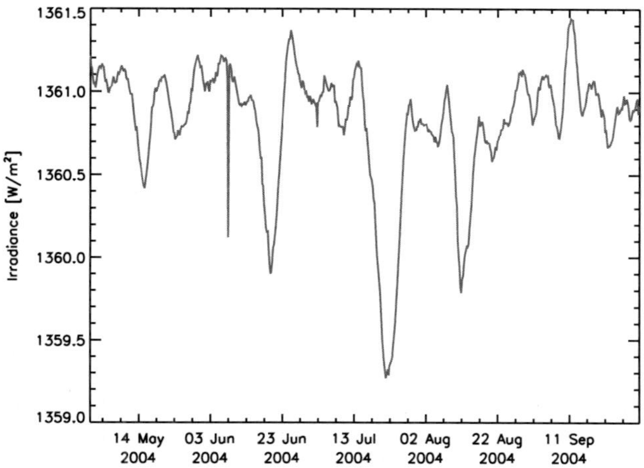


*Figure 3.* TSI data from the SORCE TIM instrument for a four-month time period in 2004. The sharp spike on 08 June is due to the transit of Venus across the disk of the Sun.

Shorter-term variations of TSI are also apparent in the observational record, and their dominant cause is the passage of dark sunspots across the disk of the Sun. These appear as dips in TSI data of about 0.1% and last for several days as the sunspots and sunspot groups traverse the center of the solar disk. The dips due to the passage of sunspots appear as "noise" in the time series of Figure 2, and of course they are far more prominent when sunspots are abundant during solar maxima.

Figure 3 provides an expanded view of TSI during a four-month period in 2004, and this more clearly shows the passage of dark sunspots and the associated enhancements due to the bright faculae. Since the faculae associated with the sunspots are extended features and more uniformly spread across the solar disk, their positive contribution to TSI does not produce intermediate- and short-term increases as striking as the sunspot dips.

## 2.2. Solar Spectral Irradiance

The measurement of solar spectral irradiance, SSI, is similar to the measurement of TSI with the important additional complexity that incoming solar radiation must first be separated by wavelength before the detector records the signal. A simple optical device might employ a filter to isolate a broad or narrow spectral band, while more precise measurements would be provided by a spectrometer using a prism, grating or other dispersive element to isolate a wavelength interval. A typical science requirement is to achieve a spectral resolution ($\Delta\lambda/\lambda$) on the order of 0.01 (resolving power greater than 100) or better. Figure 4 demonstrates the dramatic range of the solar spectrum from X-ray wavelengths to infrared – more than 3 orders

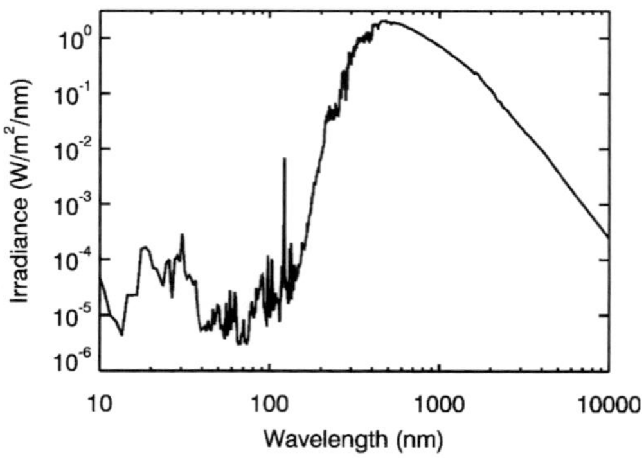


*Figure 4.* The TOA solar irradiance as shown in Figure 1, but in a log-log format that emphasizes the weaker emissions and the shorter wavelengths. More than 95% of the radiative power is in the visible and infrared. This spectrum is a composite of data taken on 21 April 2004 by the XPS, SOLSTICE and SIM on SORCE and by the EGS from TIMED.

of magnitude in wavelength and more than 5 orders of magnitude in intensity. No single instrument is capable of covering this large dynamic range and broad spectral extent, and in general the measurement requires several spectrometers each covering a specific portion of the spectrum.

Early observations of solar spectral irradiance concentrated on the ultraviolet wavelengths – the extreme ultraviolet and X-rays ($\lambda < 100$ nm) because of their importance to the ionization and heating of the Earth's upper atmosphere and the ultraviolet (100 nm $< \lambda <$ 300 nm) because of its importance to ozone photochemistry in the middle atmosphere. These observational programs advanced with steady improvement, beginning with multiple sounding rocket observations and progressing to long-duration observations from satellites. The spectral coverage was expanded and observing techniques were refined, until slowly an understanding of the more energetic ultraviolet irradiance evolved. A workshop was held in 1975 from which a compilation of review articles (White, 1977) summarized the prevailing understanding of the solar spectrum and its variations as derived from the first twenty years of space observations. Some of the conclusions made at that time remain true today, while others were incorrect probably due to overly optimistic assumptions of the observational uncertainty. For example, at that time there was conjecture that a solar cycle variation as large as 20% occurred near 300 nm, whereas subsequent observations now constrain the variations at 300 nm to no more than 1%.

Since 1975 there has been significant improvement in the precision and accuracy of spectral irradiance observations. In particular the Upper Atmosphere Research Mission, UARS, operated from September 1991 to August 2005 and carried two ultraviolet spectrometers, SOLSTICE (Rottman *et al.*, 1993) and SUSIM (Brueckner *et al.*, 1993) covering the spectral range of 120 to 420 nm. These observations cover more than one full solar cycle and have provided a reliable estimate of solar

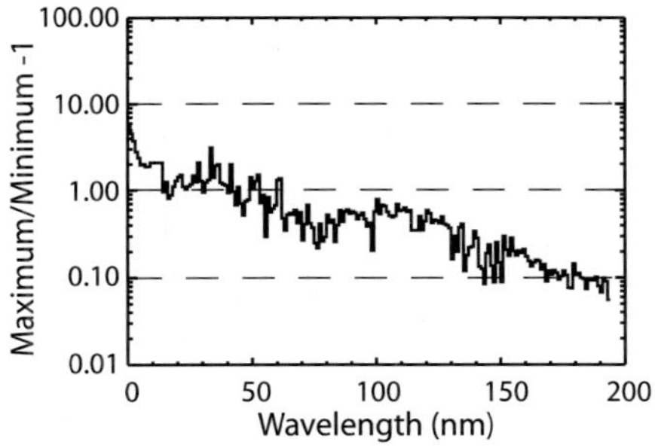


*Figure 5.* This plot shows solar variations as determined from TIMED observations during a two-year portion of solar cycle 23 (Woods *et al.*, 2005b). The log scale is used because of the large range of variability across this wavelength interval. High levels of irradiance in early 2002 are compared to lower levels in early 2004, and ratios of the two irradiance levels minus 1 (*maximum/minimum – 1*) are plotted versus wavelength. Near 200 nm the variations are 10% (i.e., *maximum/minimum* is 110% and plotted as $1.1 - 1 = 0.10$), increasing to 100% (i.e., $2 - 1 = 1.00$) near 50 nm, and approaching a factor of 10 (i.e., $10 - 1 = 9$) at the shortest X-ray wavelengths.

variations between 120 to 300 nm, and are described in publications of the first ISSI workshop on Solar Variability and Climate (Rottman, 2000).

Today there is consensus on at least three major aspects of solar cycle variations. First, emission from the solar corona and transition region dominate the very short-wavelength and energetic X-ray and EUV portions of the solar spectrum. Here the X-rays vary by factors of ten and larger and the EUV ($\lambda < 120$ nm) by factors of two (see Figure 5 and discussion in Woods *et al.*, 2005b). At the somewhat longer UV wavelengths (200 nm $< \lambda <$ 300 nm) emission from the solar chromosphere and upper photosphere dominate the spectrum, and here the solar cycle variations range from 1% to 70% (see Figure 6 and discussion in Rottman *et al.*, 2004). A second important finding is that the integrated effect of this UV ($\lambda < 300$ nm) variability accounts for roughly 30% of the solar cycle variation of TSI discussed earlier in Section 2.1, with the remaining 70% assigned to the visible and infrared portions of the spectrum (London and Rottman, 1989; Lean, 1989). The inevitable third aspect is therefore that the visible and infrared irradiance as constrained by the TSI observations (variation $\sim$0.1%) must also be extremely small and well below one percent.

Considering this final point, the measurement of such small variations of spectral irradiance in the visible and near infrared is a technical challenge. It requires that the space-based spectrometer provides a stable responsivity over many years on-orbit and that it be capable of establishing solar variability at the level well below 0.1%. The Spectral Irradiance Monitor, SIM, on SORCE launched in January 2003, was specifically designed to achieve this level of precision and accuracy. SIM is a newly developed prism spectrometer measuring solar irradiance from 200 to 2400 nm (Harder *et al.*, 2005a,b). The science objective of SIM is to make these

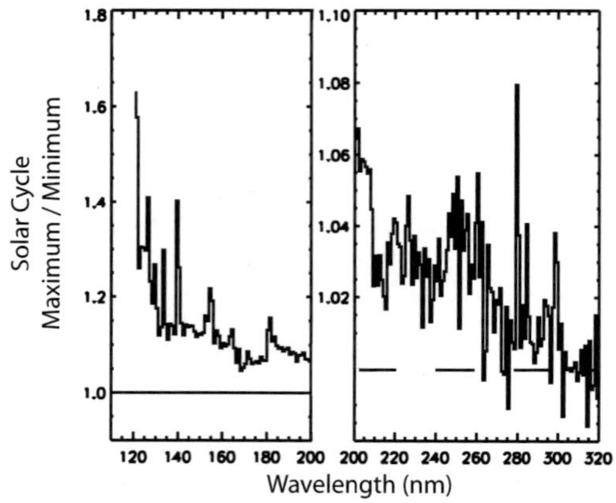


*Figure 6.* Solar cycle variations observed by UARS from solar maximum in 1992 to solar minimum in 1996. The solar maximum values are an 81-day average in early 1992 and the minimum values are an 81-day average in mid 1996. The ratio of the maximum to minimum is plotted and values range from less than 1% to almost a factor of 2.

visible to near infrared measurements with a combined standard uncertainty of less than 0.1% and precision and long-term relative accuracy of 0.03%.

Three samples of the SIM data are shown in Figure 7 at wavelengths of 656 nm, 857 nm, and 1000 nm (see Rottman *et al.*, 2005). All data are normalized to an arbitrary reference day and each panel compares the spectral data to the TSI as shown in Figure 3. It is interesting to note that in the visible SSI and TSI closely match, especially regarding darkening due to the passage of sunspots across the solar disk. Further in the infrared the darkening of sunspots is progressively less. The variations displayed in this figure span a relatively short time period (about four months), and it will remain to be seen what magnitude and what shape the SSI variations will take throughout the solar cycle. It will also be interesting to see how the long-term variations of SSI compare with TSI observations as shown in Figure 2. Of course, such comparisons will require that the SIM observations achieve long-term relative accuracy comparable with the TSI accomplishment of a few tens of parts per million (<0.01%).

## 3. Today's Measurement and Prospects for the Future

Figure 8 is a historical view of the many irradiance observations made since the 1960s. The top row contains the TSI observations, and the following rows are the spectral observations in order of decreasing wavelength. The encouraging message is that there has been nearly complete temporal coverage, often with duplicate measurements. For example, the top row showing TSI observations is more fully described by the data of Figure 2. It is easy to understand the confusion and ambiguity that would arise if only one or two of these observations were available.

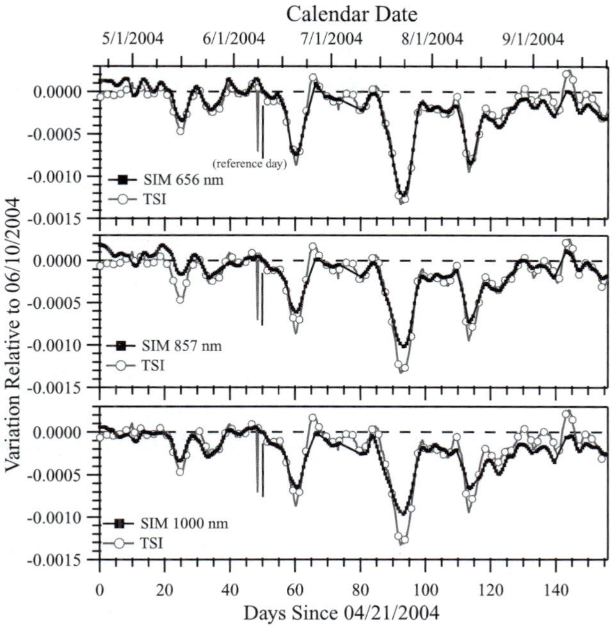


*Figure 7.* Three sample time series of SIM data from the visible to near infrared. These data are compared to the TSI data for the same period. The spectral data are daily values and do not include the Venus transit on 08 June 2004 as seen in the higher cadence TSI data.

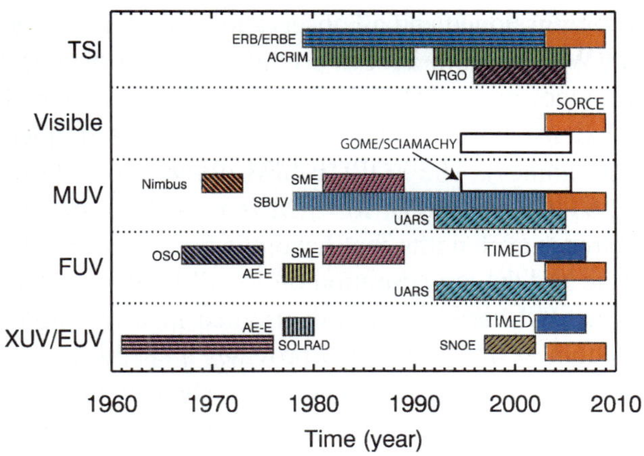


*Figure 8.* Approximate historical record of total solar irradiance observations and spectral irradiance observations. The spectral programs are sorted by wavelength.

Gaps in the TSI time series would have prohibited the construction of a composite data set described in an accompanying paper (Fröhlich, 2006), and today there would be little insight regarding the true variability of TSI. In general the UV (MUV: 200 nm $< \lambda <$ 300 nm and FUV: 100 nm $< \lambda <$ 200 nm) solar irradiance has also been well covered, with only a few gaps and with multiple observations. The EUV and X-ray irradiance observations started aggressively in the early 1960s, but then there was a long hiatus with no observations through the 1980s and most of the 1990s. This period, referred to as the EUV gap, has been detrimental to upper atmospheric studies conducted during that time. Fortunately, EUV observations were resumed with the TIMED mission launched late in 2001 (Woods *et al.*, 2005b) and continuing today.

### 3.1. On-Going Programs

TSI is presently being measured by three separate missions – ACRIMSat, VIRGO and SORCE. ACRIMSat and VIRGO are in an extended mission phase, and hopefully will continue for several additional years. The SORCE TIM data will extend the TSI record, and they have the additional goal of establishing a new level of precision and accuracy (~300 ppm). The SORCE TSI data are complementary to the ACRIMSat and VIRGO TSI data, providing mutual validation and redundancy.

The SORCE SOLSTICE observations extend the UV irradiance record, which began with SBUV in 1978 and SME in 1981. SORCE has now had two years overlap with the UARS SOLSTICE (Rottman *et al.*, 1993) and SUSIM (Brueckner *et al.*, 1993) observations, and in particular, the stellar comparison technique of SOLSTICE (McClintock *et al.*, 2005a,b) provides a direct and reliable method of tying together the UARS and SORCE data. After 14 years of successful operations, the UARS was decommissioned and all observations were ended late in 2005. The SORCE XPS data (Woods *et al.*, 2005a) provide essential overlap and validation with similar data provided by the TIMED XPS (Woods *et al.*, 2005b), but in addition, SORCE has more extensive wavelength coverage and more frequent observations (~5 minutes vs 90 minutes). The SORCE SIM observations, however, are quite different than SOLSTICE and XPS, for SIM is providing new and unique data at visible and infrared wavelengths that have not been recorded before. The full wavelength coverage and the long duration of the SIM measurements will provide insight as to how the TSI variations are distributed in wavelength. Likewise the SCIAMACHY (Skupin *et al.*, 2005) also provides a spectral irradiance data set that will be used to validate the SIM observations. The new findings of visible and near infrared irradiance variations will have important implications regarding the atmospheric and climate response to solar variability. They will also improve and constrain models that describe the radiation processes and energy balance of the Sun.

### 3.2. Future Programs

There are plans to include the TIM and SIM instruments on the NOAA/DoD/NASA National Polar-orbiting Operational Environmental Satellite System, NPOESS. This series of satellites may have a first launch opportunity after 2010. The two irradiance instruments comprise the Total Solar Irradiance System, TSIS, presently scheduled on the second platform with a launch in 2013 (all of these dates are of course subject to change). The NPOESS SIM will have extended capability in the UV with wavelengths as short as 200 nm, but will not recoup the entire spectral range of SOLSTICE down to Lyman-$\alpha$ (121.6 nm). The immediate problem is that it is unlikely that SORCE will remain operational for ten or more years, and there is consequently a real danger that a gap will occur in the irradiance record. The break in the data sets is troubling for two reasons – first, if there are missing data they can only be filled with proxy or modeled data leading to ambiguity in the long-term record; and second, the two (or more) observational records that do not overlap must rely on their individual, inherent accuracies to bridge the break in the data, more than likely compromising the entire long-term climate record.

There is a plan to partially fill a potential gap between SORCE and NPOESS by flying a TIM only (without SIM) on a NASA mission called Glory. This mission may launch in 2008 (again subject to change) and therefore, more than likely it will overlap with SORCE, but it will take a fortuitous extension to its three-year lifetime to reach all the way to NPOESS. Glory does not include SIM or any other spectral irradiance measurement capability, and will therefore leave the long-term spectral irradiance record in great jeopardy. Moreover, there is presently no plan that includes the far UV spectral irradiance after SORCE – a fact most troubling to maintaining the long-term UV data record, but also a great loss to process studies in atmospheric chemistry. An additional mission, PICARD (Thuillier, private communication, 2005) with a primary scientific objective of measuring the solar diameter will launch in 2008. This mission will also carry SOVA (Solar Variability) to measure TSI and PREMOS (Precision Monitoring of Solar Variability) to measure selected UV and visible spectral bands of SSI.

In the meantime, and at least for the next three to five years, SORCE will continue to make irradiance observations, filling out the long-term climate record, providing a fundamental measurement for atmospheric studies, and likely leading to important new discoveries related to atmospheric sciences and solar physics.

## Acknowledgements

This research is supported at the University of Colorado, Boulder, under NASA contract NAS5-97045. Special recognition is extended to the scientists, professional staff, and students at the Laboratory for Atmospheric and Space Physics whose commitment and dedication insure the quality and reliability of the SORCE and TIMED data sets. Likewise, recognition and appreciation is extended to the many other research groups participating in the careful and precise radiometric study of the Sun.

## References


Abbot, C. G.: 1948, 'The Radiation of the Planet Earth to Space', Smithsonian Miscellaneous Collections **110**, Publication 3940.

Brueckner, G. E., Edlow, K. L., Floyd, L. E., Lean, J. L., and VanHoosier, M. E.: 1993, 'The solar ultraviolet spectral irradiance monitor (SUSIM) experiment on board the upper atmospheric research satellite (UARS)', *J. Geophys. Res.* **98**, 10695–10711.

Foukal, P. and Lean, J.: 1988, 'Magnetic modulation of solar luminosity by photospheric activity', *Astrophys. J., Part I* **328**, 347–357.

Friedman, H.: 1961, 'X-ray and Ultraviolet Radiation Measurements from Rockets', in: W. Liller (ed.), *Space Astrophysics*, McGraw-Hill, N.Y., pp. 107–120.

Fröhlich, C.: 2004, 'Solar Irradiance Variability', in: J. Pap, C. Fröhlich, H. Hudson, J. Kuhn, J. McCormack, G. North, W. Sprig, and S. T. Wu (eds.), *Solar Variability and Its Effects on Climate*, Geophysical Monograph Series **141**, American Geophysical Union, Washington, D.C., pp. 97–110.

Fröhlich, C.: 2006, 'Solar irradiance variability since 1978', *Space Sci. Rev.*, this volume, doi: 10.1007/s11214-006-9045-6.

Fröhlich, C. and Lean, J.: 1998, 'The Sun's total irradiance: Cycles, trends, and related climate change uncertainties since 1976', *Geophys. Res. Lett.* **25**, 4377–4380.

Haigh, J. D.: 2001, 'Climate variability and the influence of the Sun', *Science* **294**, 2109–2111.

Harder, J., Lawrence, G., Fontenla, J., Rottman, G., and Woods, T.: 2005a, 'The spectral irradiance monitor: Scientific requirements, instrument design, and operation modes', *Solar Phys.* **203**, 141–167.

Harder, J., Fontenla, J., Lawrence, G., Woods, T., and Rottman, G.: 2005b, 'The spectral irradiance monitor: Measurement equations and calibration', *Solar Phys.* **203**, 169–203.

Hickey, J. R., Stowe, L., Jacobowitz, H., Muschoff, R., House, F., and VonderHaar, T.: 1980, 'Initial solar irradiance determinations from Nimbus 7 cavity radiometer measurements', *Science* **208**, 281.

Hickey, J. R., Alton, B. M., Kyle, H. L., and Hoyt, D. V.: 1988, 'Total solar irradiance measurements by ERB/Nimbus–7, a review of nine years', *Space Sci. Rev.* **48**, 321–342.

Kiehl, J. T. and Trenberth, K. E.: 1997, 'Earth's annual global mean energy budget', *Bull. Amer. Meteor. Soc.* **78**, 197–208.

Kopp, G. and Lawrence, G.: 2005, 'The total irradiance monitor (TIM): Instrument Design', *Solar Phys.* **203**, 91–109.

Kopp, G., Heuerman, K., and Lawrence, G.: 2005, 'The total irradiance monitor (TIM): Instrument Calibration', *Solar Phys.* **203**, 111–127.

Lean, J.: 1989, 'Contribution of ultraviolet irradiance variations to changes in the Sun's total irradiance', *Science* **244**, 197–200.

Lean, J., Rottman, G., Harder, J., and Kopp, G.: 2005, 'SORCE contributions to new understanding of global change and solar variability', *Solar Phys.* **203**, 27–53.

Lee, R. B., Barkstrom, B. R., and Cess, R. D.: 1987, 'Characteristics of the earth radiation budget experiment solar monitors', *Appl. Opt.* **26**, 3090–3096.

London, J. and Rottman, G. J.: 1989, 'The contribution of solar UV irradiance variations to variations of the solar constant', in: J. Lenoble and J. Geleyn (eds.), *IRS'88: Current Problems in Atmospheric Radiation*, A. Deepak Publishing, Hampton, Virginia, pp. 472–473.

McClintock, W. E., Rottman, G. J., and Woods, T. N.: 2005a, 'Solar stellar irradiance comparison experiment II (SOLSTICE II): Instrument concept and design', *Solar Phys.* **203**, 225–258.

McClintock, W. E., Snow, M., and Woods, T. N.: 2005b, 'Solar stellar irradiance comparison experiment II (SOLSTICE II): Pre-launch and on-orbit calibrations', *Solar Phys.* **203**, 259–294.

Menzel, D. H.: 1949, *Our Sun*, The Blakiston Co., Garden City, NY.

Rind, D.: 2002, 'The Sun's role in climate variations', *Science* **296**, 673–677.

Rottman, G. J.: 2000, 'Variations of solar ultraviolet irradiance observed by the UARS SOLSTICE – 1991 to 1999', *Space Sci. Rev.* **94**, 83–91.

Rottman, G.: 2005, 'The SORCE mission', *Solar Phys.* **203**, 7–25.

Rottman, G. J., Woods, T. N., and Sparn, T. P.: 1993, 'Solar stellar irradiance comparison experiment I: 1 instrument design and operation', *J. Geophys. Res.* **98**, 10,667–10,677.

Rottman, G. J., Floyd, L., and Viereck, R.: 2004, 'Measurement of the solar ultraviolet irradiance', in: J. Pap, C. Fröhlich, H. Hudson, J. Kuhn, J. McCormack, G. North, W. Sprig, and S. T. Wu (eds.), *Solar Variability and Its Effects on Climate*, Geophysical Monograph Series **141**, American Geophysical Union, Washington, DC, pp. 111–126.

Rottman, G., Harder, J., Fontenla, J., Woods, T., White, O. R., and Lawrence, G.: 2005, 'SIM: Early observations', *Solar Phys.* **203**, 205–224.

Skupin, J., Noël, S., Wuttke, M. W., Gottwald, M., Bovensmann, H., Weber, M., and Burrows, J. P.: 2005, 'SCIAMACHY solar irradiance observation in the spectral range from 240 to 2380 nm', *Adv. Space Res.* **35**, 370–375.

White, O. R.: 1977, *The Solar Output and Its Variation*, Colorado Assoc. University Press, Boulder, CO.

Willson, R. C.: 1984, 'Measurements of solar total irradiance and its variability', *Space Sci. Rev.* **38**, 203–242.

Willson, R. C.: 1994, *The Sun as Variable Star, Solar and Stellar Irradiance Variations*, Cambridge Univ. Press, Cambridge, UK.

Willson, R. C. and Hudson, H. S.: 1991, 'The Sun's luminosity over a complete solar cycle', *Nature* **351**, 42–44.

Willson, R. C., Gulkis, S., Janssen, M., Hudson, H. S., and Chapman, G. A.: 1981, 'Observations of solar irradiance variability', *Science* **211**, 700–702.

Woods, T. N., Rottman, G., and Vest, R.: 2005a, 'XUV photometer system (XPS): Overview and calibrations', *Solar Phys.* **203**, 345–374.

Woods, T. N., Eparvier, F. G., Bailey, S. M., Chamberlin, P. C., Lean, J., Rottman, G. J., Solomon, S. C., Tobiska, W. K., and Woodraska, D. L.: 2005b, 'Solar EUV Experiment (SEE): Mission overview and first results', *J. Geophys. Res.* **110**, doi:10.1029/2004JA010765.

# SOLAR IRRADIANCE VARIABILITY SINCE 1978

*Revision of the PMOD Composite during Solar Cycle 21*

C. FRÖHLICH

*Physikalisch-Meteorologisches Observatorium Davos, World Radiation Center, CH-7260 Davos Dorf, Switzerland*

*(E-mail: cfrohlich@pmodwrc.ch)*

(Received 22 August 2005; Accepted in final form 9 February 2006)

**Abstract.** Since November 1978 a set of total solar irradiance (TSI) measurements from space is available, yielding a time series of more than 25 years. Presently, there are three TSI composites available, called PMOD, ACRIM and IRMB, which are all constructed from the same original data, but use different procedures to correct for sensitivity changes. The PMOD composite is the only one which also corrects the early HF data for degradation. The results from the detailed analysis of the VIRGO radiometry allow a good understanding of the effects influencing the long-term behaviour of classical radiometers in space. Thus, a re-analysis of the behaviour of HF/NIMBUS-7 and ACRIM-I/SMM was indicated. For the former the situation is complicated by the fact that there are no in-flight means to determine changes due to exposure to solar radiation by comparison with a less exposed radiometer on the same spacecraft. The geometry and optical property of the cavity of HF is, however, very similar to the PMO6-type radiometers, so the behaviour of the PMO6V radiometers on VIRGO can be used as a model. ACRIM-I had to be revised mainly due to a henceforth undetected early increase and a more detailed analysis of its degradation. The results are not only important for solar radiometry from space, but they also provide a more reliable TSI during cycle 21. The differences between the revised PMOD composite and the ACRIM and IRMB are discussed by comparison with a TSI reconstruction from Kitt-Peak magnetograms. As the PMOD composite is the only one which has reliable data for cycle 21, the behaviour of the three solar cycles can now be compared and the similarities and differences discussed.

**Keywords:** total solar irradiance, solar variability, solar activity

## 1. Introduction

Since late 1978 total solar irradiance (TSI) measurements were made by different radiometers in space, HF on NIMBUS 7, ACRIM I on SMM, ACRIM II on UARS, VIRGO on SOHO, ACRIM III on ACRIMSat and since 2003 TIM on SORCE (not used in the construction of the composites). Figure 1 shows these original time series and it is clear that not only the absolute values are quite different, especially at the beginning of the series, but there are also important differences between the series. This is, e.g., obvious from comparison of early ACRIM-I with HF and is due to the fact that the original data from the HF radiometer cannot be corrected for degradation by internal means. These time series – either as they are and/or corrected for some effects not considered in the original data sets – can be used for the construction of a TSI composite by shifting each series to a common level

Space Science Reviews (2006) 125: 53–65

DOI: 10.1007/s11214-006-9046-5 

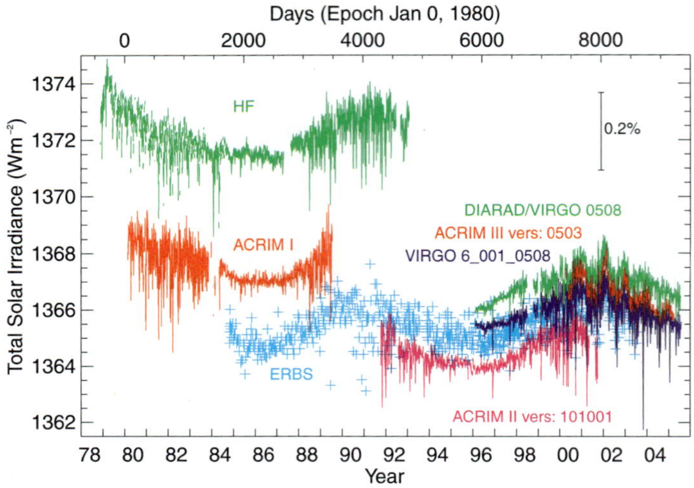


*Figure 1.* Compared are daily averaged values of the Sun's total irradiance from radiometers on different space platforms as published by the instrument teams since November 1978. Note that the VIRGO TSI is determined from both VIRGO radiometers (PMO6V and DIARAD), whereas the DIARAD TSI is only based on this one.

and merging them together. Presently there are three composites available, the first one was presented in 1997 at the IAU General Assembly in Kyoto by Fröhlich and Lean (1998b) and is now called PMOD composite. A few months later the ACRIM composite was published by Willson (1997) which has been updated in 2003 (Willson and Mordvinov, 2003). Recently a third composite, called IRMB, was presented by Dewitte *et al.* (2004).

Already in the first versions of the PMOD composite corrections for the HF degradation were introduced (Fröhlich and Lean, 1998a, b). Due to the fact that the HF radiometer is similar to the PMO6V radiometers on VIRGO/SOHO the corrections were based on early results from VIRGO and used exponential functions to fit the changes due to an early increase and the degradation, and a linear trend to account for a gradual increase of the sensitivity. The origin of the latter effect is still unclear, but it was needed to explain the behaviour of HF up to about 1986. The so corrected data set was then used before the start of ACRIM-I and during the spin mode of SMM. Similarly, the degradation of ACRIM-I during its first year was corrected for the effect of the rather short exposure time during the spin mode which was not taken into account in the original treatment by Willson and Hudson (1991). The two time series with the corrections mentioned were the basis for the composite of Fröhlich and Lean (1998a) during the period before the end of ACRIM-I in 1989 and remained unchanged up to version d40_61_0502. The other two composites use the data as published during this period: HF up to 1980 and then ACRIM-I.

Another problem for all composite constructions is how to bridge the so-called ACRIM gap between the end of ACRIM-I and the start of ACRIM-II, from June

1989 to October 1991. During this period daily values from HF and some 70 data points from ERBS with a sampling every 14 days are available. A detailed comparison of the two series by Lee III *et al.* (1995) revealed two slips in the HF data resulting in a total change of $-0.68\ \mathrm{Wm}^{-2}$ over this period. This was confirmed by comparison with a model from the San Fernando group (Chapman *et al.*, 1996). Fröhlich (2000) re-analyzed this period and the overall change was confirmed, but with a slightly different value of $-0.58\ \mathrm{Wm}^{-2}$. The date of the second slip was taken from Lee III *et al.* (1995) although it was difficult to really identify it. In Fröhlich and Lean (2002) it was first suggested that a gradual sensitivity increase of HF over the whole period together with a step at the first place would better represent the change. Moreover, it was recognized that the trend was very close to the one identified from comparison with ACRIM-I up to 1984 (Fröhlich and Lean, 1998a). The combination of a slip at 29 September 1989 after a switch-off of HF for four days ($0.417 \pm 0.043\ \mathrm{Wm}^{-2}$) and a linear trend ($0.349 \pm 0.103$ mW $\mathrm{m}^{-2}\mathrm{d}^{-1}$) was implemented in Fröhlich (2004) and yielded a total change over the gap of $-0.84 \pm 0.07\ \mathrm{Wm}^{-2}$. The standard deviation of the ratio of HF to ERBS or the proxy model decreased by 45 and 60 ppm, respectively, or up to 20% with the corrections included. All these corrections were determined from data during the period of the ACRIM gap. As shown later, the new corrections for the HF are determined for the full time series from November 1978 until January 1993 at once, after a search and correction of glitches throughout the mission. Thus, these corrections are now internally consistent and there is no longer any need to treat the period of the ACRIM gap separately for the construction of the PMOD composite. The ACRIM composite neglects the corrections of the HF during the gap and this is the main reason for the claimed upward trend of TSI over the last 25 years. The IRMB composite traces ACRIM-II to I via ERBS and thus the difference between the two minima is also not significant, but only if the minimum value is calculated without the DIARAD data.

In the following Sections we describe first the new procedure for ACRIM-I and HF, present the results and the new PMOD composite, discuss the differences to the earlier versions and compare it with the other two composites. Finally we compare the three solar cycles with a proxy model and discuss the similarities and differences between them.

## 2. Radiometric Corrections

The most important effect influencing space radiometry is what is normally called degradation. As this effect depends on the time of exposure to solar radiation it can be determined by comparison with a less exposed back-up radiometer of the same type. Only the ACRIM and VIRGO radiometers have backups. As the backup measurements have to be scarce, a way to interpolate between the ratios has to be devised. For this the model developed during the detailed analysis of the PMO6V radiometry on VIRGO/SOHO is ideally suited. It is based on hyperbolic functions which take the dose of irradiation explicitly into account (a detailed description can be found in Fröhlich, 2003).

Another effect is the early increase in sensitivity of radiometers in space which was first identified for the PMO6V radiometers on VIRGO and also for the HF, which is the main reason for the high values at the beginning of this record. This effect is important for all radiometers used in space missions which have their primary aperture directly in front of the cavity. It is due to a blackening of the primary aperture by the strong solar UV radiation in space which increases the temperature of the innermost part of the aperture and simulates an increased sensitivity by the extra IR radiation emitted into the cavity. Only apertures with a small cylindrical part are concerned, and thus, this effect is important for the HF, ACRIM, ERBS and PMO6 radiometers, but not for DIARAD which has a much larger land. During the evaluation of the PMO6V radiometers it was recognized that the early increase was followed by a short-term degradation (Fröhlich, 2003). The inspection of the SOVA-2 radiometers which were on the EURECA platform and retrieved after 10 months in space (Crommelynck *et al.*, 1993) show that the early blackening is followed by a bleaching which explains the combination of the early increase and the short-term degradation of the PMO6V.

In the following we describe the newly determined corrections of ACRIM-I and HF, and discuss the differences to the early treatment.

### 2.1. Corrections for ACRIM-I on SMM

A first result of the re-analysis of ACRIM-I was the detection of an early increase with similar amplitudes as observed for PMO6V. As there are not enough data points available, however, only the blackening effect can be determined as shown in Figure 2; the neglect of the effect of the bleaching weakens the blackening somewhat and the remaining part is included in the overall degradation.

In a next step the degradation is re-analyzed. In Fröhlich and Lean (1998a) it was already realized that there is a problem related to the fact that during the spin-mode operation of SMM (after failure of the pointing system of SMM in late 1980 until the spectacular repair in 1984 from the shuttle by astronauts) the exposure was drastically reduced. This effect was not accounted for in the original corrections by Willson and Hudson (1991) and the results of the improved degradation analysis are shown in Figure 3. The s-shape of the degradation curve illustrates the influence of the dose changing with solar activity from the maximum at the beginning through the minimum from days 2000–2600 into the ascending part of cycle 22. The flat part from days 320–1500 represents the spin-mode data with much less exposure. After the repair of SMM a further complication has been observed due to the rather long switch-off during the repair and a correction is determined by comparison with HF, ERBS and the proxy model, resulting in an exponential function with an

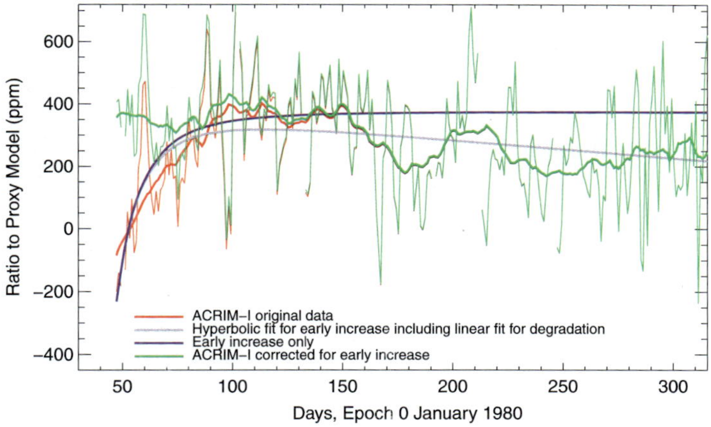


*Figure 2.* Shown are the early measurements of ACRIM-I together with a fit of a hyperbolic function describing the early increase and a linear fit taking the long-term degradation into account during the first few hundred days.

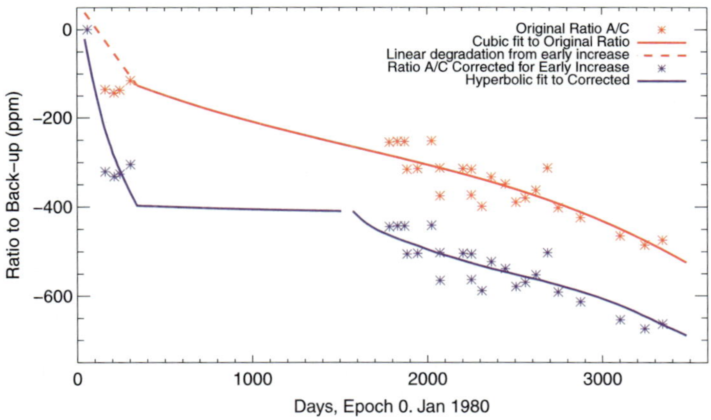


*Figure 3.* The original ratios of sensor A to sensor C (red symbols) of the ACRIM-I experiment on SMM are from Willson and Hudson (1991). The cubic fit corresponds to the correction originally applied and the dashed line includes the linear fit found by fitting the early increase. The blue symbols are corrected for the early increase and then fitted with a hyperbolic function.

amplitude of 88 ppm and a time constant of 80 days. Finally the corrected time series is presented in Figure 4.

### 2.2. Corrections for HF on NIMBUS-7

For the HF radiometer the situation is more complicated as there is no back-up instrument which can be used for in-flight corrections and there are many slips or glitches in the data which are thought to be due to operational changes on the spacecraft and have not been taken into account in the evaluation by Hoyt *et al.*

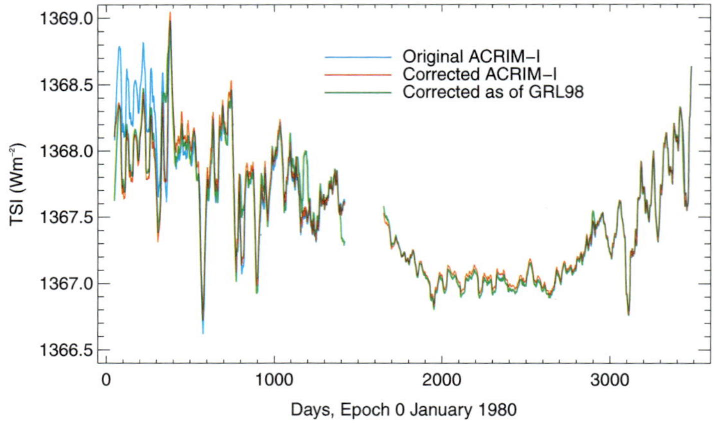


*Figure 4.* The original ACRIM-I record is compared to the corrected one as described in the text. Also plotted is the record as it was used in Fröhlich and Lean (1998a). The overall behaviour is very similar, the differences are in the details.

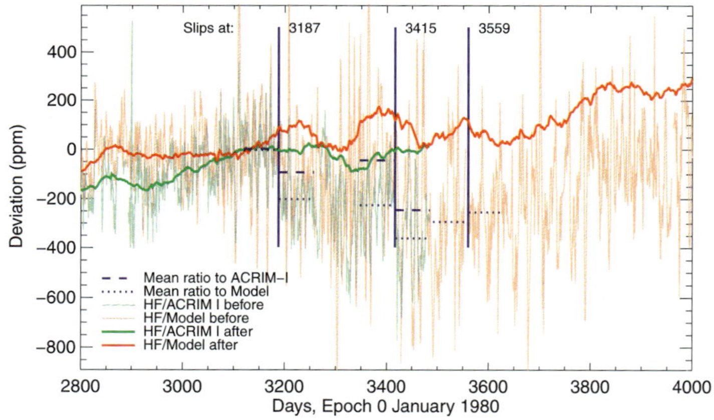


*Figure 5.* Shown is the period around the end of the ACRIM-I record with two slips of HF during ACRIM-I and one just after. The latter is the famous one responsible for the difference between the PMOD and ACRIM composite. "Before" and "After" mean before and after the corrections applied, with the "Before" plotted as daily values and the "After" as 81-day running means.

(1992). An example is presented in Figure 5. Besides the three glitches shown in the figure, more than 15 have been identified in the whole record by local comparison with a proxy model and/or ACRIM-I and were corrected. Also the famous one of 28 September 1989 (day 3559) is clearly identified in Figure 5.

After having removed the slips and glitches we need to correct for the exposure-dependent early increase, the degradation, and for an increase of the sensitivity which may not be exposure-dependent. As there is no backup instrument we need a reference for the early observations. This reference is built from the proxy model calibrated against the corrected ACRIM-I and used to extrapolate ACRIM-I back to

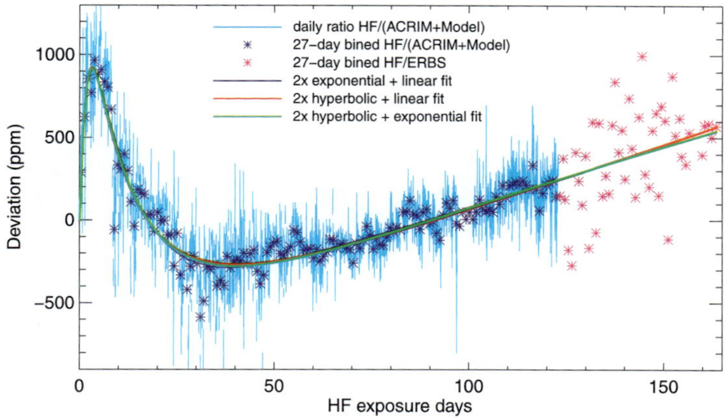


*Figure 6.* Shown is the ratio of HF, corrected for glitches, to the reference TSI. Corrections are needed for an early increase, for degradation, which may also include the bleaching of the aperture, and for a long-term increase of the sensitivity. The latter can be modelled with different functions, in the following the exponential one is preferred.

November 1978, the start of NIMBUS-7. For the period after February 1980 we use the corrected ACRIM-I data until their end and then ERBS data. Figure 6 illustrates the procedure and the results. The non-exposure dependent effect is modelled with an exponential function as it is similar to the one found for DIARAD/VIRGO by Fröhlich (2003). So the corrections during the ACRIM gap need no longer to be determined by comparison with ERBS and a proxy model as in Fröhlich (2000). The final result of the corrected HF data set is shown in Figure 7 which demonstrates how important these corrections are (up to 1 $Wm^{-2}$). Moreover, the internal consistency of the corrections prove their reliability, and also their need.

### 2.3. Corrections for ACRIM-II on UARS and VIRGO on SOHO

ACRIM-II needs also some corrections which have been determined in Fröhlich (2004) and are not changed since. Finally, we use version 6_001_0508 for the VIRGO data, updated to July 2005. The corrections to get from level-1 to level-2 data are described in Fröhlich (2003) and on `http://www.pmodwrc.ch/pmod.php?topic=tsi/virgo` from where the hourly and daily TSI data from VIRGO can be downloaded.

## 3. The PMOD Composite

Having applied all the described corrections we are almost ready to construct the composite. We need to refer ACRIM-II to ACRIM-I which is done by a weighted

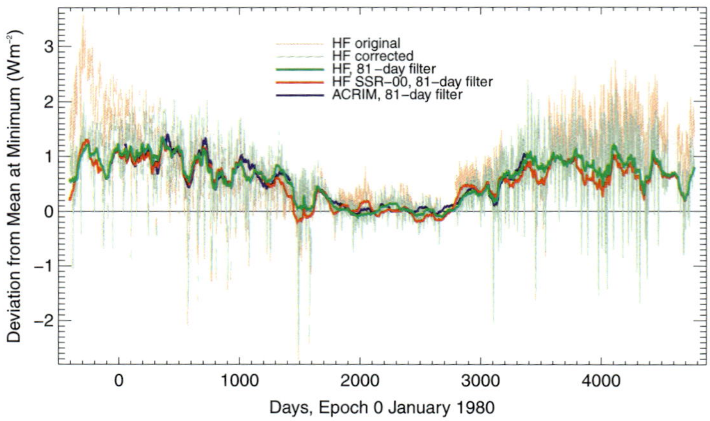


*Figure 7.* The final result of the HF after corrections is compared to the original data set and to ACRIM-I. Differences between the corrected and original data is up to 1 Wm$^{-2}$. Also plotted is the corrected HF as used in Fröhlich (2000).

average of the ratios of ACRIM-I to the corrected HF and ERBS data and the corresponding average ratios to ACRIM-II. The result is shown in Figure 8. Note, that this result depends on the corrections applied to HF and ACRIM-I and II. The correction of ACRIM-II changed by only about 30 ppm with the new HF compared to the one used by Fröhlich (2000).

Having done this scaling the rest is straightforward and consists in adjusting the HF and VIRGO measurements to the ACRIMs leaving them at their respective levels. For practical reasons the result is then scaled to SARR (Crommelynck *et al.*, 1995) and is shown in Figure 9. The differences to the earlier versions have been shown in the Figures 4 and 7 and lie more in the details than in the overall behaviour. Thus the final result does not change the trend or the amplitudes significantly. Overall it is a more reliable time series mainly due to the fact that the corrections of the different records are based on the same type of model.

In Figure 10 the PMOD, ACRIM (Willson and Mordvinov, 2003) and IRMB (Dewitte *et al.*, 2004) composites are compared to a reconstruction of TSI from Kitt-Peak magnetograms by Wenzler (2005). These reconstructions are based on the identification of sunspot (umbra, penumbra and pores) pixels from white light images and of quiet, network and facular pixels from the magnetograms (see, e.g., Wenzler *et al.*, 2005). Together with a contrast from, e.g., Fligge *et al.* (2000) the radiance of each pixel is determined and summed up to yield the irradiance. For the bright pixels a filling factor is defined which depends linearly on the magnetic field in the pixel up to a saturation field after which it is set to one. This saturation field is the only free parameter in the reconstruction which is determined by calibration against the PMOD composite as 340 G for the Kitt-Peak magnetograms (Wenzler, 2005). The best agreement is with the PMOD composite explaining 83% of the

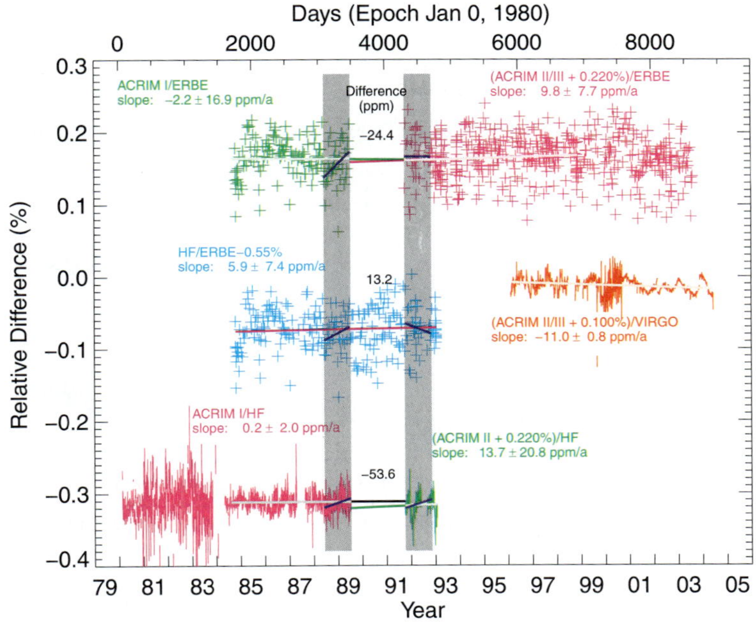


*Figure 8.* This plot illustrates the tracing of ACRIM-II to ACRIM-I. The change of the ACRIM-II level is determined by a weighted average of the comparison of ACRIM-I and II with HF and ERBS during the periods indicated by shading.

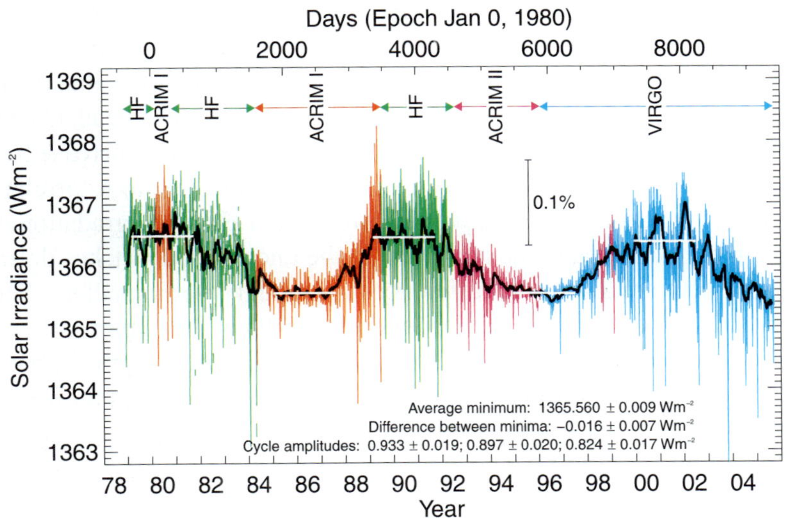


*Figure 9.* Shown is the final version of the PMOD composite. Compared to the earlier versions the maximum of cycle 21 is at about the same level as before, but has less noise, especially in the early part. This may indicate that the early HF corrections have indeed been improved. Finally, the difference between the minima has also not changed. The dates of the maxima are 26/03/1979–25/12/1981, 28/04/1989–21/02/1992 and 19/01/2000–18/02/2003 and of the minima 13/04/1985–07/06/1987 and 05/05/1995–02/08/1997.

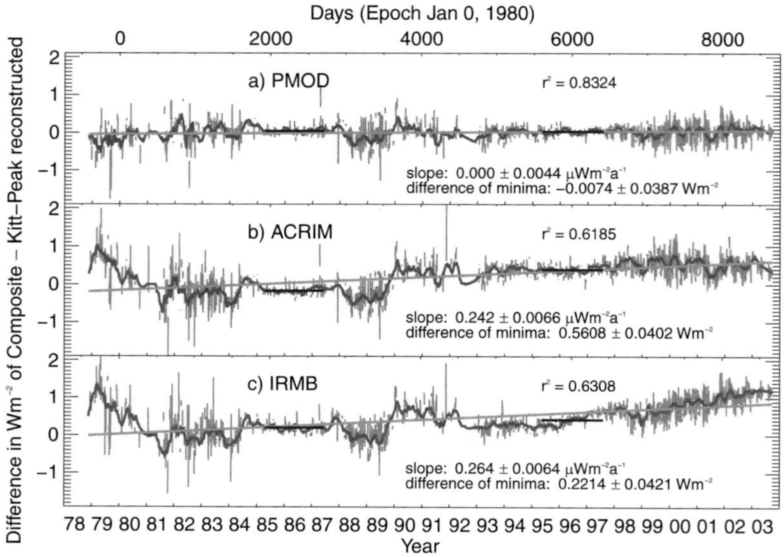


*Figure 10.* The comparison of the three composites with a reconstruction of TSI from Kitt-Peak magnetograms by Wenzler (2005).

variance, an unexpectedly high correlation for the rather simple assumptions on which the reconstruction is based. For the ACRIM and IRMB composites this reconstruction explains only 62 and 63% of the variance. For the former the low value is mainly due to the difference between the two minima due to the neglect of the HF corrections during the ACRIM gap. For the latter the main reason is that the IRMB evaluation of DIARAD/VIRGO does neglect the non-exposure dependent changes, which explains also the difference between the minima due to the increase from the beginning of the measurements in early 1996. Furthermore, the low correlation of both composites is also due to the missing correction of HF for degradation during the maximum of cycle 21 and the noise due to the uncorrected glitches during the ACRIM gap. There are also some differences – although much smaller – between the reconstruction and the PMOD composite, especially at the maxima of cycle 21 and 22 which are probably due to the noisier NSO-512 magnetograms, but could also be partly due to mistaken corrections applied to the TSI records for the PMOD composite.

## 4. Comparison with a 3-Component Proxy Model

Multiple regression of a 3-component proxy model against TSI is used to "calibrate" it and explains somewhat less than 80% of the variance. This is slightly less than what the magnetogram-based reconstruction is able to do. The three components of this proxy model are the photometric sunspot index as represented by PSI ($P_S$) and

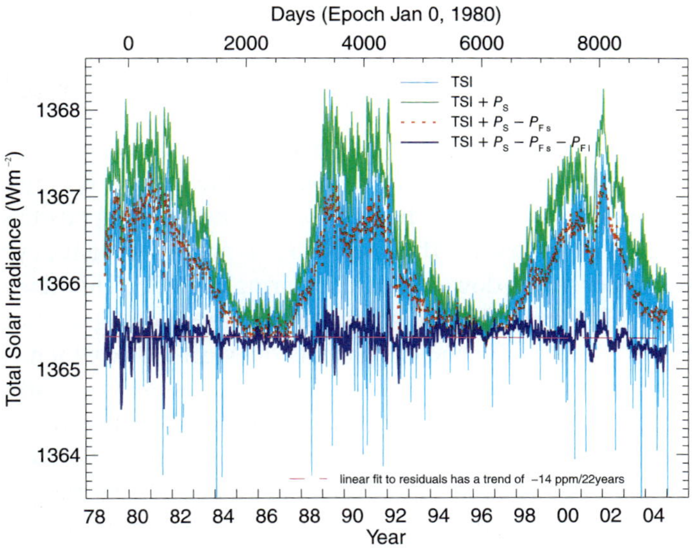


*Figure 11.* The comparison of the PMOD composite with a 3-component model (PSI $P_S$, short- $P_{Fs}$ and long-term MgII $P_{Fl}$). This model has been calibrated against the composite for each cycle separately which makes the overall trend of the residuals approximately zero.

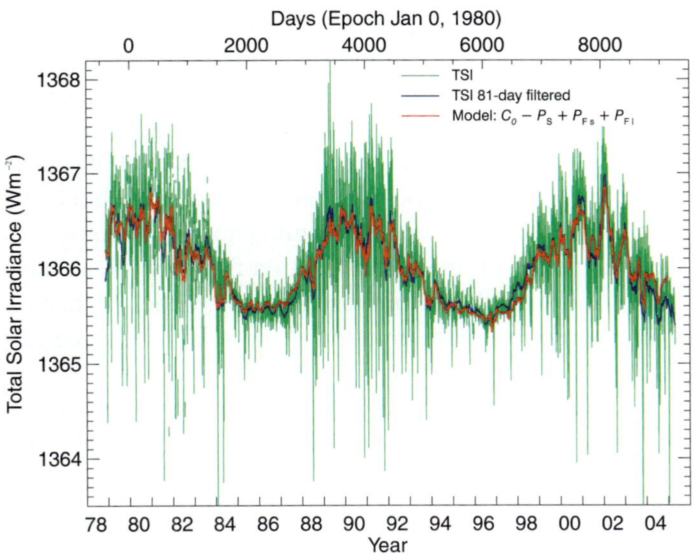


*Figure 12.* Shown is the comparison of the proxy model with the PMOD composite. Although some discrepancies are obvious, the overall fit is quite impressive. Note the rather large deviation at the end of the time series which may be related to a possible influence of sunspots on the MgII index (see text).

a short- and long-term MgII index ($P_{\mathrm{Fs,Fl}}$) as described in, e.g., Fröhlich and Lean (2004). The corresponding coefficients are determined by multiple regression with TSI. This can be done over the full period or for each cycle separately. The results of the latter method are shown in Figures 11 and 12. The coefficients for the short and long-term MgII ($P_{\mathrm{Fs,Fl}}$) vary at most by $\pm 10\%$ between the cycles which may not be significant in terms of representativeness of the model. It is interesting to note that $P_{\mathrm{Fs}}$ is lowest for cycle 21, whereas $P_{\mathrm{Fl}}$ is highest. The opposite situation is found for cycle 23. The highest coefficients for $P_{\mathrm{Fs,Fl}}$ are observed for cycle 22. The residuals show a rather strong downward trend during cycle 23. This deviation could also be due to the MgII index having some still uncorrected degradation. Another explanation may be some influence of sunspots on MgII index which confuse the facular part and should be removed. This may be indicated by the trend of the ratio of the MgII index to the 10.7 cm radio flux as indicated by Figure 6 of Viereck *et al.* (2004). Also the rather large deviations during the maxima of cycle 22 and 23 could be for the same reason. The overestimation of TSI by the proxy model towards the end of descending part of cycle 23 may also be related to such an effect, when the measurements show already values as low as those observed during the last minima and the model is still higher.

## 5. Conclusions

The presented corrections are based on the improved understanding of the short and long-term changes of classical radiometers in space, and have substantially improved the reliability of the composite. Overall, the changes are small and do not change the earlier conclusions about a non-existing long-term trend or the amplitudes of the cycles. A detailed error analysis shows that the PMOD composite has a long-term uncertainty of less than about 90 ppm per decade (Fröhlich, 2004), which makes the observed difference between the minima not significantly different from zero. The close agreement with the reconstruction from Kitt-Peak magnetograms by Wenzler (2005), and with the 3-component proxy model supports the PMOD composite as the most reliable representation of the solar irradiance variability for the last 25 years.

The PMOD composite is available from `http://www.pmodwrc.ch/pmod.php?topic=tsi/composite/SolarConstant`, the ACRIM and IRMB composites from `http://www.acrim.com/Data%20Products.htm` and `http://remotesensing.oma.be/solarconstant/sarr/SARR.txt` respectively.

## Acknowledgements

Continued support of solar irradiance research at PMOD/WRC by the Swiss National Science Foundation is greatly acknowledged. Thanks are extended to the

VIRGO team for the continuous support of the experiment on SOHO, a cooperative ESA/NASA mission, and to the ACRIM and SORCE teams for still unpublished data.

## References

Chapman, G. A., Cookson, M. A., and Dobias, J. J.: 1996, 'Variations in total solar irradiance during solar cycle 22', *J. Geophys. Res.* **101**, 13,541–13,548.

Crommelynck, D., Domingo, V., Fichot, A., Fröhlich, C., Penelle, B., Romero, J., and Wehrli, C.: 1993, 'Preliminary results from the sova experiment on board the european retrievable carrier (eureca)', *Metrologia* **30**, 375–380.

Crommelynck, D., Fichot, A., Lee III, R. B., and Romero, J.: 1995, 'First realisation of the space absolute radiometric reference (sarr) during the ATLAS 2 flight period', *Adv. Space Res.* **16**, (8)17–(8)23.

Dewitte, S., Crommelinck, D., Mekaoui, S., and Joukoff, A.: 2004, 'Measurement and uncertainty of the long-term total solar irradiance trend', *Sol. Phys.* **224**, 209–216.

Fligge, M., Solanki, S. K., and Unruh, Y. C.: 2000, 'Modelling irradiance variations from the surface distribution of the solar magnetic field', *A&A* **353**, 380–388.

Fröhlich, C.: 2000, 'Observations of irradiance variability', *Space Sci. Rev.* **94**, 15–24.

Fröhlich, C.: 2003, 'Long-term behaviour of space radiometers', *Metrologia* **40**, 60–65.

Fröhlich, C.: 2004, 'Solar irradiance variability', In: *Geophysical Monograph 141: Solar Variability and its Effect on Climate*, American Geophysical Union, Washington DC, USA, Chapt. 2: Solar Energy Flux Variations, pp. 97–110.

Fröhlich, C., and Lean, J.: 1998a, 'The sun's total irradiance: cycles and trends in the past two decades and associated climate change uncertainties', *Geophys. Res. Lett.* **25**, 4377–4380.

Fröhlich, C., and Lean, J.: 1998b, 'Total solar irradiance variations: The construction of a composite and its comparison with models'. in: F. L., Deubner, J. Christensen-Dalsgaard, and Kurtz, D. (eds.): *IAU Symposium 185: New Eyes to See Inside the Sun and Stars*, pp. 89–102.

Fröhlich, C., and Lean, J.: 2002, 'Solar irradiance variability and climate', *Astron. Nachr.* **323**, 203–212.

Fröhlich, C., and Lean, J.: 2004, 'Solar radiative output and its variability: Evidence and mechanisms', *A&A Rev.* **12**, 273–320.

Hoyt, D. V., Kyle, H. L., Hickey, J. R., and Maschhoff, R. H.: 1992, 'The NIMBUS-7 solar total irradiance: a new algorithm for its derivation', *J. Geophys. Res.* **97**, 51–63.

Lee III, R. B., Gibson, M. A., Wilson, R. S., and Thomas, S.: 1995, 'Long-term total solar irradiance variability during sunspot cycle 22', *J. Geophys. Res.* **100**, 1667–1675.

Viereck, R. A., Floyd, L. E., Crane, P. C., Woods, T. N., Knapp, B. G., and Rottman, G. *et al.*: 2004, 'A composite Mg II index spanning from 1978 to 2003', *Space Weather* **2**, doi:10.1029/2004SW000084.

Wenzler, T.: 2005, 'Reconstruction of solar irradiance variations in cycles 21–23 based on surface magnetic fields', Ph.D. thesis, ETH Nr 16199, Eidgenössische Technische Hochschule, Zürich.

Wenzler, T., Solanki, S. K., and Krivova, N. A.: 2005, 'Can surface magnetic fields reproduce solar irradiance variations in cycles 22 and 23?', *A&A* **432**, 1057–1061.

Willson, R. C.: 1997, 'Total solar irradiance trend during solar cycles 21 and 22', *Science* **277**, 1963–1965.

Willson, R. C., and Hudson, H. S.: 1991, 'The sun's luminosity over a complete solar cycle', *Nature* **351**, 42–44.

Willson, R. C. and Mordvinov, A. V.: 2003, 'Secular total solar irradiance trend during solar cycles 21–23', *Geophys. Res. Lett.* **30**, doi:10.1029/2002GL016038.

# SOLAR VARIABILITY OVER THE PAST SEVERAL MILLENNIA

J. BEER[1,*], M. VONMOOS[1] and R. MUSCHELER[2]
[1]*Swiss Federal Institute of Environmental Science and Technology (EAWAG), Ueberlandstrasse 133, CH-8600 Dübendorf, Switzerland*
[2]*NASA/Goddard Space Flight Center, Greenbelt, MD, USA*
*(*Author for correspondence: E-mail: beer@eawag.ch)*

(Received 11 November 2005; Accepted in final form 28 February 2006)

**Abstract.** The Sun is the most important energy source for the Earth. Since the incoming solar radiation is not equally distributed and peaks at low latitudes the climate system is continuously transporting energy towards the polar regions. Any variability in the Sun-Earth system may ultimately cause a climate change. There are two main variability components that are related to the Sun. The first is due to changes in the orbital parameters of the Earth's position relative to the Sun induced by the other planets. Their gravitational perturbations induce changes with characteristic time scales in the eccentricity ($\sim$100,000 years), the obliquity (angle between the equator and the orbital plane) ($\sim$40,000 years) and the precession of the Earth's axis ($\sim$20,000 years). The second component is due to variability within the Sun. A variety of observational proxies reflecting different aspects of solar activity show similar features regarding periodic variability, trends and periods of very low solar activity (so-called grand minima) which seem to be positively correlated with the emitted energy from the Sun, the total and the spectral solar irradiance. The length of these records ranges from 25 years (solar irradiance) to 400 years (sunspots). In order to establish a quantitative relationship between solar variability and solar forcing it is necessary to extend the records of solar variability much further back in time and to identify the physical processes linking solar activity and total and spectral solar irradiance. The first step, the extension of solar variability, can be achieved by using cosmogenic radionuclides such as $^{10}$Be in ice cores. After removing the effect of the changing geomagnetic field on the $^{10}$Be production rate, a 9000-year long record of solar modulation was obtained. Comparison with paleoclimatic data provides strong evidence for a causal relationship between solar variability and climate change. It will be the subject of the next step to investigate the underlying physical processes that link solar variability with the total and spectral solar irradiance.

**Keywords:** solar activity, solar influence on climate, cosmogenic radionuclides

## 1. Introduction

The Sun plays a central role for the Earth at least in two respects. Firstly, it is the gravitational centre around which the globe cruises on an elliptical orbit and secondly the Sun is by far the most important source of energy for the Earth system. Every second, the Sun looses 4.3 million tons of its mass, emitted in the form of electromagnetic radiation from the photosphere into space. This power is generated in the core by nuclear fusion and transported to the surface by radiative and convective processes. The Sun emits a total power of $4 \times 10^{26}$ W in the form of

Space Science Reviews (2006) 125: 67–79
DOI: 10.1007/s11214-006-9047-4

*Figure 1.* The Sun emits a total power of $4 \times 10^{26}$ W, $10^{17}$ W of which arrive at the Earth. When distributing this power homogeneously over the globe and taking into account that 30% is reflected (albedo), the average solar power amounts to 239 $Wm^{-2}$. Potential causes for changes in the solar radiation received by the Earth are: changes in the albedo of the Earth (cloud cover, ice and snow cover, vegetation), the atmospheric greenhouse gas concentration ($CO_2$, $CH_4$ etc.), the atmospheric aerosol and dust content (volcanic eruptions, ashes, soot), and internal variability in the energy transport on Earth.

electromagnetic radiation with a spectrum that resembles a black body of 5700 K. Because of the large distance between Sun and Earth ($150 \times 10^6$ km) and the relatively small radius of the Earth (6370 km), only about a billionth of this emitted power arrives at the top of the atmosphere (1365 $Wm^{-2}$). Taking into account that 30% of the arriving radiation is reflected back into space (albedo) and integrating over the cross section of the Earth, the total power received by the Earth system amounts to about $10^{17}$ W (Figure 1). In comparison, the contributions from other sources of energy such as cosmic ray particles ($10^{10}$ W), geothermal energy as a result of radioactive decay in the Earth interior and gravitational energy from the time of the Earth formation ($10^{13}$ W) and tidal energy ($10^{12}$ W), are negligible.

Considering the Earth as a black body and assuming steady state conditions, a simple calculation reveals a mean global temperature of −18°C which is considerably lower than the effective value of 15°C. The difference of 33°C is the result of the greenhouse gases in the atmosphere (natural greenhouse effect). As a consequence of the spherical shape of the Earth, the incoming solar power has its maximum at low latitudes (1365 $Wm^{-2}$) and decreases with the cosine of the latitude. This leads to a large energy gradient between equatorial and polar regions which the climate machine tries to remove by continuously transporting energy through the atmosphere and the ocean from low to high latitudes.

In principle, this could be a stable system consisting of a constant flow of solar energy arriving at Earth, distributed by the climate system and finally reemitted into space. However, in reality, the climate system is highly dynamic and never reaches equilibrium. As indicated in Figure 1 there are several potential causes for climate change:

– changes in the total solar irradiance (TSI) and its spectral distribution (SSI)

– changes in the albedo (cloud and snow cover, vegetation etc.)
– changes in the optical properties of the atmosphere (greenhouse gases, aerosols, volcanic ash, dust etc.)
– internal variability within and between the various components of the climate system (e.g. changes in the ocean circulation)

We are faced with two main difficulties when dealing with the climate system. The first is its complexity, with many non-linear couplings that makes it very difficult to predict its behaviour without appropriate models. The second is its dynamics, with time scales ranging from seconds (atmospheric processes) to millions of years (evolution of the Sun, continental drift etc). This implies that the climate system has the potential to respond significantly to even very weak forcings and that this response may depend strongly, not only on the amplitude, but also on the duration of the forcing with potentially larger effects for longer lasting or repeated forcings. In this paper, we will concentrate on some aspects of solar variability on mainly millennial time scales.

## 2. Solar Forcing: Orbital Parameters

As discussed above, the amount of solar radiation arriving at the top of the atmosphere depends on the energy emitted by the Sun and the distance and orientation of the Earth with respect to the Sun. Let's begin with the latter. 99.85% of the mass of the solar system is concentrated in the Sun. The remaining 0.15% is distributed among the planets with Jupiter and Saturn taking the largest share (92%). Consequently, the orbit of the Earth is dominated by the gravitational force of the Sun. However, the influence of the other planets cannot be neglected. These disturbing effects depend on the relative positions between Sun, Earth, and the other planets. Detailed calculations carried out by Milankovich (1930) and more recently by Berger (1978) and Laskar *et al.* (2004) reveal that gravitational forces of the planets influence three different orbital parameters of the Earth with different characteristic periodicities. These parameters are the eccentricity, the deviation of the Earth's orbit from a circle, the obliquity, the mean tilt angle of the Earth's axis relative to the ecliptic plane, and the precession of the Earth's rotational axis around the mean axis. The eccentricity changes with periodicities around 400,000 and 100,000 years, while changes in the obliquity occur with periodicities around 40,000 years and changes of the precession with 20,000 years. There are also shorter periodicities (Bertrand *et al.*, 2002). However, these are too weak to have any significant effect on the climate.

Orbital forcing is special in at least two respects:

1. Only changes in the eccentricity lead to changes in the absolute amount of solar radiation received by the Earth. On the other hand, obliquity and precession affect only the relative distribution of the incoming radiation.

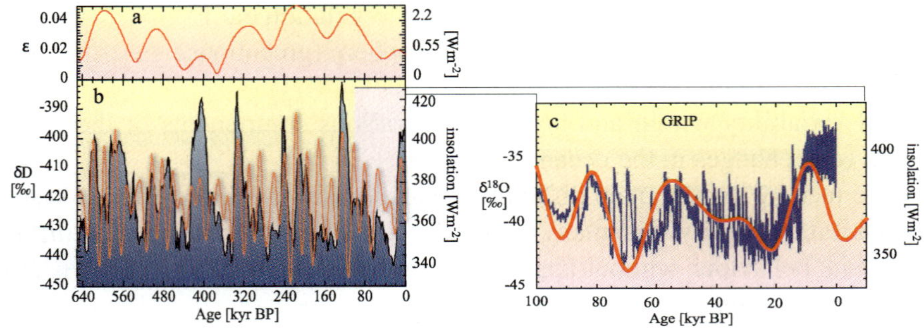


*Figure 2.* Orbital forcing and its effect on the climate. The gravitational forces of the planets (mainly Jupiter and Saturn) cause changes in the orbital parameters of the Earth that lead to changes in the total amount of solar radiation and its latitudinal distribution. Panel (a) shows the eccentricity, the deviation from a circular orbit for the past 640,000 years with a clear period of ~100,000 years which corresponds nicely with the sequence of glacial and interglacial periods found in the $\delta$D record from Dome C (Antarctica) (Spahni, 2005) that is an indicator of temperature shown in panel (b). Low $\delta$D values (blue curve) mean a depletion of the heavier hydrogen isotope D corresponding to colder climatic conditions. The red curve in panel (b) reflects the summer insolation at 65°N which includes, in addition to the eccentricity, the tilt angle (period of ~40,000 years) and the precession of the Earth's axis (period of ~20,000 years). Panel (c) shows a blow up of the past 100,000 years. Instead of $\delta$D, here the $\delta^{18}$O record (blue curve) from the GRIP ice core (Johnsen *et al.*, 1997) is plotted which also reflects the temperature. The abrupt short-term changes in $\delta^{18}$O (so-called Dansgaard-Oeschger events) are most probably due to melt-water pulses shutting down the deep-water formation in the North Atlantic. It is interesting to note that the change in the insolation for the coming 10,000 years is comparatively small.

2. Orbital forcing is the only forcing that can be calculated precisely for the past and even the future several million years because it is based purely on celestial mechanics.

As an example, Figure 2 shows the eccentricity (panel a) and the summer insolation (red curve) at 65°N, together with the $\delta$D record (in blue) from the Dome C ice core, Antarctica (Spahni, 2005) (panel b). $\delta$D, the deviation of the isotopic ratio D/H from a standard value, is a proxy of the temperature, with low values indicating cold conditions. The strongest correspondence between $\delta$D and the insolation is in the 100,000 years band caused by the eccentricity changes. Note that the mean annual global insolation changes caused by the eccentricity are very small ($<$2.5 Wm$^{-2}$) (panel a).

In panel (c) of Figure 2, the past 100,000 years and the future 10,000 years of insolation at 65°N are depicted together with $\delta^{18}$O (blue curve), another temperature proxy measured in the GRIP ice core from Greenland (Johnsen *et al.*, 1997). The comparison shows a generally good agreement between orbital forcing and climate response. However, during the glacial period from 80,000 to 10,000 years before present, the $\delta^{18}$O record is characterized by abrupt climate changes that are usually attributed to fluctuations in the thermohaline circulation of the ocean (Broecker, 1997). During such an event, the mean annual temperature drops by about 20°C in Greenland within a decade or two. Other interesting features in Figure 2c are that the decrease of the insolation during the past 10,000 years is not visible in the isotopic record and that insolation changes are not predicted to play an important role in the next 10,000 years.

We conclude this short discussion of the orbital forcing with one last interesting fact. During the past 650,000 years, by far the largest signal induced by insolation changes is the 100,000 years cycle of glacial-interglacial periods (Figure 2b). However, the corresponding mean annual forcing changes are extremely small (Figure 2a).

## 3. Solar Variability

Solar variability has only recently been considered to be a potential forcing factor of climate change. This is also reflected by the fact that the term "solar constant" is still widely used to describe the total solar irradiance (TSI). This may be somewhat surprising given the fact that the Sun provides the power which drives the climate system. Nevertheless, there were a few scientists who recognised the importance of this issue a long time ago. Among others, Abbot (1910) tried hard to detect fluctuations in the solar constant from the Earth's surface. However, he failed since changes in the optical properties of the atmosphere inhibit the detection of the relatively small TSI changes from the Earth's surface. A new era began when Eddy (1976) pointed to the striking similarity between the winter severity index of Paris and London and proxies for solar activity, as for example the sunspot record or changes of the atmospheric concentration of the cosmogenic radionuclide $^{14}C$ measured in tree rings.

But only after it was possible to operate radiometers outside the atmosphere on satellites, did it become clear that indeed the solar constant is not constant (Fröhlich, 2006). However, the observed changes of the TSI over an 11-year cycle are very small (0.1%), corresponding to an average temperature change of 1.5 K of the photosphere and, on Earth, to a global forcing change of 0.25 $Wm^{-2}$ (averaging over the globe and taking into account the albedo of 30%). This led many people to conclude that, even if the solar constant is not constant, the changes are too small to be climatically relevant without invoking additional strong amplification mechanisms. This conclusion seems to be premature, firstly because there is no doubt that there are positive feedback mechanisms in the climate system. A cooling for example, leads to growing ice sheets which increases the albedo and thus the cooling. The existence of feedback mechanisms is illustrated by the discussed glacial-interglacial cycles that are related to a very weak annual mean change in insolation. General circulation models show that a change of the TSI by 0.1% over decades to centuries is not negligible (Cubash and Voss, 2000). Last but not least, the change of 0.1% in the TSI is accompanied by far greater changes in the UV part of the solar spectrum (Rottman, 2006) causing changes in the stratosphere which can be coupled down to the troposphere (Haigh, 2006). Secondly, it could be misleading to extrapolate from just three decades of TSI observations to centennial and millennial time scales. There are no physical reasons why the emission of radiation from the Sun should not show larger fluctuations on longer time scales up to $3 \times 10^7$ years (Helmholtz-Kelvin time scale). Other stars, although not exactly of the same type as the Sun, show considerably larger fluctuations (Radick, 2001).

In the past years, large efforts have been made to model the TSI by combining brightening (faculae and magnetic networks) and darkening (sunspots) effects related to the magnetic activity on the Sun (Solanki, 2006). Although successful in reproducing the short-term variability over the past few decades, these models are unable to predict the full potential range of solar variability unless we know more about the solar conditions (TSI, open and closed magnetic flux) during a period of very low solar activity such as the Maunder Minimum (1645–1715 AD) when the sunspots were almost completely absent (Lockwood, 2006). To reach the ambitious goal of reconstructing the total and spectral solar irradiance over decadal to millennial time scales, we need to know more about the physical conditions of the Sun in the past. We also have to develop physical models of how these conditions are related to the TSI and to the SSI. Here, we will focus on the first task: the reconstruction of long-term solar variability inferred from cosmogenic radionuclide records.

## 4. Reconstruction of Long-Term Solar Variability

There are several proxies of solar activity which are all based on direct measurements or on observations and are therefore called direct proxies. They have in common that they are all related – even though in different ways – to magnetic processes on the Sun with temporal resolutions of weeks to years and that they are generally more reliable for more recent times. Depending on their length, all the records show similar features (Figure 3): the 11-year Schwabe cycle, a long-term trend from 1900–1950 with a generally increasing activity with time and some short periods (1–7 decades) with reduced solar activity. The most pronounced is the Maunder Minimum (1645–1715 AD) which is characterised by an almost complete absence of sunspots (Eddy, 1976) (pink band in Figure 3). A more detailed comparison shows that, apart from these common features, there are also clear differences, mainly due to the fact that all these proxies are related in various ways to different aspects of magnetic processes taking place on the Sun. Two records do not extend to the present. The reasons are that, to our knowledge, aurorae were no longer counted after 1955, and that the ice core which was analyzed for $^{10}$Be (see below) was drilled in 1985.

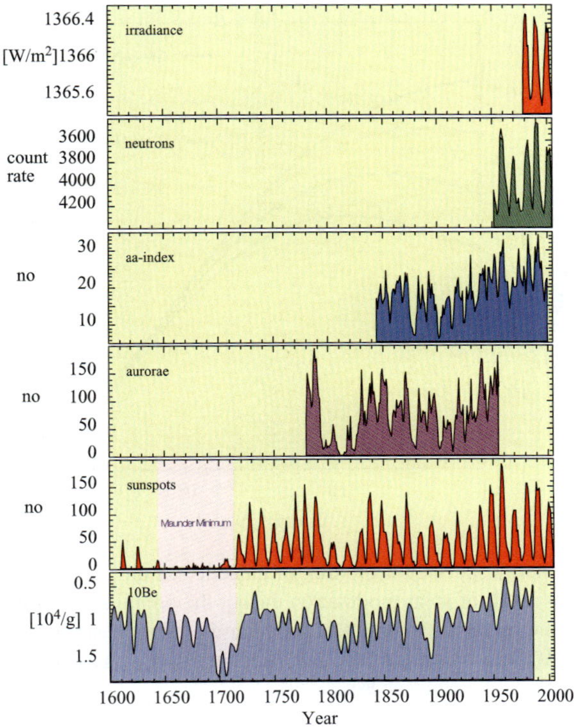


*Figure 3.* Comparison of proxies of solar activity. Each proxy reflects a specific aspect of solar activity. As a result, all proxies show some common features (11-year cycle, trends, solar minima), but also differences. The length is limited to the periods of instrumental observations and ranges from 25 years (TSI) to about 400 years (sunspots). No aurora counts could be found for the period since 1955. $^{10}$Be is the only record which is not based on direct observations and, therefore has the potential to be extended over at least 10,000 years. The Dye 3 ice core used for the $^{10}$Be analysis was drilled in 1985 (Beer *et al.*, 1990).

The only record in Figure 3, which is not based on direct observations, is the $^{10}$Be record (Beer *et al.*, 1994). $^{10}$Be and other cosmogenic radionuclides such as $^{14}$C and $^{36}$Cl offer the unique opportunity to extend the reconstruction of solar activity back to at least 10,000 years. Cosmogenic radionuclides are produced in the Earth's atmosphere by the interaction of galactic cosmic rays (GCR) with nitrogen, oxygen, and argon. Before reaching the Earth, GCR have to cross the heliosphere where they are subject to solar induced modulation effects. The propagation of GCR through the heliosphere is described by the cosmic ray transport equation formulated by Parker (1965). To reduce the complexity, several simplifying assumptions have been introduced leading to the so-called force field equation (Gleeson and Axford, 1967). It shows that the effect of solar activity on the energy spectrum of galactic cosmic rays can be described by the so-called solar modulation function $\Phi$ (Figure 4).

Figure 4 illustrates the enhanced shielding effect during periods of high solar activity, i.e. the enhanced energy loss of the low energy GCR protons due to

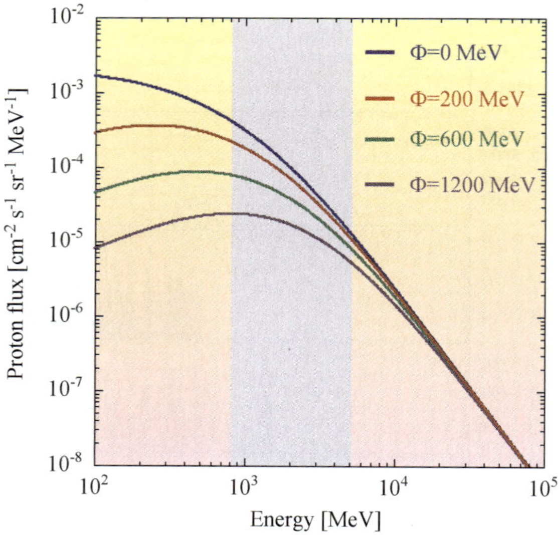


*Figure 4.* Differential energy spectra for the flux of galactic cosmic ray protons arriving at the top of the atmosphere. As a result of solar modulation during the propagation through the heliosphere, the flux is reduced depending on the intensity of the solar modulation expressed by the parameter Φ (Masarik and Beer, 1999). The light blue band indicates the energy range that is mainly responsible for the production of $^{10}$Be in the atmosphere.

the increased intensity of the magnetic field carried by the solar wind. The solar modulation affects mainly the GCR protons with energies below $10^4$ MeV. This coincides with the most efficient energy range for the subsequent production of the cosmogenic radionuclide $^{10}$Be from GCR in the Earth's atmosphere, which covers the range 0.8–5 GeV (McCraken, 2004; Webber and Higbie, 2003) (light blue band in Figure 4). This makes $^{10}$Be time series sensitive recorders of solar modulation in the past and, thus, of the temporal changes of solar variability.

The production rate of cosmogenic radionuclides also depends on the shielding effect of the geomagnetic field. As the geomagnetic modulation of cosmic rays is mainly due to the Earth's dipole field, the shielding effect is highly dependent on the geomagnetic latitude, being maximal at low latitudes. The production rate of cosmogenic radionuclides as a function of geomagnetic and heliospheric modulations is well understood and can be simulated using Monte Carlo technique (Masarik and Beer, 1999; Webber and Higbie, 2003). Figure 5 shows the dependence of the global mean $^{10}$Be production rate as a function of the geomagnetic field intensity and the solar modulation function Φ.

Subsequent to its production, $^{10}$Be becomes attached to aerosols and follows their pathways. After a mean residence time of 1–2 years, $^{10}$Be is removed from the atmosphere, mainly by wet deposition, and stored in natural archives such as

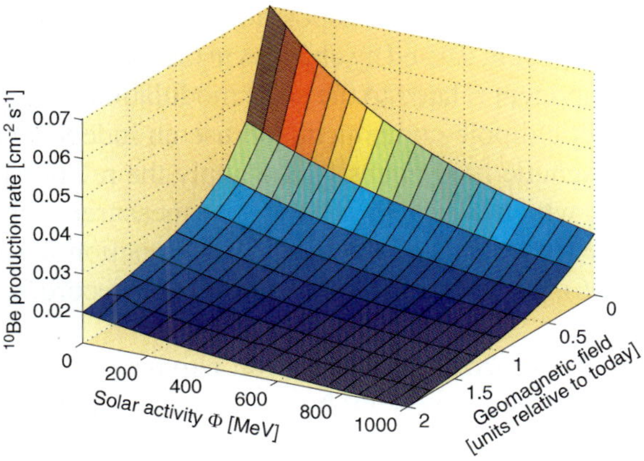


*Figure 5.* Dependence of the $^{10}$Be production rate on the intensity of the geomagnetic dipole field in relative units, and on the solar modulation function Φ (Masarik and Beer, 1999).

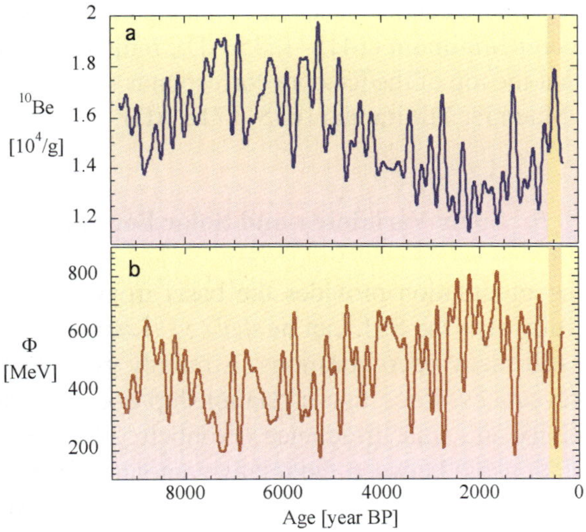


*Figure 6.* Derivation of the solar modulation function Φ from the measured $^{10}$Be concentration. *Panel (a)*: Measured $^{10}$Be concentration in the GRIP ice core from Greenland after applying a low-pass filter with a cut-off frequency of 1/(200 years) to remove system effects caused by the transport of $^{10}$Be from the atmosphere into the ice. *Panel (b)*: Reconstructed solar modulation function Φ derived from the $^{10}$Be record of panel (a). The production contribution from the geomagnetic field (not shown) has been removed using the dependence of the $^{10}$Be production rate on the geomagnetic field intensity and Φ shown in Figure 5 (Vonmoos *et al.*, 2006).

ice caps. Therefore, cosmogenic radionuclides provide information on the solar magnetic activity with a maximal temporal resolution of 1 year on a time range depending on the half-life and the age of the investigaed archive. Figure 6a depicts the measured $^{10}$Be concentration record of the GRIP ice core in Central Greenland (Muscheler *et al.*, 2004) after applying a low-pass filter with a cut-off frequency of

1/(200 years) to remove transport and deposition effects. The remaining signal is still composed of a component caused by changes in the intensity of the geomagnetic dipole field and a component related to the solar modulation we are looking for. To remove the geomagnetic dipole component, we use an independently determined record of the paleomagnetic field intensity and apply the relationship between the $^{10}$Be production rate, which is proportional to the measured $^{10}$Be concentration, the magnetic field intensity, and the solar modulation function $\Phi$ shown in Figure 5 (Vonmoos *et al.*, 2006). The resulting solar modulation record is shown in panel (b) of Figure 6.

The $\Phi$-record is characterized by a high and persistent variability throughout the Holocene. A period of relatively high solar activity around 8000 years BP is followed by a decrease reaching a minimum around 7000 years BP. After a slow but steady increase, $\Phi$ reaches another longer lasting maximum around 2000 BP (Roman period) before decreasing again. Superimposed to these slow changes are abrupt short-term troughs, some of which very pronounced, with $\Phi$ values around 200 MeV and lasting about a century. These are called grand minima. A well-known example is the Spoerer Minimum (1415–1535 AD), indicated by the yellow band in Figure 6. Note that the top of the ice core corresponds to 1645 AD and therefore does not show the Maunder Minimum (1645–1715 AD).

## 5. Solar Variability and Solar Forcing

This record of solar modulation provides the basis from which, in a next step, the TSI, and hopefully also the SSI, can be derived (Lockwood, 2006). This is a difficult task with still many open questions. A rapidly growing number of paleoclimate records reveals evidence that a relationship does indeed exist between solar magnetic activity and solar irradiance (Fröhlich, 2006). For example, Figure 7 shows the relationship between the relative amount of ice rafted debris in a sediment core from the North Atlantic (Bond *et al.*, 2001) and the reconstructed solar modulation of Figure 6 after applying a low pass filter with a cut-off frequency of 1/(900 years). This "ice rafted debris" consists of specific glass particles which originate from volcanic eruptions in Iceland. They are picked up by glaciers moving towards the coast where they break up into icebergs and drift southwards carrying the particles. During the melting of the iceberg, the particles are released, sink to the bottom, and are incorporated into the sediment. The distance an iceberg can drift south before melting depends on the prevailing climate conditions (surface winds and surface hydrography). The close correlation between the amount of glass particles produced by volcanic eruptions in Iceland and the solar modulation function $\Phi$ clearly points to a causal relationship. Some of the discrepancies are due to the dating uncertainties of the two cores. Similar relationships have been found with the regional precipitation rate by analyzing stalagmites in a cave in Oman (Neff *et al.*, 2001), with the advances and retreats of glaciers (Denton and

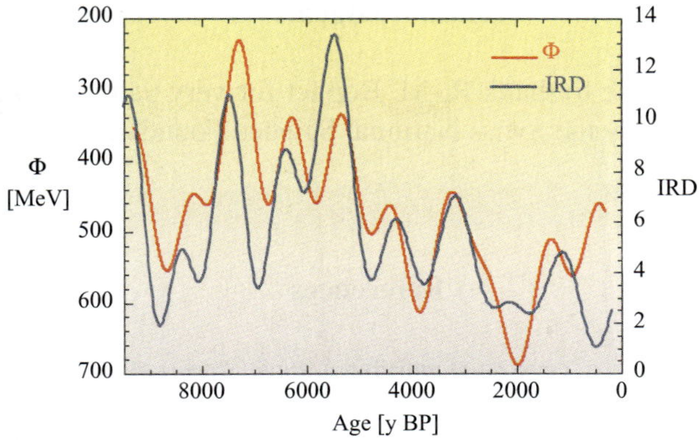


*Figure 7.* Comparison of the reconstructed solar modulation function Φ of Figure 6 with ice rafted debris (IRD) retrieved from a sediment core in the North Atlantic (Bond *et al.*, 2001). In this figure, the Φ record has been low-pass filtered with a cut-off frequency of 1/(900 years) and is plotted inversely compared to Figure 6. The ice rafted debris consists of glass particles originating from volcanic eruptions in Iceland. Incorporated into icebergs, they drift much further south during periods of low solar activity.

Karlén, 1973) and the water level of lakes in the French Jura (Holzhauser, 1997). These are just a few examples of a rapidly growing number of publications in this field.

## 6. Conclusions

It is well known that the Sun plays the fundamental role as our energy source. However, it is still an open question what role the Sun plays in climate change. Direct measurements of the total (TSI) and the spectral (SSI) irradiance using satellite born radiometers show relatively small changes in accordance with the solar magnetic activity. To answer this question, we have (1) to understand the physical mechanisms responsible for the irradiance changes, (2) to reconstruct the solar variability over the past millennia, (3) to derive from this variability record, reliable estimates of the TSI and SSI changes, and (4) to determine the response of the climate system using appropriate models. As far as the second topic is concerned we can rely on a variety of solar activity proxies based on direct observations (sunspots, *aa*-index, aurorae). However, all of them are limited to periods ranging from a few decades to a few centuries. To date, the only proxy providing information about the solar variability on millennial time scales are cosmogenic radionuclides stored in natural archives such as ice cores. They clearly reveal that the Sun varies significantly on millennial time scales and most likely plays an important role in climate change.

## Acknowledgements

The authors would like to thank R.-M. Bonnet for very valuable comments. This work was supported by the Swiss National Science Foundation.

## References

Abbot, C. G.: 1910, 'The solar constant of radiation', *Annual Report of the Smithsonian Institution*, 319 pp.

Beer, J., Blinov, A., Bonani, G., Finkel, R. C., Hofmann, J. J., *et al.*: 1990, 'Use of $^{10}$Be in polar ice to trace the 11-year cycle of solar activity', *Nature* **347**, 164–166.

Beer, J., Baumgartner, S., Hannen-Dittrich, B., Hauenstein, J., Kubik, P., *et al.*: 1994, 'Solar Variability Traced by Cosmogenic Isotopes', in J. M. Pap, C. Fröhlich, H. S. Hudson, and S. K. Solanki (eds.), *The Sun as a Variable Star: Solar and Stellar Irradiance Variations*, Cambridge University Press, Cambridge, pp. 291–300.

Berger, A.: 1978, 'Long-term variations of daily insolation and quaternary climatic changes', *J. Atmos. Sci.* **35**, 2362–2367.

Bertrand, C., Loutre, M.-F., and Berger, A.: 2002, 'High frequency variations of the Earth's orbital parameters and climate change', *Geophys. Res. Lett.* **29**, doi:10.1029/2002GL015622.

Bond, G., Kromer, B., Beer, J., Muscheler, R., Evans, M. N., *et al.*: 2001, 'Persistent solar influence on north Atlantic climate during the Holocene', *Science* **294**, 2130–2136.

Broecker, W. S.: 1997, 'Thermohaline circulation, the Achilles heel of our climate system: Will man-made $CO_2$ upset the current balance?', *Science* **278**, 1582–1588.

Cubasch, U. and Voss, R.: 2000, 'The influence of total solar irradiance on climate', *Space Sci. Rev.* **94**, 185–198.

Denton, G. H. and Karlén, W.: 1973, 'Holocene climatic variations – their pattern and possible cause', *Quat. Res.* **3**, 155–205.

Eddy, J. A.: 1976, 'The maunder minimum', *Science* **192**, 1189–1201.

Fröhlich, C.: 2006, 'Solar irradiance variability since 1978', *Space Sci. Rev.*, this volume, doi: 10.1007/s11214-006-9046-5.

Gleeson, L. J. and Axford, W. I.: 1967, 'Cosmic rays in the interplanetary medium', *Astrophys. J.* **149**, L115–L118.

Haigh, J.: 2006, 'Solar influences on dynamical coupling between the stratosphere and troposphere', *Space Sci. Rev.*, this volume, doi: 10.1007/s11214-006-9067-0.

Holzhauser, H.: 1997, 'Gletscherschwankungen innerhalb der letzten 3200 Jahre am Beispiel des grossen Aletsch- und des Gornergletschers. Neue Ergebnisse', in B. Salm (ed.), *Gletscher im ständigen Wandel*, VDF-Hochschulverlag AG, Zürich.

Johnsen, S. J., Clausen, H. B., Dansgaard, W., Gundestrup, N. S., Hammer, C. U., *et al.*: 1997, 'The $\delta^{18}$O record along the Greenland Ice Core Project deep ice core and the problem of possible Eemian climatic instability', *J. Geophys. Res.* **102**, 26,397–26,410.

Laskar, J., Robutel, P., Joutel, F., Gastineau, M., Correia, A. C. M., and Levrard, B.: 2004, 'A long-term numerical solution for the insolation quantities of the Earth', *Astron. Astrophys.* **428**, 261–285.

Lockwood, M.: 2006, 'What do cosmogenic isotopes tell us about past solar forcing of climate?', *Space Sci. Rev.*, this volume, doi: 10.1007/s11214-006-9049-2.

Masarik, J. and Beer, J.: 1999, 'Simulation of particle fluxes and cosmogenic nuclide production in the Earth's atmosphere', *J. Geophys. Res.* **104**, 12,099–12,111.

McCracken, K. G.: 2004, 'Geomagnetic and atmospheric effects upon the cosmogenic Be-10 observed in polar ice', *J. Geophys. Res.* **109**, doi:10.1029/2003JA010060.

Milankovich, M.: 1930, 'Mathematische Klimalehre und atsronomische Theorie der Klimaschwankungen', in W. Köppen and R. Geiger (eds.), *Handbuch der Klimatologie*, Gebrüder Bornträger, Berlin, pp. 1–176.

Muscheler, R., Beer, J., Wagner, G., Laj, C., Kissel, C., *et al.*: 2004, 'Changes in the carbon cycle during the last deglaciation as indicated by the comparison of $^{10}$Be and $^{14}$C records', *Earth Planet. Sci. Let.* **219**, 325–340.

Neff, U., Burns, S., Mangini, A., Mudelsee, M., Fleitmann, D., and Matter, A.: 2001, 'Strong coherence between solar variability and the monsoon in Oman between 9 and 6 kyrs ago', *Nature* **411**, 290–293.

Parker, E. N.: 1965, 'The passage of energetic charged particles through interplanetary space', *Planet. Space Sci.* **13**, 9–49.

Radick, R. R.: 2001, 'A brief survey of chromospheric and photometric variability among sunlike stars', *Adv. Space Res.* **26**, 1739-1745.

Rottman, G.: 2006, 'Measurements of total and spectral solar irradiance', *Space Sci. Rev.*, this volume, doi: 10.1007/s11214-006-9045-6.

Solanki, S. K.: 2006,'Solar variability of possible relevance for planetary climates', *Space Sci. Rev.*, this volume, doi: 10.1007/s11214-006-9044-7.

Spahni, R., Chappellaz, J., Stocker, T. F., Loulergue, L., Hausammann, G., *et al.*: 2005, 'Atmospheric methane and nitrous oxide of the late Pleistocene from Antarctic ice cores', *Science* **310**, 1317–1321.

Vonmoos, M., Beer, J., and Muscheler, R.: 2006, 'Large variations in Holocene solar activity constraints from $^{10}$Be in the GRIP ice core', *J. Geophys. Res.* **111**, doi:10.1029/2005JA011500.

Webber, W. R. and Higbie, P. R.: 2003, 'Production of cosmogenic Be nuclei in the Earth's atmosphere by cosmic rays: Its dependence on solar modulation and the interstellar cosmic ray spectrum', *J. Geophys. Res.* **108**, 1355–1365.

# SOLAR AND HELIOSPHERIC MODULATION OF GALACTIC COSMIC RAYS

B. HEBER[1,*], H. FICHTNER[2] and K. SCHERER[2]
[1]*Institut für Experimentelle und Angewandte Physik, Christian-Albrechts-Universität Kiel, Leibnizstr. 11, D-24118 Kiel, Germany; previously at: 1. Physikalisches Institut, Universität Stuttgart, Germany*
[2]*Theoretische Physik IV, Ruhr-Universität Bochum, Germany*
*(*Author for correspondence: E-mail: heber@physik.uni-kiel.de)*

(Received 10 September 2005; Accepted in final form 6 March 2006)

**Abstract.** The significance of external influences on the environment of Earth and its atmosphere has become evident during recent years. Especially, on time scales of several hundred years, the cosmogenic isotope concentration during the Wolf-, Spoerer-, Maunder- and Dalton-Minimum indicates an increased cosmic ray flux. Because these grand minima of solar activity coincide with cold periods, a correlation of the Earth climate with the cosmic ray intensities is plausible. Any quantitative study of the effects of energetic particles on the atmosphere and environment of the Earth must address their transport to Earth and their interactions with the Earth's atmosphere including their filtering by the terrestrial magnetosphere. The first problem is one of the fundamental problems in modern cosmic ray astrophysics, and corresponding studies began in the 1960s based on Parker's cosmic ray modulation theory taking into account diffusion, convection, adiabatic deceleration, and (later) the drift of energetic particles in the global heliospheric magnetic field. It is well established that all of these processes determining the modulation of cosmic rays are depending on parameters that are varying with the solar magnetic cycle. Therefore, the galactic cosmic ray intensities close to Earth is the result of a complex modulation of the interstellar galactic spectrum within the heliosphere. The modern view of this cosmic ray modulation is summarized in our contribution.

**Keywords:** cosmic rays, solar activity, modulation, heliosphere, particle transport in astrophysical plasmas, solar terrestrial relation

## 1. Introduction

Victor Hess discovered in 1912 an evidence of a very penetrating radiation (cosmic rays) coming from outside our atmosphere, which is dominated by hydrogen and helium (Simpson, 1983). At energies below a few GeV per nucleon the influence of solar and heliospheric modulation on the galactic cosmic ray energy spectra becomes important. Figure 1 shows in panel (A) the monthly averaged count rates of cosmic rays measured at 1 AU by the Kiel neutron monitor and in panel (B) the monthly sunspot number (black line) and the evolution of the maximum latitudinal extension of the heliospheric current sheet (tilt angle, red line). From the Figure two characteristics of the cosmic ray intensity history are evident:

Space Science Reviews (2006) 125: 81–93
DOI: 10.1007/s11214-006-9048-3

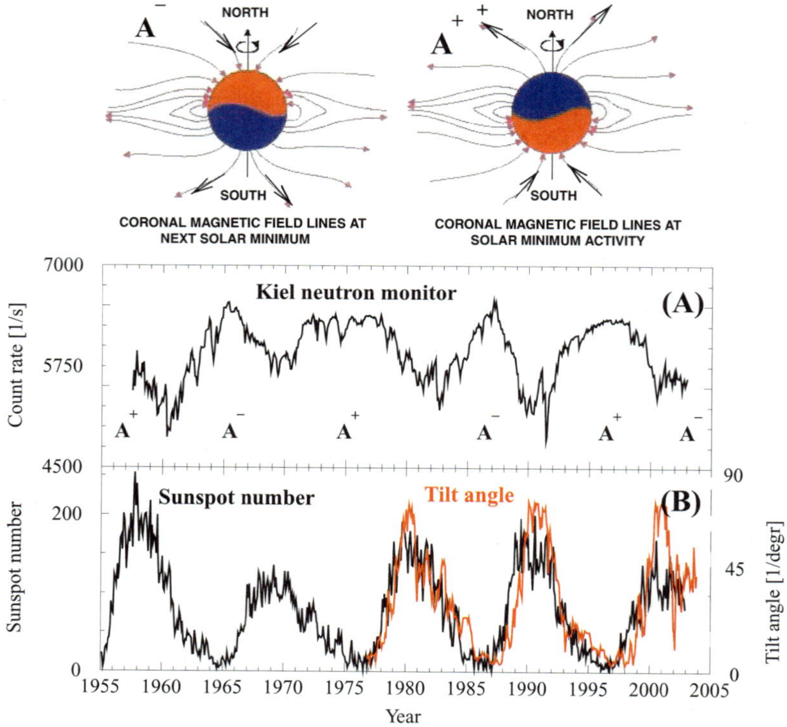


*Figure 1.* Solar modulation of galactic cosmic rays, monthly sunspot number and tilt angle $\alpha$ of the heliospheric current sheet. Marked by $A^+$ ($A^-$) are times when the solar magnetic field is directed inward (outward) from the Sun in the northern polar and outward (inward) in the southern polar region, as sketched on top (Scherer *et al.*, 2004).

1. The cosmic ray flux is varying in anti-correlation with the 11-year solar activity cycle, leading to the conclusion that galactic cosmic rays entering the regions close to the Sun are modulated as they traverse the region controlled by the Sun, called the heliosphere.
2. In the 1960s and 1980s, when the solar magnetic field is pointing towards the Sun in the northern hemisphere, the time profiles are peaked, whereas they are more or less flat in the 1970s and 1990s during the opposite solar magnetic epoch, showing a correlation with the 22-year solar magnetic cycle.

Parker (1958) introduced the concept of the supersonic expansion of the solar corona, the solar wind. Its existence was confirmed with Mariner 2 in 1962 (Snyder and Neugebauer, 1963). The solar wind is a hot, tenuous plasma surrounding the Sun, with characteristic temperatures and particle densities of about $10^6$ K and $10^{14}$ m$^{-3}$ in the corona. The interaction of the supersonic solar wind with the local interstellar medium leads to a transition from supersonic to subsonic speeds at the heliospheric termination shock, which has been observed by Voyager I at a distance of about 94 AU (Stone *et al.*, 2005; Decker *et al.*, 2005). As suggested by sophisticated models (Scherer and Ferreira, 2005), supported by radio observations (Gurnett *et al.*, 2003), the boundary layer between the local interstellar medium and the solar wind, called the heliopause, is at about 125 AU.

Although the solar wind moves out almost radially from the Sun, the rotation of the Sun gives the magnetic field the form of a three-dimensional Archemedean spiral, known as the Parker spiral (Parker, 1963). Fisk (1996) pointed out that a different correction needs to be made to the Parker spiral model for the simple reason that the Sun does not rotate rigidly but differentially, with the solar poles rotating ~20% slower than the solar equator. The interplay between the differential rotation of the magnetic field line footprints in the photosphere and the subsequent non radial expansion of the solar wind from coronal holes results in magnetic field excursions in heliographic latitude. The direction (polarity) of the field in the Sun's northern hemisphere is opposite to that of the field in the southern hemisphere, and reverses at solar maximum. A heliospheric current sheet separates the two polarities. This current sheet is tilted because of an offset between the Sun's rotational and magnetic axes. The tilt angle $\alpha$ is varying with the solar cycle from low ($<10^\circ$) to high ($>70^\circ$) values during solar maximum (Hoeksema, 1995).

## 2. Transport of Charged Particles in the Heliosphere

In order to understand the modulation of cosmic rays in the heliosphere it is vital to review the theory and observation of the particle transport in the heliosphere during periods of solar minimum. It is the variation of these transport parameters which causes the cosmic ray flux to vary with the solar cycle.

### 2.1. The Particle Transport Equation

The transport of cosmic rays in the heliosphere is described by Parker's (1965) transport equation, as discussed in detail by Potgieter (1998). Let $f(\mathbf{r}, P, t)$ be the differential cosmic ray distribution function, $\mathbf{r}$ the spatial coordinates, $P$ the particle rigidity, and $t$ the time:

$$\frac{\partial f}{\partial t} = -(\underbrace{\mathbf{V}}_{a} + \underbrace{\langle \mathbf{v}_D \rangle}_{d}) \cdot \nabla f + \underbrace{\nabla \cdot (\overleftrightarrow{\kappa} \cdot \nabla f)}_{c} + \underbrace{\frac{1}{3}(\nabla \cdot \mathbf{V}) \frac{\partial f}{\partial \ln P}}_{b} + \underbrace{Q}_{e} \tag{1}$$

where terms on the right-hand side represent:

(a) Outward convection by the solar wind.
(b) Adiabatic deceleration from the divergence of the spherically expanding solar wind: In order to illustrate the effect of adiabatic deceleration Potgieter (1984) investigated the heliospheric propagation to Earth using series of "monoenergetic" local interstellar proton spectra. The results of this computation are shown

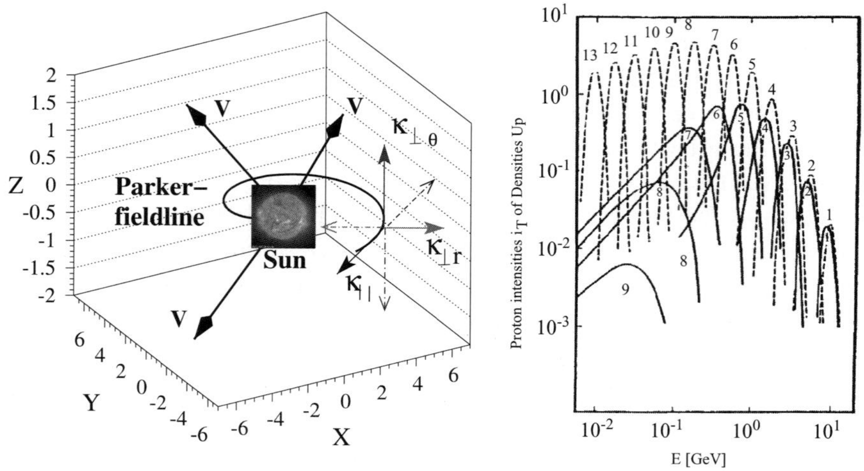

*Figure 2.* The different components of the diffusion tensor in the Parker-spiral (left). The arrows $V$ indicate the expanding solar wind. On the right the effect of adiabatic deceleration on monoenergetic "interstellar" spectra is shown (for details see text, and Potgieter, 1984).

on the right side of Figure 2. It clearly illustrates the well-known $E^1$ power-law at low energies. The gaussian spectra numbered 8 to 13, the dotted lines in the right-hand panel of Figure 2, are spectra assumed in the local interstellar medium. The corresponding spectra at Earth are given by the solid lines 8 and 9. The contribution for lower energies (lines 10 to 13) are not shown because they contribute even less to the observations at Earth (see also Moraal, 1993).

(c) Diffusion ($\overleftrightarrow{\kappa}$) through the irregular heliospheric magnetic field in response to the gradient set up by convection and deceleration: $\overleftrightarrow{\kappa}$ consists of a parallel diffusion coefficient ($\kappa_{\|}$) and the perpendicular diffusion coefficient for the radial ($\kappa_{\perp r}$) and polar direction ($\kappa_{\perp \vartheta}$). Promising progress has been made in understanding the interaction of energetic particles with the solar wind and the calculation of the diffusion tensor in recent years (Parhi *et al.*, 2001; Bieber *et al.*, 2004; Shalchi and Schlickeiser, 2004, 2005). However, applying these diffusion tensors in modulation models are not able to explain Ulysses observations, discussed below.

(d) Gradient and curvature drifts ($\langle \mathbf{v}_D \rangle$) in the global heliospheric magnetic field (Jokipii *et al.*, 1977). In an A $>$ 0 solar epoch like in the 1970s and 1990s (c.f. Figure 1) with the solar magnetic field directed outward from the Sun in the northern polar region and inward in the southern polar region, positively charged particles are expected to drift into the inner heliosphere over the solar poles and out along the heliospheric current sheet. In the same phase of the solar cycle the drift pattern of negatively charged particles is reversed. The intensity of negatively charged particles is, therefore, expected to depend on the latitudinal excursion of the heliospheric current sheet in an A $>$ 0 cycle, whereas the intensity of positively charged particles should vary significantly less (Potgieter and le Roux, 1992). The situation reverses in an A $<$ 0 magnetic cycle.

(e) Local sources like particles accelerated at the termination shock or interplanetary shocks.

## 2.2. Cosmic Ray Distribution at Recent Solar Minima

In order to understand solar and heliospheric modulation it is vital to reproduce the spatial distribution and the energy spectra of cosmic rays in the three-dimensional heliosphere around solar minimum periods. The second task is to investigate the solar cycle dependence of all propagation parameters. Keys to fulfill these tasks are to measure the cosmic ray distribution in the three-dimensional heliosphere and to model the cosmic ray transport in the heliospheric magnetic field in order to reproduce observations like the one shown in Figure 1. The combination of Ulysses, the Voyager, Pioneer and spacecraft at 1 AU constitute a unique network for studying the transport of galactic cosmic rays over a vast region of the heliosphere. The range of heliocentric distances and latitudes covered by Ulysses (U), Voyager (V1 and V2) and Pioneer 10 (P10) for the time period from mid 1993 to 2005 is shown in a semilog plot (Figure 3A).

Panel (B) of Figure 3 displays the cosmic ray spectra for protons for the 1965 and 1977 solar minimum, respectively. (1) It follows a $E^1$ law at several 10 MeV. From the Figure it is obvious that (2) the intensities are higher in the A $>$ 0 than in the A $<$ 0 solar magnetic epoch at energies below 700 MeV and (3) vice versa at energies above 700 MeV. (4) Due to the large gradients anomalous cosmic ray hydrogen can not be measured at Earth. However, Garcia-Munoz *et al.* (1973) found that the shape of the helium spectra does not follow the $E^1$ law during solar minimum. The current paradigm for the unexpected shape are anomalous cosmic rays (Fisk *et al.*, 1974). The principal ideas were further developed by Vasyliunas and Siscoe (1976), discussed in detail by Fichtner (2001). In connection with space climate it might be interesting to note that the anomalous cosmic ray helium flux is dominant at around 10 MeV per nucleon. The blue symbols in Figure 4C are the measured proton spectra at 63 AU. At energies below 100 MeV the observations are dominated by anomalous hydrogen, leading to the two peak structure.

While (1) and (4) are a consequence of the large energy losses in the heliosphere (2) and (3) can be attributed to the importance of drifts in the heliospheric magnetic field (Reinecke and Potgieter, 1994).

Another important prediction from drift-dominated modulation models is the expectation that protons will have large positive latitudinal gradients in an A $>$ 0 solar magnetic epoch (Potgieter, 1998; Heber and Marsden, 2001). The dataset from the high heliospheric latitude ESA/NASA Ulysses mission is ideally suited to investigate galactic cosmic ray flux variations with heliospheric latitude (Marsden and Wenzel, 1981). Ulysses was launched towards Jupiter in October 1990 in the

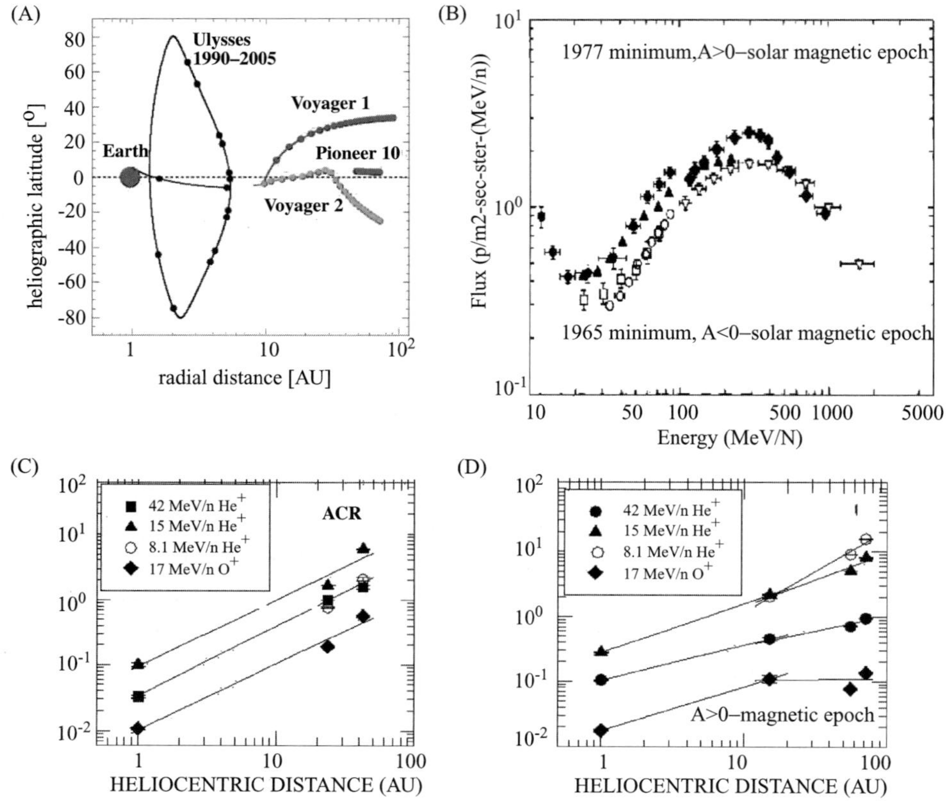


*Figure 3.* (A) Heliographic latitude and radial distance of Ulysses, V1, V2 and P10 from mid 1993 to 2005. Solid circles mark the start of each year. V1 crossed the termination shock (TS) at a distance of 94 AU in December 2004 (`http://www.nasa.gov/vision/universe/solarsystem/voyager_agu.html`). Galactic cosmic ray proton spectra (B), as measured by IMP during the 1960s and 1970s solar minima (Beatty *et al.*, 1985). Radial intensity distributions of cosmic rays for the $A < 0$ (C) and $A > 0$ solar magnetic epochs (D) have been determined by Fujii and McDonald (2005).

declining phase of solar cycle 22. The spacecraft is traveling since then in an elliptical orbit inclined at 80.2° to the solar equator and performed within 11 months a whole latitude scan of 160° up from 80°S to 80°N at solar minimum in 1994/1995 and at solar maximum in 2000/2001. Figure 4A displays the Ulysses and Earth trajectory during the minimum fast latitude scans. Marked by shading are the Ulysses polar passes, which are those periods during which the spacecraft is above 70 degrees heliographic latitude in either hemisphere. Figure 4 illustrates in its part (B) the expected variation of the proton spectrum. The expected proton spectra at 1 AU in the ecliptic and at 80° latitude are displayed together with the local interstellar spectrum (LIS). The model parameters have been chosen such that the 1 AU spectrum fits typical ecliptic 1 AU solar minimum spectra. At energies below several 100 MeV an increase by an order of magnitude was expected and the LIS

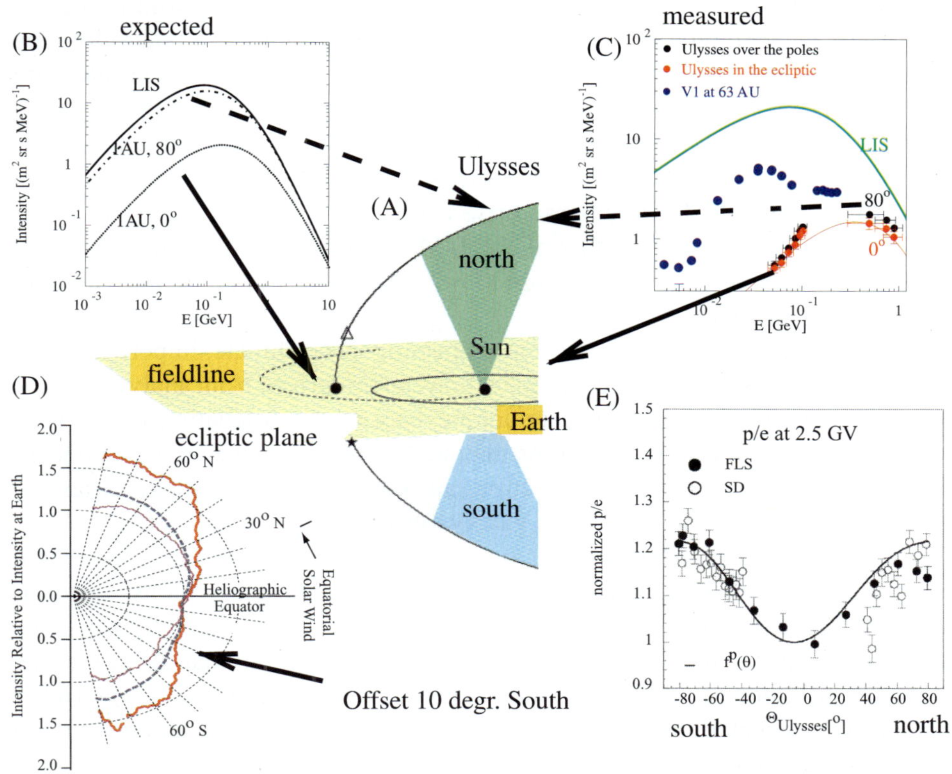


*Figure 4.* (A) Ulysses and Earth orbit during the first fast latitude scan in 1994/1995. (B) and (C) expected and measured spectra in the ecliptic and over the poles respectively. (D) Daily averages of >125 MeV protons measured at Earth and by Ulysses (red), showing a positive latitudinal gradient for galactic cosmic ray protons. (E) Proton to electron ratio as function of Ulysses heliographic latitude. The Figure is a composition from Heber and Potgieter (2000), McKibben *et al.* (1996), Heber *et al.* (1996b), and Heber *et al.* (1999).

should become almost unmodulated at polar latitudes. The Ulysses observations during solar minimum are given in panel (C) together with the Voyager observations at 63 AU. The red symbols and line correspond to the Ulysses observations and the calculation for the heliographic equator, respectively. The black symbols are Ulysses measurements above 70°. In contrast to the expectations the measured spectrum over the poles is still lower than the Voyager measurements and highly modulated. Thus Ulysses did not measure the LIS during the minimum of solar cycle 22 – with positive charged particles drifting inwards at polar regions – and led Heber *et al.* (1996a) to the conclusion that it is impossible to determine LIS in the inner heliosphere. Therefore the LIS will only be measurable by a space probe, like the Interstellar Probe (Liewer *et al.*, 2000) or an Interstellar Heliopause Explorer – investigated by ESA in 2003 (Leipold *et al.*, 2003) – to be sent far beyond the heliospheric termination shock.

Figure 4D displays nine day running averages of Ulysses to Earth-ratios of 35–70 MeV per nucleon protons, helium, and >100 MeV protons as a polar plot. The data are normalized during the equator crossing in March 1995. A constant ratio of one means a spherically symmetric cosmic ray distribution. From this Figure it is evident that 35–70 MeV protons do not show any significant latitudinal excess (pink curve), whereas >100 MeV protons (red curve) and 30–70 MeV per nucleon helium (blue curve) do. It was a big surprise when Simpson *et al.* (1996) and Heber *et al.* (1996a) reported that the flux of >100 MeV protons is not symmetric to the heliographic equator. They found a shift of $\sim 7°–10°$ of the minimum intensity into the southern hemisphere.

Although electrons are of minor importance for space climate, important information about heliospheric particle propagation can be derived from analyzing these data. Together with positrons they are unique, because (1) adiabatic energy losses are small and (2) information about the interstellar spectra can be derived independently from in situ measurements: Electrons and positrons undergo Bremsstrahlung and an inverse Compton effect in the galaxy (Webber, 1983, 1998; Strong, 2001). Since the drift pattern of electrons is opposite to the one of protons, one would expect that electrons have a different latitudinal gradient than protons. Unfortunately the Kiel Electron Telescope (Simpson *et al.*, 1992) on board Ulysses is the *only* experiment that measures GeV electrons continuously, so that the latitudinal electron distribution cannot be determined directly. Figure 4E displays how Heber *et al.* (1999) estimated the latitudinal gradient of electrons. These authors plotted the p/e-ratio as function of heliographic latitude. The line shown in panel (E) is the result of a fit to the proton distribution with latitude (not shown here). From the Figure it is obvious that the latitudinal variation of the e/p-ratio can also be approximated by that line. Therefore they concluded that the gradient of electrons is consistent with zero.

## 3. Solar Cycle Variation

Cosmic ray modulation during increased solar activity is characterized by several large steps that are easily recognized from observations at Earth and beyond, as shown in the right two panels of Figure 5 during the 1988, and 1998 period for example. These large steps are correlated with long-lasting intense magnetic fields in the outer heliosphere, called Global Merged Interaction Regions (Burlaga *et al.*, 1993). Cane *et al.* (1999) raised the question if such merging is necessary to modulate cosmic rays. They and Belov (2000) found a good correlation between the heliospheric magnetic field strength and the variation of galactic cosmic rays at Earth, indicating that the diffusion tensor should be correlated with the heliospheric magnetic field strength. Potgieter and Ferreira (2001) and Ferreira and Potgieter (2004) tested this idea using a full time-dependent numerical model with a diffusion coefficient changing with the phase of the solar cycle. They coupled the

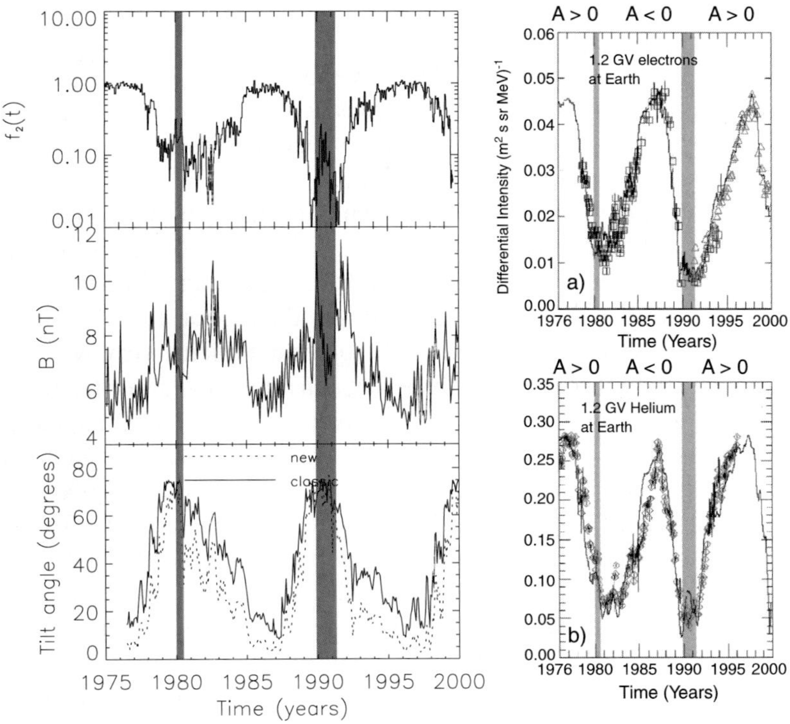


*Figure 5.* The left panels display from bottom to top the computed time history of the tilt angle, the measured magnetic field strength at 1 AU and the resulting variation of the diffusion tensor. On the right the measured and computed solar cycle modulation of 1.2 GV electron and helium are displayed (Potgieter and Ferreira, 2001).

diffusion tensor with the observed heliospheric magnetic field strength $B$ and maximum latitudinal extension of the heliospheric current sheet (tilt angle $\alpha$) so that variations are small for solar minimum, and are increasing with increasing solar activity. The time history of the tilt angle $\alpha(t)$ and of the heliospheric magnetic field $B$ at Earth are displayed in the lower two panels of Figure 5 (left). The time dependence of the diffusion tensor is then expressed by $\kappa(t) \propto B^{\frac{\alpha}{n}}$, with $n$ depending on the particle species and energy. Towards solar maximum, temporal changes in the diffusion coefficients essentially simulate step-like variations in the galactic cosmic ray intensities. In Figure 5 (right-hand panel) the results of this compound modeling approach are compared to the 1.2 GeV electron and helium observations. It is evident that the modulation amplitude, most of the steps, and the peaked and less peaked minima are reproduced, although some of the simulated steps do not have the correct magnitude and phase. It is also important to note that these authors could not use absolute identical parameters for electrons and protons, indicating that: (1) the diffusion coefficients in general may be different for the two species, (2) drifts are not yet handled correctly with increasing solar activity and/or (3) an additional charge-sign dependent processes may be contributing to modulation, e.g., magnetic helicity (Burger *et al.*, 1997).

Concerning (2) the importance of drifts over the 22-year solar magnetic cycle, measurements from the out-of-ecliptic Ulysses mission are crucial. The observed charge sign dependent latitudinal gradient at solar minimum are indicative for gradient and curvature drifts in the heliosphere.

Six years later when Ulysses performed another fast latitude scan Belov *et al.* (2003), Heber *et al.* (2003), and McKibben *et al.* (2003) reported a spherical symmetric cosmic ray distribution indicating that drifts are of minor importance around solar maximum. Ndiitwani *et al.* (2005) analyzed the e/p-ratio time profile during solar maximum and found that the no drift solution explains very well the solar maximum observations, but fails shortly before and after, indicating that the particle drift is an important factor for understanding solar modulation and that modulation models have to include all components used in Parker's transport Equation (1).

Several attempts have been made in the recent years to model the galactic cosmic ray intensities during the grand minima periods (see also Beer, 2006). These grand minima are characterized by low solar activity during several solar cycle as has been observed in the 17th century during the Maunder Minimum. Scherer and Fichtner (2004) used a dynamic multi fluid model of the global heliosphere (Fahr *et al.*, 2000). They found, that the variation of the heliospheric structure leads only to a small increase of galactic cosmic rays during the Maunder Minimum and therefore the diffusion tensor needs to be modified during such periods. In order to obtain more realistic results in future, Scherer and Ferreira (2005) combined successfully a dynamic multi fluid model of the global heliosphere with the compound particle transport code described above. Although it is beyond the scope of this paper to discuss all the effects of such an approach, it should be mentioned that for time scales longer than the length of the solar cycle several additional aspects have to be considered, like the particle transport in the heliosheath, the extrapolation of the current diffusion tensor and drift fields to different heliospheric conditions and the knowledge of the local interstellar spectra (Scherer *et al.*, 2004; Caballero-Lopez *et al.*, 2004; McCracken *et al.*, 2004).

## 4. Summary

In this paper, both the current knowledge and hypotheses about the modulation and the transport of cosmic rays in the heliosphere are reviewed. The transport theory aims at an understanding of all processes determining the propagation of cosmic rays in the solar wind plasma. Keys to fulfill this task are to measure the cosmic ray distribution in the three-dimensional heliosphere and to model the cosmic ray transport in the solar wind. The main unknowns in the models are the diffusion tensor, the importance of drifts and the local interstellar spectra. Drift effects depend on the charge sign of the particle and the polarity of the heliospheric magnetic field. Therefore it is important to analyze particles with opposite charge sign simultaneously. On the modelling side it is vital to include all modulation processes in transport models. Recent modelling has advanced to a self-consistent treatment of the dynamic heliosphere and the cosmic ray transport. This opens the unique opportunity to study long term modulation on time scales of centuries.

## References

Beatty, J. J., Garcia-Munoz, M., and Simpson, J. A.: 1985, 'The cosmic-ray spectra of H-1, H-2, and He-4 as a test of the origin of the hydrogen superfluxes at solar minimum modulation', *Astrophys. J.* **294**, 455–462.

Beer, J., Vonmoos, M., and Muscheler, R.: 2006, 'Solar variability over the past several millenia', *Space Sci. Rev.*, this volume, doi: 10.1007/s11214-006-9047-4.

Belov, A.: 2000, 'large scale modulation: View from the earth', *Space Sci. Rev.* **93**, 79–105.

Belov, A. V., Eroshenko, E. A., Heber, B., Yanke, V. G., Raviart, A., Müller-Mellin, R., and Kunow, H.: 2003, 'Latitudinal and radial variation of >2 GeV/n protons and alpha-particles at solar maximum: ULYSSES COSPIN/KET and neutron monitor network observations', *Annales Geophysicae* **21**, 1295–1302.

Bieber, J. W., Matthaeus, W. H., Shalchi, A., and Qin, G.: 2004, 'Nonlinear guiding center theory of perpendicular diffusion: General properties and comparison with observation', *Geophys. Res. Lett.* **31**, doi:10.1029/2004GL020007.

Burger, R. A., Hattingh, M., and Bieber, J. W.: 1997, 'The effect of magnetic helicity on the propagation of galactic cosmic rays', *Adv. Space Res.* **19**, 897–900.

Burlaga, L. F., Perko, J., and Pirraglia, J.: 1993, 'Cosmic-ray modulation, merged interaction regions, and multifractals', *Astrophys. J.* **407**, 347–358.

Caballero-Lopez, R. A., Moraal, H., McCracken, K. G., and McDonald, F. B.: 2004, 'The heliospheric magnetic field from 850 to 2000 AD inferred from $^{10}$Be records', *J. Geophys. Res.* **109**, doi:10.1029/2004JA010633.

Cane, H., Wibberenz, G., Richardson, I. G., and von Rosenvinge, T. T.: 1999, 'Cosmic ray modulation and the solar magnetic field', *Geophys. Res. Lett.* **26**, 565–568.

Decker, R. B., Krimigis, S. M., Roelof, E. C., Hill, M. E., Armstrong, T. P., Gloeckler, G., Hamilton, D. C., and Lanzerotti, L. J.: 2005, 'Voyager 1 in the Foreshock, Termination Shock, and Heliosheath', *Science* **309**, 2020–2024.

Fahr, H. J., Kausch, T., and Scherer, H.: 2000, 'A 5-fluid hydrodynamic approach to model the solar system-interstellar medium interaction', *Astron. Astrophys.* **357**, 268–282.

Ferreira, S. E. S., and Potgieter, M. S.: 2004, 'Long-Term Cosmic-Ray Modulation in the Heliosphere', *Astrophys. J.* **603**, 744–752.

Fichtner, H.: 2001, 'Anomalous Cosmic Rays: Messengers from the outer heliosphere', *Space Sci. Rev.* **95**, 639–754.

Fisk, L. A.: 1996, 'Motion of the footpoints of heliospheric magnetic field lines at the Sun: Implications for recurrent energetic particle events at high heliographic latitudes', *J. Geophys. Res.* **101**, 15,547–15,554.

Fisk, L. A., Kozlovsky, B., and Ramaty, R.: 1974, 'An Interpretation of the Observed Oxygen and Nitrogen Enhancements in Low-Energy Cosmic Rays', *Astrophys. J. Lett.* **190**, L35.

Fujii, Z., and F. B. McDonald: 2005, 'The spatial distribution of galactic and anomalous cosmic rays in the heliosphere at solar minimum', *Adv. Space Res.* **35**, 611–616.

Garcia-Munoz, M., Mason, G. M., and Simpson, J. A.: 1973, 'A new test for solar modulation theory: The 1972 May-July low-energy galactic cosmic-ray proton and helium spectra', *Astrophys. J. Lett.* **182**, L81.

Gurnett, D. A., Kurth, W. S., and Stone, E. C.: 2003, 'The return of the heliospheric 2–3 kHz radio emission during solar cycle 23', *Geophys. Res. Lett.* **30**, 8–1.

Heber, B. and Potgieter, M. S.: 2000, 'Galactic cosmic ray observations at different heliospheric latitudes', *Adv. Space Res.* **26**, 839–852.

Heber, B. and Marsden, R. G.: 2001, 'Cosmic Ray modulation over the poles at solar maximum: Observations.', *Space Sci. Rev.* **97**, 309–319.

Heber, B., Dröge, W., Ferrando, P., Haasbroek, L., Kunow, H., Müller-Mellin, R., Paizis, C., Potgieter, M., Raviart, A., and Wibberenz, G.: 1996a, 'Spatial variation of >40 MeV/n nuclei fluxes observed during the Ulysses rapid latitude scan', *Astron. Astrophys.* **316**, 538–546.

Heber, B., Dröge, W., Kunow, H., Müller-Mellin, R., Wibberenz, G., Ferrando, P., Ravart, A., and Paizis, C.: 1996b, 'Spatial variation of >106 MeV proton fluxes observed during the Ulysses rapid latitude scan: Ulysses COSPIN/KET results', *Geophys. Res. Lett.* **23**, 1513–1516.

Heber, B., Ferrando, P., Raviart, A., Wibberenz, G., Müller-Mellin, R., Kunow, H., Sierks, H., Bothmer, V., Posner, A., Paizis, C., and Potgieter, M. S.: 1999, 'Differences in the temporal variation of galactic cosmic ray electrons and protons: Implications from Ulysses at solar minimum', *Geophys. Res. Lett.* **26**, 2133–2136.

Heber, B., Sarri, G., Wibberenz, G., Paizis, C., Ferrando, P., Raviart, A., Posner, A., Müller-Mellin, R., and Kunow, H.: 2003, 'The Ulysses fast latitude scans: COSPIN/KET results', *Annales Geophysicae* **21**, 1275–1288.

Hoeksema, J.: 1995, 'The large-scale structure of the heliospheric current sheet during the ULYSSES epoch', *Space Sci. Rev.* **72**, 137–148.

Jokipii, J. R., Levy, E. H., and Hubbard, W. B.: 1977, 'Effects of particle drift on cosmic ray transport, I. General properties, application to solar modulation ', *Astrophys. J.* **213**, 861–868.

Leipold, M., Fichtner, H., Heber, B., Groepper, P., Lascar, S., Burger, F., Eiden, M., Niederstadt, T., Sickinger, C., Herbeck, L., Dachwald, B., and Seboldt, W.: 2003, 'Heliopause Explorer - a sailcraft mission to the outer boundaries of the solar system', In: *ESA SP-542: Low-Cost Planetary Missions*. pp. 367–375.

Liewer, P. C., Mewaldt, R. A., Ayon, J. A., and Wallace, R. A.: 2000, 'NASA's Interstellar Probe Mission', In: *AIP Conf. Proc. 504: Space Technology and Applications International Forum*. p. 911.

Marsden, R. C., and Wenzel, K. P.: 1981, 'The International Solar Polar Mission (ISPM)', In: *ESA SP-164: Solar System and Its Exploration*. pp. 51–59.

McCracken, K. G., McDonald, F. B., Beer, J., Raisbeck, G., and Yiou, F.: 2004, 'A phenomenological study of the long-term cosmic ray modulation, 850-1958 AD', *J. Geophys. Res.* **109**, doi:10.1029/2004JA010685.

McKibben, R. B., Connell, J. J., Lopate, C., Simson, J. A., and Zhang, M.: 1996, 'Observations of galactic cosmic rays and the anomalous helium during the Ulysses passage from the south pole to the north pole', *Astron. Astrophys.* **316**, 547.

McKibben, R. B., Connell, J. J., Lopate, C., Zhang, M., Anglin, J. D., Balogh, A., dalla, S., Sanderson, T. R., Marsden, R. G., Hofer, M. Y., Kunow, H., Posner, A., and Heber, B.: 2003, 'Ulysses COSPIN observations of cosmic rays and solar energetic particles from the South Pole to the North Pole of the Sun during solar maximum', *Annales Geophysicae* **21**, 1217–1288.

Moraal, H.: 1993, 'Cosmic ray modulation studies in the outer heliosphere', *Nuclear Physics B Proceedings Supplements* **33**, 161–178.

Ndiitwani, D. C., Ferreira, S. E. S., Potgieter, M. S., and Heber, B.: 2005, 'Modelling cosmic ray intensities along the Ulysses trajectory', *Annales Geophysicae* **23**, 1061–1070.

Parhi, S., Burger, R. A., Bieber, J. W., and Matthaeus, W. H.: 2001, 'Challenges for an ab initio theory of cosmic ray modulation', In: *Proceedings of the ICRC 2001*. pp. 3670–3673.

Parker, E. N.: 1958, 'Dynamics of the Interplanetary gas and magnetic fields', *Astrophys. J.* **128**, 664.

Parker, E. N.: 1963, *Interplanetary Dynamical Processes*, New York: Wiley and Sons.

Parker, E. N.: 1965, 'The passage of energetic particles through interplanetary space', *Planet. Sp. Sc.* **13**, 9–49.

Potgieter, M.: 1984, 'The modulation of galactic cosmic rays as described by a three-dimensional drift model', Ph.D. thesis, Potchefstroom University of CHE, South Africa.

Potgieter, M. S.: 1998, 'The modulation of galactic cosmic rays in the heliosphere: Theory and models', *Space Sci. Rev.* **83**, 147–158.

Potgieter, M. S., and le Roux, J. A.: 1992, 'The simulated features of heliospheric cosmic-ray modulation with time-dependent drift model I. General effects of the changing neutral sheet over the period 1985-1990', *Astrophys. J.* **386**, 336–346.

Potgieter, M., and Ferreira, S. E. S.: 2001, 'Modulation of cosmic rays in the heliosphere over 11 and 22 year cycles: A modelling perspective', *Adv. Space Res.* **27**, 481–492.

Reinecke, J. P. L. and Potgieter, M. S.: 1994, 'An explanation for the difference in cosmic ray modulation at low and neutron monitor energies during consecutive solar minimum periods', *J. Geophys. Res.* **99**, 14,761–14,768.

Scherer, K., and Fichtner, H.: 2004, 'Constraints on the heliospheric magnetic field variation during the Maunder Minimum from cosmic ray modulation modelling', *Astron. Astrophys.* **413**, L11–L14.

Scherer, K., and Ferreira, S. E. S.: 2005, 'A heliospheric hybrid model: hydrodynamic plasma flow and kinetic cosmic ray transport', *Astrophysics and Space Sciences Transactions* **1**, 17–27.

Scherer, K., Fahr, H.-J., Fichtner, H., and Heber, B.: 2004, 'Long-term modulation of cosmic rays in the heliosphere and its influence at earth', *Sol. Phys.* **224**, 305–316.

Shalchi, A., and Schlickeiser, R.: 2004, 'Quasilinear perpendicular diffusion of cosmic rays in weak dynamical turbulence', *Astron. Astrophys.* **420**, 821–832.

Shalchi, A., and Schlickeiser, R.: 2005, 'Evidence for the nonlinear transport of galactic cosmic rays', *Astrophys. J. Lett.* **626**, L97–L99.

Simpson, J. A.: 1983, 'Introduction to the galactic cosmic radiation', In: *NATO ASIC Proc. 107: Composition and Origin of Cosmic Rays*. pp. 1–24.

Simpson, J., *et al.*: 1992, 'The ulysses cosmic-ray and solar particle investigation', *Astron. Astrophys. Suppl.* **92**, 365–399.

Simpson, J., Zhang, M., and Bame, S.: 1996, 'A solar polar north-south asymmetry for cosmic ray propagation in the heliosphere: the Ulysses pole-to-pole rapid transit', *Astrophys. J. Lett.* **465**, L69.

Snyder, C. W., and Neugebauer, M.: 1963, 'Direct observations of the solar wind by the mariner ii spacecraft', *Solar Particles and Sun-Earth Relations. Proceedings from the 8th International Cosmic Ray Conference, Volume 1, p. 210* **1**, 210.

Stone, E. C., Cummings, A. C., McDonald, F. B., Heikkila, B. C., Lal, N., and Webber, W. R.: 2005, 'Voyager 1 Explores the Termination Shock Region and the Heliosheath Beyond', *Science* **309**, 2017–2020.

Strong, A. W.: 2001, 'Signatures of energetic protons and electrons in the galaxy', *Space Sci. Rev.* **99**, 167–176.

Vasyliunas, V. M., and Siscoe, G. L.: 1976, 'On the flux and the energy spectrum of interstellar ions in the solar system', *J. Geophys. Res.* **81**, 1247–1252.

Webber, W. R.: 1983, 'Cosmic ray electrons and positrons – a review of current measurements and some implications', In: *NATO ASIC Proc. 107: Composition and Origin of Cosmic Rays*. pp. 83–100.

Webber, W. R.: 1998, 'A new estimate of the local interstellar energy density and ionization rate of galactic cosmic cosmic rays', *Astrophys. J.* **506**, 329–334.

# WHAT DO COSMOGENIC ISOTOPES TELL US ABOUT PAST SOLAR FORCING OF CLIMATE?

M. LOCKWOOD
*Rutherford Appleton Laboratory, Space Science and Technology Department, Chilton, Didcot, Oxfordshire, OX11 0QX, UK; University of Southampton, Highfield, Southampton, UK*
*(E-mail: m.lockwood@rl.ac.uk)*

(Received 3 December 2005; Accepted in final form 28 April 2006)

**Abstract.** In paleoclimate studies, cosmogenic isotopes are frequently used as proxy indicators of past variations in solar irradiance on centennial and millennial timescales. These isotopes are spallation products of galactic cosmic rays (GCRs) impacting Earth's atmosphere, which are deposited and stored in terrestrial reservoirs such as ice sheets, ocean sediments and tree trunks. On timescales shorter than the variations in the geomagnetic field, they are modulated by the heliosphere and thus they are, strictly speaking, an index of heliospheric variability rather than one of solar variability. Strong evidence of climate variations associated with the production (as opposed to the deposition) of these isotopes is emerging. This raises a vital question: do cosmic rays have a direct influence on climate or are they a good proxy indicator for another factor that does (such as the total or spectral solar irradiance)? The former possibility raises further questions about the possible growth of air ions generated by cosmic rays into cloud condensation nuclei and/or the modulation of the global thunderstorm electric circuit. The latter possibility requires new understanding about the required relationship between the heliospheric magnetic fields that scatter cosmic rays and the photospheric magnetic fields which modulate solar irradiance.

**Keywords:** galactic cosmic rays, total solar irradiance, cosmogenic isotopes, paleoclimate changes

## 1. Introduction

Paleoclimate studies frequently use cosmogenic isotope abundances as indicators of past total solar irradiance (TSI) variations (see review in this volume by Beer *et al.* (2006). In Section 2 we focus on three examples by Bond *et al.* (2001), Neff *et al.* (2001) and Wang *et al.* (2005b). In these studies it is either explicitly or implicitly assumed that the inferred deposition rates of the isotopes are strongly related to either TSI or spectral solar irradiance. As discussed in the reviews by Fröhlich (2006) and Rottman (2006) in this volume, we now have solar irradiance observations covering three decades. Figure 1 shows that an anticorrelation is observed between TSI and galactic cosmic ray (GCR) fluxes (Lockwood, 2002a; Muscheler *et al.*, 2003). However, because of the large thermal capacity of the oceans, these potential decadal-scale variations in climate forcing will be heavily damped in Earth's climate response, and any century-scale changes would be much more significant (Wigley and Raper, 2005). Because we have only a partial understanding of the

Space Science Reviews (2006) 125: 95–109
DOI: 10.1007/s11214-006-9049-2

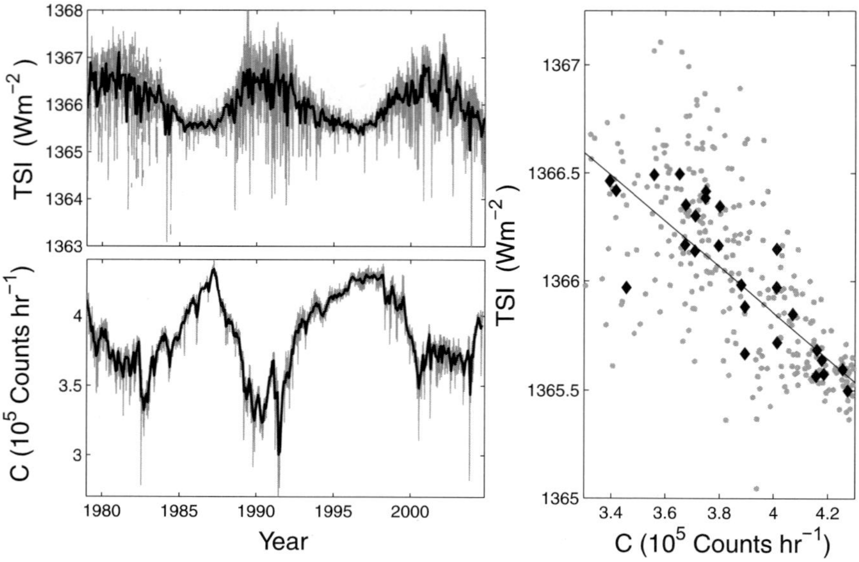


*Figure 1.* The anticorrelation of GCR fluxes with the total solar irradiance (TSI) since 1978. *Left-hand panels:* the variations of (top) the PMOD TSI composite (Fröhlich, 2006) and (bottom) the GCR counts, $C$, detected by the neutron monitor at Climax. In both cases, the grey line gives the daily values, and the black line the monthly means. It can be seen that the agreement is poor on a daily basis because of the effect of individual sunspot groups on TSI. *Right-hand panel:* Scatter plot of TSI as a function of $C$. The grey points show monthly means, the black diamonds the annual means. The best-fit linear regression to the annual data is also plotted. The correlation coefficients (and significance levels) are $-0.68$ (99.99%) and $-0.85$ (91.5%) for, respectively, monthly and annual data.

physics of the centennial and millennial changes in TSI and cosmic rays, we cannot be sure that the correlation seen on decadal timescales will also apply on century timescales. Reconstructions of TSI over the past 3 centuries do show strong similarities with records of cosmogenic isotope abundances (Lean *et al.*, 1995). However, the physics behind this connection is far from clear and quantitative scaling of TSI from cosmogenic isotope records is not yet possible. The common denominator between the two is the solar magnetic field, the TSI being determined by the field in the photosphere (see review in this volume by Solanki, 2006) and GCR fluxes and cosmogenic isotopes being modulated by the "open" field which passes through the solar corona and fills the heliosphere (see review in this volume by Heber *et al.*, 2006). However, the open solar flux is only small fraction (a few percent) of the photospheric flux and it is not immediately apparent why the former should be a reliable indicator of the latter. Furthermore, TSI variations depend not only on the total photospheric flux, but also on the distribution of flux tube sizes (see review by Lockwood, 2004).

The alternative hypothesis, discussed in Section 3, is that cosmic rays have a direct effect on climate, for example via a proposed influence on cloud formation or via modulation of the global electric circuit.

This paper addresses the issues surrounding this central question of the relationship of cosmogenic isotopes to TSI and climate, after first reviewing some of the evidence for a link between them in Section 2.

## 2. Paleoclimate Evidence for Solar Variability Effects on Climate

In recent years, a body of evidence has emerged that solar variations have had a clear effect on climate throughout the Holocene, the warm period that has prevailed throughout the past ten thousand years. These studies have been made in many parts of the world and employ a wider variety of paleoclimate indicators (Davis and Shafer, 1992; Jirikowic *et al.*, 1993; Davis, 1994; Van Geel *et al.*, 1998; Yu and Ito, 1999; Bond *et al.*, 2001; Neff *et al.*, 2001; Sarnthein *et al.*, 2003; Hu *et al.*, 2003; Prasad *et al.*, 2004; Wei and Wang, 2004; Christla *et al.*, 2004; Wang *et al.*, 2005b; Maasch *et al.*, 2005; Mayewski *et al.*, 2005). Here we discuss just three of the many examples.

The average abundance of ice-rafted debris has been measured in cores of ocean-bed sediment throughout the middle and North Atlantic. These glasses, grains and crystals are gouged out in known glaciers, from which they are carved off in icebergs and deposited in the sediment when and where the icebergs melt. The sediment is dated using microfossils found at the same level in the core. The abundances of this "ice-rafted debris" are very sensitive indicators of currents, winds and temperatures in the North Atlantic and reveal high, and highly significant, correlations with both the $^{10}$Be and $^{14}$C cosmogenic isotopes (Bond *et al.*, 2001).

A second example of such paleoclimate evidence has been obtained from the oxygen isotope ratio $\delta^{18}$O, as measured in stalagmites in Oman (see Figure 2). This has been found to show an exceptional correspondence with the cosmogenic isotopes on timescales between decades and several thousand years (Neff *et al.*, 2001). U-Th (Uranium-Thorium series) dating is used on the stalagmite and the limits to allowed temporal "wiggle-matching" are set by experimental uncertainties and have been rigorously adhered to. The $\delta^{18}$O is, in this case, a proxy for local rainfall and reveal enhanced rainfall caused by small northward migrations of the inter-tropical convergence zone. Large effects are seen because the latitudinal gradient around the site is large. These correlations are between cosmogenic isotopes and local climate indicators. However, Wang *et al.* (2005b) have recently repeated this analysis and found the same result on the edge of the monsoon belt in China. Therefore there is strong evidence for global shifts in the latitude of the monsoon belt associated with cosmic ray variations.

Importantly, these correlations are found for both the $^{14}$C and $^{10}$Be isotopes. Both are spallation products of GCRs hitting atmospheric O, N and Ar atoms. However, there the similarities end because their transport and deposition into the reservoir where they are detected (ancient tree trunks for $^{14}$C and ice sheets or

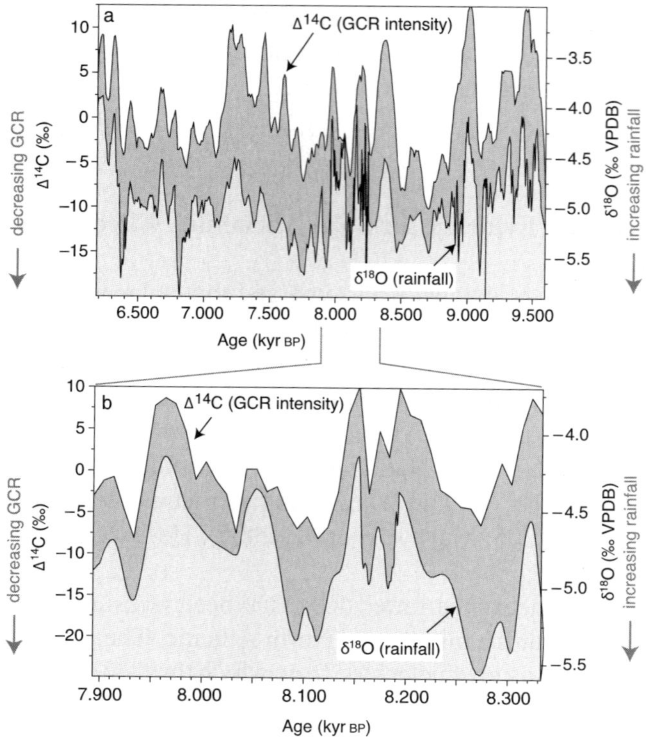


*Figure 2.* Comparison of $\delta^{18}O$, an indicator of local rainfall, from stalagmite analysis of a cave in Oman and $\Delta^{14}C$ from global tree ring data. The bottom panel shows a higher time-resolution analysis allowed by a period of rapid stalagmite growth. The correlation is seen to apply on timescales between decades and several thousand years (Neff *et al.*, 2001).

ocean sediments in the case of $^{10}Be$) are vastly different in the two cases. We can discount the possibility that the isotope abundances in their respective reservoirs are similarly influenced by climate during their terrestrial life-history because the transport and deposition of each is so vastly different. Thus we can conclude that the correlation is found for both isotopes because of the one common denominator in their production, namely the incident GCR flux.

The flux of the GCRs that generate the cosmogenic isotopes is modulated by three influences (Beer *et al.*, 2006): (1) the interstellar flux of GCRs incident on the heliosphere; (2) the GCR shielding by the heliosphere; and (3) the GCR shielding by the geomagnetic field. The spatial scale of interstellar GCR flux variation in our galaxy is sufficiently large compared to distances moved by our solar system through the galaxy, that we can neglect incident GCR variations on timescales of Myr and smaller. The geomagnetic field shield has varied on timescales of 10 kyr. This variation has, in the main, been gradual during the Holocene (Tric, 1992; Baumgartner *et al.*, 1998), although there have been shorter-lived weakenings of the field (which may be geomagnetic reversal onsets that did not develop) such as the Laschamp event around 40 kyr BP (Laj *et al.*, 2001). These events complicate the cosmogenic isotope record (Bhattacharyya and Mitra, 1997; Damon *et al.*, 1978) but are not consistent with the variations seen on timescales of order 1 kyr and less seen, for example by Bond *et al.* (2001), Neff *et al.* (2001) and Wang *et al.* (2005b). This being the case, most of the variations on these timescales arise from heliospheric shielding.

With these considerations, the correlations between these cosmogenic isotopes and paleoclimate indicators allow just two possible classes of explanation:

A. Air ions produced by GCRs have a direct effect on climate.
B. The GCR flux is anticorrelated with another factor that influences climate to the extent that it is a good proxy for that factor.

The most controversial suggestion under category A, is that cosmic rays directly modulate the formation of clouds. However, this is not the only possibility because GCRs are the source of electrical conductivity in the sub-ionospheric gap and thus are vital to the global thunderstorm electric circuit, which also may have a solar cycle variation (Schlegel *et al.*, 2001). The most likely factors for which GCRs could be a proxy (category B) are the total solar irradiance or the UV spectral irradiance. Note that paleoclimate studies usually (explicitly or implicitly) assume a category B explanation.

## 3. Direct Cosmic-Ray Effects: Clouds and the Global Electric Circuit

The most controversial suggestion for a direct effect of cosmic rays on climate is that they directly modulate the formation of clouds (Friis-Christensen and Svensmark, 1997; Svensmark, 1998; Marsh and Svensmark, 2000, 2003; Udelhofen and Cess, 2001; Kristjánsson and Kristiansen, 2000; Carslaw *et al.*, 2002; Arnold and Neubert, 2002). This idea was first proposed by analogy to cloud chamber particle detectors – an analogy that is not valid because natural atmospheric supersaturations are much smaller than those needed to make a cloud chamber work. The idea has been revived by observed correlations over recent solar cycles between GCRs counts and the global composite of satellite cloud cover observations compiled by the International Satellite Cloud Climatology Project, ISCCP (Rossow *et al.*, 1996). Udelhofen and Cess (2001) found a solar cycle signal in ground-based data from 90 weather stations across the North American Continent. Instrument relocation and changes mean that a long-term drift in these ground-based data cannot be determined, but de-trended data show a clear and persistent solar cycle variation in coastal cloud cover in data that extends back to 1900. The best correlations between GCRs and global cloud cover have been obtained by Marsh and Svensmark (2000) from the infrared observations of clouds (10–12 $\mu$m) that make up the "D2" set compiled by ISCCP: these authors find that it is primarily liquid, maritime clouds, away from regions of an El Niño (ENSO) event, that correlate well with GCR fluxes. Other authors argue that the results are still influenced by ENSO events (Farrar,

2000). Correlations on shorter timescales due to Forbush decreases in GCR fluxes, have been reported in localised datasets by Veretenenko and Pudovkin (1997).

The arguments against such direct cosmic ray-cloud connection have been:

1. Given the atmospheric supersaturations, there is no known mechanism that can cause the effect.
2. The inter-calibrations involved in the composite ISCCP dataset render it unsuitable for this type of analysis.
3. The data sequences are too short and so the significance of the correlations is low or marginal.
4. Periods of low geomagnetic field, particularly the Laschamp event, gave enhanced GCR fluxes in the Earth's atmosphere but did not influence climate in the Greenland area (Beer, 2001).

Of these objections, only 4 argues that such a mechanism is not effective: 1–3 all argue that the evidence in its favour is inadequate but cannot be used as arguments that it is not active. In relation to 4, there is some evidence that the failure to see an effect of the Laschamp event may have been a local characteristic of the climate in the Greenland area (Christla *et al.*, 2004). Nevertheless, objection 4 remains the key debate.

Concerning the potential mechanisms (objection 1), there are now some viable proposals. Air ions are produced in the atmosphere, broadly in proportion to the flux of incident GCRs (Aplin *et al.*, 2005) and there are plausible mechanisms now known, by which these can grow into cloud condensation nuclei (CCNs) at natural atmospheric supersaturations (Carslaw *et al.*, 2002). These include the growth of air ions into aerosols (Yu and Turco, 2000; Eichkorn *et al.*, 2002) and the effects of atmospheric electric fields on the microphysical connection between aerosols and clouds (Tinsley *et al.*, 2000; Tripathi and Harrison, 2002). Given these mechanism appear to be viable, the major debate now is if such an effect would be significant compared to the many other sources of CCNs (Carslaw *et al.*, 2002).

In relation to objections 2 and 3, recent studies using longer series of homogeneous data are providing new and strong evidence that there could indeed be a direct cosmic ray-cloud effect in certain regions (Harrison and Stephenson, 2005). Specifically, the "diffuse fraction" (the ratio of the intensity of daylight under direct illumination and in the shadow of an obscuring shield) is a simple and repeatable measure of local cloudiness and aerosols that has been made at many weather stations for many years. Figure 3 summarises some of these results from stations in the UK. Part (a) shows that the diffuse fraction is 2% lower on the 14% of days when the Climax GCR count is less than $3.6 \times 10^5$ hr$^{-1}$, compared to the days when it exceeds this threshold. This result is found at all the stations studied and is highly statistically significant (usually at or exceeding the 99.9% level) in almost all cases. Note that it is possible, or even probable, that the effect is only for stations in clean, maritime air (Wilding and Harrison, 2005) and appears to be less significant in regions of high rainfall.

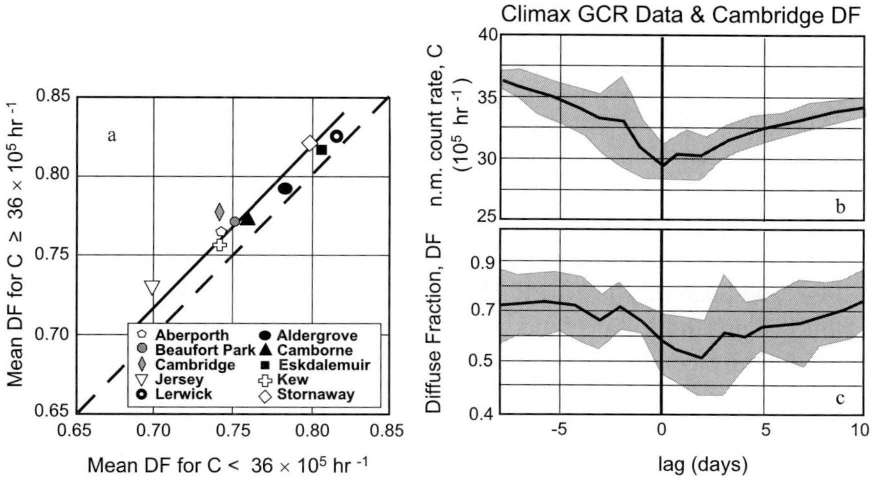


*Figure 3.* The effect of GCR fluxes on the diffuse fraction (*DF*) observed at a variety of weather stations in the UK. The data are compared to the counts, $C$, from the neutron monitor at Climax, Colorado (detecting cosmic rays of rigidity exceeding 3 GV). *Left-hand panel:* (a) The mean diffuse fraction on days when $C$ exceeds $3.6 \times 10^5$ hr$^{-1}$ as a function of the corresponding value at the same station when $C$ is less than this threshold. In all cases, *DF* is lower for the days with lower $C$: the black line shows a 2% decrease. The stations used are: Aberporth (data available from 1957–2000; probability that the difference of *DF* arrived by chance, $p < 0.1\%$); Aldergrove (1976–2000, $p = 8\%$); Beaufort Park (1865–2000, $p < 0.1\%$); Camborne (1981–2000, $p = 1\%$); Cambridge (1957–1971, $p < 0.1\%$); Eskdalemuir (1957–2000, $p = 20\%$); Jersey (1968–1994, $p < 0.1\%$); Kew (1947–1980, $p < 2\%$); Lerwick (1952–2000, $p = 0.6\%$); and Stornaway (1982–2000, $p < 0.1\%$). *Right-hand panels:* An example of a superposed epoch analysis of Forbush decreases (in this case for the Cambridge station) showing mean values as a function of the time relative to the time of the minimum $C$ in the event. The black lines in panels (b) and (c) show, respectively, the mean variation in $C$ and the corresponding variation in mean *DF*. In both cases the 95% confidence limits are shown by the grey band (Harrison and Stephenson, 2005).

The studies reviewed above appear to show a solar influence on cloud cover. However, even if these prove to be robust, we must be cautious in ascribing this variation purely to the direct effect of cosmic rays. Some (but by no means all) GCM climate simulations have revealed an effect of TSI variations on cloudiness (see review by Haigh, 2003), via the distribution of atmospheric heating and its effect on winds and circulations, and TSI is well (anticorrelated) with GCR fluxes. Lockwood (2001b, 2002b) showed that the peak correlation coefficient on decadal timescales for the global cloud cover anomaly was +0.654 with the GCR data but was −0.741 with the TSI (see also Kristjánsson *et al.*, 2002). These correlations are significant at the 99.8% and 99.6% levels, respectively. Although the correlation is marginally higher for TSI than for the GCRs, application of the Fisher-Z test shows that the difference between these two correlations is not significant (the significance level of the difference being only 30%, i.e., the probability that the difference arose by chance is 0.7).

The production of cloud condensation nuclei by cosmic rays is not the only possibility in category A, because GCRs are the source of electrical conductivity in the sub-ionospheric gap and thus are vital to the global electric thunderstorm circuit (Markson, 1981; Bering *et al.*, 1998; Harrison, 2002a). Atmospheric electric field changes are linked to changes in global temperature, as they modulate global changes in air ion concentrations (Aplin *et al.*, 2005) and, potentially, non-thunderstorm clouds. Thunderclouds charge by collisions between ice and water moving vertically at different velocities in convective activity: in most cases, the top of the cloud becomes positively charged, the base negatively charged. Current flows from the ground to the cloud in the form of lightning and up from the cloud to the horizontally-conducting ionosphere above about 80 km. The latter is made possible by the conductivity in the sub-ionospheric gap due to the air ions produced by GCRs and give rise to optical signatures such as Sprites, Elves, and Blue Jets. In this way, thunderstorms charge the ionosphere up to a positive potential of order 300 kV. Away from the thunderclouds that power the circuit, return downward current is driven by the ionospheric potential and is made possible by the GCR-induced conductivity and, at the lowest altitudes, ionisation caused by the release of radioactive gases from the ground. The fair-weather electric field corresponds to this return current and has been shown to have fallen by about 3% per decade over the 20th century at a number of sites (Harrison, 2002b, 2004; Märcz and Harrison, 2005). Given that lightening is known to be influenced by GCRs and the solar cycle (Schlegel *et al.*, 2001; Arnold and Neubert, 2002), this may be consistent with long-term modulation of sub-ionospheric conductivity caused by a predicted drop in GCRs fluxes associated with an observed rise in the heliospheric field (Carslaw *et al.*, 2002). To complicate the picture even further, changes in atmospheric electricity may be a factor in a direct effect of GCRs on clouds (Tinsley *et al.*, 2000; Tripathi and Harrison, 2002).

## 4. The Connection Between Cosmic Rays and TSI on Centennial Time Scales

Lockwood *et al.* (1999) devised a method for deriving the open solar flux $F_S$ since 1868 using the *aa* geomagnetic index. They found that $F_S$ had approximately doubled over the twentieth century. This $F_S$ value is very well anticorrelated with both GCR fluxes and the abundance of the $^{10}$Be cosmogenic isotope (Lockwood, 2001a). This variation has been reproduced by a continuity model (see the review in this volume by Solanki, 2006) and by numerical models of magnetic flux transport over the Sun (Wang *et al.*, 2005a; Schrijver *et al.*, 2002).

As discussed in the review in this volume by Heber *et al.* (2006), full understanding of GCR shielding is complex. However, the simple anti-correlations found by Lockwood (2001a) show that much of the variation (specifically, about 75%) of the GCR flux at Earth was explained by the open flux. Bonino *et al.* (2001) fitted the modulation parameter $M$ with a quadratic expression in $F_S$ and so were able to estimate the variation in the GCR spectrum from the variation in the open flux. Similarly, Usoskin *et al.* (2002) used a power-law variation of $M$ with $F_S$. The modulation parameter (Cini-Castagnoli and Lal, 1980) is a simple way of describing heliospheric GCR shielding, based on the so-called "force-free" approximation. Whilst it is not going to describe all features of all GCR masses, charge states and energies in all parts of the heliosphere, it has had good success in describing the variation of the spectrum and flux of proton GCRs at Earth over recent solar cycles (Bonino *et al.*, 2001; Masarik and Beer, 1999; Usoskin *et al.*, 2002).

The modulation parameter was also used by Masarik and Beer (1999) to predict the heliospheric modulation of cosmogenic isotope production, using GEANT high-energy particle simulations. These simulations also predict the counts observed by neutron monitors. Thus the above papers established a clear and quantitative connection between the open flux, $F_S$, the modulation parameter $M$, GCR counts seen by a neutron monitor (e.g., $C$ for Climax) and the global mean production rate of the $^{10}$Be cosmogenic isotope, $P[^{10}\text{Be}]$. (N.B. this connection was exploited by Usoskin *et al.*, 2003, and Solanki *et al.*, 2004, to estimate sunspot numbers over recent millennia using cosmogenic isotopes).

The triangular grey points in all panels of Figure 4 show annual means of the observed TSI since 1978 (the PMOD composite, Fröhlich, 2006), as a function of $P[^{10}\text{Be}]$, which has been scaled from the observed annual mean Climax n.m. count rate $C$ (as shown in Figure 1) using the results of Masarik and Beer (1999).

Recently Foster (2004) and Lockwood (2004) have used a homogeneous composite of sunspot group data extending back to 1874 (derived from the Greenwich, Polokov and Mt Wilson datasets) to reconstruct the total solar irradiance. There is one key parameter which influences these reconstructions and which is unknown: that is the average quiet-Sun photospheric field (in pixels of a set size) at sunspot-minimum, $\langle B_r \rangle_{QS}$, during the Maunder minimum. (N.B. because these authors make use of relationships between the intensity and field seen by the MDI instrument on SoHO, they use MDI-sized pixels throughout). To make the reconstructions, the authors made three assumptions: assumption A is that $\langle B_r \rangle_{QS}$ in the Maunder minimum was the same as the present-day value; assumption B is that $\langle B_r \rangle_{QS}$ in the Maunder minimum was zero (a magnetically clean Sun at MDI resolution); and assumption C is that $\langle B_r \rangle_{QS}$ in the Maunder minimum was half the present-day value. Note that the resulting reconstruction A is very similar to that proposed by Foukal *et al.* (2004), whereas that for B is very similar to those by Lean (2000) and Lockwood and Stamper (1999). All three reconstructions have significantly less century-scale drift than that by Lean *et al.* (1995), which is frequently employed in General Circulation Model (GCM) simulations of climate change over the past century (see review by Lockwood, 2004).

The reconstructions also assumed that irradiance effects are entirely due to surface magnetic fields. Over recent cycles, the success of the SATIRE modelling

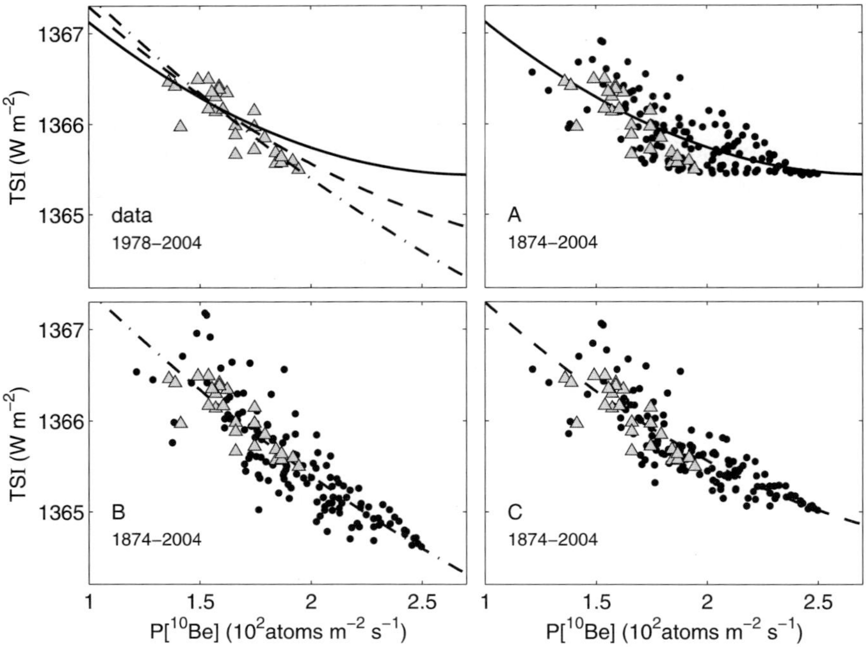


*Figure 4.* Scatter plots of annual means of the total solar irradiance $I_{TS}$ against the production rate of the $^{10}$Be isotope, $P[^{10}\text{Be}]$. *Top left:* the observed data since 1978: the PMOD TSI composite (Fröhlich, 2006) is plotted as a function of the flux of >3 GV GCRs (the data shown in Figure 1), scaled in terms of $P[^{10}\text{Be}]$ using the results of the GEANT simulations by Masarik and Beer (1999). These data are also shown as grey triangular points in the other panels. *The other 3 panels:* Black points show the annual $I_{TS}$ values since 1874, reconstructed by Foster (2004) and Lockwood (2004) for assumptions A, B and C (see text for details), plotted against $P[^{10}\text{Be}]$ derived from the open flux $[F_S]_{aa}$ (Lockwood *et al.*, 1999). The lines give the best 2nd-order polynomial fits to the reconstructed data. All the polynomial fits are also reproduced in the first panel. From Lockwood *et al.* (2006).

(Solanki, 2006) means that "shadow" effects of fields deep in the convection zone are negligible. It is most likely that any such effects on century timescales would be associated with solar radius changes and these could add to the long-term drift in TSI.

The other three panels of Figure 4 (from Lockwood *et al.*, 2005) compare the reconstructed TSI for these three assumptions with the $^{10}$Be production rate, $P[^{10}\text{Be}]$, which is here scaled from the open solar flux, as derived from the *aa* geomagnetic index, $[F_S]_{aa}$. In each case a best quadratic fit to the data is given, and all three fits are also shown in the top left panel (as solid, dot-dash, and dashed lines for assumptions A, B and C, respectively).

It can be seen that the predicted variation of TSI with $P[^{10}\text{Be}]$ is monotonic for all three assumptions. This means that cosmogenic isotopes can be used as a quantitative indicator of TSI, irrespective of the actual value of $\langle B_r \rangle_{QS}$ in the

Maunder minimum. However for more quantitative analysis, we do need an estimate of this value. All three assumptions are valid within the scatter of the points. However, at the lowest TSI a divergence between the best fit for assumption A and the data is becoming apparent. The variation is, in general, non-linear: the linear assumption used by Lockwood and Stamper (1999) being a good approximation only for assumption B. Given that cosmogenic isotopes continue to show a decadal oscillation in the Maunder minimum (Beer *et al.*, 1998) (implying some magnetic flux emergence, evolution and loss continued in the Maunder minimum), assumption B appears to be less likely than the other two.

## 5. Conclusions

Cosmogenic isotopes do tell us about past solar variations relevant to climate (Beer *et al.*, 2006). There are two potential relationships. Firstly, as we have gained a better understanding of the origin and evolution of open solar flux, a firm link between it and total and/or spectral solar irradiance is becoming apparent. Simulations reproduce this link but disagree on the scaling required (Solanki *et al.*, 2001; Wang *et al.*, 2005a). Lockwood *et al.* (2006) have shown that this scaling depends on the residual quiet-Sun photospheric magnetic flux in the Maunder minimum. With quantification of this value, the way is open for "quantitative palaeoclimatology" in which cosmogenic isotopes are used to quantify the solar irradiance input into GCMs which are then used in the forward modelling of proxies, for example predicting the ice rafted debris observed by Bond *et al.* (2001) or the stalagmite $\delta^{18}O$ observed by Neff *et al.* (2001) and Wang *et al.* (2005b).

The second relationship is the putative direct effect of cosmic rays on clouds and/or the global thunderstorm circuit. This effect remains highly controversial, but even the weak effect revealed by Harrison and Stephenson (2005) may be highly significant. This effect is seen in relation to short-term variations of GCRs (Forbush decreases, lasting only a few days) and Figure 1 shows that on these timescales the link between TSI and cosmic rays is almost non-existent. Thus these data do argue for a direct effect and not just via the anticorrelation of TSI with cosmic rays which is most apparent on timescales of a year or greater.

## Acknowledgements

The author is grateful to a large number of scientists for many discussions and preprints relevant to this review.

## References

Aplin, K. L., Harrison, R. G., and Bennett, A. J.: 2005, 'Effect of the troposphere on surface neutron counter measurements', *Adv. Space Res.* **35**, 1484–1491.

Arnold, N. and Neubert, T.: 2002, 'The electric Earth: Cosmic influences on the atmosphere', *Astron. and Geophys.* **43**, 6.9–6.12.

Baumgartner, S., Beer, J., Masarik, J., Wagner, G., Meynadier, L., and Synal, H.-A.: 1998, 'Geomagnetic Modulation of the 36Cl Flux in the GRIP Ice Core Greenland', *Science* **279**, 1330–1331.

Beer, J.: 2001, 'Ice core data on climate and cosmic ray changes', in: J. Kirkby and S. Mele (eds.), *Ion-Aerosol-Cloud Interactions*, Proc. Workshop on Ion-Aerosol-Cloud Interactions, CERN Yellow Report, **2001(007)**, ISSN 0007-8328, ISBN 92-9083-191-0, CERN, Geneva, Switzerland, pp. 3–11.

Beer, J., Tobias, S., and Weiss, N. O.: 1998, 'An active Sun throughout the Maunder minimum', *Sol. Phys.* **181**, 237–249.

Beer, J., Vonmoos, M., and Muscheler, R.: 2006, 'Solar variability over the past several millennia', *Space Sci. Rev.*, this volume, doi: 10.1007/s11214-006-9047-4.

Bering, E. A., Few, A. A., and Benbrook, J. R.: 1998, 'The global electric circuit', *Physics Today* **51(10)**, 24–30.

Bhattacharyya, A. and Mitra, B.: 1997, 'Changes in cosmic ray cut-off rigidities due to secular variations of the geomagnetic field', *Annales Geophys.* **15**, 734–739.

Bond, G., Kromer, B., Beer, J., Muscheler, R., Evans, M. N., Showers, W., Hoffmann, S., Lotti-Bond, R., Hajdas, I., and Bonani, G.: 2001, 'Persistent solar influence on North Atlantic climate during the Holocene', *Science* **294**, 2130–2136.

Bonino, G., Cini-Castognoli, G., Cane, D., Taricco, C., and Bandahri, N.: 2001, 'Solar modulation of the galactic cosmic ray spectra since the Maunder minimum', in: *Proceedings of the ICRC*, R. Schlickeiser (ed.), International Cosmic Ray Conference, **27**, Copernicus Gesellschaft, Katlenburg-Lindau, Germany, pp. 3769–3772.

Carslaw, K. S., Harrison, R. G., and Kirkby, J.: 2002, 'Cosmic Rays, Clouds and Climate', *Science* **298**, 1732–1737.

Christla, M., Manginia, A., Holzkäampera, S., and Spötlb, C.: 2004, 'Evidence for a link between the flux of galactic cosmic rays and Earth's climate during the past 200,000 years', *J. Atmos. Sol.-Terr. Phys.* **66**, 313–322.

Cini-Castagnoli, G. and Lal, D.: 1980, 'Solar modulation effects in terrestrialv production of Carbon-14', *Radiocarbon* **22**, 133–158.

Damon, P. E., Lerman, J. C., and Long, A.: 1978, 'Temporal fluctuations of atmospheric $^{14}$C: Causal factors and implications', *Ann. Rev. Earth Planet. Sci.* **6**, 457.

Davis, O. K.: 1994, 'The correlation of summer precipitation in the southwestern U.S.A. with isotopic records of solar activity during the Medieval Warm Period', *Clim. Change* **26**, 271–287.

Davis, O. K. and Shafer, D. S.: 1992, 'An early-Holocene maximum for the Arizona monsoon recorded at Montezuma Well, central Arizona', *Palaeogeogr. Palaeoclimatol. Palaeoecol.* **92**, 107–119.

Eichkorn, S., Wilhelm, S., Aufmhoff, H., Wohlfrom, K. H., and Arnold, F.: 2002, 'Cosmic ray-induced aerosol-formation: First observational evidence from aircraft-based ion mass spectrometer measurements in the upper troposphere', *Geophys. Res. Lett.* **29**, 1698–1701.

Farrar, P. D.: 2000, 'Are cosmic rays influencing oceanic cloud coverage – or is it only El Niño?', *Clim. Change* **47**, 7–15.

Foster, S. S.: 2004, 'Reconstruction of solar irradiance variations, for use in studies of global climate change: Application of recent SoHO observations with historic data from the Greenwich observations', *Ph.D. Thesis*, School of Physics and Astronomy, University of Southampton, Southampton, UK.

Foukal, P., North, G., and Wigley, T.: 2004, 'A stellar view on solar variations and climate', *Science* **306**, 68–69.

Friis-Christensen, E. and Svensmark, H.: 1997, 'What do we really know about the sun-climate connection?', *Adv. Space Res.* **20**, 913–920.

Fröhlich, C.: 2006, 'Solar irradiance variability since 1978 – Revision of the PMOD composite during cycle 21', *Space Sci. Rev.*, this volume, doi: 10.1007/s11214-006-9046-5.

Haigh, J. D.: 2003, 'The effects of solar variability on the Earths climate', *Phil. Trans. Roy. Soc. Lond.* **361**, 95–111.

Harrison, R. G.: 2002a, 'Radiolytic particle production in the atmosphere', *Atmos. Environ.* **36**, 169–160.

Harrison, R. G.: 2002b, 'Twentieth century secular decrease in the atmospheric electric circuit', *Geophys. Res. Lett.* **29**, 1600–1603.

Harrison, R. G.: 2004, 'The global atmospheric electrical circuit and climate', *Surv. Geophys.* **25**, 441–484.

Harrison, R. G. and Stephenson, D. B.: 2005, 'Empirical evidence for a non-linear effect of galactic cosmic rays on clouds', *Phil. Trans. Roy. Soc. Lond.*, in press.

Heber, B., Fichtner, H., and Scherer, K.: 2006, 'Solar and heliospheric modulation of galactic cosmic rays', *Space Sci. Rev.*, this volume, doi: 10.1007/s11214-006-9048-3.

Hu, F. S., Kaufman, D., Yoneji, S., Nelson, D., Shemesh, A., Huang, Y., Tian, J., Bond, G., Clegg, B., and Brown, T.: 2003, 'Cyclic variation and solar forcing of holocene climate in the Alaskan subarctic', *Science* **301**, 1890–1893.

Jirikowic, J. L., Kalin, R. M., and Davis, O. K.: 1993, 'Tree-Ring 14C as an indicator of climate change', *Climatic Change in Continental Isotopic Records, AGU Geophysical Monograph* **78**, 353–366.

Kristjánsson, J. E. and Kristiansen, J.: 2000, 'Is there a cosmic ray signal in recent variations in global cloudiness and cloud radiative forcing?', *J. Geophys. Res.* **105**, 11,851–11,863.

Kristjánsson, J. E., Staple, A., Kristiansen, J., and Kass, E.: 2002, 'A new look at possible connections between solar activity, clouds and climate', *Geophys. Res. Lett.* **29(23)**, 2107–2110.

Laj, C., *et al.*: 2001, 'North Atlantic paleointensity stack since 75 ka (NAPIS-75) and the duration of the Laschamp event', *Phil. Trans. R. Soc., London* **358**, 1009–1025.

Lean, J.: 2000, 'Evolution of the Suns spectral irradiance since the Maunder minimum', *Geophys. Res. Lett.* **27**, 2425–2428.

Lean, J., Beer, J., and Bradley, R.: 1995, 'Reconstruction of solar irradiance since 1610: Implications for climate change', *Geophys. Res. Lett.* **22**, 3195–3198.

Lockwood, M.: 2001a, 'Long-term variations in the magnetic fields of the Sun and the heliosphere: Their origin, effects and implications', *J. Geophys. Res.* **106**, 16,021–16,038.

Lockwood, M.: 2001b, 'Long-term variations in cosmic ray fluxes, total solar irradiance and global climate', in: J. Kirkby and S. Mele (eds.), *Ion-Aerosol-Cloud Interactions*, Proc. Workshop on Ion-Aerosol-Cloud Interactions, CERN Yellow Report, **2001(007)**, ISSN 0007-8328, ISBN 92-9083-191-0, CERN, Geneva, Switzerland, pp. 12–23.

Lockwood, M: 2002a, 'An evaluation of the correlation between open solar flux and total solar irradiance', *Astron. Astrophys.* **382**, 678–687.

Lockwood, M.: 2002b, 'Long-term variations in the open solar flux and links to variations in Earths climate', *From Solar Min to Max: Half a Solar Cycle with SoHO*, Proc. SoHO 11 Symposium, Davos, Switzerland, ESA-SP **508**, pp. 507–522.

Lockwood, M.: 2004, 'Solar outputs, their variations and their effects of Earth', in: I. Rüedi, M. Güdel, and W. Schmutz (eds.), *The Sun, Solar Analogs and the Climate*, Proceedings of Saas-Fee Advanced Course **34**, Springer, ISBN 3-540-23856-5, pp. 107–304.

Lockwood, M. and Stamper, R.: 1999, 'Long-term drift of the coronal source magnetic flux and the total solar irradiance', *Geophys. Res. Lett.* **26**, 2461–2464.

Lockwood, M., Stamper, R., and Wild, M. N.: 1999, 'A doubling of the Suns coronal magnetic field during the last 100 years', *Nature* **399**, 437–439.

Lockwood, M., Foster, S. S., and Stamper, R.: 2006, 'The Use of Cosmogenic Isotopes as an Indicator of Total Solar Irradiance in Paleoclimate Research', *Annales Geophys.*, submitted.

Maasch, K., Mayewski, P. A., Rohling, E., Stager, C., Karlen, K., Meeker, L. D., and Meyerson, E.: 2005, 'Climate of the past 2000 years', *Geografiska Annaler, A* **87**, 7–15.

Märcz, F. and Harrison, R. G.: 2005, 'Further signatures of long-term changes in atmospheric electrical parameters observed in Europe', *Annales Geophys.* **23**, 1987–1995.

Markson, R.: 1981, 'Modulation of the Earth's electric field by cosmic radiation', *Nature* **291**, 304–308.

Marsh, N. and Svensmark, H.: 2000, 'Cosmic rays, clouds and climate', *Space Sci. Rev.* **94**, 215–230.

Marsh, N. and Svensmark, H.: 2003, 'Galactic cosmic ray and El Niño-Southern Oscillation trends in International Satellite Cloud Climatology Project D2 low-cloud properties', *J. Geophys. Res.* **108**, 4195–4203.

Masarik, J. and Beer, J.: 1999, 'Simulation of particle fluxes and cosmogenic nuclide production in the Earth's atmosphere', *J. Geophys. Res.* **104**, 12,099–12,112.

Mayewski, P. A., Maasch, K., Yan, Y., Kang, S., Meyerson, E., Sneed, S., Kaspari, S., Dixon, D., Morgan, V., van Ommen, T., and Curran, M. : 2005, 'Solar forcing of the polar atmosphere', *Ann. Glaciol.*, in press.

Muscheler, R., Beer, J., and Kromer, B.: 2003, 'Long-term climate variations and solar effects', in: *Solar Variability as an Input to the Earth's Environment, ESA SP* **535**, 23–28.

Neff, U., Burns, S. J., Mangini, A., Mudelsee, M., Fleitmann, D., and Matter, A.: 2001, 'Strong coherence between solar variability and the Monsoon in Oman between 9 and 6 kyr ago', *Nature* **411**, 290–293.

Prasad, S., Vos, H., Negendank, J. F. W., Waldmann, N., Goldstein, S. L., and Stein, M.: 2004, 'Evidence from Lake Lisan of solar influence on decadal to centennial climate variability during marine oxygen isotope stage 2. Geology', *Space Sci. Rev.* **32**, 581–584.

Rossow, W. B., Walker, A. W., Beuschel, D. E., and Roiter, M. D.: 1996, 'Documentation of new datasets', *International Satellite Cloud Climatology Project (ISCCP)*, **WMO/TD 737**, World Meteorological Organization, Geneva, Switzerland.

Rottman, G.: 2006, 'Measurement of total and spectral solar irradiance', *Space Sci. Rev.*, this volume.

Sarnthein, M., Van Kreveld, S., Erlenkeuser, H., Grootes, P. M., Kucera, M., Pflaumann, U., and Schulz, M.: 2003, 'Centennial-to-millennial-scale periodicities of Holocene climate and sediment injections off the western Barents shelf, 75° N', *Boreas* **32**, 447–461.

Schlegel, K., Diendorfer, G., Thern, S., and Schmidt, M.: 2001, 'Thunderstorms, lightening and solar activity – Middle Europe', *J. Atmos. Sol.-Terr. Phys.* **63**, 1705–1713.

Schrijver, C. J., DeRosa, M. L., and Title, A. M.: 2002, 'What is missing from our understanding of long-term solar and Heliospheric activity?', *Astrophys. J.* **577l**, 1006–1012.

Solanki, S.: 2006, 'Solar variability of possible relevance for planetary climates', *Space Sci. Rev.*, this volume, doi: 10.1007/s11214-006-9044-7.

Solanki, S. K., Scüssler, M., and Fligge, M.: 2001, 'Secular evolution of the Suns magnetic flux', *Astron. Astrophys.* **383**, 706–712.

Solanki, S. K., Usoskin, I. G., Kromer, B., Schüssler, M., and Beer, J.: 2004, 'Unusual activity of the Sun during recent decades compared to the previous 11,000 years', *Nature* **431**, 1084–1087.

Svensmark, H.: 1998, 'Influence of cosmic rays on Earths climate', *Phys. Rev. Lett.* **81**, 5027–5030.

Tinsley, B. A., Rohrbaugh, R. P., Hei, M., and Beard, K. V.: 2000, 'Effects of image charges on the scavenging of aerosol particles by cloud droplets, and on droplet charging and possible ice nucleation processes', *J. Atmos. Sci.* **57**, 2118–2134.

Tric, E.: 1992, 'Paleointensity of the geomagnetic field during the last 80,000 years', *J. Geophys. Res.* **97**, 9337–9351.

Tripathi, S. N. and Harrison, R. G.: 2002, 'Enhancement of contact nucleation by scavenging of charged aerosol', *Atmos. Res.* **62**, 57–70.

Udelhofen, P. M. and Cess, R. D.: 2001, 'Cloud cover variations over the United States: An influence of cosmic rays or solar variability', *Geophys. Res. Lett.* **28**, 2617–2620.

Usoskin, I. G., Mursula, K., Solanki, S. K., Schüssler, M., and Kovaltsov, G. A.: 2002, 'A physical reconstruction of cosmic ray intensity since 1610', *J. Geophys. Res.* **107**, doi:10.1029/2002JA009343.

Usoskin, I. G., Solanki, S. K., Schüssler, M., Mursula, K., and Alanko, K.: 2003, 'Millennium-scale sunspot number reconstruction: Evidence for an unusually active Sun since the 1940s', *Phys. Rev. Lett.* **91**, doi:10.1103/PhysRevLett.91.211101.

Van Geel, B., van der Plicht, J., Kilian, M. R., Klaver, E. R., Kouwenberg, J. H. M., Renssen, H., Reynaud-Farrera, I., and Waterbolk, H. T.: 1998, 'The sharp rise of 14C around 800 cal BC: Possible causes, related climatic teleconnections and the impact on human environments', *Radiocarbon* **40**, 535–550.

Veretenenko, S. V. and Pudovkin, M. I.: 1997, 'Effects of the galactic cosmic ray variations on the solar radiation input in the lower atmosphere', *J. Atmos. Sol.-Terr. Phys.* **59**, 1739–1746.

Wang, Y.-M., Lean, J., and Sheeley, N. R. Jr.: 2005a, 'Modeling the Sun's magnetic field and irradiance since 1713', *Astrophys. J.* **625**, 522–538.

Wang, Y., Cheng, H., Edwards, R. L., He, Y., Kong, X., An, Z., Wu, J., Kelly, M. J., Dykoski, C. A., and Li, X.: 2005b, 'Strong coherence between solar variability and the monsoon in Oman between 9 and 6 kyr ago', *Science* **308**, 854–857.

Wei, J. and Wang, H.: 2004, 'A possible role of solar radiation and ocean in the mid-holocene east asian monsoon climate', *Adv. Atmos. Sci.* **21**, 1–12.

Wigley, T. M. L. and Raper, S. C. B.: 1990, 'Climatic change due to solar irradiance changes', *Geophys. Res. Lett.* **17**, 2169–2172.

Wilding, R. J. and Harrison, R. G.: 2005, 'Aerosol modulation of small ion growth in coastal air', *Atmos. Env.* **39(32)**, 5876–5883.

Yu, Z. and Ito, E.: 1999, 'Possible solar forcing of century-scale drought frequency in the northern Great Plains', *Geology* **27**, 263–266.

Yu, F. and Turco, R. P.: 2000, 'Ultrafine aerosol formation via ion-mediated nucleation', *Geophys. Res. Lett.* **27**, 883–886.

# OBSERVED LONG-TERM VARIATIONS OF SOLAR IRRADIANCE AT THE EARTH'S SURFACE

A. OHMURA
*Institute for Atmospheric and Climate Science, Swiss Federal Institute of Technology (ETH), Universitätstr. 16, CH-8092 Zurich, Switzerland; Scott Polar Research Institute, University of Cambridge, UK*
*(E-mail: ohmura@env.ethz.ch)*

(Received 14 November 2005; Accepted in final form 27 January 2006)

**Abstract.** The variation of global radiation (sum of direct solar and diffuse sky radiation) at the Earth's surface is examined based on pyranometer measurements at about 400 sites. The period of the study covers in general the last 50 years. For Europe the study is extended to the beginning of observations in the 1920s and 1930s. Global radiation generally increased in Europe from the 1920s to the 1950s. After the late 1950s and early 1960s global radiation began to decrease in most areas of the world at a mean rate of 0.7 $Wm^{-2}a^{-1}$ until 1980s, thereafter 75% of the stations showed a recovery at a mean rate of 0.7 $Wm^{-2}a^{-1}$. All stations in the Polar region, which are far from aerosol sources, also show this pattern of change. At the remaining 25% of the stations the decrease has continued to present. These regions are a part of China, most of India, and Central Africa. Both during the declining and recovering phases global radiation observed under the cloudless condition also followed the same tendency, indicating the simultaneous and parallel changes of aerosol and cloud conditions. Long-term observations of total zenith transmittance of the atmosphere indicate a decrease in transmittance to the mid 1980s and an increase after this period. Since the brighter and darker periods correspond to relatively warmer and colder periods, the present study offers the possibility to quantitatively evaluate the mutual relationships between the solar irradiance, atmospheric transmittance, cloud conditions and air temperature.

**Keywords:** climate change, radiation, irradiance, aerosol, atmospheric transmission

## 1. Introduction

The awareness that radiation in the climate system might change is not old and even today not fully grasped by many professionals. An example of the acceptance difficulty is the term, solar constant, obviously conceived under the unquestioned assumption that the emission of the Sun is invariable. The same applies to irradiances at the Earth's surface. Budyko (1982) demonstrated that global radiation is stable over a long period and several years' observations suffice to characterize a site with a stable mean value. One of the first attempts to question the long-term stability of a radiative component was that by Kessler (1985) with respect to global radiation in Hamburg. Ohmura and Lang (1989) presented observational evidence that global radiation over Europe decreased systematically by about 8 $Wm^{-2}$ during the 30 years following the International Geophysical Year (IGY). The variation was

Space Science Reviews (2006) 125: 111–128
DOI: 10.1007/s11214-006-9050-9

explained by the change in cloud cover. This work was based on the radiation and heat balance data accumulated in the Global Energy Balance Archive (GEBA). Stanhill and Moreshet (1992) investigated solar radiation for Israel, to find a gradual but clear decline of global radiation. Liepert *et al.* (1994) presented a decreasing trend in global radiation over Germany, attributing the change to aerosol. By this time the global dimension of the decreasing tendency in global radiation began to be recognized by more scientists and was coined as global dimming. During the 2002 Radiation Panel meeting of the Global Energy and Water Cycle Experiment (GEWEX) in Zurich, William B. Rossow of the Goddard Institute of Space Studies (GISS/NASA) presented two diagrams portraying the time series of total cloud amount for the globe and for the Tropics, based on the results of International Satellite Cloud Climatology Project (ISCCP) encompassing 20 years from 1983 to 2002. The diagrams indicated an unmistakably increasing trend in cloud amount towards the end of the 1980s followed by a rather sharp decrease by as much as 5% over these 15 years. These diagrams gave the author an important hint to examine global radiation by supplementing GEBA with the post-1985 data. In 2003 the author started to compile the post-1985 global radiation data based on the sources in the World Radiation Data Center (WRDC), St. Petersburg, its duplicate of data for earlier years at National Renewable Energy Laboratory (NREL), Golden, CO, and contributions from a number of national meteorological services and research institutions of the world. The present article is intended to present the results of the analyses of global radiation from the earliest observation to the present. Emphasis is, however, given to the post-1985 period.

## 2. Data Compilation and Quality Control

The best dataset for the present subject is GEBA which contains monthly mean fluxes relevant to the surface energy balance for about 1600 sites (Ohmura and Gilgen, 1992; Gilgen *et al.*, 1998). The data in this archive are instrumentally measured and quality tested. The quality test is made up of four stages organized in ascending order of strictness, and known today as the GEBA quality scheme (Ohmura *et al.*, 1989). This set contains data for the majority of the sites only up to 1985. The updating of the dataset for the post-1985 period was an urgent requirement and has been completed at least for global radiation in the occasion of this work. Presently GEBA stores about 250,000 month site fluxes.

A major effort was made to collect the monthly mean global radiation, and in limited cases the total zenith transmittance or pyrheliometric data. Global radiation data are available for many sites at the NREL file that stores the data of the WRDC up to 1990. The data for the period starting 1991 are available from the WRDC in St. Petersburg. This file however often lacks the data of the early 1990s, and those of 1992 are entirely missing.

Many organizations such as national meteorological services possess much of the data missing at WRDC. There are also many public and private research institutions and observatories that possess large amounts of radiation data that may not have been forwarded to WRDC or other equivalent international data archives. A new campaign of requesting data from these institutions was a modest success, bringing the data for the missing period from 600 stations, of which about 400 stations were judged to be of an acceptable quality for the present purpose.

A large amount of quality data is available from the Baseline Surface Radiation Network (BSRN) for the period from 1992 to present. This short but very important period for detecting changes compensates for the missing data for a number of sites in WRDC. In addition, the BSRN data are usually of the highest quality, and offer direct and diffuse sky radiation with auxiliary meteorological data observed at the same sites.

The uncertainty of the measurements by commercially available pyranometers is the order of $\pm 3\ \mathrm{Wm}^{-2}$ for a short period of observation such as one hour. The systematic error of the long-term observations can be kept much smaller ($\pm 1.5\ \mathrm{Wm}^{-2}$), when the instruments are well maintained and regular recalibrations are applied twice a year.

Through these efforts a substantial amount of monthly mean global radiation data have been added to the existing GEBA to complete the updating work. This dataset is accessible at `http://bsrn.ethz.ch`.

## 3. Results

The variation of global radiation will be discussed in the following Sections with respect to geographic regions followed by a Section summing up for the globe. For regions with a sufficient number of stations with continuous records, an arithmetic regional mean was calculated in addition to individual station time series. If the region lacks a continuous series, the arithmetic mean excluding the missing years often introduces an artifact fluctuation. In such a case, the regional mean is calculated after the series for each station are normalized with respect to their long-term means.

### 3.1. Europe and the British Isles

The longest observation series are available in Europe. At five sites, Stockholm, Wageningen, Davos, Potsdam and Locarno-Monti the observations started before 1940. Although the distances between Stockholm, Wageningen, Potsdam and Locarno are the order of 1000 km, the development of the decadal change in radiation is amazingly similar (Figure 1). At all locations global radiation increased throughout the 1940s arriving at a peak sometime between the late 1940s and early

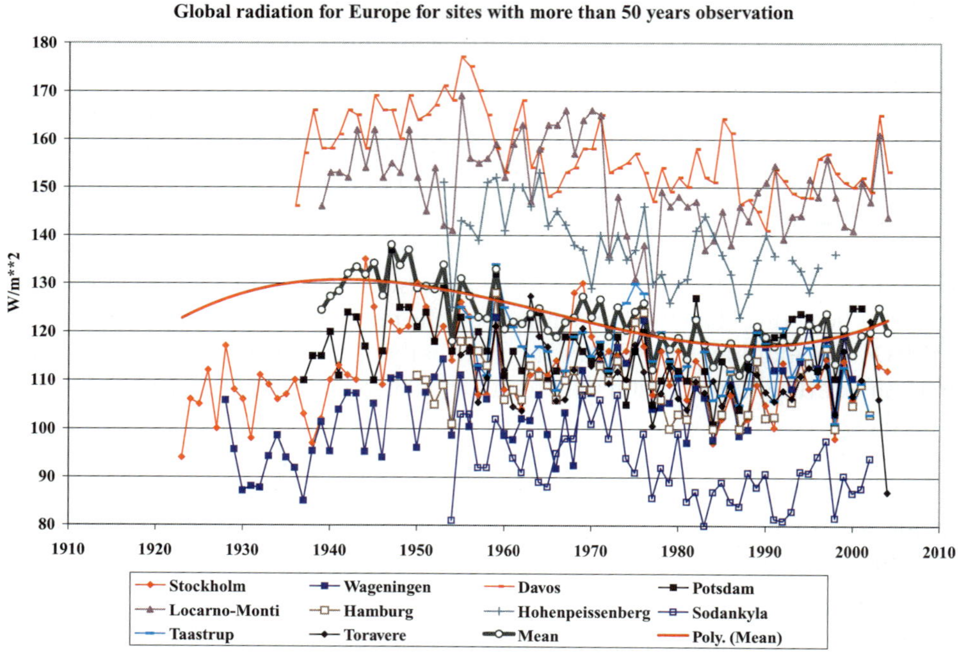


*Figure 1.* Annual mean global radiation for ten sites with more than 50 years observation in Europe. Regional mean is expressed in a thick green line with open circles. The polynomial trend is given in a thick red line.

1950s. The longer two series at Stockholm and Wageningen indicate that the increasing trend was already set in the 1920s. Global radiation increased between 1922 and 1952 by 20 $Wm^{-2}$. The mean rate of increase during the 1940s for Stockholm was 1.4 $Wm^{-2}a^{-1}$, while the regional mean was 0.5 $Wm^{-2}a^{-1}$. After the peak was reached in the late 1940s, global radiation decreased during the following 40 years until the late 1980s. Global radiation showed a recovery after late 1980s but the level remains well below the peak value of the late 1940s at all stations. The number of the sites increased rapidly after 1950 and we can add five more sites (Hamburg, Hohenpeissenberg, Sodankyla, Taastrup and Toravere) for this period, allowing a more detailed and accurate analysis. Because of the short nature of the observed period, however, this period of 50 years gives a possibility of estimating the amplitude of the fluctuation observed only for the declining period. Examining the northernmost site of Sodankyla to the southernmost site of Locarno, the absolute magnitudes of the decline from the peak to the bottom increase from north to south with the range of 10 to 15 $Wm^{-2}$ over the 40 years, which corresponds to about 10% of the annual mean global radiation at these sites. This value is larger than the decline proposed by Ohmura and Lang (1989) of 8 $Wm^{-2}$ over 30 years from the late 1950s to the late 1980s. The subsequent recovery registered an increase of about 6 $Wm^{-2}$ from 1990 to 2003.

The British Isles was earlier characterized as a region showing a prolonged increasing phase of global radiation in comparison with the rest of Europe (Ohmura and Lang, 1989). The British Isles is also an area lacking a good set of long-term observations. Therefore, for this region a mean of the normalized irradiance was calculated with 7 stations (Aberporth, Jersey, Belfast, Eskdalemuir, Lewick, Bracknell, and London). Global radiation indeed appeared to be increasing if the data only up to the mid 1980s were considered. The addition of the post-1985 data resulted in a clearer picture in which global radiation reached a peak in the late 1950s and continued to decline in the following 30 years by −0.04 in the normalized scale or −4 $Wm^{-2}$. The recovery has been very steep at 0.06 or 7 $Wm^{-2}$ during the last 10 years. Therefore, unlike the earlier characterization, the region followed the general trend of the global dimming and the subsequent brightening seen in the rest of Europe. The phase of the brightening, however, seems to have been delayed by about 5 years.

### 3.2. Israel

In Mediterranean Europe and the Near East, the only region for which the present analysis can be applied is Israel. Stanhill and Ianetz (1997) reexamined the data used earlier and concluded that global radiation decreased during the period of 40 years from 1954 to 1994, at a mean rate of 1.02 $Wm^{-2}a^{-1}$, despite a slight decrease in cloud amount. They pointed at the possible effect of aerosol including biogenic aerosol. The analysis of the data of the last 10 years will give an important clue as to if this region is also recovering from the trend of long dimming.

### 3.3. Japan

Japan is probably the best region to examine whether the secular variation discovered in Europe is local or global. It is far from Europe and has a well-run network of radiometry with a long history. Presently the series for 102 sites are available of which 49 stations have observational records longer than 40 years starting in 1961. The older data going back to the 1930s are located at the Japan Meteorology Agency, but the homogenization of the time series is ongoing. There are 26 sites among them that are of first class quality and continuation. Annual mean global radiation for these 26 sites and the regional means are presented in Figure 2. All locations share a declining trend from the early 1960s to late 1980s that is followed by a recovery at about the same rate. The mean of the 26 sites indicates a decline by 25 $Wm^{-2}$ from 1961 to 1990 at a rate of −0.8 $Wm^{-2}a^{-1}$. The recovery from 1990 to 2002 is 10 $Wm^{-2}$ over 12 years at 0.8 $Wm^{-2}a^{-1}$. Although this set of data does not include the earlier brightening phase of the pre-1960s, the declining trend from the early 1960s to late 1980s and the brightening phase from the early 1990s to the present are similar to the European sequence.

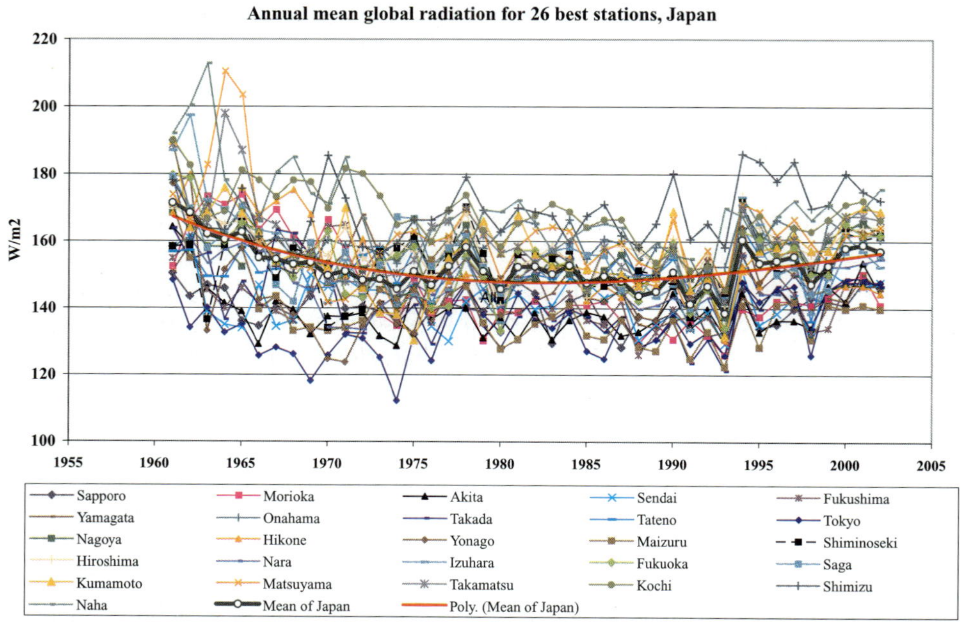


*Figure 2.* Annual mean global radiation for 26 best stations in Japan for the period 1961–2002. Regional mean is expressed in a thick green line with open circles. The polynomial trend is given in a thick red line.

## 3.4. China and Adjacent Countries

The Chinese National Meteorological Administration operates a network of 121 stations with observations of global radiation for a period of 40 years from 1961 to 2000. The author selected 66 sites considered to be of long lasting and better quality. These stations show two distinctive types of the variation: the majority (46 sites) show an initially declining tendency up to the 1980s, then a clear recovery starting in the early 1990s, while a small number of sites (20) show a trend of continued dimming to present. As an example of the majority of sites, 6 stations with the best records in southeast China are presented in Figure 3. Although the scatter of individual annual means is large, the regional means indicate a decline of 30 $Wm^{-2}$ from 1960 to the late 1980s with a rate of 1.2 $Wm^{-2}a^{-1}$. This rate belongs to the steepest rate witnessed in the world, besides Northern and Central Africa. The recovery after the late 1980s to 2000 is 10 $Wm^{-2}$. Zhang *et al.* (2004) reports a continued dimming for the period from 1961 to 2000 for this region, based on the observations at Shanghai, Nanjing and Hangzhou. This work applied only a linear trend analysis. If the time series are carefully analyzed, however, they show an unmistakable reversal of the trend in the late 1980s. A unique variation is seen in northwestern China where more than half of the stations show a continuous decline of global radiation (Figure 4). In this region a 25 $Wm^{-2}$ decrease was seen as the

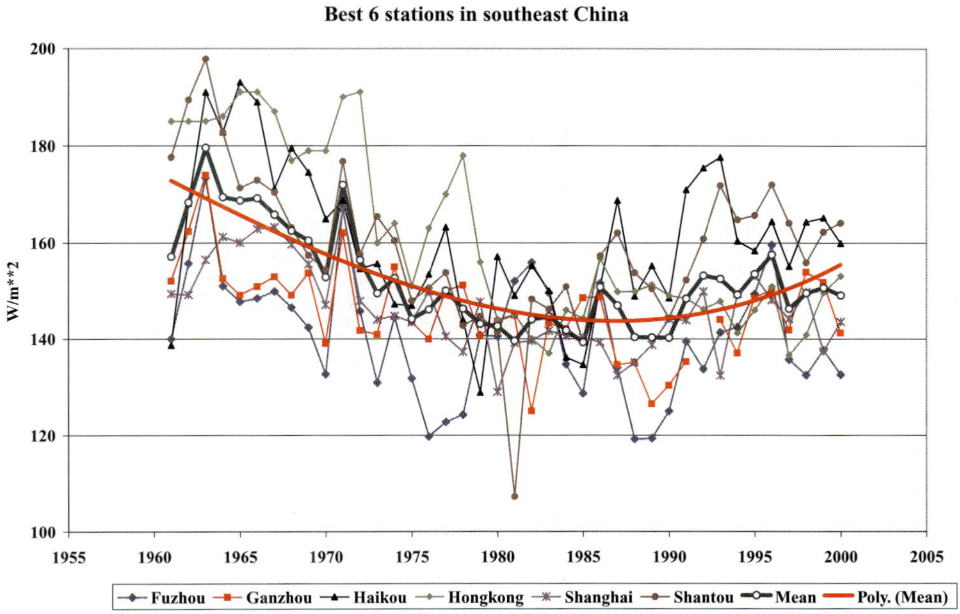


*Figure 3.* Annual mean global radiation for 6 best stations in southeast China (Fuzhou, Ganzhou, Haikou, Hongkong, Shanghai and Shantou). Regional mean is expressed in a thick green line with open circles. The polynomial trend is given in a thick red line.

regional mean in the 40 years from 1960 to 2000. There is no indication of an increase at present. Figure 5 shows a regional distribution of the two types of the radiation changes. In the immediately adjacent areas, four sites in Mongolia, two sites in Kazakhstan, one site each in Uzbekistan and in South Korea are added. The sites with continued dimming are located mostly north of the Yangtze River and in northern China. Further to the north in Mongolia, however, there is not a single site with continued dimming. The mean condition in Mongolia is a steady decrease from 1965 till the mid 1980s of 15 $Wm^{-2}$, followed by a steep recovery of 10 $Wm^{-2}$ from 1990 to 2002.

## 3.5. Indian Sub-Continent

Although India reports global radiation for 15 sites to WRDC, 5 stations are considered to be useful for the present purpose with respect to the length of observation and quality of data. The annual means and their averages for the entire India are presented for the period from 1964 to 2002 in Figure 6. While other stations report similar annual means within a narrow range from 200 to 230 $Wm^{-2}$, Calcutta shows systematically 20 $Wm^{-2}$ lower values, indicating an influence of acute local pollution. All sites, however, show basically the same trend of decrease amounting to 30 $Wm^{-2}$ over the 39 years. There is no sign of recovery.

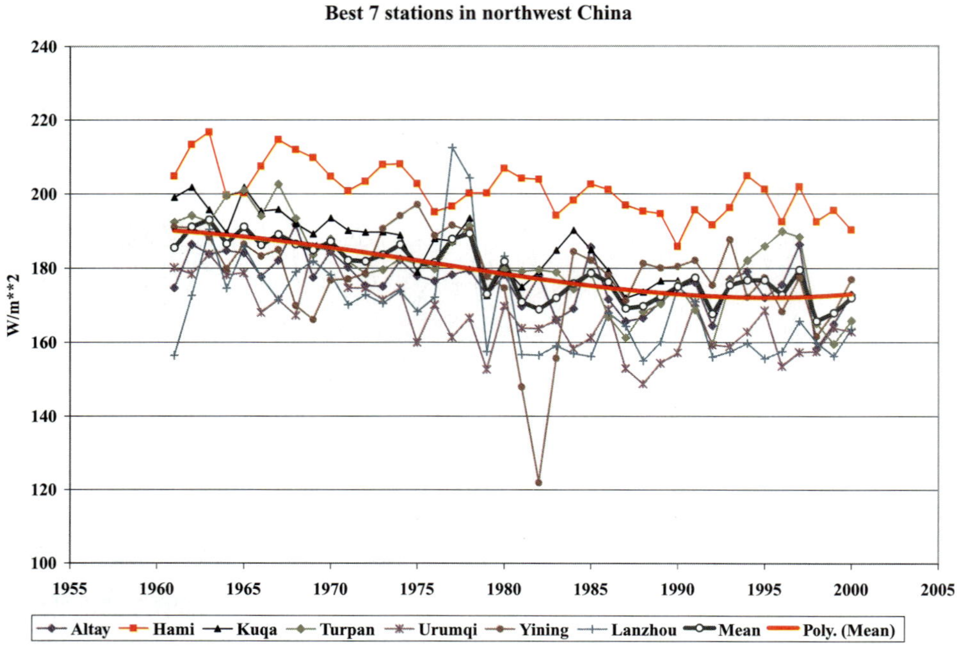


*Figure 4.* Annual mean global radiation for 7 best stations in northwest China (Altay, Hami, Kuqa, Turpan, Urumqi, Yining and Lanzhou). Regional mean is expressed in a thick green line with open circles. The polynomial trend is given in a thick red line.

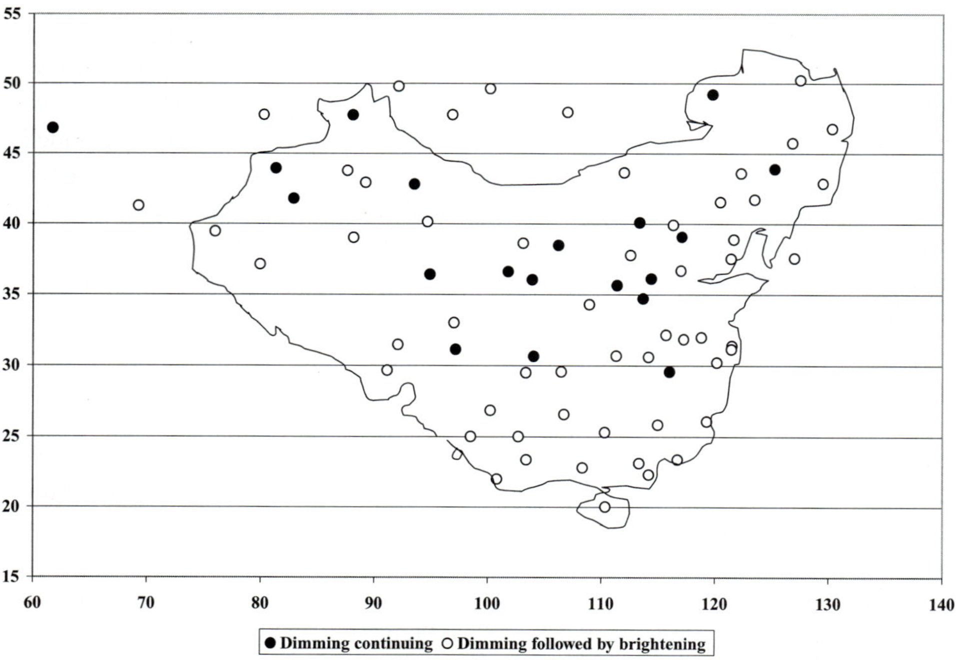


*Figure 5.* Distribution of sites with and without recent brightening trends in and around China.

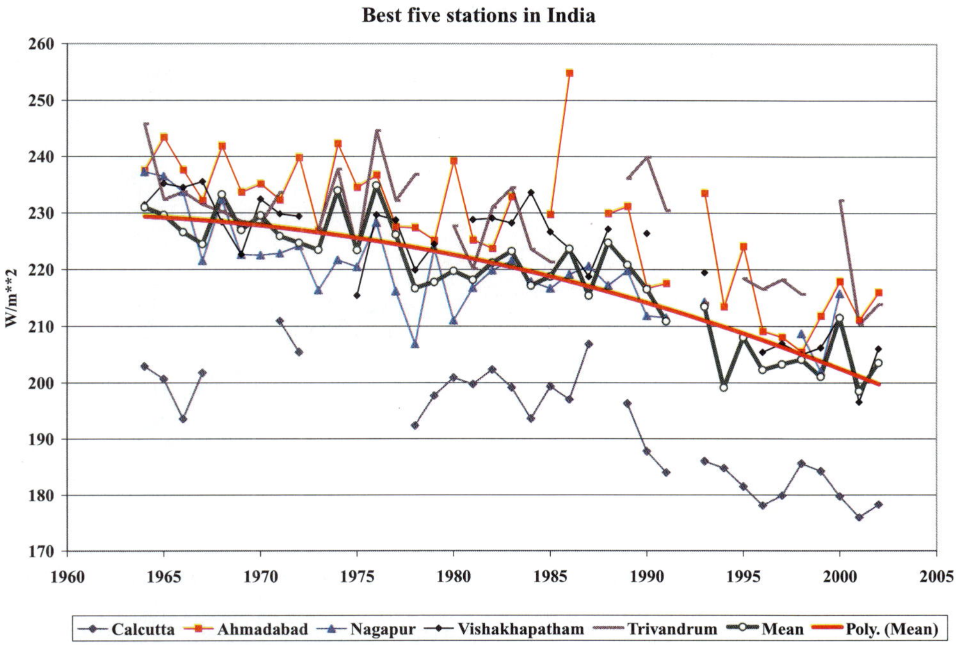


*Figure 6.* Annual mean global radiation for 5 sites in India (Calcutta, Ahmadabad, Nagapur, Vishakhapatham and Trivandrum). Regional mean is expressed in a thick green line with open circles. The polynomial trend is given in a thick red line.

## 3.6. Africa

The only region in Africa, north of the Sahara, with systematic radiation measurements is Egypt and Tunisia. There are broken records at 4 sites, Tahrir, Cairo, Bahtim and Aswan in Egypt. The records from all these sites indicate the initial dimming and recent recovery, but it appears that the time of the recovery was somewhat delayed in comparison with other regions. Since none of these records cover 30 years continuously, and the absolute values among stations differ considerably, the author has only presented the mean of normalized values with respect to the long-term means at each site (Figure 7). The mean curve shows global radiation reaching its peak in the late 1960s and decreasing from the early 1970s to early 1990s. The decline over the 20 years is 0.11 in the normalized scale, which corresponds to 26 $Wm^{-2}$. The recovery starting in the mid 1990s is a modest 8 $Wm^{-2}$ but has clearly been increasing over the past 8 years. Sidi Bouzid in Tunisia shows basically the same pattern as Egypt, but it shows the initial increasing phase more clearly than Egypt, and reaches its peak in the early 1970s. The range of the subsequent dimming is 13 $Wm^{-2}$ and is much smaller than in Egypt. The appearances of peaks and valleys in the Mediterranean North African series seem to drag behind Europe by 15 to 20 years.

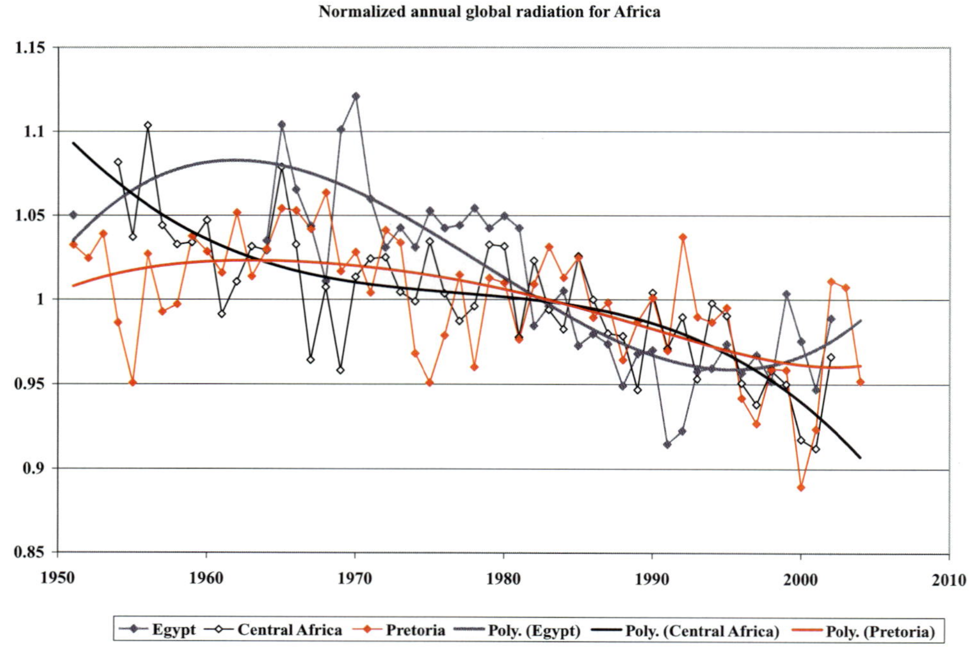


*Figure 7.* Normalized annual mean global radiation for Africa (Cairo, Tahrir, Bahtim and Aswan in Egypt; Kinshasa, Boende, Harare and Bulawayo in Central Africa; Pretoria in South Africa). Trend lines for each region are expressed in the same colours.

Two countries in Central Africa supply two records each, Kinshasa and Boende in Zaire, and Harare and Bulawayo in Zimbabwe. The records covering almost 50 years from 1954 to 2002 are, however, discontinuous. Therefore, as in the case of North Africa, normalized means have been calculated based on these 4 records and added in Figure 7. Global radiation in this region continued to decrease from the mid 1950s up to present, by 0.14, which corresponds to a decrease of 30 $Wm^{-2}$ over 49 years. Kenya also provides global radiation data for 3 sites, Nairobi, Mombasa and Voi, but these data are not of sufficient quality to be considered.

Maputo in Mozambique offers some of the better data in Africa, starting in the 1950s and continuing to present with only few breaks. The variation in this series is unique in that it shows a decline from the mid 1950s to mid 1970s by as much as 30 $Wm^{-2}$, followed by a slight brightening. The last 10 years show again a clear decline by 20 $Wm^{-2}$. The phase is exactly the opposite of what was seen in the Northern Hemisphere. At the moment this pattern does not show any similarity to the other regions. The author believes, however, that this time series may reflect a regionally important situation that is worthy of detailed investigation.

The best radiation records in Africa are available in South Africa. In Pretoria, for example, the observation of global radiation has continuously been made since 1951. The quality of the data is excellent. Global radiation increased gradually to the late 1960s, followed by a decline to 1990s by 15 $Wm^{-2}$, and then by a clear increase. The pattern of the change is similar to Europe with a phase lag of about 10 years (Figure 7).

### 3.7. South America

It is unfortunate that there is not a single site in South America that can be used for the present analysis. Nevertheless, global radiation records from Caracas and Coro, Venezuela are analyzed. The data series are frequently broken and the calibration is uncertain. There is, however, no concrete reason to reject these data. If we trust the records, global radiation in this part of South America decreased over 35 years from 1963 to 1998 by about 40 $Wm^{-2}$, which makes an annual rate of $-1.1$ $Wm^{-2}a^{-1}$. This is a very rapid decline in global radiation but similar declining rates are found in Central Africa and southeastern China. More recent data are not available.

### 3.8. North America (Contiguous) Including Alaska

North America taking the contiguous USA and Canada together does not have a good record of long-term observations. Therefore, the annual global radiation only for three better series, Bismarck, Boulder and Toronto has been normalized and the mean series was constructed (Figure 8). In the early 1950s global radiation was still on the rise culminating in the late 1950s. It fell for the next 40 years by 0.08 in the normalized scale or 13 $Wm^{-2}$ and took a recovery course in the mid 1990s. The absolute magnitude of the recovery is very small at 5 $Wm^{-2}$ at this stage. The pattern of the change is similar to Europe with a delayed phase of about 10 years.

There is only one site in Alaska that provides more than 50 years of global radiation records with some breaks. Barrow in Figure 9 indicates an increasing trend lasting until the early 1960s, followed by a 30 years declining phase until the early 1990s. The magnitude of the decline is 8 $Wm^{-2}$. The recovery during the last 10 years is very small but clearly visible. It appears as if the pattern is similar to Europe with damped amplitude but the phase is delayed by 10 years. The other site with a good prospect of the study is Fairbanks, which needs still some work to clarify.

### 3.9. Ocean Surface, Oceania, Arctic and Antarctic

The observation for this vast area is still insufficient. To cover major ocean surfaces, there are only two possibilities, either to rely on island-based observations or to use satellites. The short history of satellite data excludes the latter possibility. Island-based measurements are scarce, mostly short lived and prone to lose accurate calibration. Nevertheless, there are some results worthy of consideration. Figure 10 presents two sites available from the tropical Pacific, one in Fiji and the other in

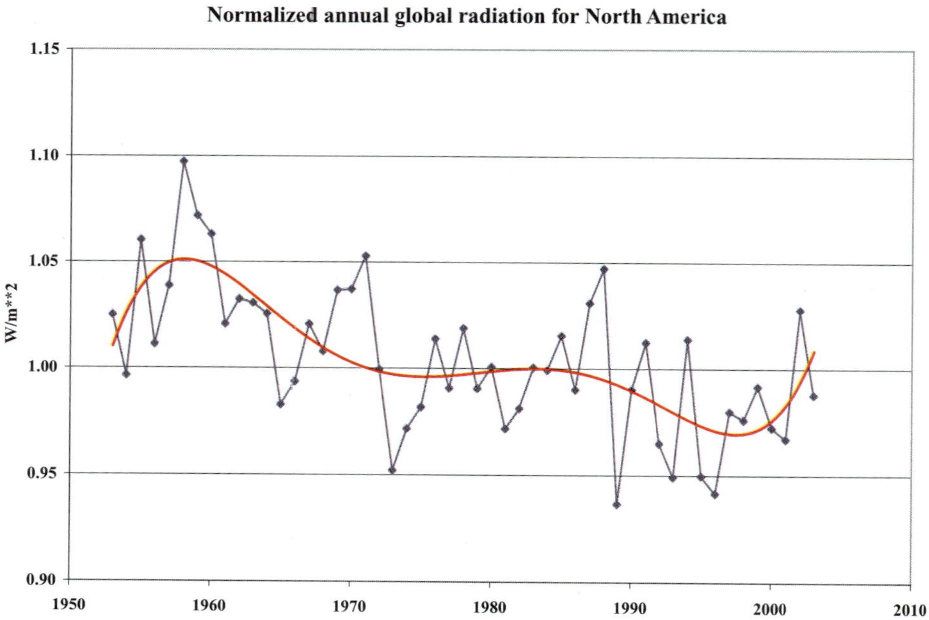


*Figure 8.* Normalized annual mean global radiation for North America (Bismarck, Boulder and Toronto). The polynomial trend is expressed in the smooth curve.

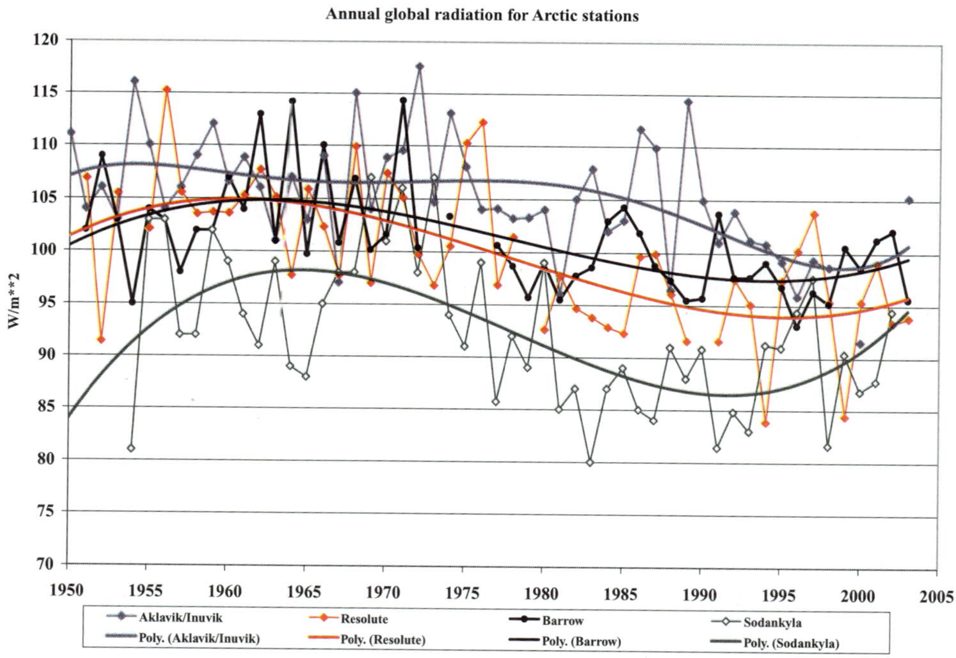


*Figure 9.* Annual mean global radiation for Arctic stations (Aklavik/Inuvik, Resolute, Barrow and Sodankyla). Trend lines for each station are expressed in the same colours.

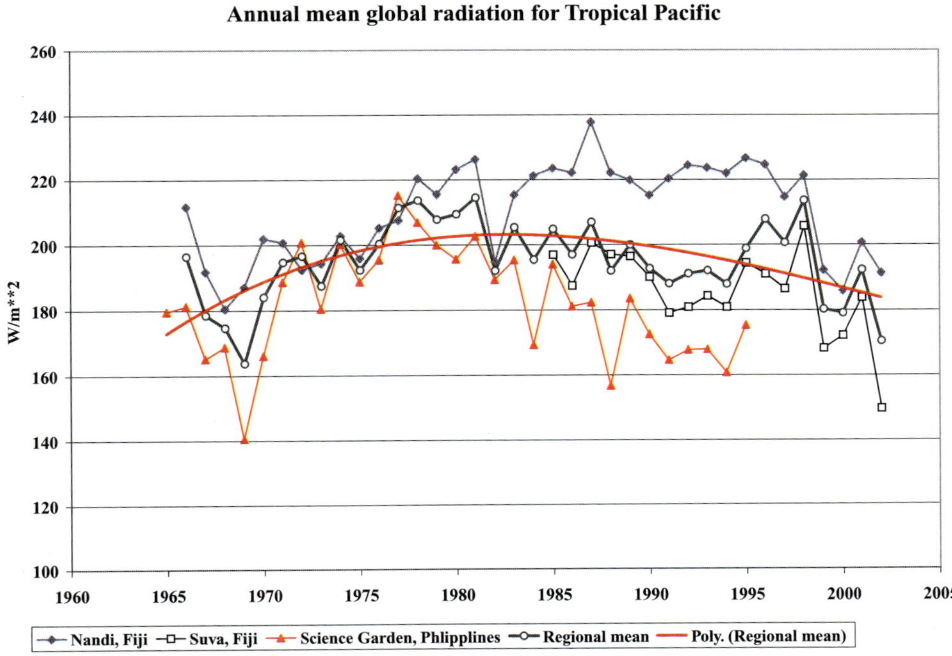


*Figure 10.* Annual mean global radiation for the tropical Pacific (Nandi and Suva in Fiji; Science Garden in the Philippines). Regional mean is expressed in a thick green line with open circles. The polynomial trend is given in a thick red line.

the Philippines. Although there are no similarities between these series and the others, it is worthy to note that these two sites in the tropical Pacific show quite large fluctuations with peaks from the late 1970s and 1980s, and are presently losing radiation. In the case of Fiji, the increase from the 1960s to the end of 1980s is 30 $Wm^{-2}$, and the decline during the last 15 years is 35 $Wm^{-2}$ at Nandi based on 40 years of continuous observation. We have little material to disprove the credibility of these data. Another station in Fiji at Suva also shows a similar sequence after 1985. The only other location offering a series of more than 30 years in the tropical Pacific is the Science Garden in the Philippines. This spot shows a range of increase amounting to 35 $Wm^{-2}$ between 1965 and the peak time in the late 1970s. The presently continuing dimming trend shows already a 35 $Wm^{-2}$ lower value than in the late 1970s. It is difficult to prove if these large ranges of variability in the tropical regions are real or artifact. The definite answer rests on having long series of accurate radiometry at a number of places in the tropics.

For the high latitude ocean climate region, Iceland was chosen. Figure 11 for Reykjavik shows the variation for 39 years from 1964 to 2002. What is common to all ocean stations discussed above is the recent decline of radiation starting in late 1980s. This is the time when the majority of land-based stations started to show signs of brightening.

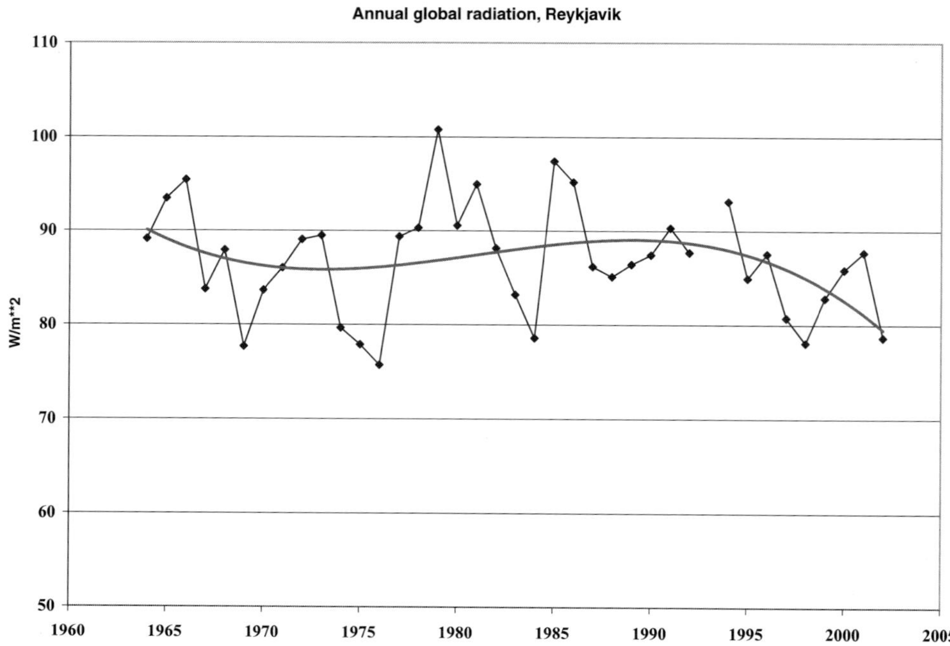


*Figure 11.* Annual mean global radiation for Reykjavik, Iceland. The polynomial trend is expressed by the smooth curve.

Australian series do not show any significant trends, not because global radiation lacked in long-term variation (Stanhill and Kalma, 1994; Stanhill, 2005), but the Bureau of Meteorology adjusted long-term averages to a fixed value to compensate for a possible drift of calibration (B. Fogan, private communication). A re-analysis based on the history of calibration is presently under way.

In the Arctic and Antarctic, there are not sufficient long-term observations of reliable quality on which the calculation of the decadal variation can be justified. The main reason is the frequent atmospheric disruptions to measurements such as frost condensation, rime accretion and snowfall. Most time series are interrupted by long breaks. Nevertheless, Stanhill (1995) and Stanhill and Cohen (1997) reported significant decline in global radiation in both polar regions up to the mid 1990s. The rate of decrease is 0.36 $Wm^{-2}a^{-1}$ and 0.28 $Wm^{-2}a^{-1}$ for the Arctic and Antarctic, respectively. Since the technology to avoid these problems has been available for some time, the observations made in recent years must supply a valuable addition to the post-1990 period. Besides Barrow, three stations (Aklavik/Inuvik, Resolute and Sodankyla) in the Arctic have been added in Figure 9. All Arctic stations show a broad peak in the late 1950 and early 1960s, followed by about 30 years of decline. All sites also show a recent brightening tendency. All polar stations of BSRN (Ny Alesund, Barrow, Syowa, von Neumayer and South Pole) show an increasing trend for the period after 1992. The mean rate of the increase for the Arctic and Antarctic are 0.40 $Wm^{-2}a^{-1}$ and 0.73 $Wm^{-2}a^{-1}$, respectively. The fact that these remote sites show a clear dimming trend only to the late 1980s, and a recovering tendency after 1990 suggests the global nature of the switch from the dimming to brightening at this time.

### 3.10. Global Condition

The observations of global radiation at 380 stations lasting more than 40 years have been analyzed. The longest series is the measurement in Stockholm starting in July 1922. The majority (75%) of stations show a similar sequence of radiation change over the last 40 years. They show a declined of radiation from the early 1960s to the end of the 1980s, followed by a recovery during the last 10 to 15 years. Much longer observations available only in Europe show an initially increasing trend from 1920s to 1950s, thereafter a decline as mentioned above. Six sites in the Arctic and Antarctic also share this trend during the last 50 years. The trends in the polar regions are considered to represent the changes in a large region of the world, as the region is far from major aerosol sources. In 4 regions, middle China, the entire Indian subcontinent, Central Africa and Venezuela global radiation has continued to decrease during the last 40 years of the observation period. It is not possible to investigate the situation in these regions for the period prior to 1960. Although represented by only small number of sites, Maputo (Mozambique), Nandi (Fiji), Science Garden (the Philippines) and Reykjavik (Iceland) show a rather rare succession of the change. Radiation increased towards the late 1980s and early 1990s and then started to decrease to present. This is a trend completely opposite to what the majority of the above-discussed sites show. This sequence is also different from the stations with continuous dimming, as they reveal a clear brightening phase in 1970s. This peculiar trend of a small number of stations should be further studied in detail, as all these sites are located in oceanic regions and may represent a large area of the world.

## 4. The Cause of the Secular Variation

For the period after January 1992, as BSRN data became available, the global brightening has been recognized at the majority of stations in the network ranging from Svalbard to the South Pole (Wild *et al.*, 2004). Further it was found that global radiation observed under cloudless conditions also showed the same increasing tendency. This means that the basic tendency of global radiation is shaped not only by the change in cloud conditions as previously believed but also by constituents of the cloudless atmosphere. This consideration had led the author to investigate the long-term variation of the total zenith transmittance of the atmosphere. Ideally, this investigation should aim at establishing the long history of the optical depth

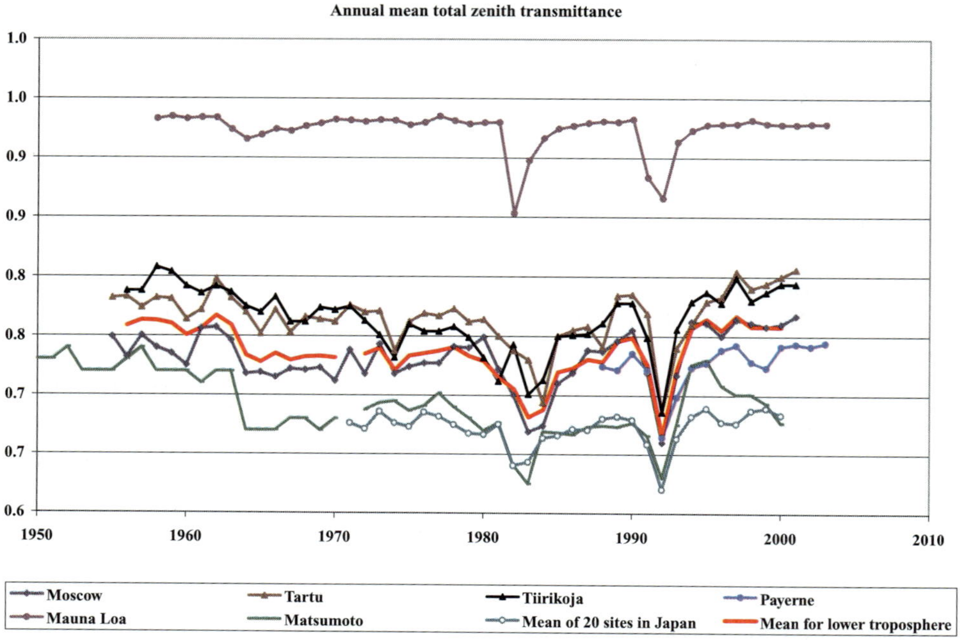


*Figure 12.* Long-term observations of annual mean zenith transmittance of the cloudless atmosphere.

due to aerosol at a number of sites. However, it is at present very difficult to obtain long-term observations with which the aerosol optical depth can be calculated, with the exception of several sites. For Table Mountain, California and Mt Montezuma, Chile, Roosen and Angione (1984) reported the constant nature of the aerosol optical depth for the period of 32 years from 1926 to 1957 for the former, and for the period of 23 years from 1932 to 1954 for the latter. If the presented diagrams are carefully analyzed, however, both sites show a clear trend of increasing optical depth for the period after the mid 1940s. Makhotkina *et al.* (2005) analyzed pyrheliometric data for the territory of the Russian Federation for the period of 1976 and 2002, and concluded that the aerosol optical depth over the vast region of Russia increased to the mid 1980s followed by a period of decrease.

The total transmittance is a useful substitute for the aerosol optical depth under the present circumstances. Total zenith transmittance for 6 sites with long observation records based on well-calibrated pyrheliometers is presented in Figure 12. All sites are located in the Northern Hemisphere. All but Mauna Loa are low altitude stations showing a decreasing transmittance through the second half of the 1950s to 1980s. The other 5 transmittance-series indicate a clear trend of recovery after the mid 1980s, despite momentary depressions caused by El Chichón and Mt Pinatubo eruptions. The trend of the total transmittance indicates a parallel change of the aerosol optical depth and global cloud amount. Since this trend is not seen in the Mauna Loa observations, the trend discussed here is a phenomenon in the lower troposphere. This general trend in the aerosol load in the lower troposphere is the basis to understand the above-discussed course of global dimming to the 1980s and the following brightening tendency observed in 75% of the regions investigated in the present article. These sites are all located in advanced nations who shared the quick industrial recovery during the first twenty years after the end of the world war. Many of these nations also launched a systematic campaign to calve air pollution in the 1980s. Global dimming and brightening may be the direct result of these activities. There is however the remaining 25% of the regions with continued dimming. Unfortunately, there are no long-term pyrheliometric observations in these regions, or BSRN stations. The recently increasing economic activities and biomass burning resulting in increased emission of aerosol may be the cause of the continued dimming in these regions. The changes of global radiation and total transmittance discussed above offer important insights into the interaction of aerosols, optical depth, clouds and solar radiation.

## Acknowledgements

The present article is a part of the summary of the author's work on atmospheric radiation that was carried out during the last twenty years. He has been greatly supported by colleagues in his group at ETH, notably Drs. Hans Gilgen, Andreas Roesch, Peter Tschuck and Martin Wild. The present article was written while the author was a visiting scholar at the Scott Polar Research Institute, and Brenda Ryman Visiting Fellow at Girton College at the University of Cambridge. The author wishes to thank Professor Julian A. Dowdeswell, the director of the Scott Polar Research Institute, and Dr. Roland Randall, a fellow of Girton College. The author was helped by many individuals who organised and supplied valuable radiation data. A special thanks is due to Professor Guangyu Shi of the Chinese Academy of Sciences, Messrs Kohei Honda at Japan Meteorological Agency and Danie Esterhuyse, South African Weather Service. The author is indebted to Professor Bert Holtslag of Wageningen University for notifying the author of the existence of the world's second longest observation series at Wageningen and Mr. Bert Heusinkveld of Wageningen University for supplying the entire observation.

## References

Budyko, M. I.: 1982, *The Earth's Climate: Past and Future*, Intern. Geophys. Ser. **29**, Academic Press, N.Y.

Gilgen, H., Wild, M., and Ohmura, A.: 1998, 'Means and trends of shortwave irradiance at the surface estimated from Global Energy Balance Archive Data', *J. Clim.* **11**, 2042–2061.

Kessler, A.: 1985, *Heat Balance Climatology*, World Survey of Climatology, General Climatology, Vol. 1A. Elsevier, Amsterdam.

Liepert, B., Fabian, P., and Grassl, H.: 1994, 'Solar radiation in Germany – observed trends and an assessment of their causes. Part 1: Regional approach', *Beitr. Phys. Atmosph.* **67**, 15–29.

Makhotkina, E. L., Plakhina, I. N., and Lukin, A. B.: 2005, 'Some features of atmospheric turbidity variations in Russian territory during the last quarter of the Twentieth Century (in Russian)', *Meteorologia i Hydrologia* **2005**, 28–36.

Ohmura, A. and Lang, H.: 1989, 'Secular variation of global radiation in Europe', in: J. Lenoble and J.-F. Geleyn (eds.): *IRS'88: Current Problems in Atmospheric Radiation*, A. Deepak Publ., Hampton, VA, pp. 298–301.

Ohmura, A., Gilgen, H., and Wild, M.: 1989, *Global Energy Balance Archive GEBA, World Climate Program – Water Project A7, Report 1: Introduction*, Verlag der Fachvereine, Zuerich, Zuercher Geografische Schriften **34**, 62 pp.

Ohmura, A. and Gilgen, H.: 1992, 'Global energy balance archive (GEBA, WCP-Water A7) and new aspects of global radiation distribution on the earth's surface', in: S. Keevallik and O. Kärner (eds.): *IRS'92: Current Problems in Atmospheric Radiation*. A. Deepak Publ., Hampton, VA, pp. 271–274.

Roosen, R. G. and Angione, R. J.: 1984, 'Atmospheric transmission and climate: Results from Smithsonian measurements', *Bull. Am. Met. Soc.* **65**, 950–957.

Stanhill, G. and Moreshet, S.: 1992, 'Global radiation climate changes: The world network', *Clim. Change* **21**, 57–75.

Stanhill, G. and Kalma, J. D.: 1994, 'Secular variation of global irradiance in Australia', *Australian Meteorological Magazine* **43**, 81–86.

Stanhill, G.: 1995, 'Global irradiance, air pollution and temperature changes in the Arctic', *Phil. Trans.: Phys. Sc. Engin.* **A352**, 247–258.

Stanhill, G. and Cohen, S.: 1997, 'Recent changes in solar irradiance in Antarctica', *J. Clim.* **10**, 2078–2086.

Stanhill, G. and Ianetz, A.: 1997, 'Long term trends in, and spatial variation of global irradiance in Israel', *Tellus* **49**, 112–122.

Stanhill, G.: 2005, 'Global dimming: A new aspect of climate change', *Weather* **60**, 11–14.

Wild, M., Ohmura, A., and Gilgen, H.: 2004, 'On the consistency of trends in radiation and temperature records and implications for the global hydrological cycle', *Geophys. Res. Let.* **31**, 10.1029/2003GL019188.

Zhang, Y. L., Qin, B. Q., and Chen, W. M.: 2004, 'Analysis of 40 years records of solar radiation data in Shanghai, Nanjing and Hangzhou in eastern China', *Theor. Appl. Climatol.* **78**, 217–227.

# AEROSOL EFFECTS ON CLOUDS AND CLIMATE

U. LOHMANN
*Institute for Atmospheric and Climate Science, ETH Zurich, Universitätsstr. 16, CH-8092 Zurich, Switzerland*
*(E-mail: ulrike.lohmann@env.ethz.ch)*

(Received 23 August 2005; Accepted in final form 3 November 2005)

**Abstract.** Aerosols affect the climate system by changing cloud characteristics in many ways. They act as cloud condensation and ice nuclei, they may inhibit freezing and they could have an influence on the hydrological cycle. While the cloud albedo enhancement (Twomey effect) of warm clouds received most attention so far and traditionally is the only indirect aerosol forcing considered in transient climate simulations, here I discuss the multitude of effects.

**Keywords:** aerosols, clouds, climate

## 1. Introduction

The burning of fossil fuels and biofuels due to human activities has greatly increased the amount of particular matter in the atmosphere. The major aerosol components are mineral dust, sea salt, sulfates, nitrates, black carbon (also termed soot) and particulate organic matter (POM). The natural aerosol species, mineral dust and sea salt, dominate the mass concentration in the atmosphere. On average they contribute 39 mg $m^{-2}$ and 13 mg $m^{-2}$ whereas the anthropogenic components, sulfate, POM and black carbon only contribute 3.9, 3.3 and 0.4 mg $m^{-2}$ to the annual global average as deduced from 20 different global models (Kinne *et al.*, 2006). So far, nitrate is not included in most models, because of its semi-volatile nature.

Optically, mineral dust and sea salt are less important because of their larger size. Thus, mineral dust and sea salt each contribute only as much to the aerosol optical depth as sulfate does (25%). Black carbon, which contributes only 3% to the optical depth, is the main aerosol type that absorbs solar radiation and can lead to a warming of the surrounding air. This warming can prevent cloud formation because the atmosphere becomes more stable or even lead to an evaporation of cloud droplets (e.g., Koren *et al.*, 2004). This so-called semi-direct effect thus counteracts some of the negative aerosol forcings from scattering aerosols, such as sea salt and sulfate, at the top-of-the atmosphere (TOA) (e.g., Lohmann and Feichter, 2005).

Aerosols also act as condensation centers for cloud droplets and ice crystals, thereby changing cloud properties. If more aerosol particles compete for the uptake of water vapor, the resulting cloud droplets do not grow as large. More smaller cloud droplets have a larger surface area than fewer larger cloud droplets for the

Space Science Reviews (2006) 125: 129–137
DOI: 10.1007/s11214-006-9051-8

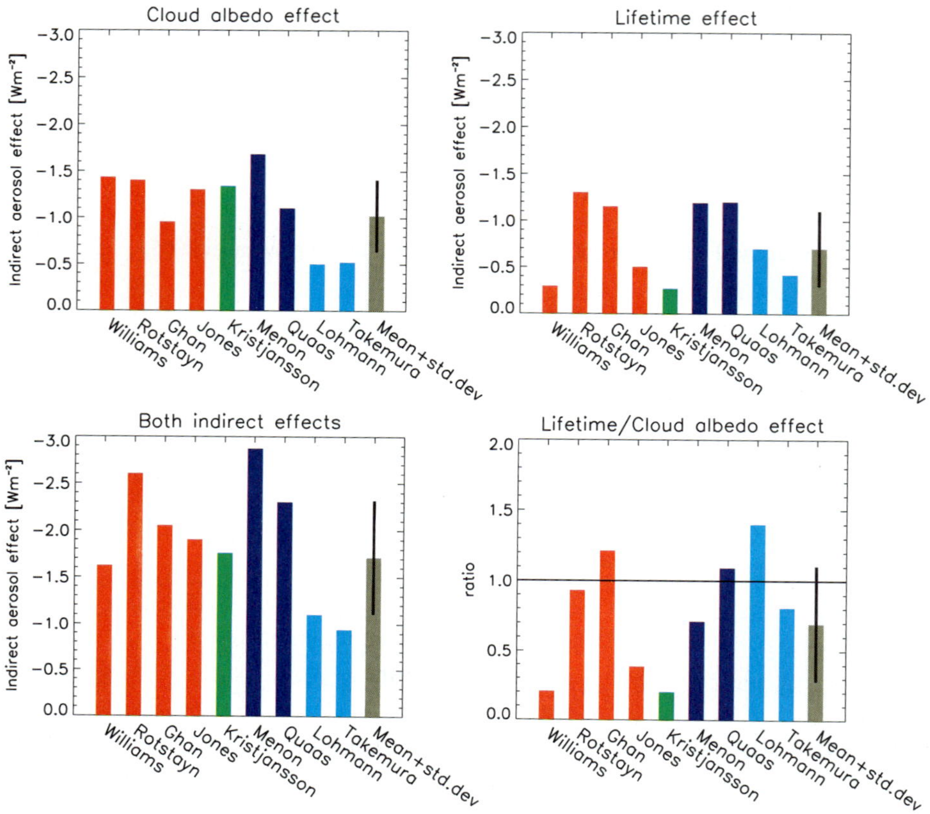


*Figure 1.* Global mean cloud albedo effect, lifetime effect, both effects combined and the ratio lifetime effect/cloud albed effect of anthropogenic sulfate aerosols (red bars) from Williams *et al.* (2001), Rotstayn and Penner (2001), Ghan *et al.* (2001) and Jones *et al.* (2001), of anthropogenic sulfate and black carbon (green bars) from Kristjánsson (2002), of anthropogenic sulfate and organic carbon (blue bars) from Menon *et al.* (2002) and Quaas *et al.* (2004), of anthropogenic sulfate and black and organic carbon (turquoise bars) from Lohmann *et al.* (2000) and Takemura *et al.* (2005), and the mean plus standard deviation from all simulations (olive bars). The results from Menon *et al.* (2002) and Ghan *et al.* (2001) are taken to be the averages of the simulations for only the cloud albedo effect and for both effects.

same amount of cloud water. Thus, a polluted cloud reflects more solar radiation back to space, resulting in a negative radiative forcing at TOA (cloud albedo effect). In addition, these more numerous but smaller cloud droplets collide less efficiently with each other, which reduces the precipitation efficiency of polluted clouds and presumably prolongs their lifetime (cloud lifetime effect). It also implies more scattering of solar radiation back to space, thus reinforcing the cloud albedo effect. Whether the cloud lifetime or the cloud albedo effect is more important, is still an open question. Whereas some models predict that the cloud albedo effect is four times as important as the cloud lifetime effect, other models predict that the cloud lifetime effect dominates over the cloud albedo effect (Figure 1).

The global mean magnitude of the cloud albedo effect since pre-industrial times is estimated between $-0.5$ and $-1.9$ Wm$^{-2}$ from different climate models and the cloud lifetime effect to be between $-0.3$ and $-1.4$ Wm$^{-2}$ (Lohmann and Feichter, 2005). The semi-direct effect, which could in principle counteract part of this negative forcing at TOA, is predicted to be only between $-0.5$ and $+0.1$ Wm$^{-2}$, where the negative values result from black carbon being located above the cloud. If the individual indirect effect values are summed up, the indirect effect could amount to almost $-3$ Wm$^{-2}$. This exceeds estimates from simple inverse models, that start from the observed land temperature raise and increased ocean heat uptake in the 20th century, which bracket the overall indirect aerosol effect to be between 0 and $-2$ Wm$^{-2}$ (Anderson *et al.*, 2003). Thus, either climate model predictions of the cloud albedo and/or cloud lifetime effect are too large, or a counteracting effect is missing.

A proposed counteracting effect could include the ice phase (glaciation effect). Here increases in ice nuclei in the present-day climate result in more frequent freezing of supercooled clouds. As the precipitation formation in ice clouds is faster than in water clouds, this would increase the overall amount of precipitation. A climate model prediction including this effect resulted in reduced cloud cover at mid and high latitudes of the Northern Hemisphere and more of solar radiation absorbed within the Earth-atmosphere system, thus partly offsetting the indirect effects on warm clouds (Lohmann, 2002). Lohmann and Diehl (2006) extended this approach by introducing new parameterizations of contact freezing and immersion freezing in stratiform mixed-phase clouds for black carbon and mineral dust assumed to be composed of either kaolinite (simulation KAO) or montmorillonite (simulation MON) in the ECHAM4 general circulation model from a compilation of laboratory studies. The rather subtle differences between these sensitivity simulations in the present-day climate have significant implications for the anthropogenic indirect aerosol effect. The decrease in net radiation in these sensitivity simulations at the top-of-the-atmosphere varies from $1 \pm 0.3$ Wm$^{-2}$ to $2.1 \pm 0.1$ Wm$^{-2}$ depending on whether dust is assumed to be composed of kaolinite or montmorillonite. In simulation KAO, black carbon has a higher relevancy as an ice nucleus than in simulation MON, because kaolinite is not freezing as effectively as montmorillonite. In simulation KAO, the addition of anthropogenic aerosols results in a larger ice water path, a slightly higher precipitation rate and a reduced total cloud cover. On the contrary, in simulation MON the increase in ice water path is much smaller and globally the decrease in precipitation is dominated by the reduction in warm-phase precipitation due to the indirect cloud lifetime effect (Figure 2).

Aerosols may also influence convective clouds (Rosenfeld and Woodley, 2000). Using a detailed cloud microphysics model Khain *et al.* (2005) found that smaller cloud droplets, such as originating from human activity, reduce the production of drizzle drops in convective clouds. When these droplets freeze, the associated latent heat release results in more vigorous convection. In a clean cloud, on the other hand,

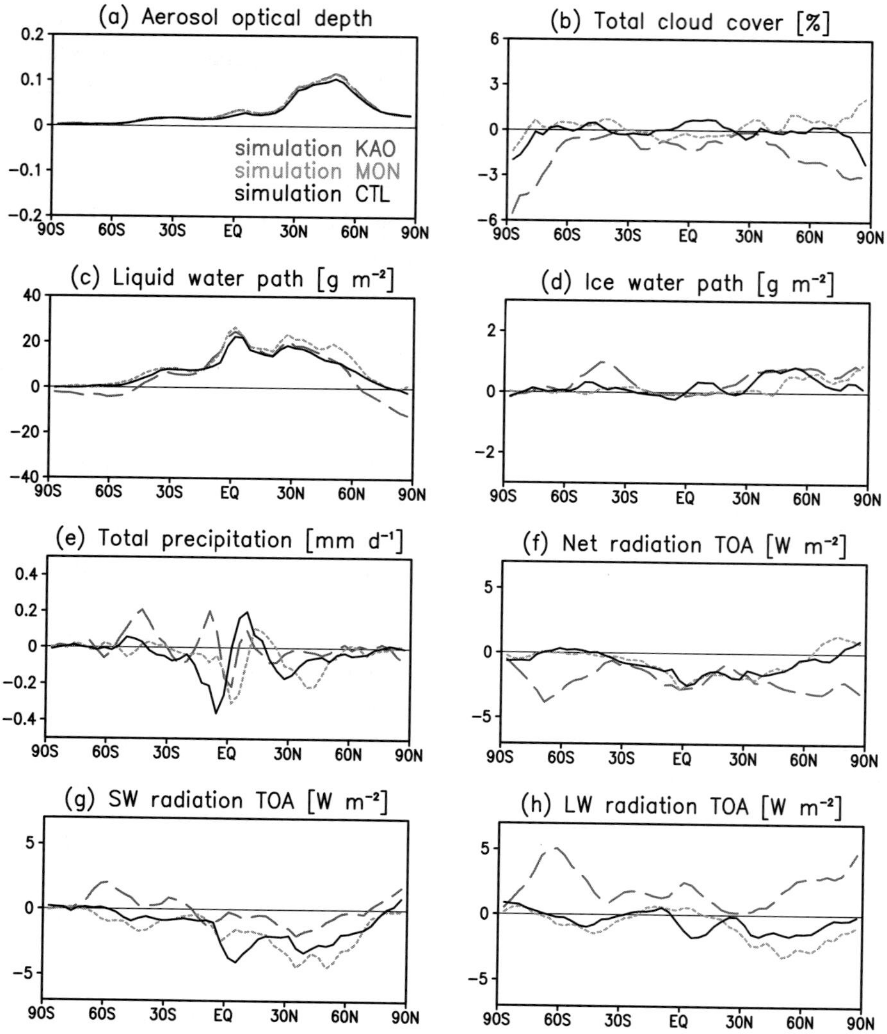


*Figure 2.* Zonal, annual mean changes in (a) aerosol optical depth, (b) total cloud cover, (c) liquid water path (g $m^{-2}$), (d) ice water path (g $m^{-2}$), (e) total precipitation (mm $d^{-1}$), (f) TOA net, (g) shortwave (SW), and (h) longwave (LW) radiation between pre-industrial and present day times for the reference simulation CTL (solid line), simulation KAO (dashed line) and MON (dotted line). For more details, please refer to Lohmann and Diehl (2006).

drizzle would have left the cloud so that less latent heat is released when the cloud glaciates resulting in less vigorous convection. Therefore, no squall line is formed with maritime aerosol concentrations, but the squall line arises under the influence of higher continental aerosol concentrations and results in more precipitation after two hours of simulations.

A smaller indirect aerosol effect is also obtained when constraining the total indirect aerosol effect by taking the difference in the slope of the cloud droplet effective radius-aerosol index relationship between the POLDER satellite data (Bréon *et al.*, 2002) and the ECHAM GCM results taken into account (Lohmann and Lesins, 2002). This reduces the total global mean aerosol effect from $-1.4$ Wm$^{-2}$ to $-0.85$ Wm$^{-2}$. Along the same lines, the cloud albedo effect is reduced from $-0.7$ to $-0.4$ Wm$^{-2}$, when the empirical relationship between sulfate aerosol mass and the cloud droplet number concentration (Boucher and Lohmann, 1995) is constrained to match the relationship between the cloud droplet radius versus aerosol index from the POLDER and MODIS satellite data (Figure 3 and Quaas and Boucher, 2005).

At the Earth surface, however, both scattering and absorbing aerosols work in the same direction of reducing the amount of solar radiation reaching the surface. Since pre-industrial times, increasing emissions of aerosols from human activity and their precursors have caused a reduction of solar radiation at the surface ("solar dimming") by increasing aerosol and cloud optical depth. Such a reduction of 1.3% per decade over the land surfaces from 1961 to 1990 has been observed in many regions worldwide (e.g., Liepert, 2002; Stanhill and Cohen, 2001; Wild *et al.*, 2004).

Likewise equilibrium simulations with a global climate model coupled to a mixed-layer ocean model with increasing aerosol particles and greenhouse gases due to human activity from pre-industrial times to present-day (Liepert *et al.*, 2004) and transient simulations (Roeckner *et al.*, 1999) showed that the decrease in solar radiation at the surface resulting from the increases in optical depth due to the direct and indirect anthropogenic aerosol effects is more important for controlling the surface energy budget than the greenhouse gas induced increase in surface temperature. The conductive flux from below the surface is negligible in the long-term mean surface energy budget. The other components of the surface energy budget (thermal

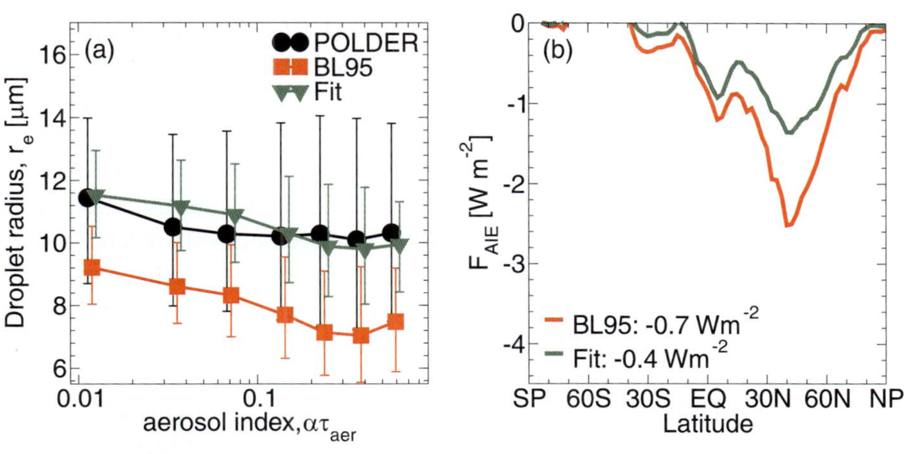


*Figure 3.* Fit of the cloud droplet radius vs aerosol index relationship (black: POLDER data, red: original (Boucher and Lohmann, 1995) parameterization, green: adaptation using parameters that match the POLDER data). For more details, please refer to Quaas and Boucher (2005).

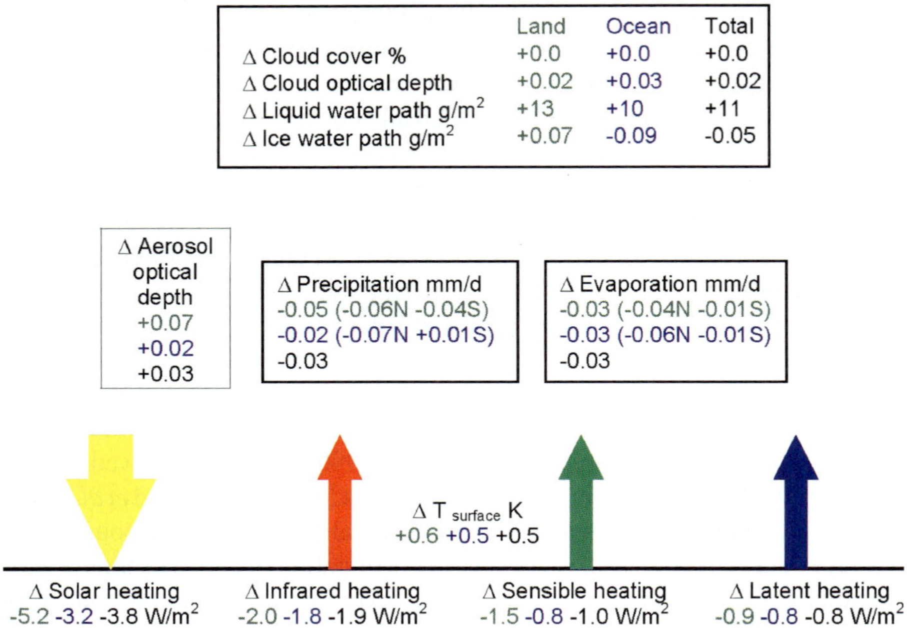


*Figure 4.* Simulated changes in atmospheric variables and surface energy and water fluxes. Changes are calculated as differences for a present day minus pre-industrial climate equilibrium experiments with a general circulation model where anthropogenic aerosol and greenhouse gas concentrations are modified. The arrows indicate direction and relative strengths of the energy fluxes. For more details, please refer to Liepert *et al.* (2004).

radiative flux, sensible and latent heat fluxes) decreased in response to the reduced input of solar radiation (Figure 4). This mechanism could explain the observations of decreased pan evaporation over the last 50 years (Roderick and Farquhar, 2002) and it agrees with the findings by Wild *et al.* (2004). As evaporation must equal precipitation on the global scale in these equilibrium climate simulations, a reduced latent heat flux led to a reduction in precipitation.

Recent surface observations, however, show that the long term decline in solar radiation at land surfaces turned into an increase in surface solar radiation during the 1980s (Wild *et al.*, 2005; Pinker *et al.*, 2005), in agreement with recent emission trends in the "old" industrial regions in the northern hemisphere (Krüger and Graßl, 2002) as well as with long-term black carbon trends in the Canadian Arctic (Sharma *et al.*, 2004) and sulfate deposition declines over Europe and North America since 1978 (E. Holland, private communication). Thus, the increasing greenhouse effect may no longer be masked by an aerosol-induced decline in solar radiation, resulting in the enhanced warming observed since the 1990s (Wild *et al.*, 2005).

Climate model simulations suggest that the decrease in global mean precipitation from pre-industrial times to the present day may reverse into an increase in global mean precipitation of about 1% in 2021–2050 as compared with 1961–1990. This is caused by the increased warming due to black carbon and greenhouse gases that will dominate over the sulphate cooling in the mid 21st century (Roeckner *et al.*, 2006). In South Asia, absorbing aerosols in atmospheric brown clouds may have played a major role in the observed South Asian climate and hydrological cycle changes and may have masked as much as 50% of the surface warming due to the global increase in greenhouse gases (Ramanathan *et al.*, 2005). Their simulations raise the possibility that, if current trends in emissions continue, the South Asian subcontinent may experience a doubling of the drought frequency in future decades.

## 2. Conclusions

The following conclusions regarding the various aerosol effects on clouds and climate can be reached:

– The range of model estimates for the cloud albedo indirect radiative forcing varies from $-0.5$ to $-1.9$ $Wm^{-2}$. All models indicate a negative forcing, which on average is $-1$ $Wm^{-2}$, with a standard deviation of 0.4 $Wm^{-2}$ (Lohmann and Feichter, 2005).
– The cloud lifetime effect varies considerably between the different models (from $-0.3$ to $-1.4$ $Wm^{-2}$), resulting in an average forcing of $-0.7$ $Wm^{-2}$ and a standard deviation of 0.5 $Wm^{-2}$ (Lohmann and Feichter, 2005).
– Estimates of the semi-direct effect range from $+0.1$ to $-0.5$ $Wm^{-2}$. The variations arise from different locations of black carbon with respect to the cloud.
– Aerosol effects on stratiform or convective mixed-phase clouds are too uncertain until now to even provide an estimate.

In summary, it is not possible to obtain a best estimate of the total indirect aerosol effect from pre-industrial times to present-day solely from observations. The satellite record is not long enough and other existing long-term records do not provide the aerosol and cloud microphysical properties needed for such an assessment. Climate models by themselves have weaknesses that could bias the indirect effect. Thus, to obtain a best estimate of the indirect aerosol effect, measurements and models should be combined. Combining satellite measurements with climate model estimates offline, Lohmann and Lesins (2002) obtained a total indirect effect of $-0.85$ $Wm^{-2}$, which falls within the range of the indirect effect as estimated from inverse simulations (Anderson *et al.*, 2003). By applying the cloud droplet number versus fine mode aerosol optical depth relationship from MODIS directly within climate models, the total indirect effect is reduced even further to $-0.3$ to $-0.5$ $Wm^{-2}$ (Quaas *et al.*, 2006).

## Acknowledgements

Ulrike Lohmann thanks Johannes Quaas for providing Figure 3.

## References

Anderson, T. L., Charlson, R. J., Schwartz, S. E., Knutti, R., Boucher, O., Rodhe, H., and Heintzenberg, J.: 2003, 'Climate forcing by aerosols – a hazy picture', *Science* **300**, 1103–1104.

Boucher, O. and Lohmann, U.: 1995, 'The sulfate-CCN-cloud albedo effect: A sensitivity study with two general circulation models', *Tellus* **Ser. B, 47**, 281–300.

Bréon, F.-M., Tanré, D., and Generoso, S.: 2002, 'Aerosol effect on cloud droplet size monitored from satellite', *Science* **295**, 834–838.

Ghan, S. J., Easter, R. C., Chapman, E., Abdul-Razzak, H., Zhang, Y., Leung, R., Laulainen, N., Saylor, R., and Zaveri, R.: 2001, 'A physically-based estimate of radiative forcing by anthropogenic sulfate aerosols', *J. Geophys. Res.* **106**, 5279–5293.

Jones, A., Roberts, D. L., and Woodage, M. J.: 2001, 'Indirect sulphate aerosol forcing in a climate model with an interactive sulphur cycle', *J. Geophys. Res.* **106**, 20,293–30,310.

Khain, A., Rosenfeld, D., and Pokrovsky, A.: 2005, 'Aerosol impact on the dynamics and microphysics of convective clouds', *Q. J. R. Meteorol. Soc.* **131**, 2639–2663.

Kinne, S., Schulz, M., Textor, C., Guibert, S., *et al.*: 2006, 'An AeroCom initial assessment – optical properties in aerosol component modules of global models', *Atmos. Chem. Phys.* **6**, 1815–1834.

Koren, I., Kaufman, Y. J., Remer, L. A., and Martins, J. V.: 2004, 'Measurements of the effect of smoke aerosol on inhibition of cloud formation', *Science* **303**, 1342–1345.

Kristjánsson, J. E.: 2002, 'Studies of the aerosol indirect effect from sulfate and black carbon aerosols', *J. Geophys. Res.* **107**, doi: 10.1029/2001JD000887.

Krüger, O. and Graßl, H.: 2002, 'The indirect aerosol effect over Europe', *Geophys. Res. Lett.* **29**, doi: 10.1029/2001GL014081.

Liepert, B. G.: 2002, 'Observed reductions of surface solar radiation at sites in the United States and worldwide from 1961 to 1990', *Geophys. Res. Lett.* **29**, doi: 10.1029/2002GL014910.

Liepert, B. G., Feichter, J., Lohmann, U., and Roeckner, E.: 2004, 'Can aerosols spin down the water cycle in a warmer and moister world?', *Geophys. Res. Lett.* **31**, doi:10.1029/2003GL019060.

Lohmann, U.: 2002, 'A glaciation indirect aerosol effect caused by soot aerosols', *Geophys. Res. Lett.* **29**, doi: 10.1029/2001GL014357.

Lohmann, U. and Lesins, G.: 2002, 'Stronger constraints on the anthropogenic indirect aerosol effect', *Science* **298**, 1012–1016.

Lohmann, U. and Diehl, K.: 2006, 'Sensitivity studies of the importance of dust ice nuclei for the indirect aerosol effect on stratiform mixed-phase clouds', *J. Atmos. Sci.* **63**, 968–982.

Lohmann, U. and Feichter, J.: 2005, 'Global indirect aerosol effects: A review', *Atmos. Chem. Phys.* **5**, 715–737.

Lohmann, U. Feichter, J., Penner, J. E., and Leaitch, W. R.: 2000, 'Indirect effect of sulfate and carbonaceous aerosols: A mechanistic treatment', *J. Geophys. Res.* **105**, 12,193–12,206.

Menon, S., DelGenio, A. D., Koch, D., and Tselioudis, G.: 2002, 'GCM simulations of the aerosol indirect effect: Sensitivity to cloud parameterization and aerosol burden', *J. Atmos. Sci.* **59**, 692–713.

Pinker, R. T., Zhang, B., and Dutton, E. G.: 2005, 'Do satellites detect trends in surface solar radiation?', *Science* **308**, 850–854.

Quaas, J. and Boucher, O.: 2005, 'Constraining the first aerosol indirect radiative forcing in the LMDZ GCM using POLDER and MODIS satellite data', *Geophys. Res. Lett.* **32**, doi: 10.1029/2005GL023850.

Quaas, J., Boucher, O., and Bréon, F.-M.: 2004, 'Aerosol indirect effects in POLDER satellite data and the Laboratoire de Météorologie Dynamique-Zoom (LMDZ) general circulation model', *J. Geophys. Res.* **109**, doi: 10.1029/2003JD004317.

Quaas, J., Boucher, O., and Lohmann, U.: 2006, 'A new estimate of the aerosol indirect radiative forcing by constraints of global climate models using satellite datasets', *Atmos. Chem. Phys.* **6**, 947–955.

Ramanathan, V., Chung, C., Kim, D., Bettge, T., Buja, L., Kiehl, J. T., Washington, W. M., Fu, Q., Sikka, D. R., and Wild, M.: 2005, 'Atmospheric brown clouds: Impacts on South Asian climate and hydrological cycle', *Proc. Nat. Acad. Sc.* **102**, 5326–5333.

Roderick, M. L. and Farquhar, G. D.: 2002, 'The cause of decreased pan evaporation over the past 50 years', *Science* **298**, 1410–1411.

Roeckner, E., Bengtsson, L., Feichter, J., Lelieveld, J., and Rodhe, H.: 1999, 'Transient climate change simulations with a coupled atmosphere-ocean GCM including the tropospheric sulfur cycle', *J. Climate* **12**, 3004–3032.

Roeckner, E., Stier, P., Feichter, J., Kloster, S., and Esch, M.: 2006, 'Impact of carbonaceous aerosol emissions on regional climate change', *Clim. Dyn.*, in press.

Rosenfeld, D. and Woodley, W. L.: 2000, 'Deep convective clouds with sustained supercooled liquid water down to $-37.5°C$', *Nature* **405**, 440–442.

Rotstayn, L. D. and Penner, J. E.: 2001, 'Indirect aerosol forcing, quasi-forcing, and climate response', *J. Climate* **14**, 2960–2975.

Sharma, S., Lavoué, D., Cachier, H., Barrie, L. A., and Gong, S. L.: 2004, 'Long-term trends of the black carbon concentrations in the Canadian Arctic', *J. Geophys. Res.* **109**, doi: 10.1029/2003JD004331.

Stanhill, G. and Cohen, S.: 2001, 'Global dimming a review of the evidence for a widespread and significant reduction in global radiation with discussion of its probable causes and possible agricultural consequences', *Agric. For. Meteorol.* **107**, 255–278.

Takemura, T., Nozawa, T., Emori, S., Nakajima, T. Y., and Nakajima, T.: 2005, 'Simulation of climate response to aerosol direct and indirect effects with aerosol transport-radiation model', *J. Geophys. Res.* **110**, doi:10.1029/2004JD00502.

Wild, M., Ohmura, A., Gilgen, H., and Rosenfeld, D.: 2004, 'On the consistency of trends in radiation and temperature records and implications for the global hydrological cycle', *Geophys. Res. Lett.* **31**, doi: 10.1029/2003GL019188.

Wild, M., Gilgen, H., Roesch, A., Ohmura, A., Long, C. N., Dutton, E. G., Forgan, B., Kallis, A., Russak, V., and Tsvetkov, A.: 2005, 'From dimming to brightening: Decadal changes in solar radiation at Earth's surface', *Science* **308**, 847–850.

Williams, K. D., Jones, A., Roberts, D. L., Senior, C. A., and Woodage, M. J.: 2001, 'The response of the climate system to the indirect effects of anthropogenic sulfate aerosols', *Clim. Dyn.* **17**, 845–856.

# SATELLITE OBSERVATIONS OF NATURAL AND ANTHROPOGENIC AEROSOL EFFECTS ON CLOUDS AND CLIMATE

Y. J. KAUFMAN
*NASA/Goddard Space Flight Center, 613.2, Greenbelt, MD 20771, USA*
*(E-mail: lorraine.a.remer@nasa.gov)*

(Received 13 September 2005; Accepted in final form 3 November 2005)

**Abstract.** Anthropogenic aerosols affect the climate system and the hydrological cycle. The net effect of aerosols is to cool the climate system, directly by reflecting sunlight to space, and indirectly by increasing the brightness and cover of clouds that in turn also reflect more sunlight to space. The uncertainty in the aerosol effect on climate is 5 times greater than that of the greenhouse gases. The reason for this is the short aerosol lifetime and chemical complexity, that makes it difficult to represent the global aerosol budget from surface or aircraft measurements. Satellites provide daily global information about the aerosol content, generating large statistics with excellent regional and global representation of the aerosol column concentration, and differentiating fine from coarse aerosol. Here we use observations performed with the MODIS instrument onboard the Terra and Aqua satellites to differentiate natural from anthropogenic aerosols, and to measure the aerosol effect on cloud properties and on the reflectivity of sunlight.

**Keywords:** anthropogenic, aerosols, cloud aerosol interaction, MODIS

## 1. Introduction

Anthropogenic aerosols affect the climate system and the hydrological cycle. The net effect of aerosols is to cool the climate system by directly reflecting sunlight to space, and indirectly by increasing the brightness and cover of clouds that in turn also reflect more sunlight to space. However, depending on their composition, atmospheric aerosols can also absorb incoming sunlight, thus further cooling the surface but warming the atmosphere in the process. Understanding the aerosol effect on the climate system is important in order to better assess the amplitude of the current man-made radiative forcing of climate. As pointed out by Andreae *et al.* (2005), the opposed effects of aerosol cooling and greenhouse gases warming imply that the amplitude of estimated future greenhouse warming will depend on the present-time aerosol cooling. In other words, the stronger is the effective present-time aerosol cooling, the stonger will be the implied future warming by greenhouse gases. Hence, if last century's 0.6°C Earth's surface temperature increase (IPCC, 2001) is attributed to greenhouse gases alone, the sensitivity of the climate system would be of 0.6°C per $2.4 \pm 0.3$ W/m$^2$ of greenhouse gas warming or $0.25 \pm 0.03$°C per W/m$^2$. However, if the warming was in fact partially counterbalanced by an


DOI: 10.1007/s11214-006-9052-7 

aerosol-induced cooling of the order of $1.8 \pm 1$ W/m$^2$, then the actual climate sensitivity to anthropogenic aerosol is 4 times larger, with a very large uncertainty.

Why is it so much more difficult to calculate the actual aerosol forcing versus that of the greenhouse gases? Greenhouse gases, such as carbon dioxide and methane, have a lifetime of up to 100 years in the atmosphere and a rather homogeneous distribution around the globe. This is in contrast to the heterogeneous spatial and temporal distribution of tropospheric aerosols, which results from their short lifetime of about a week (Kaufman *et al.*, 2002). Therefore it was possible to measure the global increase in the $CO_2$ concentration of 1–2 ppm/yr already half a century ago using a single ground-based instrument (Keeling, 1960), while daily satellite observations (Kaufman *et al.*, 2002) and continuous *in situ* measurements (Holben *et al.*, 2001; Prospero and Nees, 1986) are needed to observe the emission and transport of dense aerosol plumes downwind of populated and polluted regions (urban haze), regions with vegetation fires (smoke), and deserts (dust).

Assessing the aerosol effect on climate and the hydrological cycle is particularly difficult because aerosols take a multitude of chemical compositions, ranging from desert dust to urban pollution and smoke from fires with their organic, sulfate and black carbon composition, each with corresponding different optical properties; and because aerosol interaction with clouds and precipitation increases the aerosol variability over time and space. To accurately study aerosol distribution and composition therefore requires continuous observations from satellites, networks of ground-based instruments and dedicated field experiments that can study regionally the aerosol properties, distribution and impacts (Ramanathan *et al.*, 2001; Dubovik *et al.*, 2002).

## 2. Natural Versus Anthropogenic Aerosol

Satellite instruments do not measure the aerosol chemical composition needed to discriminate anthropogenic from natural aerosol components. However the ability of new satellite instruments to distinguish fine (submicron) from coarse (supermicron) aerosols over the oceans (Tanré *et al.*, 1997), can be used to identify the signature of the anthropogenic component. This technique allows us to estimate the fraction of anthropogenic aerosols with an uncertainty of ±30% (Kaufman *et al.*, 2005a). Figure 1 shows an example of such satellite data, taken by the MODIS (Moderate Resolution Imaging Spectroradiometer) on board the Terra satellite. In this case the wide spectral range of the MODIS measurements over the ocean between 0.53 $\mu$m and 2.1 $\mu$m is clearly distinguishing fine smoke from coarse dust (Remer *et al.*, 2005). The method is further illustrated in Figure 2 that shows a 3-month composite image of the average aerosol distribution over the Atlantic Ocean. The colors (green for coarse and red for fine aerosol particles) clearly differentiate the dust regions west of West Africa, from the pollution and smoke regions to the north and south. The method allowing us to distinguish anthropogenic from natural

*Figure 1.* Spectral aerosol reflectance measured by the MODIS instrument onboard the Terra satellite. *Left panels:* Color composites of channels in the visible spectrum. *Right panels:* Composites in the near infrared spectrum. *Top panels*: Coarse dust particles emitted from West Africa (7 January 2002), which are visible in both panels over the ocean. *Bottom panels*: Fine smoke particles from fires in Australia (25 December 2001), which are invisible over the ocean in the near infrared.

aerosol is based on estimates of the fraction of optical thickness induced by fine aerosol particles in aerosol dominated by natural components, e.g., dust and marine aerosol, and of the fraction of optical thickness induced by fine aerosol in anthropogenic aerosol, e.g., pollution and smoke from prescribed fires. The method is described in Kaufman *et al.* (2005a).

An application of this method to two years of global MODIS data is shown in Figure 3. The analysis shows that $21 \pm 7\%$ of the aerosol optical thickness over the oceans has an anthropogenic origin. The MODIS measurements were compared with 3 chemical transport models: SPRINTARS (Takemura *et al.*, 2002), GOCART (Chin *et al.*, 2002), and LMDZ (Reddy *et al.*, 2005). Although the results disagree as to the total aerosol optical thickness, we found that the three models, used to compute global estimates of the aerosol forcing of climate, calculate a global average anthropogenic optical thickness over the ocean between 0.030 and 0.036, in

*Figure 2.* Spatial distribution of aerosol column concentration (expressed as the optical thickness) and type (given by the fraction of the aerosol in the sub-micron mode) over the Atlantic Ocean from MODIS data for June–August 2002. The optical thickness is represented by the brightness of the image, the aerosol type by the color. Red: dominance by submicron particles – smoke from central Africa and pollution from Europe and North America. Green: dominance by dust from Africa or sea salt in regions with high winds.

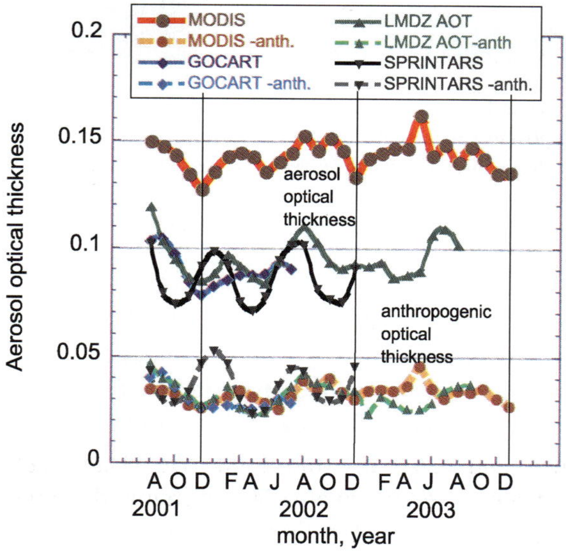


*Figure 3.* Global average aerosol optical thickness over the cloud-free ocean measured by MODIS (top red curve), and simulated (center curves) by the GOCART (blue), LMDZ (green) and SPRINTARS (black) models. Bottom curves: anthropogenic aerosol optical thickness for MODIS (red), and models. MODIS mostly agrees with models regarding the anthropogenic component, with average optical thicknesses of 0.033, 0.032, 0.030, and 0.036 for MODIS, LMDZ, GOCART, and SPRINTARS, respectively.

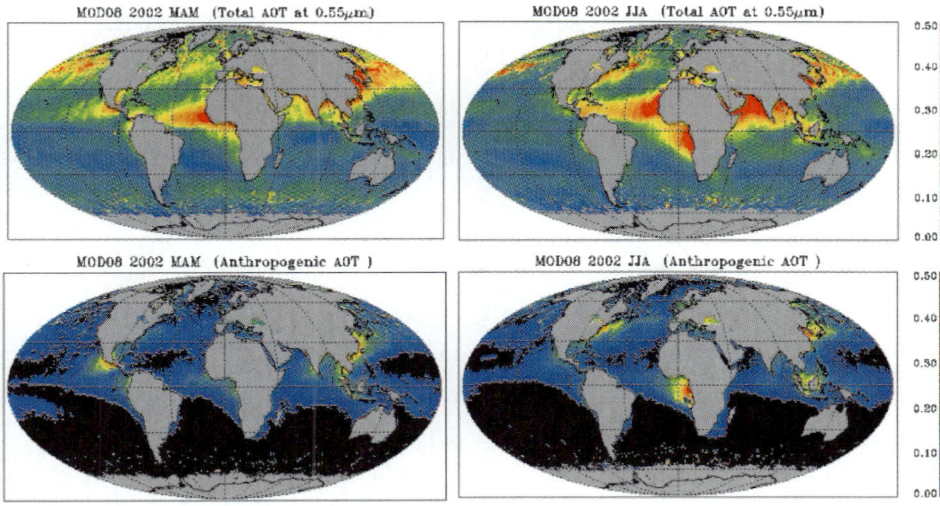


*Figure 4.* Global distribution over the oceans of the total aerosol optical thickness (top) and the anthropogenic component (bottom) derived from the MODIS data. Left panels for March–May, right panels for June–August, 2002.

line with the present MODIS assessment of 0.033. This increases our confidence in model assessments of the aerosol direct forcing of climate. The MODIS-estimated aerosol forcing over cloud free oceans is therefore $-1.4 \pm 0.4$ W/m$^2$. The spatial distribution of the total and anthropogenic optical thickness is shown in Figure 4 for two seasons. Note that the anthropogenic panels show the smoke and polluted regions, while regions affected by dust and marine aerosol appear as black – no anthropogenic aerosols.

## 3. Aerosol Effect on Shallow Cloud Development

Clouds developing in a polluted environment tend to have more numerous, but smaller droplets. This may lead to suppression of precipitation and longer cloud lifetime (Gunn and Phillips, 1957). Absorption of incoming solar radiation by aerosols, on the other hand, can reduce the cloud cover (Hansen *et al.*, 1997; Koren *et al.*, 2004). The net aerosol effect on clouds is currently the largest uncertainty in evaluating climate forcing (IPCC, 2001). Using large statistics of 1 km resolution MODIS satellite data, we study the aerosol effect on shallow water clouds in 4 distinct regions of the Atlantic Ocean, for June through August 2002 (see Figure 2): marine aerosol (30°S–20°S), smoke (20°S–5°N), mineral dust (5°N–25°N) and pollution aerosols (30°N–60°N). The main results are shown in Figure 5 and Table I Kaufman *et al.* (2005b). As expected from previous results found in the literature (e.g., Kaufman *et al.*, 2002) all the 4 aerosol types affect the cloud droplet size.

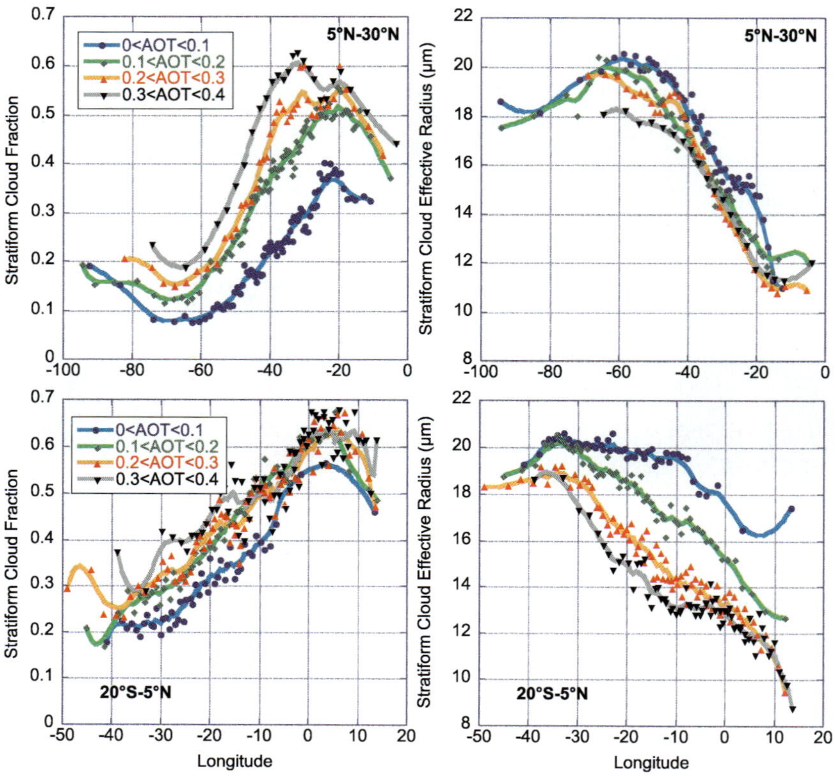


*Figure 5.* Longitudinal dependence of the shallow cloud fraction (left panels) and droplet effective radius (right panels) for the Northern tropical Atlantic with dust intrusions (top panels) and Southern tropical Atlantic with smoke intrusion (bottom panels). The results are shown for the 4 ranges of aerosol optical thickness (AOT) given in the figure. The dots are averages of 50 to 200 $1^\circ \times 1^\circ$ grid boxes located at similar longitude locations and for the same aerosol optical thickness range.

However, we also found that the coverage of shallow clouds increases in all the cases by 0.2–0.4 from clean to polluted, smoky, or dusty aerosol conditions. In Figure 5 we show the results for the dust and smoke regions as a function of the location. The cloud fraction varies as a function of longitude due to variation in the conditions for shallow cloud developments, but also as a function of the aerosol optical thickness. In some locations, dust doubles the cloud cover. The figure also shows variation of the cloud droplet size with longitude and aerosol concentration.

A natural question arises here: is this correlation of the presence of aerosol and cloud cover an indication of aerosol effect on clouds? Coincidental correlation between aerosol concentration and cloud cover due to changes in atmospheric circulation that can affect both aerosol and clouds is an alternative explanation. We used co-variability analysis with meteorological parameters most relevant to the transport of aerosol and cloud formation (Kaufman *et al.*, 2005b). The analysis

TABLE I

Results of the analysis for 4 regions in the Atlantic Ocean.

| Region and dominant aerosol | Fraction of region | Shallow cloud cover | AOT Range & Average | $\Delta$cl-aer | $\delta$cl-aer | Change in $R_{eff}$ (%) | Change in LWP (%) | Change in CLTP (hPa) | Radiative effects (W/m$^2$) due to $\Delta N_c$ | $\Delta N_c$ + $\Delta$LWP | $\Delta N_c$ + $\Delta$LWP + $\delta$cl | Total forcing TOA | $\Delta$abs |
|---|---|---|---|---|---|---|---|---|---|---|---|---|---|
| 30°N–60°N Pollution | 0.17 | 0.07 | 0.03–0.19 *0.102* | $0.20 \pm 0.06$ | $0.19 \pm 0.03$ | $-12 \pm 10$ | $6 \pm 34$ | $-39 \pm 20$ | −1.0 | −1.1 | −4.5 | −8.0 | 0.7 |
| 5°N–30°N Saharan Dust | 0.26 | 0.11 | 0.03–0.46 *0.174* | $0.36 \pm 0.12$ | $0.25 \pm 0.04$ | $-12 \pm 13$ | $9 \pm 34$ | $-66 \pm 13$ | −0.7 | −0.9 | −6.8 | −14.0 | 0.7 |
| 30°S–5°N Biomass Burning | 0.53 | 0.29 | 0.03–0.43 *0.152* | $0.31 \pm 0.07$ | $0.31 \pm 0.04$ | $-32 \pm 3$ | $-21 \pm 8$ | $-55 \pm 11$ | −1.5 | −1.0 | −9.5 | −11.3 | 2.9 |
| 30°S–20°S Marine | 0.47 | 0.27 | 0.02–0.24 *0.085* | $0.45 \pm 0.10$ | $0.45 \pm 0.04$ | $-19 \pm 7$ | $35 \pm 22$ | $-72 \pm 18$ | — | — | — | — | — |

Columns from left: location; fraction of the region classified as shallow clouds; shallow cloud fraction; range of the aerosol optical thickness (AOT) in the analysis (5th-clean and 95th-hazy percentile) and average value; change in the cloud cover from the clean and hazy conditions ($\Delta$cl-aer); partial change in the cloud cover associated with aerosol from the multiple regression ($\delta$cl-aer); percent change in the cloud effective radius ($R_{eff}$) from the clean and hazy conditions; percent change in the cloud liquid water content (LWP); change in the cloud top pressure. For each value the variability within the 3 months of analysis (June–August) is given. Also given are the average radiative effects due to change in the aerosol optical thickness from the base oceanic value of 0.06, associated with: increase in cloud droplet concentration ($\Delta N_c$) due to reduction in $R_{eff}$; plus change in the column cloud water content ($\Delta$LWP); plus change in the cloud cover ($\delta$cl); plus direct aerosol radiative effect; absorption of sunlight by aerosol $\Delta$abs. Note that the sum of the last 2 columns is the total aerosol radiative forcing at the surface. The radiative effect was calculated as half of the effect for 60° solar zenith angle, only for the $1° \times 1°$ latitude grid boxes characterized as shallow clouds. The uncertainty in the aerosol measurements from MODIS is ~10%, ~3% for the cloud fraction and 20% for the cloud effective radius. The 95th percentile confidence limit of the multiple regression is 4–8% of the stated values. The overall error in the radiative effects calculations is therefore ~20%. Absorption computations depend on the validity of the assumed single scattering albedo with uncertainty of 50%. No calculations are given for the marine region, since the average AOT is too close to the baseline value.

associates most of the change in the cloud fraction to variability of aerosols, for each of the 4 regions and 3 months studied. In our opinion, there is low probability that the net aerosol effect can be explained by coincidental, unresolved changes in meteorological conditions that also accumulate aerosol, or errors in the data, though further in situ measurements and model developments is needed to fully understand the processes.

The radiative effect at the top of the atmosphere resulting from the aerosol effect on the shallow clouds and solar radiation is: $-11 \pm 3$ W/m$^2$ for the 3 months studied, 2/3 of it is due to the aerosol-induced cloud changes, and 1/3 due to aerosol direct radiative effect. A parallel study of the aerosol effect on convective clouds in the same region and season, found a strong correlation between the presence of aerosols and the structural properties of convective clouds. These correlations suggest systematic invigoration of convective clouds by pollution, desert dust and biomass burning aerosols. On average an increase in the aerosol concentration of 100–150% from baseline to average values is associated with a $5 \pm 1\%$ increase in the cloud fraction and a $40 \pm 5$ mb decrease in the cloud top pressure (Koren *et al.*, 2005).

## 4. Summary

The satellite point of view, some 800 km above the Earth surface, provides a unique perspective for the investigation of natural and anthropogenic aerosol effects on clouds and climate. Satellites can neither measure the detailed aerosol chemical composition and size distribution, nor the clouds vertical profile. But they can provide daily global coverage measurements, generating large statistics with excellent regional and global representation of the aerosol column concentration (expressed here by the optical thickness), differentiating fine from coarse aerosol. They can also provide a measure of the main properties of clouds: cloud cover, cloud top pressure and effective droplet radius. Combining this information, in our case from the MODIS instrument on the Terra platform, we are able to produce the first measurement-based estimates of the anthropogenic aerosol fraction and of the impact of aerosol on the cloud cover and height, leading to a better understanding of the impact of the aerosol effect on climate forcing.

## References

Andreae, M. O., Jones, C. D., and Cox, P. M.: 2005, 'Strong present-day aerosol cooling implies a hot future', *Nature* **435**, doi:10.1038/nature03671.

Chin, M., *et al.*: 2002, 'Tropospheric aerosol optical thickness from the GOCART model and comparisons with satellite and Sun photometer measurements', *J. Atmos. Sci.* **59**, 461–483.

Dubovik, O., *et al.*: 2002, 'Variability of absorption and optical properties of key aerosol types observed in worldwide locations', *J. Atmos. Sci.* **59**, 590–608.

Gunn, R. and Phillips B. B.: 1957, 'An experimental investigation of the effect of air pollution on the initiation of rain', *J. Meteor.* **14**, 272–280.

Hansen, J., Sato, M., Ruedy, R., *et al.*: 1997, 'Radiative forcing and climate response', *J. Geophys. Res.* **102**, 6831–6864.

Holben, B. N., Tanré, D., Smirnov, A., Eck, T. F., Slutsker, I., Abuhassan, N., *et al.*: 2001, 'An emerging ground-based aerosol climatology: Aerosol optical depth from AERONET', *J. Geophys. Res.* **106**, 12,067–12,097.

IPCC: 2001, 'Climate change 2001: The scientific basis', Contribution of Working Group I to the Third Assessment Report of the Intergovernmental Panel on Climate Change, J. T. Houghton, Y. Ding, D. J. Griggs, M. Noguer, P. J. van der Linden, X. Dai, K. Maskell, and C. A. Johnson (eds.). Cambridge University Press, Cambridge, UK, 881 pp.

Kaufman, Y. J., Tanré, D., and Boucher, O.: 2002, 'A satellite view of aerosols in the climate system', *Nature* **419**, 215–223.

Kaufman, Y. J., Boucher, O., Tanré, D., Chin, M., Remer, L. A., and Takemura, T.: 2005a, 'Aerosol anthropogenic component estimated from satellite data', *Geophys. Res. Lett.* **32**, doi:10.1029/2005GL023125.

Kaufman, Y. J., Koren, I., Remer, L. A., Rosenfeld, D., and Rudich, Y.: 2005b, 'The effect of smoke, dust, and pollution aerosol on shallow cloud development over the Atlantic Ocean', In: *Proc. Nat. Acad. Sci.* **102**, 11,207–11,212.

Keeling, C. D.: 1960, 'The concentration and isotopic abundances of carbon dioxide in the atmosphere', *Tellus* **12**, 200–203.

Koren, I., Kaufman, Y. J., Remer, L. A., and Martins, J. V.: 2004, 'Measurement of the effect of Amazon smoke on inhibition of cloud formation', *Science* **303**, 1342–1344.

Koren, I., Kaufman, Y. J., Rosenfeld, D., Remer, L. A., and Rudich, Y.: 2005, 'Aerosol invigoration and restructuring of Atlantic convective clouds', *Geophys. Res. Lett.* **32**, doi:10.1029/2005GL023187.

Prospero, J. M., and Nees, R. T.: 1986, 'Impact of the North African drought and El Niño on mineral dust in the Barbados trade wind', *Nature* **320**, 735–738.

Ramanathan, V., *et al.*: 2001, 'The Indian Ocean Experiment: An integrated analysis of the climate forcing and effects of the great Indo-Asian haze', *J. Geophys. Res.* **106**, 28,371–28,398.

Reddy, M. S., *et al.*: 2005, 'Global multi-component aerosol optical depths and radiative forcing estimates in the LMDZT general circulation model', *J. Geophys. Res.*, in press.

Remer, L. A., et al.: 2005, 'The MODIS aerosol algorithm, products and validation', *J. Atmos. Sci.* **62**, 947–973.

Takemura, T., *et al.*: 2002, 'Single-scattering albedo and radiative forcing of various aerosol species with a global three-dimensional model', *J. Climate* **15**, 333–352.

Tanré, D., Kaufman, Y. J., Herman, M., and Mattoo, S.: 1997, 'Remote sensing of aerosol over oceans from EOS-MODIS', *J. Geophys. Res.* **102**, 16,971–16,988.

# AEROSOL-CLOUD INTERACTIONS CONTROL OF EARTH RADIATION AND LATENT HEAT RELEASE BUDGETS

D. ROSENFELD
*Institute of Earth Sciences, The Hebrew University of Jerusalem, Jerusalem 91904, Israel*
*(E-mail: daniel.rosenfeld@huji.ac.il)*

(Received 6 November 2005; Accepted in final form 2 February 2006)

**Abstract.** Aircraft observations and model simulations show that cloud development is strongly modulated by the impact of cloud-aerosol interactions on precipitation forming processes. New insights into the mechanisms by which aerosols dominate the cloud cover of marine shallow clouds suggest that feedbacks between the cloud microstructure and cloud dynamics through precipitation processes play a major role in determining when a solid cloud cover will break up into a field of trade wind cumulus. Cloud-aerosol interactions dominate not only the dynamics of marine shallow clouds, but also the lifetime and the vertical disposition of latent heat of deep convective clouds over ocean and even more strongly over land. Recent coincident satellite measurements of aerosols and cloud properties quantify the aerosol effects on cloud cover and radiative forcing on regional and global scales. The shapes of the satellite retrieved relations between aerosols and cloud properties are consistent with the suggested ways by which aerosols affect clouds via precipitation processes, particularly by affecting the intensity of the cloud vertical air motions and its vertical development.

**Keywords:** climate change, cloud-aerosol-interactions, precipitation

## 1. The Major Role of Clouds and Precipitation in the Earth's Energy Budget

Clouds play a major role in energizing the climate system: they reflect much of the solar radiation back to space and so modulate the global temperature. Nearly half of the remaining radiation that reaches the surface is consumed for evaporation; only when the reverse to evaporation occurs, i.e., precipitation, the latent heat in the vapor is released, warms the air and drives the global winds and weather systems. Here we will briefly review the key role of cloud-aerosol-precipitation interactions and the ways by which aerosols affect precipitation, which in turn plays a major role in the atmospheric energy budget. This can be appreciated by reviewing the components of the Earth energy budget as published in the IPCC report (IPCC, 2001): clouds and aerosols reflect back to space 22.5% of the solar radiation (Figure 1). An additional 8.8% is reflected by the surface. Out of the remaining 68.7% only 19.6% are absorbed by the atmosphere and heats it directly and 49.1% are absorbed by the surface. Surface evaporation consumes 22.8% of the solar energy that subsequently heats the atmosphere when condensation and precipitation occurs, whereas 18.7% is used for direct heating of the atmosphere

Space Science Reviews (2006) 125: 149–157
DOI: 10.1007/s11214-006-9053-6

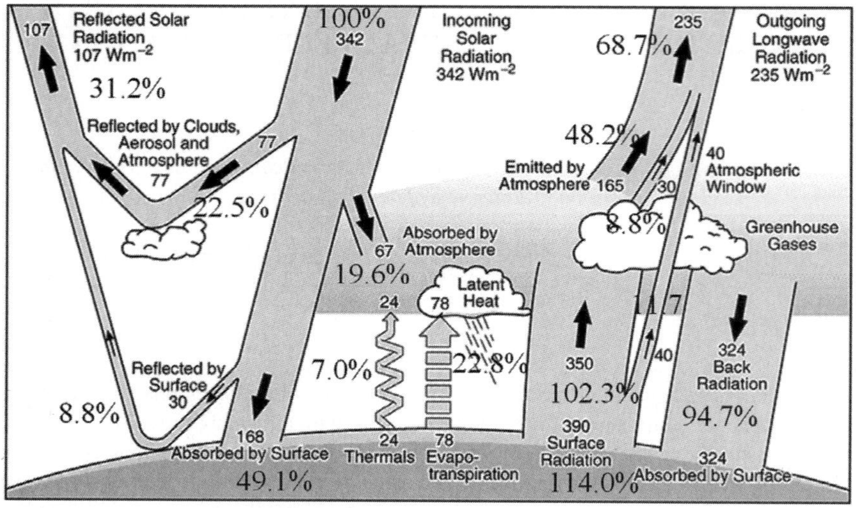


*Figure 1.* The Earth's energy budget. From IPCC (2001), Chapter 1.2.1.

by conduction and rising air (7%) and by thermal radiation (11.7%). The rest, 7.6%, escapes from the surface to space as thermal radiation.

This means that 37.3% of the atmospheric heating results where and when precipitation occurs and deposits in the atmosphere the 23% of the solar energy that was consumed for evaporation from the surface, whereas the atmospheric heating from all sources combined is 61.1% of the solar energy.

The role of aerosols was considered important mainly for shallow clouds, by acting as cloud condensation nuclei (CCN) and so determining the initial cloud drop concentrations. The same amount of cloud water is redistributed to a larger number of smaller droplets in more polluted and CCN-laden air. Twomey (1977) showed that by this mechanism aerosols increase the cloud albedo (namely the fraction of sunlight reflected back to space), and so cool the climate system. This effect is confined to shallow clouds because deep clouds reflect most of the solar radiation regardless of their drop size (Twomey, 1977). Albrecht (1989) observed that the aerosol-induced smaller droplets are slower to coalesce into raindrops and so less depletion of cloud water occurs. Based on this, Albrecht suggested that aerosols should increase cloud lifetime and area coverage. Again, this effect was considered only for shallow clouds because deep clouds were thought to precipitate always anyway.

The clouds albedo effect received much attention, while the existence of the cloud lifetime effect was acknowledged but its quantification was elusive. The next Sections provide some recent measurements suggesting that the cloud albedo effect is dwarfed compared to the cloud lifetime effect, which is a results of the aerosols suppressing precipitation and leaving the water floating in the air for longer in the form of cloud droplets.

The change in precipitation affects the latent heat release and redistributes the timing and location of this substantial component of the atmospheric heat source.

Therefore, not taking into account the cloud-aerosol-precipitation interactions misses a key component that regulates the climate system. These cloud-aerosol-precipitation processes are not taken into account as they should, not because we do not recognize their importance, but rather because we know too little on how to quantify them accurately in the weather and climate models.

## 2. Quantifying the Relative Magnitudes of the Albedo and Cloud Cover Effects

Recent satellite comparisons of aerosols with the albedo and cover of shallow clouds revealed that the radiative forcing at the top of the atmosphere (TOA) due to increased cloud cover, $F_C$, is more than 4 times greater than the forcing due to enhanced cloud albedo, $F_A$ (Sekiguchi *et al.*, 2003). New Moderate Resolution Imaging Spectroradiometer (MODIS)-based detailed satellite observations over the Atlantic Ocean, using higher spatial resolution and therefore including the aerosol effect on broken clouds, show that the overall radiative forcing at the top of the atmosphere incurred by the aerosol effect on the shallow clouds and solar radiation is $-11 \pm 3$ Wm$^{-2}$ for the 3 months studied; two thirds of it is due to the aerosol-induced cloud changes, and one third is due to aerosol direct radiative effect (Kaufman *et al.*, 2005). That study showed three times as strong effect on cloud fraction and radiative forcing over that Atlantic Ocean compared to the results of Sekiguchi *et al.* (2003). Kaufman *et al.* (2005) showed that $F_A$ was <20% of the combined cloud forcing ($F_C + F_A$). Only about half of $F_A$, i.e., ~10% of the combined cloud forcing, can be explained by the Twomey effect (Twomey, 1977), which was previously believed to be the dominant effect of aerosols on clouds. The other half of $F_A$ can be explained by decreasing loss of cloud water to precipitation with higher aerosol concentrations.

Figure 2 (from Rosenfeld *et al.*, 2006) shows that the cloud cover increases with the satellite retrieved aerosol optical depth (AOD) up to AOD = 0.3 in a remarkably similar way at four very different climatic and aerosol regimes. In parallel the satellite retrieved cloud drop effective radius ($R_{\mathrm{eff}}$) decreases at AOD = 0.3 to below 15 $\mu$m. At lower AOD the retrieved $R_{\mathrm{eff}}$ increases to values >15 $\mu$m, which is indicative for precipitating clouds (Rosenfeld and Gutman, 1994; Gerber, 1996). This relation suggests that decreased cloud cover with larger cloud droplets is related to increased precipitation associated with cloud water removal.

## 3. The Mechanism by which Aerosols Dominate the Convective Regime of the MBL

The dependence of cloud cover on AOD is much stronger than on any other potential meteorological factor that was tested. The best correlated meteorological

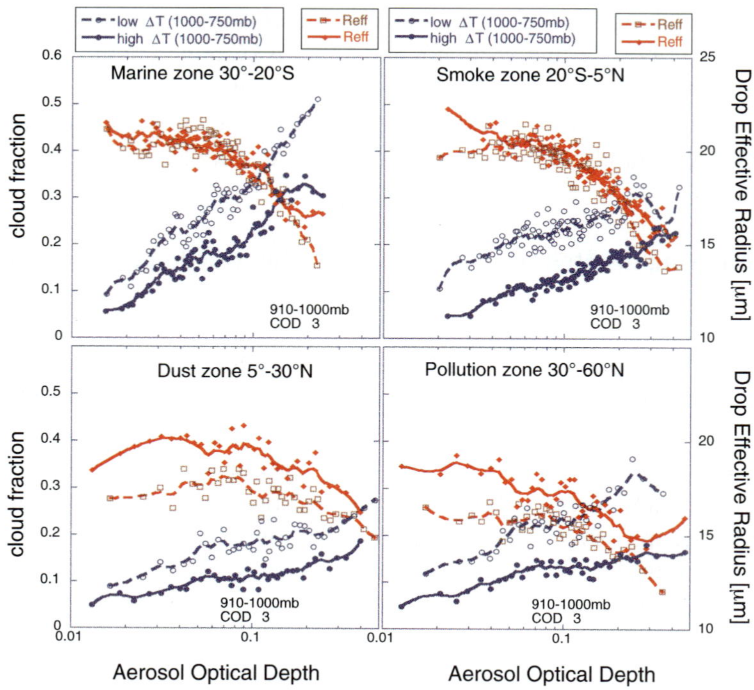


*Figure 2.* Cloud cover fraction (CF, blue lines with circles) and drop effective radius ($R_{\text{eff}}$, red lines with rhombs) of shallow clouds as a function of aerosol optical depth (AOD) for 4 zones over the Atlantic Ocean during August 2002, as defined by Kaufman *et al.* (2005). The clouds were separated based on stability. Greater stability, which is indicated by smaller ΔT(1000–750 mb) and denoted by the broken lines, promotes greater shallow cloud cover. Greater stability also produces shallower clouds with smaller $R_{\text{eff}}$ for the same AOD. Remarkably, the CF increases with AOD as much as with stability. Note that $R_{\text{eff}}$ increases and CF decreases strongly at the small values of AOD, most clearly at the pristine subtropical latitudes of the south Atlantic, supporting the runaway rainout effect of clouds and aerosols. From Rosenfeld *et al.* (2006).

parameter is the difference between surface and 750 mb temperatures, ΔT(1000–750 mb). Smaller difference means stronger static stability and better conditions for layer clouds at the marine boundary layer (MBL). Therefore ΔT(1000–750 mb) is correlated positively with greater cloud cover of marine shallow clouds. Even so, the AOD appears to have stronger effect on the cloud cover than the low level static stability. This is a remarkable observation demonstrating the dominant role that the internal cloud processes have in determining the properties of the marine shallow clouds. This is underlined by the statement of Wood and Hartmann (2006), who wrote:

> *"In regions where mesoscale cellular convection (MCC) is prevalent, we found a lack of sensitivity of the MCC type (open or closed) to the large scale meteorology, suggesting a mechanism internal to the MBL may be important in determining MCC type. The results*

*indicate that a knowledge of the physics of mesoscale cellular convection (MCC) will be required to completely understand and predict low cloud coverage and variability in the subtropics."*

Rosenfeld *et al.* (2006) suggested that this missing mechanism involves the transition from open to closed Benard cells. When precipitation is suppressed a full cloud cover can persist. Under such conditions most of the radiative cooling occurs from the cloud tops. This sets in motion an inverse cellular convection that is energized by cloud top cooling. In such convection the downward motion occurs on the polygons that define the borders of the cells, whereas the compensating rising air occurs at the rest of the area, which includes the center of the cells. This way most of the area is occupied by the rising branch, and hence most of the area is cloudy. This contributes to continuation of the radiative cooling from the cloud tops and self-maintenance of this regime of closed Benard cells.

When aerosol concentrations decreases such that cloud droplet $R_{\mathrm{eff}}$ increases above the precipitation threshold, the cloud water is lost to precipitation and it can no longer support a full cloud cover even under closed Benard convective regime. The radiative cooling is no longer concentrated at the cloud top altitude, but rather comes from deeper in the MBL. The cooled air becomes colder than the sea surface temperature to the extent that heat flux from the surface to the bottom of the MBL dominates over the radiative cooling of the top of the MBL. Under such conditions the surface heating dominates the propulsion of the convection, and the regime inverses from closed to open Benard cells, where the rising motion occurs at the polygons that form the border of the cells and the compensating downward motion occurs at the center of the cells. Therefore most of the area is occupied with downward air motion, which maintains the partly cloudy conditions and allows continuation of the radiative cooling from within the MBL. This so self maintains the regime of open Benard cells. Furthermore, the more concentrated updrafts over smaller areas of the open cells allow more efficient precipitation from the clouds and so further maintaining of this regime.

The self cleansing and cloud clearing feedback processes can lead to a runaway effect, in which cloud-free areas are maintained because of lack of CCN as their removal occurs at a rate that exceeds their replacement. It is in those regions that a large sensitivity of cloud cover to cosmic rays-generated CCN might potentially exist. This might provide the missing physical link between cosmic rays and the climate system that was suggested by Svensmark and Friis-Christensen (1997) and Carslaw *et al.* (2002). See Rosenfeld *et al.* (2006) for additional detail.

## 4. Aerosols Forcing of Deep Convective Clouds

Aerosol effects similar to those documented for marine stratocumulus were not anticipated for deep clouds because deep convection was believed to be dominated by atmospheric dynamics and not by aerosols. This perception was challenged by the author of this paper in observational (Williams *et al.*, 2002; Andreae *et al.*, 2004) and simulation (Khain *et al.*, 2005) studies, where he suggested that in very clean air, the fast formation of precipitation would cause early rainout from the lower levels of the clouds and so deprive the water and the respective latent heat of freezing from the upper portions of the clouds. Early rain formation also replaces the updrafts, which feed the clouds with heat and moisture, with downdrafts. This process leads to early maturation and dissipation of clouds, and it explains how aerosols affect cloud lifetime and cover mainly as a response to the dynamic feedbacks of the cloud to the aerosol-induced changes in the precipitation forming processes. The other side of the same coin is observed in CCN-rich clouds where the water ascends to higher altitude and freezes before precipitating. This results in additional latent heat release at greater heights, inducing higher and more extensive clouds and anvils. This additional ice precipitation melts at the lower elevations and takes up there additional heat. This added cooling enhances the downdrafts and so further invigorates and organizes of the storm system. The overall effect is stronger convective overturning for the same atmospheric lapse rate, which results in greater consumption of the instability. Energetically, greater upward transport of heat occurs for the same amount of precipitation reaching the surface. This is a fundamental change in the precipitation amount that is required to compensate a given radiative cooling of the troposphere. This must result in changes in the atmospheric lapse rate and overall Earth energy budget and global circulation. This can be measured in units of $Wm^{-2}$, so that the anthropogenic component of this energy can be defined as *latent heat* forcing.

A cloud-resolving climate model over a tropical ocean planet (Grabowski, 2003) simulated the aerosol effects by prescribing smaller cloud drops as surrogate for higher CCN concentrations. The model reproduced the large observed aerosol-induced increase in cloud cover, which occurred equally with height, except for a maximum in the anvil levels, similar to the observations of Koren *et al.* (2005) shown in Figure 3. The model also replicates the increase in cloud top height for the clouds with smaller drops. This lends credibility to other predictions of this model, suggesting that enhanced aerosols reduce the lapse rate (i.e., the convective available potential energy) and the surface fluxes of sensible and latent heat. This in turn must result in reduced precipitation and slowing down of the hydrological cycle. This occurs in addition to the similar effect of absorbing aerosols that block part of the solar radiation from reaching the surface and so further reducing evaporation and precipitation (Ramanathan *et al.*, 2001).

Key elements of the proposed mechanism were quantified using MODIS measurements by Koren *et al.* (2005). The results reproduced previous observations indicating that more polluted clouds are composed of smaller particles, and for the first time showed that such clouds grow to higher altitudes, cover more area, and have a larger ice anvil portion (see Figure 3).

These satellite observations of Koren *et al.* (2005) and Kaufman *et al.* (2005) were made over ocean because aerosols can be measured by MODIS with much

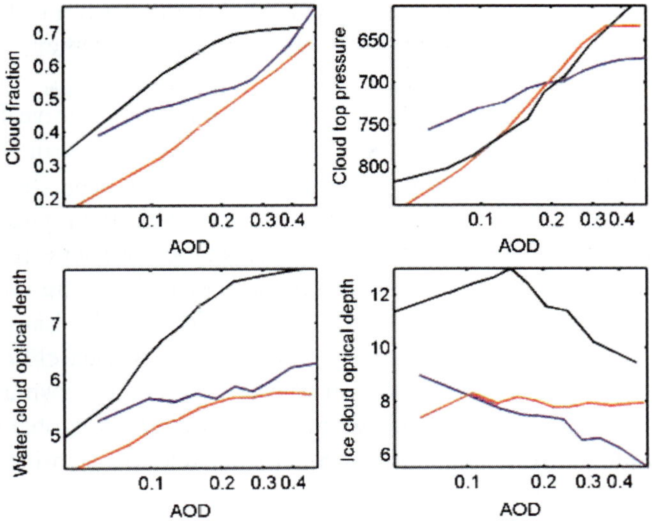


*Figure 3.* Cloud properties as a function of AOD for the three regions in the Atlantic Ocean. Blue: tropics (ITCZ); red: subtropics (STC); black: mid-latitudes (MLC). For each zone the data are sorted as a function of the AOD and averaged into 10 sequential subgroups. The upper left plot shows the cloud fraction as function of the AOD (in logarithmic scale). The upper right plot shows the cloud top pressure vs AOD. The left and right plots on the lower row show the water cloud optical depth (COD) and the ice COD as function of AOD respectively. The water optical depth increases with AOD but the ice COD decreases or remains constant, suggesting formation of more anvils that increase the ice cloud fraction but decrease the average COD. From Koren *et al.* (2005).

higher accuracy than over land. This does not mean that the profound effects that are first quantified using MODIS measurements over ocean are limited to the ocean atmosphere. Aerosol precipitation effects over land are expected to be even greater for deep clouds due to the lack of large salt CCN generated by sea spray that nucleate large drops and so enhances the precipitation in polluted clouds (Rosenfeld *et al.*, 2002). On the other hand, the effect on shallow clouds is expected to be greater over ocean, because very shallow clouds are susceptible to rainout only in very clean air that rarely occurs over land.

## 5. Conclusions

The emerging picture is of aerosols strongly regulating the cloud cover and depth by suppressing rainout and its dynamical feedbacks, which are potent mechanisms for dissipating clouds of all depths. This effect dominates the cloud-mediated radiative forcing at TOA over the north Atlantic, which amounts to about $-5$ $Wm^{-2}$ between the baseline AOD of 0.06 and the observed aerosol conditions. Unlike the previous belief that almost all of this forcing comes from shallow clouds, much is due to the aerosol cloud cover effect on scenes containing deep clouds. Furthermore, less than 20% of this forcing comes from the Twomey effect (Twomey, 1977), which operates mainly on shallow clouds.

The ramifications of the findings go much beyond the radiative forcing. The aerosol control of the precipitation processes affects in turn the vertical distribution of latent heat release and the tropospheric lapse rate, which drive the atmospheric circulation and feedbacks to the distribution of clouds and precipitation.

The mechanism of runaway cleansing via rainout processes of the marine boundary layer creates super-clean cloud free areas in which the cloud cover is likely extremely susceptible to additions of CCN, as evident by the long-lasting ship tracks in the otherwise cloud-free areas (see Figure 4 of Rosenfeld *et al.*, 2006). We suggest that this is where the addition of CCN induced by cosmic rays might make a big difference in the cloud cover, and so modulate the global temperature via affecting cloud cover as suggested by Svensmark and Friis-Christensen (1997) and by Carslaw *et al.* (2002).

It appears that we cannot get the climate system right without properly accounting for the aerosol-cloud-precipitation processes, the dynamic response of the clouds and the cascade of feedbacks. However, the representation of these aerosol effects is still very crude or mostly missing in numerical weather prediction and global climate models. It is time to go back to the drawing board in designing the next field campaigns and climate simulations.

## References


Albrecht, B. A.: 1989, 'Aerosols, cloud microphysics, and fractional cloudiness', *Science* **245**, 1227–1230.

Andreae, M. O., Rosenfeld, D., Artaxo, P., Costa, A. A., Frank, G. P., Longo, K. M., and Silva-Dias, M. A. F.: 2004, 'Smoking rain clouds over the Amazon', *Science* **303**, 1337–1342.

Carslaw, K. S., Harrison, R. G., and Kirkby, J.: 2002, 'Cosmic rays, clouds, and climate', *Science* **298**, 1737–1739.

Gerber, H.: 1996, 'Microphysics of marine stratocumulus clouds with two drizzle modes', *J. Atmos. Sci.* **53**, 1649–1662.

Grabowski, W. W.: 2003, 'Impact of cloud microphysics on convective–radiative quasi equilibrium revealed by cloud-resolving convection Parameterization', *J. Clim.* **16**, 3463–3475.

IPCC: 2001, 'Climate change 2001: The scientific basis', Contribution of Working Group I to the Third Assessment Report of the Intergovernmental Panel on Climate Change, J. T. Houghton, Y. Ding, D. J. Griggs, M. Noguer, P. J. van der Linden, X. Dai, K. Maskell, and C. A. Johnson (eds.), Cambridge University Press, Cambridge, UK, 881 pp.

Khain, A., Rosenfeld, D., and Pokrovsky, A.: 2005, 'Aerosol impact on the dynamics and microphysics of convective clouds', *Q. J. R. Meteorol. Soc.* **131**, 1–25.

Kaufman, Y. J., Koren, I., Remer, L. A., Rosenfeld, D., Rudich, Y.: 2005, 'Smoke, dust and pollution aerosol clouding the atlantic atmosphere', *Proc. Natl. Acad. Sci. USA* **102**, 11,207–11,212.

Koren, I., Kaufman, Y. J., Rosenfeld, D., Remer, L. A., and Rudich, Y.: 2005, 'Aerosol invigoration and restructuring of Atlantic convective clouds', *Geophys. Res. Lett.* **32**, doi:10.1029/2005GL023187.

Ramanathan, V., Crutzen, P. J., Kiehl, J. T., and Rosenfeld, D.: 2001, 'Aerosols, climate and the hydrological cycle', *Science* **294**, 2119–2124.

Rosenfeld, D., and Gutman, G.: 1994, 'Retrieving microphysical properties near the tops of potential rain clouds by multispectral analysis of AVHRR data', *Atmos. Res.* **34**, 259–283.

Rosenfeld, D., Lahav, R., Khain, A. P., and Pinsky, M.: 2002, 'The role of sea-spray in cleansing air pollution over ocean via cloud processes', *Science* **297**, 1667–1670.

Rosenfeld, D., Kaufman, Y. J., and Koren, I.: 2006, 'Switching cloud cover and dynamical regimes from open to closed Benard cells in response to the suppression of precipitation by aerosols', *Atmos. Chem. Phys.* **6**, 2503–2511.

Sekiguchi, M., Nakajima, T., Suzuki, K., *et al.*: 2003, 'A study of the direct and indirect effects of aerosols using global satellite data sets of aerosol and cloud parameters', *J. Geophys. Res.* **108**, doi:10.1029/2002JD003359.

Svensmark, H., and Friis-Christensen, E.: 1997, 'Variation of cosmic ray flux and global cloud coverage – a missing link in solar-climate relationships', *J. Atmos. Terr. Phys.* **59**, 1225–1232.

Twomey, S. A.: 1977, 'The influence of pollution on the shortwave albedo of clouds', *J. Atmos. Sci.* **34**, 1149–1152.

Williams, E., Rosenfeld, D., Madden, M., *et al.*: 2002, 'Contrasting convective regimes over the Amazon: Implications for cloud electrification', *J. Geophys. Res.* **107**, doi:10.1029/2001JD000380.

Wood, R., and Hartmann, D. L.: 2006, 'Spatial variability of liquid water path in marine low cloud: The importance of mesoscale cellular convection', *J. Clim.* **19**, 1748–1764.

# ATMOSPHERIC ION-INDUCED AEROSOL NUCLEATION

J. CURTIUS[1,*], E. R. LOVEJOY[2] and K. D. FROYD[3]
[1]*Institut für Physik der Atmosphäre, Universität Mainz, Germany*
[2]*NOAA ESRL Chemical Sciences Division, Boulder, CO, USA*
[3]*Cooperative Institute for Research in the Environmental Science, University of Colorado, Boulder, CO, USA*
(**Author for correspondence: E-mail: curtius@mail.uni-mainz.de*)

(Received 13 September 2005; Accepted in final form 16 November 2005)

**Abstract.** Ion-induced nucleation has been suggested to be a potentially important mechanism for atmospheric aerosol formation. Ions are formed in the background atmosphere by galactic cosmic rays. A possible connection between galactic cosmic rays and cloudiness has been proposed. However, the predictions of current atmospheric nucleation models are highly uncertain because the models are usually based on the liquid drop model that estimates cluster thermodynamics based on bulk properties (e.g., liquid drop density and surface tension). Sulfuric acid ($H_2SO_4$) and water are assumed to be the most important nucleating agents in the free troposphere. Measurements of the molecular thermodynamics for the growth and evaporation of cluster ions containing $H_2SO_4$ and $H_2O$ were performed using a temperature-controlled laminar flow reactor coupled to a linear quadrupole mass spectrometer as well as a temperature-controlled ion trap mass spectrometer. The measurements were complemented by quantum chemical calculations of the cluster ion structures. The analysis yielded a complete set of $H_2SO_4$ and $H_2O$ binding thermodynamics extending from molecular cluster ions to the bulk, based on experimental thermodynamics for the small clusters. The data were incorporated into a kinetic aerosol model to yield quantitative predictions of the rate of ion-induced nucleation for atmospheric conditions. The model predicts that the negative ion-$H_2SO_4$-$H_2O$ nucleation mechanism is an efficient source of new particles in the middle and upper troposphere.

**Keywords:** ion-induced nucleation, aerosol, clusters ions, particles, atmosphere, sulfuric acid

## 1. Introduction

Formation of new aerosol particles from gas phase precursors has been observed frequently in the atmosphere (Kulmala *et al.*, 2004). However, a general quantitative understanding of the aerosol nucleation mechanisms has not been achieved. Field measurements indicate sulfuric acid and water to be important nucleating agents in the atmosphere, especially in the free troposphere (e.g., Weber *et al.*, 1999).

Ions are ubiquitous in the lower atmosphere, mainly produced by galactic cosmic rays (GCR) forming 1–30 ion pairs $cm^{-3}s^{-1}$, with maximum formation rates at ~10 km (Rosen *et al.*, 1985). The ion lifetime with respect to recombination of positive and negative ions is typically hundreds of seconds. Other ion sources, such as radioactive decay, lightning, power transmission lines and combustion

Space Science Reviews (2006) 125: 159–167
DOI: 10.1007/s11214-006-9054-5

can enhance the ion concentrations locally. Nucleation mechanisms involving ions have been proposed to be important for aerosol formation in the atmosphere (e.g., Arnold, 1980; Yu and Turco, 2001). Recently, Eichkorn *et al.* (2002) detected large ions in the upper troposphere providing evidence for ion-mediated nucleation processes.

Marsh and Svensmark (2000) found a close correlation between the variation of galactic cosmic rays caused by the 11-year solar cycle and low-level cloud coverage above oceans. This correlation led to speculations about a connection between solar variability and climate via GCR, ion-induced aerosol nucleation and cloud formation (e.g., Carslaw *et al.*, 2002).

It is interesting to note that out of the hundreds of trace substances present in the atmosphere apparently only very few substances are able to form new particles. As a precondition for the formation of new particles from the gas phase, low volatility products must form in the gas phase from higher volatility precursors in sufficiently large amounts. Sulfuric acid, produced in the gas phase in a series of reactions from sulfur dioxide ($SO_2$), has a low vapor pressure which is further reduced in the presence of water vapor and ammonia. Other potential substances able to form new particles in the atmosphere are iodine oxides produced from the photooxidation of iodine compounds emitted by algae in coastal areas during low tide (O'Dowd *et al.*, 2002), as well as highly oxidized higher organics, such as oxidation products of sesquiterpenes which are emitted by vegetation (Bonn and Moorgart, 2003).

An atmospheric system such as an air parcel at a certain temperature and pressure is always striving to reach thermodynamic equilibrium conditions by lowering its Gibbs free energy towards a minimum. If a substance is supersaturated in the gas phase then the Gibbs free energy of the system is lowered by a transfer of molecules into the condensed phase. Supersaturated gases will preferentially condense on preexisting surfaces. The formation and spontaneous growth of new particles is generally hindered by a nucleation barrier, as the formation of a new particle means that a new surface has to be formed which consumes energy. For clusters smaller than the critical cluster (see Figure 1) the addition of a molecule costs more energy to enlarge the surface than energy is won from the phase change, and the cluster will evaporate. The nucleation barrier therefore usually prevents significant particle formation in the atmosphere, even when substances are supersaturated.

Ions can act as nucleation cores for the attachment of the supersaturated gas phase molecules. In this case, the nucleation barrier is reduced due to the stabilizing electrostatic interaction between the ionic charge and the ligand molecules. The reduction in barrier height makes the ion-induced process thermodynamically more favorable. A schematic representation of the neutral and ion-induced nucleation processes in the atmosphere is shown in Figure 1 (top). The thermodynamics for these processes are illustrated in Figure 1 (bottom). For simplicity, the system is reduced to just one chemical component. The cumulative Gibbs free energy change to transfer $i$ molecules from the gas phase to condense

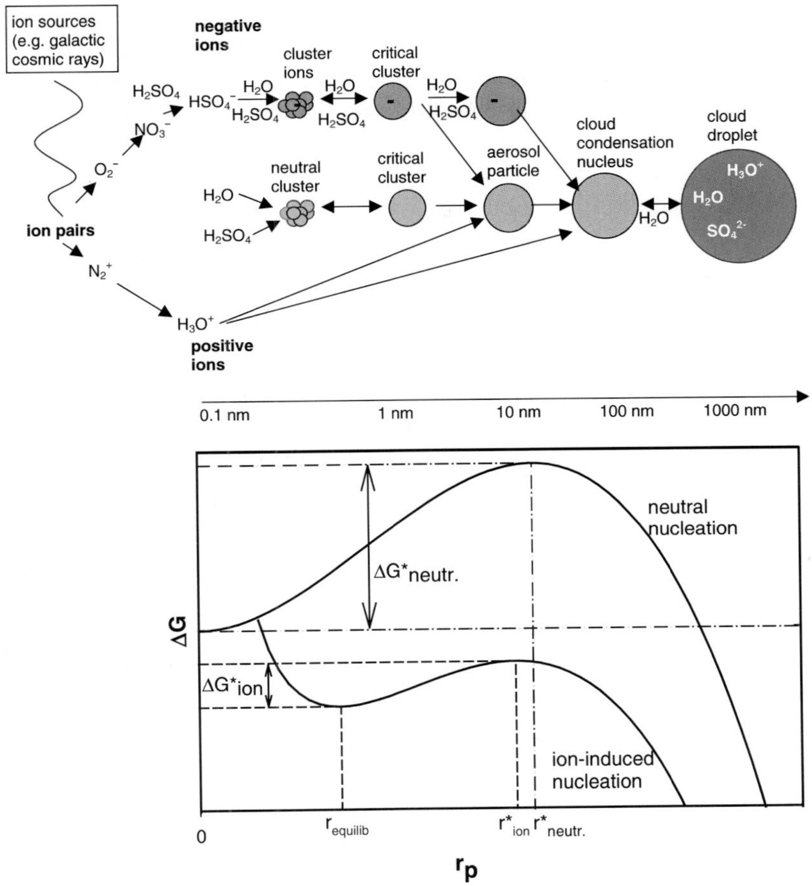


*Figure 1. Top panel:* Schematic representation of the formation and growth of aerosol particles by neutral nucleation of sulfuric acid and water and by ion-induced nucleation via negative ions. *Bottom panel:* Schematic representation of the cumulative change in Gibbs free energy as a function of cluster radius. Nucleation barriers ($\Delta G^*$) and radius of the critical cluster ($r^*$) for the neutral and ion-induced nucleation pathway are indicated.

around an ion is represented in classical nucleation theory by the Thomson equation:

$$\Delta G = -i\,k\,T\,\ln(S) + 4\pi\,r_i^2\,\sigma + \frac{q^2}{2}\left(1 - \frac{1}{\varepsilon}\right)\left(\frac{1}{r_i} - \frac{1}{r_0}\right)$$

Here $i$ is the number of molecules in a particle of radius $r_i$, $k$ is the Boltzmann constant, $T$ is the temperature, $S$ is the saturation ratio with respect to the bulk phase, $\sigma$ is the surface tension, $q$ is the unit charge, $\varepsilon$ is the dielectric constant and $r_0$ is the radius of the core ion (e.g., Laaksonen *et al.*, 1995). The first term describes the difference in $\Delta G$ between gas phase and the bulk condensed phase, the second term represents the surface energy and the third term gives the electrostatic interaction between the core ion and the ligand molecules. Several quantities derived from bulk phase properties enter this classical equation, e.g., the density $\rho$ (to calculate the number of molecules $i$ per volume) and the surface tension $\sigma$. As clusters grow in size to resemble bulk-phase liquid droplets, the Thomson model should accurately predict cluster thermodynamics. However, because these bulk phase properties are generally not suitable for the description of clusters consisting of just a few molecules, large discrepancies between the predictions of the classical nucleation theory and experimental nucleation measurements occur frequently, (see, e.g., Ball *et al.*, 1999; Wyzlozil *et al.*, 1991, for a comparison for neutral binary nucleation of $H_2SO_4$ and $H_2O$).

To improve the modeling of ion-induced nucleation, we determined the thermodynamic standard changes in enthalpy $\Delta H^0$, entropy $\Delta S^0$ and Gibbs free energy $\Delta G^0$ for each molecular association/dissociation step of the bisulfate ion-sulfuric acid-water system for up to 6 sulfuric acid molecules and up to 10 water molecules by laboratory measurements and quantum chemical calculations (Curtius *et al.*, 2001; Lovejoy and Curtius, 2001; Froyd and Lovejoy, 2003). From these measurements the $\Delta G$ curve shown in Figure 1 (bottom) can be assembled without the use of the classical Thomson equation for the smallest clusters. The thermodynamic dataset was then incorporated in a kinetic nucleation model (Lovejoy *et al.*, 2004) to obtain predictions on the relevance of ion-induced nucleation processes in the atmosphere.

## 2. Measurements

The enthalpy changes for the clustering reactions of the bisulfate ion with sulfuric acid

$$HSO_4^-(H_2SO_4)_{s-1} + H_2SO_4 \leftrightarrow HSO_4^-(H_2SO_4)_s$$

for $s \leq 5$ were derived from a master equation analysis of the temperature and pressure dependence of the thermal decomposition reactions measured in a quadrupole ion trap mass spectrometer (Curtius *et al.*, 2001; Lovejoy and Curtius, 2001). The enthalpy and entropy changes for the clustering reactions of water molecules

$$HSO_4^-(H_2SO_4)_s(H_2O)_{w-1} + H_2O \leftrightarrow HSO_4^-(H_2SO_4)_s(H_2O)_w$$

for $s \leq 6$ and $w \leq 10$ were derived from the temperature dependence of the equilibrium constants measured in an ion-molecule flow reactor coupled to a quadrupole mass spectrometer (Froyd and Lovejoy, 2003). The measurements were accompanied by quantum chemical calculations to obtain information about the ion cluster structures and to calculate the cluster's vibrational frequencies, rotational moments and the reaction entropies. Calculations were performed at the HF level using 6−31+G(d) basis sets. The remaining reaction enthalpies and entropies for sulfuric

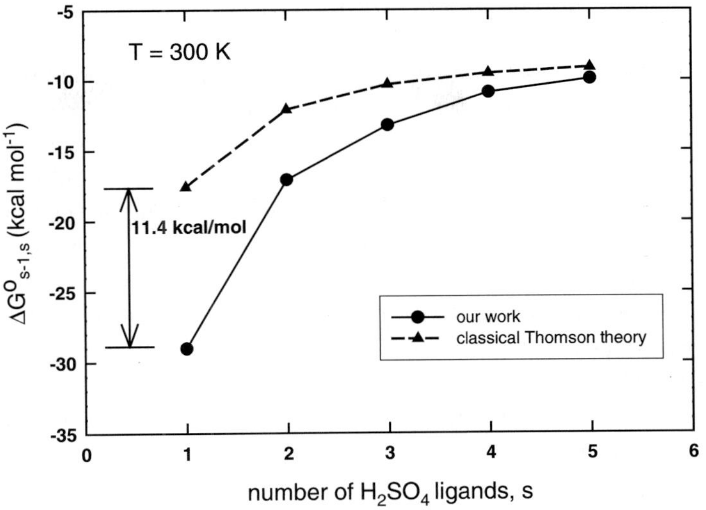


*Figure 2.* Comparison of $\Delta G^0_{s-1,s}$ values for binding of an $H_2SO_4$ ligand to the cluster ion $HSO_4^-(H_2SO_4)_s$ as determined from measurements and quantum chemical calculations (solid line), and derived from classical Thomson theory (dashed line).

acid clustering to the hydrated clusters were derived from the experimental results by using thermodynamic cycles

$$HSO_4^-(H_2SO_4)_{s-1}(H_2O)_w + H_2SO_4 \leftrightarrow HSO_4^-(H_2SO_4)_s(H_2O)_w$$

for $s \leq 6$ and $w \leq 10$. The set of measured cluster ion thermodynamics was connected to classical predictions for larger clusters by adding small terms to the Thomson equation that decayed exponentially with cluster size. This analysis yielded a complete set of $H_2SO_4$ and $H_2O$ binding thermodynamics extending from molecular cluster ions to the bulk, based on experimental thermodynamics for the small clusters.

Figure 2 shows the $\Delta G^0$ values at 300 K derived from the measurements and quantum chemical calculations for the pure sulfuric acid clusters in comparison to predictions from the classical Thomson equation. It can be seen that large differences exist for the smallest clusters that can lead to differences of several orders of magnitude in the predicted nucleation rates. On the other hand, already at a cluster size of 5 ligand molecules, Thomson theory predictions and experimentally determined molecular thermodynamics agree to better than 1 kcal/mol, justifying the assumption of an adjusted Thomson theory for the large clusters ($s > 6$, $w > 10$). For the binary ion clusters the sulfuric acid binding was generally under predicted by the Thomson equation whereas the water binding was over predicted by up to 7 kcal $mol^{-1}$.

## 3. Kinetic Model of Ion-Induced Nucleation

A numeric nucleation model (Lovejoy *et al.*, 2004) was developed that treats the kinetics of growth and evaporation of neutral and ionic clusters explicitly. Positive ions are treated as a single species whereas for the neutral and negative clusters, the model uses 20–40 bins that increment by one sulfuric acid molecule, representing hydrated $(H_2SO_4)_n$ and $HSO_4^-(H_2SO_4)_{n-1}$, respectively. The molecular resolution of these bins gives a full kinetic treatment of nucleation based on the measured thermodynamics. In the next 40–60 bins the number of sulfuric acid molecules increases geometrically to account for particles up to about 1 $\mu$m diameter. All clusters are assumed to equilibrate immediately with water vapor while they grow and evaporate by addition and loss of $H_2SO_4$. This assumption is justified as atmospheric water vapor concentrations exceed gaseous sulfuric acid concentrations by several orders of magnitude. Coagulation of ionic and neutral clusters, ion recombination as well as competing processes such as uptake of gas phase $H_2SO_4$ on a preexisting aerosol distribution are considered in the model explicitly when solving the set of stiff differential equations. The neutral nucleation pathway is modeled by applying classical nucleation theory as thermodynamic data are not available. For the neutral nucleation the classical theory was adjusted to be consistent with the experimental results of Ball *et al.* (1999). Using these settings, growing neutral clusters could not surmount the nucleation barrier over the entire range of atmospheric background conditions but it has to be emphasized that the predictions for the neutral system have very large uncertainties. However, negative cluster ions can react with positive ions to produce a neutral cluster larger than the critical cluster, which can then continue to grow along the neutral pathway and therefore it is important to include the neutral clusters in the model (Figure 1, top).

## 4. Results

The model predicts particle nucleation rates in the atmosphere as a function of $H_2SO_4$ production rate, relative humidity, ion production rate, preexisting aerosol surface area and temperature. The $H_2SO_4$ production rate is usually given by a semi-sinusoidal cycle, defined by the daily maximum $H_2SO_4$ production rate and the length of day. The daily maximum $H_2SO_4$ concentration and relative humidity needed to produce 1000 particles cm$^{-3}$ >3 nm within 12 hours is shown for a series of temperatures in Figure 3. The contour lines were derived from model calculations using a preexistent aerosol distribution providing surface areas ranging from 0 to 9 $\mu$m$^2$cm$^{-3}$ and midday $H_2SO_4$ production rates ranging from 100 to 18,000 molecule cm$^{-3}$s$^{-1}$. Relative humidity is with respect to ice for T < 273 K. Ion production in the atmosphere varies with altitude, ion pair production rates of 34/32/17/6 cm$^{-3}$s$^{-1}$ at 220/240/260/280 K were derived from the measurements

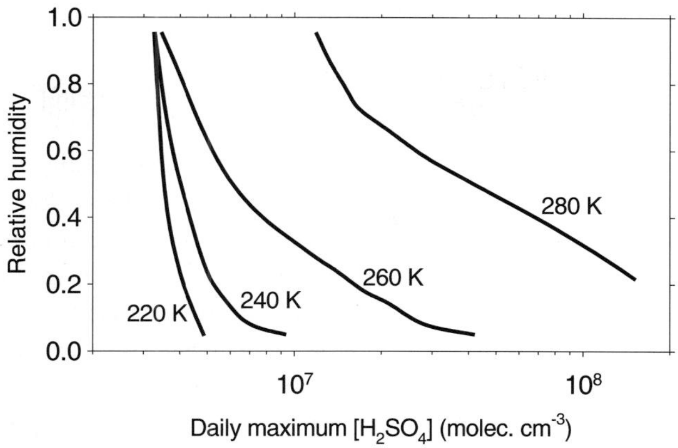


*Figure 3.* Modeled ion-induced nucleation of 1000 particles $cm^{-3}$ larger than 3 nm in diameter formed within 12 hours for different atmospheric temperatures (adapted from Lovejoy *et al.*, 2004).

by Rosen *et al.* (1985) assuming a standard atmosphere temperature and pressure profile. Typical atmospheric $H_2SO_4$ concentrations range at $1 \times 10^6$ to $2 \times 10^7$ $cm^{-3}$ (e.g., Weber *et al.*, 1999). The model calculations show that significant particle production can be expected at conditions typical for the middle and upper troposphere where more than 1000 particles $cm^{-3}$ larger than 3 nm (lower size detection limit of condensation nucleus counters) are produced within a day (Lovejoy *et al.*, 2004).

The model was compared with several observations of nucleation events in the troposphere. As an example, for the case of ACE 1 flight 27 (Clarke *et al.*, 1998) the model predicts up to 35,000 particle $cm^{-3}$ greater than 3 nm in diameter for the conditions measured during this flight (T = 253 K, altitude = 4 km, RH = 45–65%, $H_2SO_4$ = 1.2–1.9 × $10^7$ $cm^{-3}$), in fair agreement with the observed concentration of up to 6,000 ultrafine particles $cm^{-3}$. In general the model reproduces observed particle production for events with atmospheric temperatures below about 270 K, but under predicts the production for most of the lower altitude, higher temperature nucleation events. These lower altitude events may be due to an efficient ionic mechanism involving additional species, for example $NH_3$, or to homogeneous nucleation processes without significant involvement of ions (Lovejoy *et al.*, 2004).

Yu (2002) and Kazil and Lovejoy (2004) have studied the sensitivity of ultrafine particle production to variations in ionization by GCR due to solar variability. The numeric nucleation model by Kazil and Lovejoy (2004) showed for the simplified situation of an average atmosphere that the concentrations of freshly nucleated particles >3 nm are expected to be up to 10% lower during solar maximum than for solar minimum conditions. These calculations describe an average atmosphere situation. It is likely that variations in gaseous $H_2SO_4$ concentration will have a larger influence on the strength of observed nucleation events than the relatively small variation in ion concentration due to variation of GCR by solar activity.

## 5. Summary and Outlook

Ion-induced aerosol nucleation was studied for the sulfuric acid/water system. The thermodynamic properties for attachment/detachment of $H_2SO_4$ and $H_2O$ molecules to $HSO_4^-$ core ions were determined for clusters containing up to 6 sulfuric acid molecules and up to 10 water molecules. Based on these thermodynamics a kinetic model of ion-induced nucleation was developed. The model predicts that ion-induced nucleation processes are expected to be important in the atmosphere, especially in the middle and upper troposphere where temperature is low, ion concentrations are relatively large, pre-existent aerosol surface area is small and gaseous sulfuric acid concentrations are sufficiently large.

Further experimental studies and quantum chemical calculations of the thermodynamic properties of neutral sulfuric acid/water clusters are needed. Thereby, nucleation modeling of the neutral pathway for sulfuric acid and water could be improved. The kinetic ion-induced nucleation model should be validated by further laboratory investigations, chamber studies and field experiments. The role of ternary ion-induced and homogeneous nucleation including ammonia should be addressed in the future. Nucleation around positive ions might also be an efficient aerosol formation mechanism. The role of aerosol nucleation for CCN activation and cloud characteristics needs to be investigated.

## References


Arnold, F.: 1980, 'Multi-ion complexes in the stratosphere – implications for trace gases and aerosol', *Nature* **284**, 610–611.

Ball, S. M., Hanson, D. R., and Eisele, F. L.: 1999, 'Laboratory studies of particle nucleation: Initial results for $H_2SO_4$, $H_2O$, and $NH_3$ vapors', *J. Geophys. Res.* **104**, 23,709–23,718.

Bonn, B. and Moortgart, G. K.: 2003, 'Sesquiterpene ozonolysis: Origin of atmospheric new particle formation from biogenic hydrocarbons', *Geophys. Res. Lett.* **30**, 10.1029/2003GL017000.

Carslaw, K. S., Harrison, R. G., and Kirby, J.: 2002, 'Cosmic rays, clouds, and climate', *Science* **298**, 1732–1737.

Clarke, A. D., Varner, J. L., Eisele, F., Mauldin, R. L., Tanner, D., and Litchy, M.: 1998, 'Particle production in the remote marine atmosphere: Cloud outflow and subsidence during ACE 1, *J. Geophys. Res.* **103**, 16,397–16,409.

Curtius, J., Froyd, K. D., and Lovejoy, E. R.: 2001, 'Cluster ion thermal decomposition (I): Experimental kinetics study and ab intio calculations for $HSO_4^-(H_2SO_4)_x(HNO_3)_y$', *J. Phys. Chem. A* **105**, 10,867–10,873.

Eichkorn, S., Wilhelm, S., Aufmhoff, H., Wohlfrom, K. H., and Arnold, F.: 2002, 'Cosmic ray-induced aerosol-formation: First observational evidence from aircraft-based ion mass spectrometer measurements in the upper troposphere', *Geophys. Res. Lett.* **29**, 10.1029/2002GL015044.

Froyd, K. D., and Lovejoy, E. R.: 2003, 'Experimental thermodynamics of cluster ions composed of $H_2SO_4$ and $H_2O$. 2. Negative ion measurements and ab initio structures', *J. Phys. Chem. A* **107**, 9812–9824.

Kazil, J., and Lovejoy, E. R.: 2004, 'Tropospheric ionization and aerosol production: A model study', *J. Geophys. Res.* **109**, 10.1029/2004JD004852.

Kulmala, M., Vehkamäki, H., Petäjä, T., Dal Maso, M., Lauri, A., Kerminen, V.-M., Birmili, W., and McMurry, P. H.: 2004, 'Formation and growth rates of ultrafine atmospheric particles: A review of observations', *J. Aerosol Sci.* **35**, 143–176, 10.1016/j.jaerosci.2003.10.003.

Laaksonen, A., Talanquer, V., and Oxtoby, D. W.: 1995, 'Nucleation: Measurements, theory, and atmospheric applications', *Ann. Rev. Phys. Chem.* **46** , 489–524.

Lovejoy, E. R., and Curtius, J.: 2001, 'Cluster ion thermal decomposition (II): Master equation modeling in the low pressure limit and fall-off regions. Bond Energies for $HSO_4^-(H_2SO_4)_x\ (HNO_3)_y$', *J. Phys. Chem. A* **105**, 10,874–10,883.

Lovejoy, E. R., Curtius, J., Froyd, K. D.: 2004, 'Atmospheric ion-induced nucleation of sulphuric acid and water', *J. Geophys. Res.* **109**, 10.1029/2003JD004460.

Marsh, N. D., and Svensmark, H.: 2000, 'Low cloud properties influenced by cosmic rays', *Phys. Rev. Lett.* **85**, 5004–5007.

O'Dowd, C. D., Jimenez, J. L., Bahreini, R., Flagan, R. C., Seinfeld, J. H., Hämeri, K., Pirjola, L., Kulmala, M., Jennings, S. G., and Hoffmann, T.: 2002, 'Marine aerosol formation from biogenic iodine emissions', *Nature* **417**, 632–636, 10.1038/nature00775.

Rosen, J. M., Hofmann, D. J., and Gringel, W.: 1985, 'Measurements of ion mobility to 30 km', *J. Geophys. Res.* **90**, 5876–5884.

Weber, R. J., McMurry, P. H., Mauldin III, R. L., Tanner, D. J., Eisele, F. L., Clarke, A. D., and Kapustin, V. N.: 1999, 'New particle formation in the remote troposphere: A comparison of observations at various sites', *Geophys. Res. Lett.* **26**, 307–310.

Wyzlozil, B. E., Seinfeld, J. H., Flagan, R. C., and Okuyama, K.: 1991, 'Binary nucleation in acid-water systems. II. Sulfuric acid-water and a comparison with methanesulfonic acid-water', *J. Chem. Phys.* **94**, 6842–6850.

Yu, F.: 2002, 'Altitude variations of cosmic ray induced production of aerosols: Implications for global cloudiness and climate', *J. Geophys. Res.* **107**, 10.1029/2001JA000248.

Yu, F., and Turco, R. P.: 2001, 'From molecular clusters to nanoparticles: Role of ambient ionization in tropospheric aerosol formation', *J. Geophys. Res.* **106**, 4797–4814.

# ATMOSPHERIC AEROSOL AND CLOUD CONDENSATION NUCLEI FORMATION: A POSSIBLE INFLUENCE OF COSMIC RAYS?

F. ARNOLD

*Atmospheric Physics Division, Max Planck Institute for Nuclear Physics (MPIK), P.O. Box 103980, D-69029 Heidelberg, Germany*

*(E-mail: frank.arnold@mpi-hd.mpg.de)*

(Received 6 February 2006; Accepted in final form 31 May 2006)

**Abstract.** A physical mechanism which may have a potential to connect climate with cosmic rays (CR) involves aerosol particle formation by CR generated atmospheric ions followed by new particle growth. Only grown particles can scatter sunlight efficiently and can eventually act as cloud condensation nuclei (CCN) and thereby may influence climate. Moreover grown particles live longer as they are less rapidly scavenged by pre-existing larger particles. The present paper discusses aerosol particle formation and growth in the light of new measurements recently made by our MPIK Heidelberg group. Emphasis is placed upon the upper troposphere where very low temperatures tend to facilitate new particle formation by nucleation. The new measurements include: laboratory measurements of cluster ions, aircraft measurements of ambient atmospheric ions, and atmospheric measurements of the powerful nucleating gas $H_2SO_4$ and its precursor $SO_2$. The discussion also addresses model simulations of aerosol formation and growth. It is concluded that in the upper troposphere new aerosol formation by CR generated ions is a frequent process with relatively large rates. However new particle formation by homogeneous nucleation (HONU) which is not related to CR also seems to be efficient. The bottleneck in the formation of upper troposphere aerosol particles with sizes sufficiently large to be climate relevant is not nucleation but growth of small particles. Our recent upper troposphere $SO_2$ measurements suggest that particle growth by gaseous sulphuric acid condensation is at least occasionally efficient. If so CR mediated formation of CCN sized particles should at least occasionally be operative in the upper troposphere.

**Keywords:** aerosols, clouds, cosmic rays

## 1. Introduction

Do cosmic rays influence climate? This question originally addressed by Ney (1959) finds considerable current interest stimulated by several recent reports claiming observational evidence for a cosmic ray climate connection hypothesis (Marsh and Svensmark, 2000; Neff *et al.*, 2001; Shaviv, 2002; Shaviv and Veizer, 2004; Harrison and Stephenson, 2006). In particular observations were reported suggesting that low clouds are correlated with cosmic rays (Marsh and Svensmark, 2000). However the current discussion of the cosmic ray climate connection hypothesis is highly controversial. A major point of criticism is the apparent lack of an obvious physical mechanism linking climate with CR. Several potential mechanisms have been proposed, some of which being only poorly understood. Among the proposed


DOI: 10.1007/s11214-006-9055-4

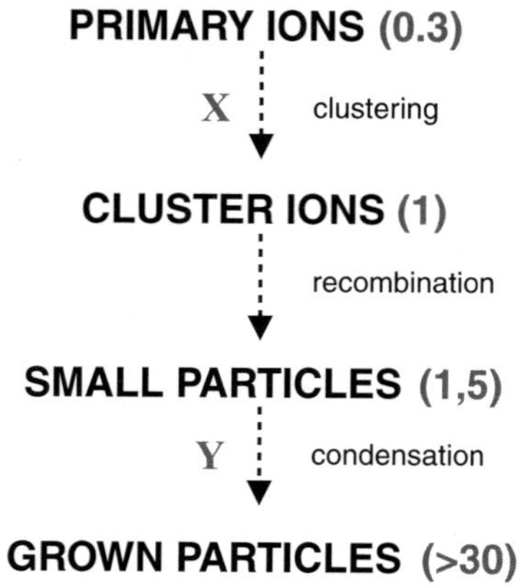


*Figure 1.* Simplified scheme of atmospheric ion-induced formation of aerosol particles followed by particle growth to CCN size. Numbers denote approximate diameters (in nm) of ions and particles. X denotes nucleating gas molecules and Y denotes condensing gas molecules.

mechanisms, the probably most often considered and perhaps least speculative one involves ion induced formation of aerosol particles followed by new particle growth to CCN size (Arnold, 1980a,b, 1981a,b; Yu and Turco, 2001; Eichkorn *et al.*, 2002; Carslaw *et al.*, 2002; Harrison and Carslaw, 2003; Lee *et al.*, 2003; Lovejoy *et al.*, 2004; Kazil and Lovejoy, 2004). However, ion induced CCN formation is itself not fully understood and it is uncertain whether it can significantly contribute to the atmospheric CCN population and thereby influence climate.

Aerosols influence climate in numerous direct (WMO, 2001) and indirect ways (WMO, 2001; Lohmann and Feichter, 2005; Chen and Penner, 2005). If sufficiently large, aerosols scatter sunlight and eventually act as CCN. If CR mediated aerosol formation was indeed contributing significantly to the atmospheric budget of grown aerosol particles. CR would have a potential to influence climate. CR are modulated by solar activity variations, and so might also be their potential effect on aerosols and CCN. An analysis of Marsh and Svensmark (2000) identified a signature of the 11-year sunspot cycle in low clouds (cloud-coverage).

The present paper focuses on processes that lead to the formation of CCN from CR generated ions (Figure 1). These processes involve clustering of atmospheric trace gas molecules X (particularly $H_2SO_4$) to ions, mutual recombination (neutralization) of large cluster ions leading to stable but still very small aerosol particles, and growth of fresh aerosol particles preferably by condensation of atmospheric trace gases Y ($H_2SO_4$ and possibly in addition also certain organic trace gases photochemically formed from organic precursor gases). The first two steps leading from ions to stable aerosol particles are termed ion induced nucleation (hereafter INU).

On the first glance the CR induced CCN formation seems to be similar to what is taking place in the famous "Wilson cloud chamber" which had an important role in the development of high-energy particle physics. Tracks of ionizing high energy particles traversing the chamber are visualized by converting ions to light scattering water droplets. This is achieved by building up a very high water vapour super-saturation (about 400%) which allows efficient INU and efficient small particle growth. In the Wilson chamber $X = Y = H_2O$.

In the atmosphere water vapour super-saturation builds up occasionally leading to cloud droplet formation. However, these super-saturations remain very small and only very rarely exceed 1% but almost never reach 2%. In contrast with the situation in a Wilson cloud chamber the minimum water droplet diameter of a stable water droplet in the atmosphere is larger than about 100 nm. Hence ions cannot grow sufficiently to become stable water droplets. Ion-induced aerosol particle formation can take place only in the presence of highly supersaturated atmospheric trace gases X. Since such trace gases rapidly (within less than about one hour) condense on pre-existing aerosols, they cannot be efficiently transported in the atmosphere and must therefore be produced locally by photochemical processes.

Sulphuric acid ($H_2SO_4$) seems to be the most important atmospheric nucleating trace gas X. Being very hygroscopic, it tends to co-cluster with water vapour molecules, eventually leading to new stable aerosol particles (bi-molecular nucleation). However, in ground level air the gaseous sulphuric acid concentration mostly does not become large enough for homogeneous binary nucleation. At least over the continents the atmospheric trace gas ammonia seems to promote homogeneous sulphuric acid nucleation (ternary $H_2SO_4/NH_3/H_2O$ nucleation). Due to their strongly hygroscopic nature, sulphuric acid aerosol particles may act as atmospheric CCN already when their diameter exceeds only about 30 nm (at a relative humidity of 101% corresponding to 1% super-saturation; see below).

Low temperatures favour nucleation and therefore the cold upper troposphere represents a region of primary interest, at the condition that gases of type X are present and that their concentrations are large enough. In fact, so far only a single X-gas has been detected in the upper troposphere, namely, $H_2SO_4$ (Heitmann and Arnold, 1983; Viggiano and Arnold, 1995). It is also the only Y-gas known to be present in the upper troposphere. Another feature of the upper troposphere which tends to facilitate new particle formation and growth is the relatively small surface of pre-existing aerosol particles. After their formation in the upper troposphere, new particles may undergo downward transport into the middle and lower troposphere where they may act as CCN. Alternatively new particles formed in the upper troposphere may also experience upward transport to the lower stratosphere. If this cross-tropopause transport takes place in the tropics, it is likely that it is followed by further transport from the lower to the middle stratosphere. Hence particles formed in the upper troposphere may also experience transport into the stratosphere and thereby feed the stratospheric aerosol layer.

The present paper critically discusses CR induced formation of atmospheric secondary sulphuric acid aerosol particles and CCN in the light of new laboratory experiments and new atmospheric measurements recently made by our MPIK

Heidelberg group. Emphasis is placed upon the upper troposphere where low temperatures and low abundances of pre-existing aerosol particles tend to facilitate new particle formation and growth. The present paper also places CR mediated INU into a broader perspective by comparing INU with new particle formation via homogeneous sulphuric acid nucleation (HONU), a mechanism that does not require ions.

## 2. Atmospheric Ions

Primary ions formed by CR are the simple molecular and atomic ions $N_2^+$, $O_2^+$, $N^+$, $O^+$, and $O_2^-$. In fact $O_2^-$ is rapidly formed by free electron attachment to $O_2$, the most abundant atmospheric gas, which possesses an appreciable electron affinity. Hereafter primary positive ions and $O_2^-$ undergo ion-molecule-reactions leading to secondary positive and negative ions, mostly complex cluster ions. Ultimately ions are lost by ion-ion recombination, and in aerosol rich air masses, also by attachment to aerosol particles. In the upper troposphere the aerosol load is relatively low and a steady state treatment of ion formation and ion loss by ion-ion recombination yields a total ion concentration

$$n_i = (Q/\alpha)^{1/2} \tag{1}$$

Below about 12 km the ion-ion recombination coefficient $\alpha$ is about $1.7 \times 10^{-6}\,\mathrm{cm}^{-3}\mathrm{s}^{-1}$ (Bates, 1985), and hence $n_i = 0.0013 \times Q^{1/2}$.

The rate $Q$ of CR induced ionization is largest at about 14 km altitude. At middle latitudes where the local tropopause is located at about 12 km, the maximum $Q$ is found just above the tropopause in the lowermost stratosphere. Since CR experience partial shielding by the magnetic fields of the Sun and the Earth, $Q$ varies with geomagnetic latitude and solar activity. It is larger at the Earth's poles compared to the equator. However, the variation of $Q$ over a solar sunspot cycle is relatively small. At middle latitudes and around 14 km altitude $Q$ decreases with increasing solar activity (from sunspot minimum to sunspot maximum) only from about 40 to $30\,\mathrm{cm}^{-3}\mathrm{s}^{-1}$. According to expression (1), at middle latitudes the corresponding variation of $n_i$ is only 4200–4850 $\mathrm{cm}^{-3}$ over a solar cycle.

Therefore, a very sensitive process must be involved if the faint increase in $n_i$ induced by the 11-year solar activity cycle was to lead to a significant increase of CCN formation via INU.

The vertical $n_i$ profile (Figure 2) has two pronounced maxima, one in the ionosphere (around 100–200 km), and one in the stratosphere (around 10–30 km). The ion lifetime with respect to ion-ion recombination is $t_r = 1/(\alpha \cdot n_i)$. Below 12 km altitude $t_r$ increases with increasing height from about 120 s (12 km) to 400 s (3 km) (Viggiano and Arnold, 1995).

Therefore the time span ($t_r$) available for the ion chemical evolution is relatively short which implies that atmospheric trace gases involved in the conversion

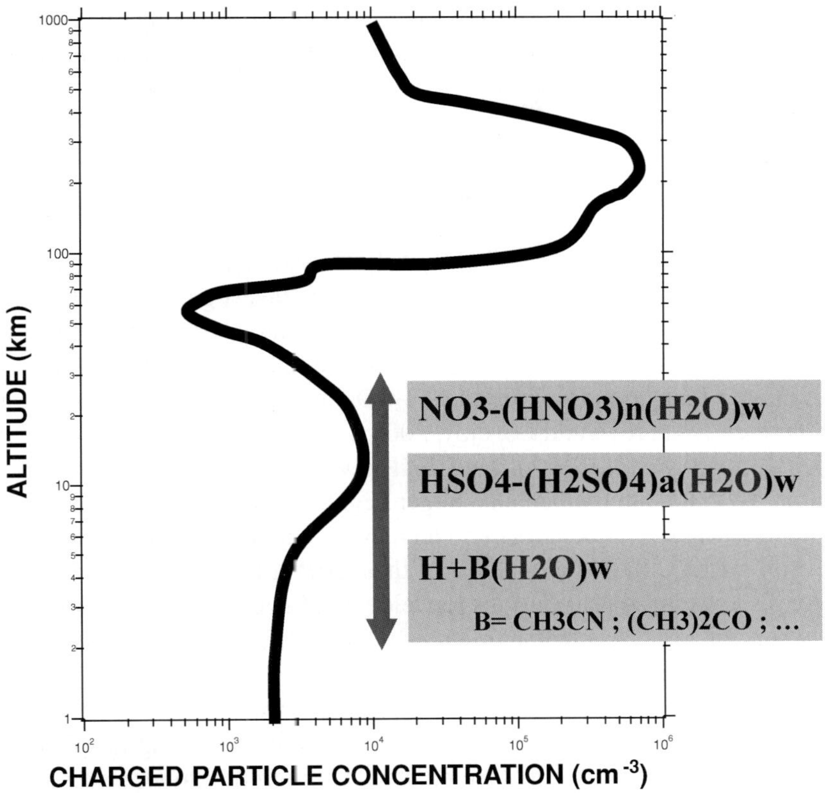


*Figure 2.* Vertical profile of the total concentration of atmospheric positive ions for daytime conditions. Also indicated are the most abundant positive and negative ion species measured by MPIK Heidelberg.

of primary to secondary ions and in the growth of secondary ions must be sufficiently abundant. For example for an altitude of 10 km where $t_r = 200$ s the trace gas concentration must be at least about $3 \times 10^6$ cm$^{-3}$ (corresponding to an atmospheric mole fraction of about $3 \times 10^{-13}$). Here a typical ion molecule collision rate coefficient of $2 \times 10^{-9}$ cm$^3$s$^{-1}$ was considered.

The most abundant ion species observed in the free troposphere and the stratosphere (Figure 2) have originally been detected by the MPIK Heidelberg group (Heitmann and Arnold, 1983; Möhler and Arnold, 1992; Viggiano and Arnold, 1995) using ion mass spectrometers on flying platforms including aircraft, balloons, and rockets. These observed major ions are complex cluster ions containing $H_2SO_4$, $H_2O$, $HNO_3$, $(CH_3)_2CO$ and $CH_3CN$ molecules attached to core ions. These cluster ions are formed from precursor ions via ion-molecule-reactions involving the atmospheric trace gases $H_2SO_4$, $H_2O$, $HNO_3$, $(CH_3)_2CO$, and $CH_3CN$. Of these gases $H_2SO_4$ is particularly interesting as it represents a powerful nucleating agent (see above). An interesting question is whether large $H_2SO_4$ cluster ions involved in INU can in fact build up in the atmosphere. Another interesting question is why the observed major positive ions do not contain $H_2SO_4$. Both questions will be answered below.

## 3. Ion Induced Sulphuric Acid Nucleation

This Section addresses steps 1 and 2 of the scheme shown in Figure 1, focussing on the case $X = H_2SO_4$. During recent time laboratory investigations of $H_2SO_4$ cluster ions have been made by two laboratories (Aeronomy Laboratory in Boulder, Colorado, and MPIK Heidelberg). The Aeronomy Lab investigations focussed on negative cluster ions containing up to 6 $H_2SO_4$ molecules and were made at different temperatures (Froyd and Lovejoy, 2003; Lovejoy and Curtius, 2001; Curtius *et al.*, 2001; Lovejoy *et al.*, 2004). The MPIK investigations were made at only one temperature (295 K), but include also larger negative and positive cluster ions containing up to 24 $H_2SO_4$ molecules (Wiedner, 2000; Eichkorn, 2001; Wilhelm *et al.*, 2004; Sorokin *et al.*, 2006). Here some of these MPIK results will be addressed.

Figure 3 shows mass distributions (envelopes) of negative cluster ions obtained in a flow reactor-mass spectrometer experiment of MPIK (Wilhelm *et al.*, 2004; Sorokin *et al.*, 2006). The ion mass spectrometer used was a large ion-mass spectrometer (LIOMAS) with a mass range of up to 10,000 amu (atomic mass units). This instrument was developed by MPIK for measurements of atmospheric large cluster ions on aircraft platforms (see below). Through the flow reactor synthetic

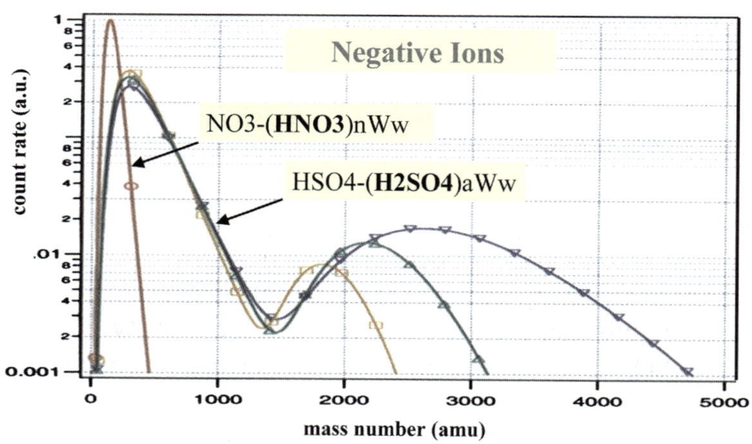


*Figure 3.* Mass distributions of negative cluster ions measured in a laboratory flow reactor experiment of MPIK Heidelberg (Wilhelm *et al.*, 2004; Sorokin *et al.*, 2006). The four mass distributions correspond to different gaseous $H_2SO_4$ concentrations (see insert). For the lowest [$H_2SO_4$] the ions are $NO_3^-(HNO_3)_a(H_2O)_w$ cluster ions and for the 3 cases with elevated $H_2SO_4$ the ions are $HSO_4^-(H_2SO_4)_a(H_2O)_w$ cluster ions. See text for details. From Wilhelm *et al.* (2004).

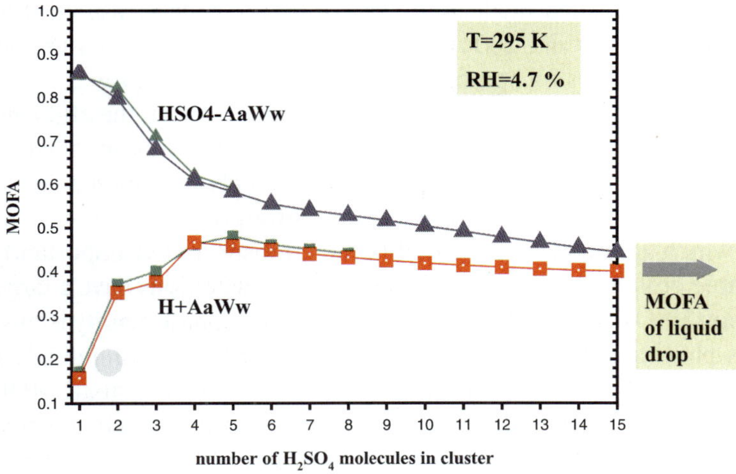


*Figure 4.* Sulphuric acid mole fraction MOFA of $HSO_4^-(H_2SO_4)_a(H_2O)_w$ and $H{+}(H_2SO_4)_a(H_2O)_w$ cluster ions measured in the laboratory flow reactor experiment of MPIK (Sorokin *et al.*, 2006). Also given are model calculations for cluster ions, electrically neutral $H_2SO_4/H_2O$ droplets, and droplets carrying a single elementary electrical charge. See text for details. From Sorokin *et al.* (2006).

air was passed and traces of $H_2SO_4$ vapour and $H_2O$ vapour (relative humidity: 4.1–6.2%) were added. Reagent ions introduced into the flow reactor were of the type $NO_3^-(HNO_3)_n(H_2O)_w$ which are the same ions acting in the atmosphere as precursors of $HSO_4^-(H_2SO_4)_a(H_2O)_w$ ions (see above). The four mass spectra (Figure 3) correspond to different gaseous $H_2SO_4$ concentrations. For the lowest $H_2SO_4$ concentration the spectrum is narrow, peaking around 200 amu and the major ions are $NO_3^-(HNO_3)_n(H_2O)_w$ reagent ions with $n = 1$ and $n = 2$ and mostly $w = 0$. These ions were introduced into the flow reactor and unambiguously identified using a second ion mass spectrometer with higher mass resolution.

As the $H_2SO_4$ concentration is increased to $[H_2SO_4] = 5 \times 10^9$ cm$^{-3}$, the spectrum becomes bi-modal and the second mode extends to 2400 amu. Now the major ions are $HSO_4^-(H_2SO_4)_a(H_2O)_w$ cluster ions (hereafter termed $HSO_4^- A_a W_w$) containing numerous $H_2O$ molecules. As $[H_2SO_4]$ is further increased the second mode shifts to larger mass numbers extending up to 4700 amu. In an analogous experiment with positive ions, cluster ions of the form $H^+A_aW_w$ were detected when gaseous $H_2SO_4$ was present. Again a second mode developed but remained smaller (3000 amu) compared to negative ions.

Figure 4 shows the measured $H_2SO_4$ mole fraction (MOFA) of $HSO_4^- A_a W_w$ and $H^+A_aW_w$ cluster ions versus the parameter $a$ (number of $H_2SO_4$ molecules in the cluster ion). Also shown are model simulations for cluster ions. For a small parameter $a$, MOFA is small for positive ions but large for negative ions. As the parameter $a$ increases, the MOFA of positive and negative ions approach each other. Note that neutral $A_aW_w$ clusters which are also important in theories of homogeneous bi-molecular nucleation have so far not been measured (Laaksonen *et al.*, 1995; Kulmala *et al.*, 2004). Therefore the cluster ion measurements also help to constrain theories on homogeneous nucleation.

The above laboratory investigations of cluster ions confirm the efficiency of co-clustering of gaseous $H_2SO_4$ and $H_2O$. This co-clustering relies on strong hydrogen bonds forming between the clustering molecules. The $H_2SO_4$ molecule has a large gas-phase acidity and therefore tends to form strong hydrogen bonds with $H_2O$ molecules which are not only abundant in the atmosphere but importantly posses an appreciable proton affinity. Small $HSO_4^-A_aW_w$ cluster ions have a large MOFA as the core ion $HSO_4^-$ prefers $H_2SO_4$ ligands whose gas-phase acidity is much larger than the gas-phase acidity of $H_2O$ ligands. In other words $H_2SO_4$ ligands bond much more strongly to $HSO_4^-$. Small $H^+A_aW_w$ cluster ions have a small MOFA since $H_2SO_4$ and $H_2O$ have approximately the same proton affinity but in the flow reactor gaseous $H_2O$ was much more abundant than gaseous $H_2SO_4$. For larger cluster ions ligand-ligand bonding becomes increasingly important and therefore the MOFA of positive and negative cluster ions approach each other.

The above laboratory experiments also explain why the major positive cluster ions observed in the atmosphere (see Figure 2) do not contain $H_2SO_4$. In the atmosphere similar to the laboratory experiment gaseous $H_2O$ is far more abundant than gaseous $H_2SO_4$. Therefore the most abundant positive cluster ions can not contain $H_2SO_4$. Even worse the atmosphere contains trace gas molecules (like $(CH_3)_2CO$ and $CH_3CN$) which possess substantially larger proton affinities than $H_2O$ and therefore are the preferred inner ligands of small atmospheric positive cluster ions.

The behaviour seen in Figure 4 indicates that under the conditions of the laboratory experiment A-attachment to $HSO_4^-A_aW_w$ ions becomes faster than thermal A-detachment for mass numbers >1300–1400 amu corresponding to $a > 6$. It also indicates that for $a > 6$ A-attachment occurs on every collision of a gaseous A-molecule with a cluster ion (Sorokin *et al.*, 2006). The ultimate decrease of the second mode simply reflects a kinetic limitation resulting from the limited time span ($t_{res} = 0.8$ s) an ion resides in the flow reactor. When gaseous $H_2SO_4$ is added, the ion life time $t_c$ with respect to collision with a gaseous A-molecule ranges between about 0.025 s and about 0.06 s depending on the gaseous A-molecule concentration. Hence the ratio $t_{res}/t_c$ ranges from about 8 to 32. This implies that on average during $t_{res}$ an ion collides with 8–32 gaseous $H_2SO_4$ molecules depending on the gaseous $H_2SO_4$ concentration. If each of these collisions would lead to A-attachment the average mass numbers of the ions would be about 1600–6000 amu (including equilibrium $H_2O$ uptake as well). In comparison the second modes of the mass spectra of negative ions extend to largest mass numbers of 2400 to 4700 amu. Our laboratory experiments with $H^+A_aW_w$ ions also revealed the presence of a second mode. However, compared to the second mode $HSO_4^-A_aW_w$ ions, the second mode $H^+A_aW_w$ ions are smaller. Probably this reflects more efficient A-detachment from $H^+A_aW_w$ ions, induced either by thermal collisions or by switching with $H_2O$ molecules.

In the atmosphere at 10 km altitude where temperatures are very low (around 223 K at middle latitudes), thermal A-detachment from ions is expected to be very slow. Hence ion growth is expected to be limited by the number of gaseous $H_2SO_4$ molecule collisions an ion experiences during its life time $t_r = 200$ s. On average this number is equal to $t_r/t_c$ where $t_c = 1/k[H_2SO_4]$ is the time needed for an ion to collide with a gaseous $H_2SO_4$ molecule. Here $k = 2 \times 10^{-9}$ cm$^3$s$^{-1}$ is the ion-molecule-collision rate coefficient and $[H_2SO_4]$ is the gaseous $H_2SO_4$ concentration. Hence $t_c = t_r = 200$ s corresponds to $[H_2SO_4] = 3 \times 10^6$ cm$^{-3}$. For efficient formation of $HSO_4^- A_a W_w$ and $H^+ A_a W_w$ cluster ions at 10 km altitude $[H_2SO_4]$ needs to be larger than $3 \times 10^6$ cm$^{-3}$.

Are $H_2SO_4$ concentrations sufficiently high in the upper troposphere? This question will be addressed in the following Section.

## 4. Atmospheric Sulphuric Acid

Gaseous sulphuric acid is present in the upper troposphere. It was originally detected via ambient atmospheric ion measurements by Arnold and Fabian (1980); see also review by Viggiano and Arnold (1995). Moreover, laboratory measurements (Reiner and Arnold, 1993, 1994; Kolb *et al.*, 1994; Lovejoy *et al.*, 1996) explored the mechanism by which gaseous sulphuric acid is formed in the atmosphere.

Figure 5 shows a simplified schematic of atmospheric sulphuric acid processes. The sulphur bearing precursor gas from which gaseous $H_2SO_4$ is formed is $SO_2$. Its sources are mostly combustion of sulphur containing fossil fuels, volcanism, and atmospheric photochemical $SO_2$ formation from oceanic plankton generated

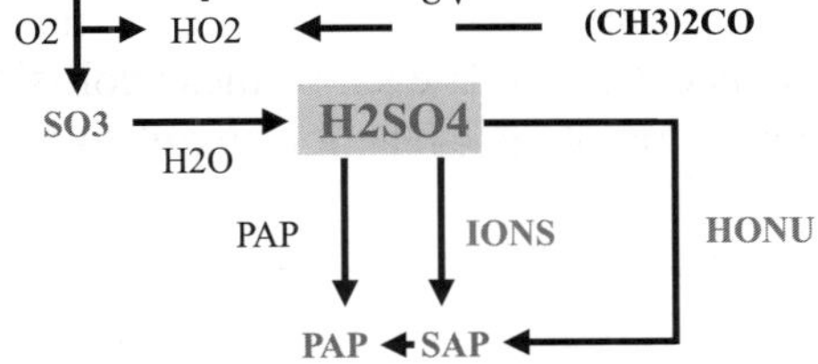


*Figure 5.* Simplified scheme of atmospheric gaseous $H_2SO_4$ processes. HONU denotes homogeneous nucleation, PAP denotes pre-existing aerosol particles, and SAP denotes secondary aerosol particles formed by HONU and ion-induced nucleation.

$(CH_3)_2S$. Of these sources fossil fuel combustion is clearly dominant. The $SO_2$ sinks are mostly deposition at the Earth's surface, cloud processes, and the gas-phase reaction with the hydroxyl radical OH. The OH reaction represents also the first and rate limiting step of $SO_2$ conversion to gaseous $H_2SO_4$. Importantly, OH is recycled and therefore acts merely as a catalyst. In the upper troposphere OH is formed via $O_3$ photolysis and also via the photolysis of acetone ($(CH_3)_2CO$) and other organic trace gases. Once formed gaseous $H_2SO_4$ rapidly condenses on pre-existing aerosol particles PAP (mostly within less than 1000 s). Hence a steady-state treatment yields as a first approximation $k[SO_2][OH] = [H_2SO_4]/t_s$ where $k = 1.5 \times 10^{-12}\,cm^3s^{-1}$ is the rate coefficient of the OH-reaction and $t_s$ is the gaseous $H_2SO_4$ lifetime with respect to condensation. For example, for $t_s = 1000$ s and a typical noon-time $[OH] = 2 \times 10^6\,cm^{-3}$, one obtains $[H_2SO_4] = 0.003 \times [SO_2]$ for cloud-free conditions around noon. Hence for such conditions, gaseous $H_2SO_4$ is for a large part controlled by $SO_2$.

If gaseous $H_2SO_4$ reaches a sufficiently large super-saturation, nucleation takes place involving INU and eventually also HONU leading to secondary aerosol particles SAP. These SAP grow by Y-condensation and mutual coagulation. Ultimately SAP may suffer from coagulation scavenging by PAP. Importantly, the SAP lifetime increases very markedly with increasing SAP size. Therefore SAP growth has a strong influence on SAP concentrations (see below). Whether INU and HONU become efficient critically depends on the gaseous sulphuric acid concentration $[H_2SO_4]$.

Figure 6 shows in a somewhat schematic form a compilation of the vertical distribution of daytime atmospheric gaseous sulphuric acid number concentration ranges measured by MPIK Heidelberg using mass spectrometers on the ground and on aircraft, balloons, and rockets. At ground-level and for cloud-free daytime conditions the measured $[H_2SO_4]$ range from about $1 \times 10^5$–$1 \times 10^7\,cm^{-3}$. This substantial range of variability reflects to a large part a variability of $[SO_2]$. Data obtained in the free troposphere on a central European mountain site at 2300 m altitude range from $4 \times 10^5$–$4 \times 10^6\,cm^{-3}$.

Data obtained in the upper troposphere by aircraft-based instruments (box in Figure 6) range from about $6 \times 10^5$–$6 \times 10^6\,cm^{-3}$. Finally data obtained in the stratosphere by balloon-based and rocket-based instruments (only an average daytime profile is shown) are around $1 \times 10^5$–$3 \times 10^6\,cm^{-3}$ and exhibit a pronounced layer with a peak around 35 km.

Also included in Figure 6 is an equilibrium curve representing the $H_2SO_4$ saturation concentration of an $H_2SO_4$-$H_2O$ mixture which is in equilibrium with atmospheric water vapour. Whenever the experimental data fall to the right of the equilibrium curve, gaseous $H_2SO_4$ is super-saturated. The largest $H_2SO_4$ super-saturations (saturation ratios > 1000) are found in the upper troposphere where temperatures are lowest. These very high super-saturations imply that thermal A-detachment from $HSO_4^-A_aW_w$ and $H^+A_aW_w$ cluster ions can be neglected and therefore under these conditions INU should have no thermodynamic limitation.

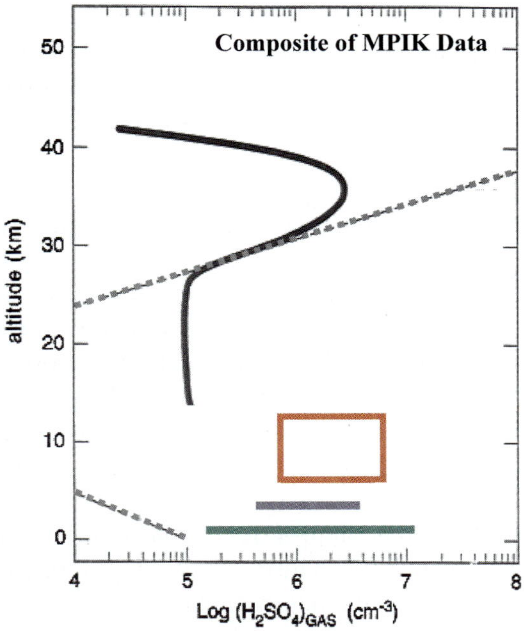


*Figure 6.* Compilation of the ranges of atmospheric gaseous sulphuric acid concentrations measured by MPIK Heidelberg using ground-based, aircraft-based, balloon-based, and rocket-based mass spectrometers. Light green: ground-level data range. Deep blue: lower free troposphere data range. Red: upper troposphere data range. Black: stratospheric daytime average profile. The dotted curve represents an equilibrium saturation curve for gaseous sulphuric acid (see text). Light blue: $H_2SO_4$ inferred from large cluster ion measurements (see text).

The upper tropospheric $H_2SO_4$ concentrations correspond to $t_c = 100$–$1000$ s. In comparison, the ion lifetime $t_r$ with respect to ion-ion recombination (around 200 s) falls in between the above $t_c$. This implies that in many situations a severe kinetic limitation ($t_r > t_c$) exists which prevents INU from becoming efficient. However occasionally $t_c$ becomes smaller than $t_r$ which should allow cluster ions to grow and INU to become efficient.

The major cause of these INU promoting conditions seems to be elevated $SO_2$. Hence, $SO_2$ seems to be the major key to upper tropospheric INU and probably also HONU. However, our knowledge of upper tropospheric $SO_2$ is rather limited since only few data have previously been obtained by advanced instruments. Only very recently a number of advanced $SO_2$ measurements have been made by our group (Speidel *et al.*, 2006) which revealed that $SO_2$ concentrations have a high degree of spatial variability. Around 10 km altitude $SO_2$ mole fractions range mostly from about 30–100 pptV but occasionally have a much larger variability from 10–3000 pptV. This implies highly variable concentrations of gaseous $H_2SO_4$ which in turn implies highly variable nucleation rates and particle growth rates. Importantly formation and growth of sulphuric acid aerosol particles in the upper troposphere should be operative particularly in air masses containing elevated $SO_2$. Therefore a decisive prerequisite of upper troposphere modelling of aerosol and CCN formation is a precise knowledge of $SO_2$ which relies on in situ $SO_2$ measurements. Current models cannot predict $SO_2$ with the required accuracy.

## 5. Observations of Large Cluster Ions in the Upper Troposphere

Building on the laboratory experiments with $H_2SO_4$ cluster ions and the atmospheric gaseous $H_2SO_4$ measurements one expects large $HSO_4^- A_a W_w$ and $H^+ A_a W_w$ cluster ions to be at least occasionally present in the cold upper troposphere. If so, such ions would induce new aerosol particle formation and therefore would represent fingerprints of INU. Using our aircraft-based large ion-mass spectrometer LIOMAS we have made a search for such large cluster ions in the upper troposphere (Eichkorn *et al.*, 2002; Arnold *et al.*, 2006).

Figure 7 shows as an example time-series of ion composition data obtained by LIOMAS in cloud-free upper troposphere air during one aircraft flight (Arnold *et al.*, 2006). Given are flight altitude and fractional abundances of positive ions with mass numbers larger than 200, 300, and 600 amu. While the aircraft was cruising mostly around 8000 m altitude it occasionally intercepted air masses containing large negative and positive cluster ions with mass numbers larger than 600 amu. Negative and positive ions behave rather similar which suggests that one or several trace gases X are present which cluster to negative and positive ions. From the measured ions the concentrations of X were inferred. These range between 1–4 × $10^6$ $cm^{-3}$ which is very similar to the gaseous $H_2SO_4$ concentrations previously measured by our group in the upper troposphere around 8000 m altitude (see Figure 6). This

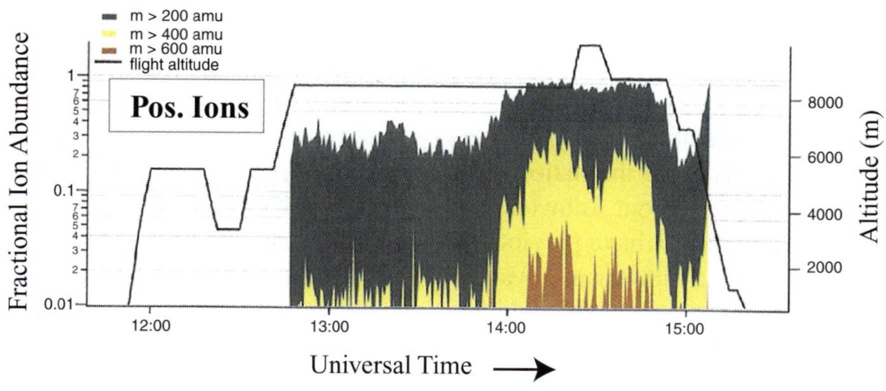


*Figure 7.* Fractional abundances of positive ions with mass numbers larger than 200 (green), 400 (yellow), and 600 (red) amu measured by MPIK Heidelberg during an aircraft mission over Central Europe. Also given is the flight altitude profile. From Arnold *et al.* (2006).

suggests at least that X is not more abundant than gaseous $H_2SO_4$ which implies that most likely X can be identified as $H_2SO_4$.

Building on the inferred X and assuming X to be gaseous $H_2SO_4$ the rate $J_i$ of ion-induced nucleation was calculated (Arnold *et al.*, 2006). Around 8000 m altitude $J_i$ ranges mostly from 3–25 $cm^{-3}s^{-1}$. The latter value is equal to the maximum possible $J_i = Q$ where $Q$ is the ionization rate. When $J_i = Q$, all ions become stable aerosol particles. Also calculated was the rate $J_h$ of homogeneous bi-molecular nucleation (which does not require ions, Arnold *et al.*, 2006). For the flight sections with low concentration of X, $J_i$ tends to exceed $J_h$ whereas for the section with high x concentrations $J_h$ exceeds $J_i$. Even the highest $J_h$ of about 100 $cm^{-3}s^{-1}$ are much smaller than the highest possible $J_h$ dictated by the rate at which an $H_2SO_4$ molecule collides with another $H_2SO_4$ molecule. Note that a nucleation rate of 3 $cm^{-3}s^{-1}$ is already relatively high allowing the formation of more than 10,000 aerosol particles $cm^{-3}hr^{-1}$.

The above example of the atmospheric situation encountered during a single upper troposphere flight demonstrates that in the cold upper troposphere CR mediated INU can be very efficient. However this example also demonstrates that HONU is also efficient and may become even more efficient than INU.

## 6. Aerosol Growth

Growth of newly formed, still very small aerosol particles is important for at least three reasons. Only sufficiently grown particles become long lived, scatter sunlight efficiently, and eventually act as CCN. Particle growth proceeds by Y condensation and coagulation. Growth by Y condensation may involve various types of different trace gases $Y_i$. Our gaseous $H_2SO_4$ measurements made in ground level air (Fiedler *et al.*, 2005) indicate that gaseous $H_2SO_4$ contributes on average only about 5% to the growth of particles with diameters larger than 3 nm. Probably growth is mostly due to so far unidentified condensable organic trace gases. These condensable organics must be formed by chemical reactions involving relatively short lived organic precursor gases.

However, in the upper troposphere the situation may be different. Here the short lived organic precursor gases may be less important since they probably become destroyed already during transport from the surface to the upper troposphere. Also the atmospheric trace gas ammonia which promotes sulphuric acid nucleation in continental ground level air may not be sufficiently abundant in the upper troposphere. Due to its large solubility in liquid water ammonia should experience very efficient wet removal and therefore should not be efficiently transported to the upper troposphere. In contrast $SO_2$, the precursor of gaseous $H_2SO_4$ is less soluble in liquid water. Its photochemical lifetime is about 10–20 days (see above) and therefore in the upper troposphere gaseous $H_2SO_4$ formation can continue for about 10–20 days.

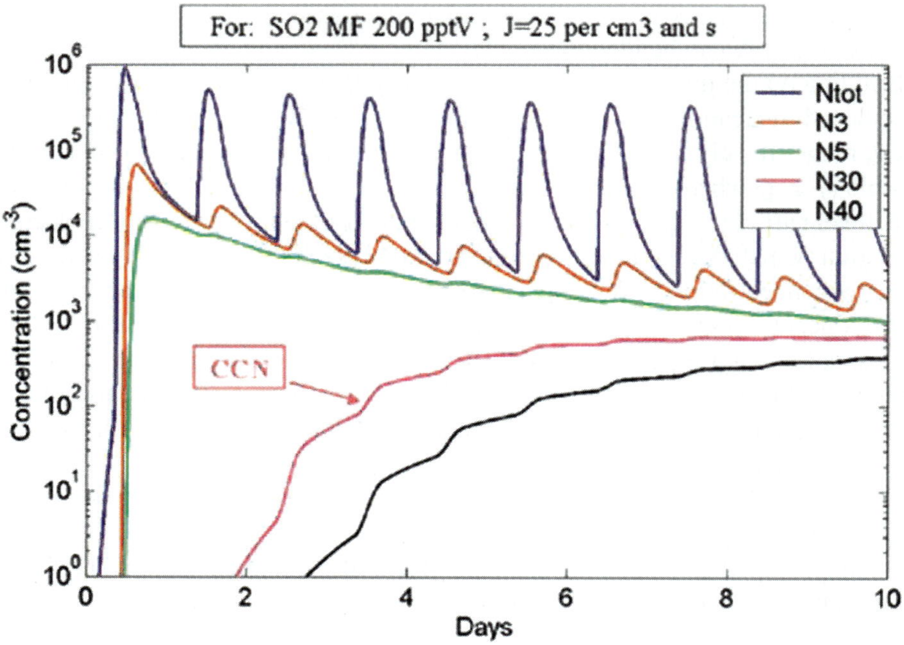


*Figure 8.* Results of a 10-day model simulation of atmospheric gaseous $H_2SO_4$, and aerosol particle number concentrations at 10 km altitude. The air mass under consideration was assumed to ascend to 10 km altitude and reside there for a period of 10 days in cloud-free conditions. The initial $SO_2$ mole fraction was assumed to be 200 pptV. See text for details. From Arnold *et al.* (2006).

Hence new particles generated in the upper troposphere by INU or HONU can grow at least by gaseous $H_2SO_4$ condensation. Note that gaseous $H_2SO_4$ condensation does not require particularly low temperatures (see Figure 6) and therefore can take place also in the middle and lower troposphere. This implies that particles formed in the upper troposphere may still continue to grow by gaseous $H_2SO_4$ condensation after downward transport and can act as CCN at lower altitudes.

Figure 8 shows as an example a 10 day model simulation (Arnold *et al.*, 2006) considering an air mass which has ascended to 10 km and resided there for 10 days at clear-sky conditions. After arrival at 10 km the initial $SO_2$ concentration was 200 pptV. The diurnal maximum OH concentration is $1 \times 10^6$ cm$^{-3}$ which implies an e-folding lifetime for $SO_2$ of about 23 days.

On day 1, as the sun is rising, OH increases, leading to $H_2SO_4$ production. As a consequence $H_2SO_4$ rises to about $2 \times 10^7$ cm$^{-3}$s$^{-1}$ leading to HONU which in turn leads to a rise of the total aerosol particle number concentration $N_{tot}$ to nearly $1 \times 10^5$ cm$^{-3}$. Also the number concentrations $N4$ and $N6$ of particles with diameters larger than 4 and 6 nm rise steeply maximizing on day 1. In the afternoon as the solar elevation decreases OH, $H_2SO_4$ decrease steeply again. Also $N_{tot}$ and to a lesser extent also $N4$ and $N6$ decrease due to coagulation. On the following days, the diurnal $H_2SO_4$ maximum decreases to about $4 \times 10^6$ cm$^{-3}$ which is due to the decrease of OH and the increase of the aerosol surface leading to a decrease of the condensational $H_2SO_4$ lifetime. After day 1 the diurnal $N_{tot}$ maxima nearly vanish, reflecting decreased HONU and increased scavenging of new particles via coagulation. The $N12$, $N20$, and $N30$ curves maximize on days 2, 4, and 6.

Note that the model results of particle formation and growth depend very critically on temperature, relative humidity, $SO_2$, OH, and the initial aerosol surface. For example, when increasing the initial aerosol abundance, an increasing fraction of the photochemically formed $H_2SO_4$ will condense on pre-existing particles rather than nucleate.

In the atmosphere water vapour super-saturations are usually less than 1% and almost never exceed 2%. The corresponding minimum diameters of $H_2SO_4/H_2O$ aerosol particles acting as CCN are about 30 nm and 20 nm respectively. An inspection of Figure 8 reveals that within about 6 days the concentration $N30$ of CCN sized particles increases very substantially from 60 cm$^{-3}$ to nearly 1000 cm$^{-3}$.

Particle growth is not controlled by temperature but is controlled by the rate of gaseous sulphuric acid formation which in turn is proportional to $[SO_2] \times [OH]$. The concentration of $SO_2$ is controlled by emissions of $SO_2$ and $SO_2$-precursors at the Earth surface, by atmospheric transport of these species, and by $SO_2$ removal by clouds.

Also, OH concentrations may be higher than considered by the above model case. This is particularly true for summer and for low latitudes. Higher OH will reduce the $SO_2$ lifetime and increase the rate of gaseous $H_2SO_4$ formation. In turn this will increase new particle formation and growth.

Variations of the CR ionization rate $Q$ induce a variation of the INU rate $J_i$. However, the $N30$ reached after 10 days substantially depends on $J_i$ only if $J_i$ is sufficiently larger than $J_h$ and only if $J_i$ is less than about 1 cm$^{-3}$s$^{-1}$ (Arnold *et al.*, 2006).

## 7. Conclusions

The preceding discussion suggests that ion induced nucleation INU is a frequent process in the cold upper troposphere. However, homogeneous nucleation HONU is also taking place. According to the model simulations reported by Arnold *et al.* (2006), upper troposphere INU and HONU rates are relatively high even for typical $SO_2$ mole fractions. Therefore the bottleneck of upper troposphere CR mediated formation of CCN sized particles seems to be new particle growth and not new particle formation by INU and HONU. In fact, the gaseous $[H_2SO_4]$ required to allow new particle growth to become efficient is mostly more than sufficient for INU.

Particle growth by gaseous $H_2SO_4$ condensation is to a large part controlled by $SO_2$. In fact spatial and temporal variations of upper troposphere $SO_2$ should have a much stronger effect on $N30$ than the small variations of the CR ionization rate.

Therefore more information on upper troposphere $SO_2$ and its spatial and temporal variability is needed. Furthermore efforts should be made to detect additional condensable gases Y which may eventually be present in the upper troposphere. The upper troposphere seems to be a source region of new sulphuric acid aerosols which may experience transport to the lower troposphere as well as the stratosphere. Their transport to the marine lower stratosphere would be potentially relevant for climate since marine clouds distribute markedly to the earths albedo.

## Acknowledgements

The work presented here took place in close collaboration with DLR-IPA and the University of Helsinki. Help and comments by Yasmine Calisesi and comments by Joachim Curtius are greatly acknowledged. Financial support by the Max Planck Society is greatly appreciated.

## References

Arnold, F.: 1980a, 'Multi-ion complexes in the stratosphere – Implications for trace gases and aerosol', *Nature* **284**, 610–611.

Arnold, F.: 1980b, 'Ion-induced nucleation of atmospheric water vapour at the mesopause', *Planet Space Sci.* **28**, 1003.

Arnold, F.: 1981a, 'Solvated electrons in the upper atmosphere', *Nature* **294**, 732.

Arnold, F.: 1981b, 'Ion nucleation – a potential source for stratospheric aerosols', *Nature* **299**, 134.

Arnold, F. and Fabian, R.: 1980, 'First measurements of gas phase sulfuric acid in the stratosphere', *Nature* **282**, 55.

Arnold, F., Wilhelm, S., and Pirjola, L.: 2006, 'Cosmic ray induced formation of aerosol particles and cloud condensation nuclei: First detection of large negative and positive cluster ions in the upper troposphere', manuscript in preparation.

Bates, D. R.: 1985, 'Ion-ion recombination in an ambient gas, *Adv. Atom. Molec. Phys.* **20**.

Carslaw, K. S., Harrison, R. G., and Kirkby, J.: 2002, 'Cosmic rays, clouds and climate', *Science* **298**, 1732–1737.

Chen, Y. and Penner, J. E.: 2005, 'Uncertainty analysis of the first indirect aerosol effect', *Atmos. Chem. Phys.* **5**, 2935–2948.

Curtius, J., Froyd, K. D., and Lovejoy, E. R.: 2001, 'Cluster ion thermal decomposition (I): Experimental kinetics study and ab initio calculations for $HSO_4^-(H_2SO_4)_x(HNO_3)_y$', *J. Phvs. Chem. A* **105**, 10,867–10,873.

Eichkorn, S.: 2001, 'Development of an aircraft-based ion mass spectrometer with a large mass range: Measurements in the laboratory, aircraft exhaust plumes and the upper troposphere', Ph. D. thesis, Univ. Heidelberg.

Eichkorn, S., Wilhelm, S., Aufmhoff, H., Wohlfrom, K. H., and Arnold, F.: 2002, 'Cosmic ray-induced aerosol-formation: First observational evidence from aircraft-based ion mass spectrometer measurements in the upper troposphere', *Geophys. Res. Lett.* **29**, 10.1029/2002GL015044.

Fiedler, V., Dal Maso, M., Boy, M., Aufmhoff, H., Hoffmann, J., Schuck, T., Birmili, W., Arnold, F., and Kulmala, M.: 2005, 'The contribution of suphuric acid to atmospheric particle formation and growth: a comparison between boundary layers in Northern and Central Europe', *Atmos. Chem. Phys. Disc.* **5**, 1–33.

Froyd, K. D., and Lovejoy, E. R.: 2003, 'Experimental thermodynamics of cluster ions composed of $H_2SO_4$ and $H_2O$. 2. Negative ion measurements and ab initio structures', *J. Phys. Chem. A* **107**, 9812–9824.

Harrison, R. G., and Carslaw, K. S.: 2003, 'Ion-aerosol-cloud processes in the lower atmosphere', *Rev. Geophys.* **41**, 1012, doi:10.1029/2002RG000114.

Harrison, R. G. and Stephenson, D. B.: 2006, 'Empirical evidence for a nonlinear effect of cosmic rays on clouds', *Proc. R. Soc. A*, in press.

Heitmann, H. and Arnold, F.: 1983, 'Composition measurements of tropospheric ions', *Nature* **306**, 747.

Kazil, J. and Lovejoy, E. R.: 2004, 'Tropospheric ionization and aerosol production: A model study', *J. Geophys. Res.* **109**, 10.1029/2004JD004852.

Kolb, C. E., Jayne, J. T., Wornshop, D. R., Molina, M. J., Meads, R. F., and Viggiano, A. A.: 1994, 'Gas phase reaction of sulfur trioxide with water vapour', *J. Atmos. Chem. Soc.* **116**, 10,314–10, 315.

Kulmala, M., Vehkamäki, H., Petäjä, T., Dal Maso, M., Lauri, A., Kerminen, V.-M., Birmili, W., and McMurry, P. H.: 2004, 'Formation and growth rates of ultrafine atmospheric particles: A review of observations', *J. Aerosol Sci.* **35**, 10.1016/j.jaerosci.2003.10.003.

Laaksonen, A., Talanquer, V., and Oxtoby, D. W.: 1995, 'Nucleation: Measurements, Theory, and Atmospheric Applications', *Ann Rev. Phys. Chem.* **46**, 489–524.

Lee, S. H., *et al.*: 2003, 'Particle formation by ion nucleation in the upper troposphere and lower stratosphere', *Science* **301**, 1886–1889.

Lohmann U. and Feichter, J.: 2005, 'Global indirect aerosol effects: a review', *Atmos. Chem. Phys. Discuss.* **5**, 715–737.

Lovejoy, E. R. and Curtius, J.: 2001, 'Cluster Ion Thermal Decomposition (II): Master Equation Modeling in the Low Pressure Limit and Fall-Off Regions. Bond Energies for $HSO_4^-(H_2SO_4)_x(HNO_3)_y$', *J. Phys. Chem. A* **105**, 10,874–10,883.

Lovejoy, E. R., Hanson, D. R., and Huey, G. G.: 1996, 'Kinetics and products of gas-phase reactions of SO3 with water', *J. Phys. Chem.* **100**, 19.911–19.916.

Lovejoy, E. R., Curtius, J., and Froyd, K. D.: 2004, 'Atmospheric ion-induced nucleation of sulphuric acid and water', *J. Geophys. Res.* **109**, 10.1029/2003JD004460.

Marsh, N. D. and Svensmark, H.: 2000, 'Low cloud properties influenced by cosmic rays', *Phys. Rev. Lett.* **85**, doi:10.1103/PhysRevLett.85.5004.

Möhler, O. and Arnold, F.: 1992, 'Gaseous sulphuric acid and sulfur dioxide measurements in the arctic troposphere and lower stratosphere: Implications for hydroxyl radical abundances', *Geophys. Res. Lett.* **19**, 1763–1766.

Neff, U., *et al.*: 2001, 'Strong coincidence between solar variability and the monsoon in Oman between 9 and 6 kyr ago', *Nature* **411**, 290-293.

Ney, E. P.: 1959, 'Cosmic radiation and the weather', *Nature* **183**, 451–452.

Reiner, T. and Arnold, F.: 1993, 'Laboratory flow reactor measurements of the reaction $SO_2 + H_2O + M \rightarrow H_2SO_4 + M$: Implications for gaseous $H_2SO_4$ and aerosol formation in the plume of jet aircraft', *Geophys. Res. Lett.* **20**, 2659–2662.

Reiner, T. and Arnold, F.: 1994, 'Laboratory investigations of gaseous sulfuric acid formation via $SO_2 + H_2O + M \rightarrow H_2SO_4 + M$: Measurements of the rate constant and products identification', *J. Chem. Phys.* **101**, 7399–7407.

Shaviv, N. J.: 2002, 'Cosmic ray diffusion from the galactic spiral arms, iron meteorites, and a possible climatic connection', *Phys. Rev. Lett.* **89**, doi: 10.1103/PhysRevLett.89.051102.

Shaviv, N. R. and Veizer, J.: 2004, 'Celestial driver of phanerozoic climate?', *GSA Today* **13**, 7.

Sorokin, A., Arnold, F., and Wiedner, D.: 2006, 'Flow reactor experiments and model calculations of sulfuric acid-water cluster ion formation and ion-induced nucleation', *Atmos. Env.*, in press.

Speidel *et al.*: 2006, 'Sulfur dioxide measurements in the lower, middle and upper troposphere: Deployment of an aircraft-based chemical ionization mass spectrometer with permanent in-flight calibration', manuscript in preparation.

Viggiano, A. A. and Arnold, F.: 1995, *Ion Chemistry and Composition of the Atmosphere*, Handbook of Atmospheric Electrodynamics, Vol. 1, CRC Press.

Wiedner, D.: 2000, 'Flow reactor investigations of aerosol particle formation by ion induced nucleation: The H2SO4/H2O system', Diploma thesis, Univ. Heidelberg.

Wilhelm, S., Eichkorn, S., Wiedner, D., Pirjola, L., and Arnold, F.: 2004, 'Ion-induced aerosol formation: new insights from laboratory measurements of mixed cluster ions $HSO_4^-(H_2SO_4)_a(H_2O)_w$ and $H^+ (H_2SO_4)_a(H_2O)_w$', *Atmos. Env.* **38**, 1735–1744.

WMO: 2001, in J. T. Houghton, *et al.* (eds.), *Climate change 2001: The Scientific Basis*, Cambridge University Press.

Yu, F. and Turco, R. P.: 2001, 'From molecular clusters to nanoparticles: Role of ambient ionisation in tropospheric aerosol formation', *J. Geophys. Res.* **106**, doi: 10.1029/2000JD900539.

# ON THE RESPONSE OF THE CLIMATE SYSTEM TO SOLAR FORCING

*Introductory Paper*

L. BENGTSSON

*Max-Planck-Institut für Meteorologie, Bundesstrasse 55, D–20146 Hamburg, Germany;*
*Environmental System Science Center, University of Reading, UK*
*(E-mail: bengtsson@dkrz.de)*

(Received 22 March 2006; Accepted in final form 30 May 2006)

**Abstract.** The climate response to changes in radiative forcing depends crucially on climate feedback processes, with the consequence that solar and greenhouse gas forcing have both similar response patterns in the troposphere. This circumstance complicates significantly the attribution of the causes of climate change. Additionally, the climate system displays a high level of unforced intrinsic variability, and significant variations in the climate of many parts of the world are due to internal processes. Such internal modes contribute significantly to the variability of climate system on various time scales, and thus compete with external forcing in explaining the origin of past climate extremes. This highlights the need for independent observations of solar forcing including long-term consistent observational records of the total and spectrally resolved solar irradiance. The stratospheric response to solar forcing is different from its response to greenhouse gas forcing, thus suggesting that stratospheric observations could offer the best target for the identification of the specific influence of solar forcing on climate.

**Keywords:** solar forcing, climate change, climate variability

## 1. Introduction

Traditionally, long-term observations of the sunspot number have been combined to surface temperature records to provide evidence for a solar forcing-climate relationship (Eddy, 1976). The coincidence of the Maunder minimum with a period of very cold winters in Europe in the late 17th century (Lean, 2000; and others) suggests the existence of such a relationship, which is also supported by numerous model studies (Crowley, 2000; Cubasch and Voss, 2000; Shindell *et al.*, 2001; Rind, 2002). The practice of using estimates of global temperature records, combined to proxy-based reconstructions of the total solar irradiance, TSI, to infer the sensitivity of the climate system to long-term solar variations is, however, questionable. Changes in the solar input associated with the 11-year cycle of solar activity are nowadays well documented, but the actual range of TSI variations over longer time scales is still a subject of debate (see different contributions in this issue) as no direct measurements exist prior to the advent of satellite-based observations, in 1978 (Fröhlich, 2006). Present reconstructions of the TSI over millennial timescales have been subject to several revisions (Hoyt and Schatten, 1993; Lean *et al.*, 1995, 2002; and others), and latest results (Wang *et al.*, 2005) suggest that the amplitude of past

Space Science Reviews (2006) 125: 187–197
DOI: 10.1007/s11214-006-9056-3

solar irradiance changes could in fact be significantly smaller than had been previously estimated. Similarly, long-term instrumental records of past climate data are limited in time and space and very few records exists before the middle of the 18th century (Jones and Moberg, 2003). Surface temperature variations in Europe, where most long-term temperature records are found (e.g., Luterbacher *et al.*, 2004), are only weakly correlated with the global surface temperatures (Bengtsson *et al.*, 2006).

Another shadow on the discussed evidence for the solar-climate variability relationship is cast by a possible model underestimation of the reproduced system internal variability (Oldenborgh *et al.*, 2005), which would force the ascription of extreme past climate events to external causes such as changes in TSI. Recent high-resolution models tend indeed to generate larger internal system variability, thus suggesting that an episode of extreme climate may simply result from the combination of internal system processes, independently of changes in external forcing. As outlined in the next paragraph, climate in Europe is indeed markedly influenced by atmospheric circulation patterns (e.g., Hurrell *et al.*, 2003), which are in turn largely determined by non-linear atmospheric processes, unpredictable on longer time-scales (Gerber and Vallis, 2005).

At middle and high latitudes the variability of the atmosphere is dominated by synoptic weather patterns, with maximum variance on time-scales of a few days to a few weeks. However, because these high-amplitude patterns are essentially non-periodic, considerable high variance contributions can be found in the low frequency part of the signal power spectrum (Hasselmann, 1976; Manabe and Stouffer, 1996). The North Atlantic Oscillation (NAO) (Hurrell *et al.*, 2003) for example, which corresponds to an oscillation in the strength of the geostrophic wind in the eastern North Atlantic (or in the amplitude of the positive pressure gradient between Iceland and the Azores), is presumably generated by changes in the distribution of transient storms (Loptien and Ruprecht, 2005). According to these results, a northerly storm track leads to a high index in the NAO (stronger than average pressure gradient) while a southerly storm track leads to a weaker than average pressure gradient. Other dominant circulation patterns like the Pacific-North America pattern (PNA) are presumably also driven by altered distributions of high frequency weather systems. The most dominant internal mode of variability in the climate system, ENSO (El Niño-Southern Oscillation), is a coupled atmosphere/ocean mode of the tropical Pacific driven by the feedback between surface wind and ocean temperature distribution. ENSO has a major influence in the tropics, but influence also the climate in the extra-tropics, particularly in the western hemisphere. Although ENSO has a dominant time-scale of some four years, the signal has considerable variability. The effect of ENSO is also supported by model studies (e.g., Zebiak and Cane, 1987; Oldenborgh *et al.*, 2005).

As a consequence, the fundamental question of the identification of specific causes of climate change has been found to be a more complex issue than hitherto generally anticipated. Two main issues stand out.

First, the climate system is characterized by internal modes of variability, with time-scales from days to several decades and possibly longer. These modes are characterized by considerable amplitudes, they dominate atmospheric circulation on virtually all time scales and are, as far as we know, largely chaotic and unpredictable beyond a certain time. *The determination of the influence of external forcing on the climate system does thus require a preliminary assessment of the influence of forcing on these modes of variability, as well as of the relative influence of these modes on the global climate system.*

Second, recent studies have demonstrated that the response to external forcing, such as from the sun, aerosols and greenhouse gases, is largely controlled by regional feedback processes, rather than directly by the forcing itself (Boer and Yu, 2003). A consequence of this is that the pattern of response is potentially uncorrelated with the pattern of forcing, making it difficult to separate different forcings as solar forcing and greenhouse forcing may generate the same climate response.

In Section 2 and 3 we will discuss in some detail the possible mechanisms involved in the climate response to external forcing. In Section 4 we will report of a recent study of possible mechanisms explaining the European climate in the period 1500–1900. Finally, general conclusions will be drawn in Section 5.

## 2. The Mechanisms of Climate Forcing and Response: Geographically-Dependent Feedbacks

External forcing of the climate system is commonly expressed as the difference between incoming and outgoing radiation in $Wm^{-2}$ at the tropopause level (IPCC, 2001, Chapter 6). This definition is justified by the fact that adjustment to altered radiative conditions, such as enhanced solar irradiation, is fast in the stratosphere (of the order of a month), but very slow in the troposphere (of the order of decades) due to the huge heat capacity of the oceans. An increase (decrease) in the solar irradiation at the top of the atmosphere will create an immediate imbalance between incoming and outgoing radiation, thus gradually leading to a warming (cooling) of the troposphere and the Earth's surface. In turn, and because of enhanced thermal radiation from a warmer planet, increased solar input will lead to a slow increase in outgoing infrared radiation that will gradually offset the radiation imbalance at the top of the atmosphere. This is a very slow process, and several centuries will be needed before a complete radiation balance is restored. The effect is analogue in case of warming resulting from increased greenhouse gases concentration levels, but in this case the immediate imbalance is due to *reduced outgoing radiation*. Hansen *et al.* (1997) have undertaken a series of model experiments to explore the response of climate to various forcing factors, including increases in $CO_2$ levels and in solar irradiance. Although the model used by Hansen *et al.* (ibid) was significantly simpler than present General Circulation Models (GCMs), the latter nevertheless respond rather similarly (e.g., Boer and Yu, 2003). Following

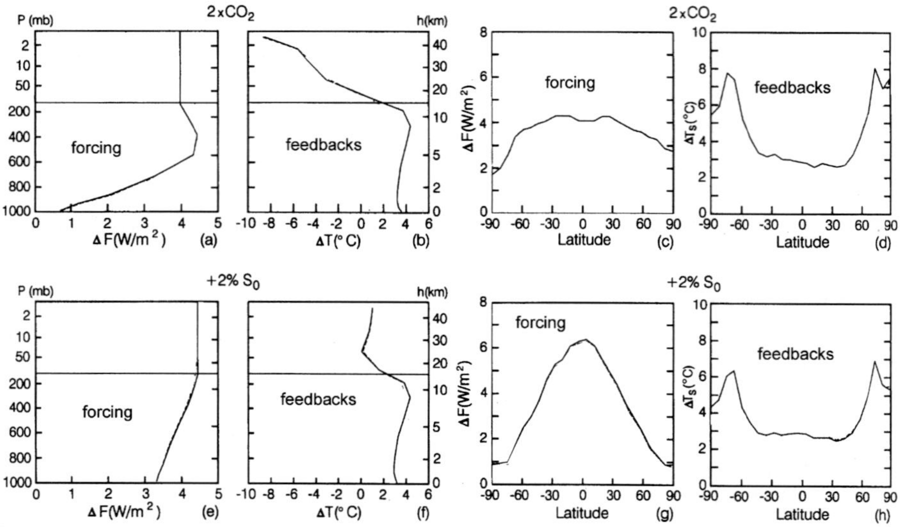


*Figure 1.* Forcing and feedback in 2 × $CO_2$ and +2% increase in TSI. (a) and (e) shows the altitude distribution of the net radiative flux and (b) and (f) are the corresponding equilibrium temperature changes. (c) and (g) are the zonal mean forcing versus latitude. (d) and (h) are the corresponding zonal mean surface temperature response. From Hansen *et al.* (1997).

Hansen *et al.* (ibid) we show the climate response to greenhouse gases and solar forcing in Figure 1.

If solar irradiance increases by 2%, the instantaneous flux change at the top of the atmosphere is 4.7 $Wm^{-2}$. A fraction of this energy is absorbed in the stratosphere and radiated back to space without affecting tropospheric temperatures. After convective radiative adjustment of the stratosphere, lasting a few months or so, the adjusted forcing entering the troposphere amounts to 4.5 $Wm^{-2}$. This adjusted forcing is comparable in amplitude to the adjusted forcing corresponding to a doubling of $CO_2$. Restoration of the energy balance without alteration of the atmospheric lapse rate implies a warming of the Earth's surface by about 1.3 K. As indicated by Hansen *et al.* (ibid), feedback processes in the climate system increase the temperature further to between 3 and 4 K, although the range of this interval is strongly model-dependent and is likely to be even larger. While the equilibrium temperature for the doubling of $CO_2$ and increasing the solar irradiation by 2% is the same in the troposphere, the situation is however quite different in the stratosphere.

In the case of forcing by greenhouse gases, the stratosphere cools because of increased radiation to space (higher concentration of greenhouse gases). The cooling of the stratosphere will not be compensated by radiation from below as this radiation is absorbed in the troposphere and only a weak thermal radiation from the temperature minimum at the tropopause enters the stratosphere. However, the stratosphere responds quite differently to solar warming. In contrast to the case with greenhouse gases, a slight warming occurs instead in the stratosphere as a result of increased solar input, partly due to increased absorption by stratospheric ozone.

Both solar and greenhouse gases forcing amplitudes are largest at lower latitudes (Figure 1). This feature is easily understandable in the case of solar forcing, as solar input to the Earth is a function of latitude. Although less marked, a similar behavior is also observed in the case of greenhouse gases forcing. In contrast, understanding the latitudinal system response to the different forcings cannot be achieved without an analysis of the model results. The immediate warming increases the amount of water vapor in the troposphere (the atmosphere tends to conserve relative humidity), thus enhancing the warming as water vapor is a dominant greenhouse gas. The amount of water vapor in the atmosphere is not uniform as the response will naturally be largest in the areas where the relative humidity is high. This occurs in the Intertropical Convergence Zone (ITCZ) and in the storm track regions. Changes in cloud cover and cloud distribution will strongly affect both short- and long-wave radiation. While low clouds have been shown to cool the atmosphere because of increased reflection of solar radiation, high clouds warm the atmosphere because of enhanced absorption of terrestrial radiation. The total effect of clouds leading to an overall net cooling of some 20 $Wm^{-2}$ in the present climate (Stephens, 2005). Whether changes in clouds will give rise to a warming or a cooling of the climate system is however still unclear, and present models behave differently in this respect. In addition, atmospheric and ocean circulation do transport heat and water vapor to higher latitudes, thus affecting climate and the temperature distribution even further. As a general result, *surface warming is actually larger at high latitudes than at low latitudes in spite of the fact that the forcing has a maximum at low latitudes* (Boer and Yu, 2003).

Figure 1 showed results obtained with a simplified mixed layer model experiment. More realistic models accounting for the full heat exchange of the atmosphere with the oceans however yield different results (Figure 2). Because of the response to the dominant westerlies over the Southern Oceans, an active exchange of heat takes place between the ocean and the atmosphere, leading to an efficient mixing of heat into the deep ocean (Held, 2005). This slows down the warming process, so at the Southern Hemisphere with its strong persistent winds ocean surface warming is much slower than at northern high latitudes. At the Northern Hemisphere, reduced surface warming is restricted to limited parts of the North Atlantic. As a result, most transient climate change models provide a response pattern as shown in Figure 2 (see Räisänen, 2002 for a more complete discussion).

## 3. Dynamical Contributions

The second problem related to attribution of the causes of climate change is the chaotic variability of atmospheric and ocean circulation patterns over longer time scales. This contribution is considerable, and best illustrated by repeated general

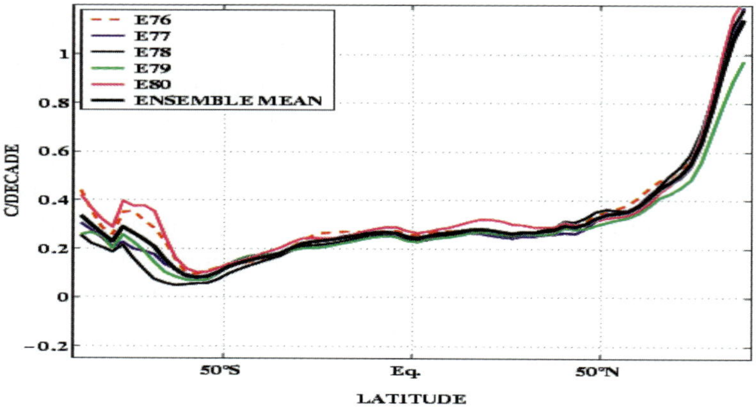


*Figure 2.* 80-year mean of a five different transient climate change experiments with the Bergen climate model showing the zonal average temperature increase during a time when the $CO_2$ forcing is increasing by 1%/year compared to a case when $CO_2$ is constant. The ensemble mean is also indicated. Courtesy H. Drange, Bergen Climate Centre.

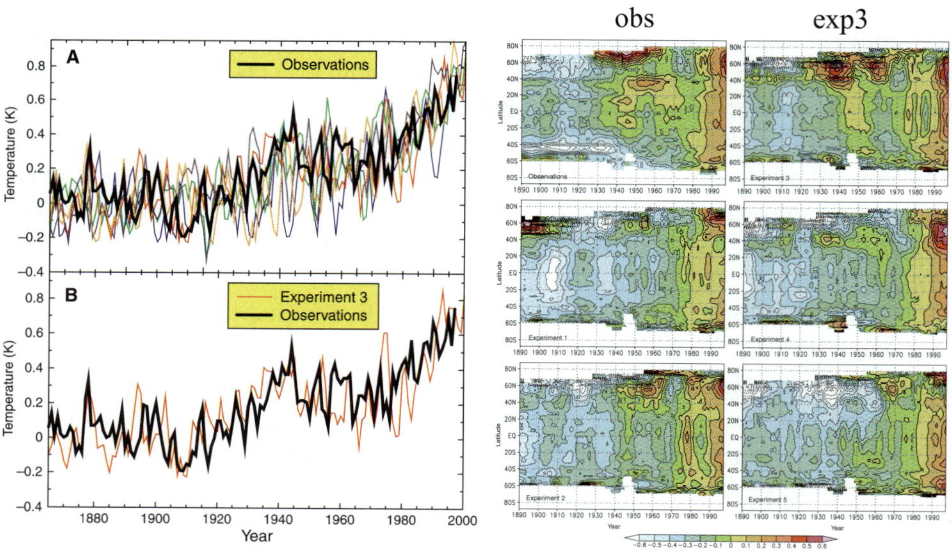


*Figure 3.* Monte-Carlo simulation with a couple climate model forced by observed greenhouse gases and sulphate aerosols. One out of five simulations almost perfectly reproduces the observed global temperature variability for the period 1860–2000. From Delworth and Knutson (2000).

circulation models simulations using slightly different initial conditions but exposed to identical forcing. In such an ensemble experiment, consisting of five simulations for the period 1860–2000 (Delworth and Knutson, 2000), one of the simulations agreed almost perfectly with the observed global surface temperature (Figure 3). This realization reproduced the warm events in the 1930s and 1940s, and as in observational records, the warming was most pronounced in the Arctic. The other four simulations also reproduced similar anomalies, but at other times. However, the underlying slow warming trend due to the forcing induced by observed greenhouse gases and aerosols increases could be found in all the ensemble experiment members. Considering the above discussed similarities between the system response to solar and greenhouse gases forcing, it is to be expected that a similar result would be achieved if a gradual increase in solar forcing had been considered in the model experiment instead of greenhouse gas forcing. In the next Section, we report the results of a modeling study that do highlight the characteristics of internal climate variability in Europe.

## 4. A Recent Model Experiment

Recently, Bengtsson *et al.* (2006) investigated the possible causes of observed surface temperature variations in Europe for the period 1500–1900. During this period, only minor changes in greenhouse gases concentrations and in other anthropogenic influences on climate occurred, so that temperature variations could with reasonable confidence be attributed to natural variations. These include solar forcing, and the effect of increased stratospheric aerosol concentration levels following intense volcanic eruptions. The reconstructed observational record of surface temperature variations over the European land area for this period is probably one of the most reliable and best documented in the world, with reconstructions achieving even seasonal resolution (Luterbacher *et al.*, 2004; Xoplaki *et al.*, 2005). It is interesting to note that extreme values of seasonally averaged temperature occurred at different periods. The coldest winter and spring temperatures over the considered time period occurred at the beginning of the 18th century and at the end of the 18th century , respectively, while the coldest summer and autumn took place at the beginning of the 20th century (Table I). Intuitively, this does not suggest major external forcing as the cause of the noted extreme temperatures, but rather natural variations in the climate system.

This interpretation is strongly supported by the result of a climate simulation performed with a coupled general circulation model (Jungclaus *et al.*, 2006) exposed to

TABLE I

Compiled data from Luterbacher *et al.* (2004) and Xoplaki *et al.* (2005) covering the period 1500–2003. From Bengtsson *et al.* (2006).

| | Coldest year | Coldest decades | Warmest year | Warmest decades |
|---|---|---|---|---|
| Winter | 1708/09 | End 1600 | 1990 | 1990s |
| Spring | 1785 | Around 1700 | 1989 | 1990s |
| Summer | 1902 | Beginning 1900 | 2003 | 1990s |
| Autumn | 1912 | End 1800<br>Early 1900 | 1938 | 1930s |

TABLE II

Mean value, standard deviation, and extreme temperature for model and observed data.

| | (a) Luterbacher (2005) | | | (b) Model | | |
|---|---|---|---|---|---|---|
| | Annual | DJF | JJA | Annual | DJF | JJA |
| Mean | 8.13 | −2.31 | 16.83 | 7.39 | −2.30 | 16.96 |
| Stand. dev. | 0.41* | 1.15 | 0.47 | 0.55 | 1.23 | 0.55 |
| Min. | 6.61 (1695) | −5.69 (1695/96) | 15.62 (1821) | 5.41 (16) | −7.45 (15) | 15.55 (170) |
| Max. | 9.40 (1822) | −0.09 (1842/43) | 18.18 (1757) | 9.41 (78) | 1.47 (52) | 19.08 (459) |
| Max.–Min. | 2.79 | 5.60 | 2.56 | 4.00 | 8.92 | 3.53 |

(a) Observed land surface temperature 1500–1900 for the European land area (35°N–70°N, 25°W–40°E) (b) the same for the ECHAM5/ MPI-OM integrated and initialized under pre-industrial forcing. Units in C. From Bengtsson *et al.* (2006). Years of extreme temperatures are indicated within the parentheses. Model years are taken from the 500-year calculation using pre-industrial greenhouse forcing only.

*Standard deviation for the last 100 years is 0.53, suggesting that the variance is underestimated in the earlier data records.

an atmospheric composition characteristic of the pre-industrial period, but without any changes in TSI nor in atmospheric aerosol concentrations (Bengtsson *et al.*, 2006). Irrespective of this, the model simulated the same average, standard deviation and extreme seasons for winter and summer (Table II) as had been derived from the observational temperature reconstruction (Luterbacher, 2005, private communication), thus supporting the idea that long-term variability of climate in Europe is dominated by natural processes. Further inspection of the model results shows that the simulated extremely cold winters and extremely warm summers have large similarities with observational reconstructions, including 30-year long periods of colder and warmer temperatures (not shown). The only difference of significance was that the occurrence of longer periods of extreme summer temperature was slightly higher in the observational reconstruction. This suggests a possible external influence, either from volcanic aerosols or from solar forcing. We shall emphasize here that these results do not imply the exclusion of any influence from low-frequency TSI variations on the climate system, but do rather highlight the difficulty faced when attempting to infer the relative influence of concurring forcing on the climate system from observed temperatures only.

## 5. Concluding Remarks

The present contribution highlights some of the principal issues that greatly complicate the attribution of climate change and climate variations to a specific forcing. One of the reasons for this difficulty is the large difference observed between the places of forcing and response patterns, the latter being mainly determined by feedback processes in the climate system. These feedback processes explain why warming due to both higher TSI or greenhouse gases is largest at high latitudes, in spite of the fact that forcing there is smaller than at low latitudes. There are also clear indications from model studies (e.g., Boer and Yu, 2003) that the climate response patterns are *additive in their geographical distribution even if the forcing patterns are different.* This stresses the fact that it is the feedback mechanisms which dominate the pattern of response irrespective of the forcing patterns.

Another difficulty is related to the high level of unforced variability in the climate system, resulting from the combination of internal processes. Such processes are chaotic in nature and unpredictable beyond a certain period of time, as predictability is conditioned by the growth of small errors in the initial system state.

The general conclusion of the present discussion is that a reliable attribution of the lower atmospheric response to TSI changes is not achievable on the basis of past temperature records only, as internal modes of variability of the climate system, at least over the last several hundred years, can give rise to variations with amplitudes similar to the ones expected from solar input changes (Bengtsson *et al.*, 2006). We are therefore inclined to conclude that the effect of solar forcing in the troposphere is still an open issue.

The results of this study highlight the need for long term monitoring of both total and spectrally resolved solar irradiance measurements in the upper atmosphere, where the time scale of response to changes in the solar input is significantly shorter than in the troposphere. Furthermore, the stratospheric response to external forcing factors is different depending on the type of considered forcing, thus providing a discrimination criterion among the various forcings of climate.

The following volume chapters, grouped under the title "Detection and attribution of climate change," will underline different aspects of the problem discussed in the present paper. We purposely did not address here the question of how forced stratospheric changes in turn affect the troposphere. This is an area of great interest and importance, which is explored in detail in Section IV of the present volume. A well documented example of such processes is the role of ozone depletion and its influence on the stratospheric circulation over Antarctica (Thompson *et al.*, 2005). Recent studies have demonstrated how this circulation also has affected the troposphere (Thompson *et al.*, ibid), thus contributing to the large warming observed on the Antarctic Peninsula during winter and spring.

## References

Bengtsson, L., Hodges, K. I., Roeckner, E., and Brokopf, R.: 2006, 'On the natural variability of the pre-industrial European climate', *Climate Dynamics*, in press.

Boer, G., and Yu, B.: 2003, 'Climate sensitivity and response', *Climate Dyn.* **20**, doi:10.1007/s00382-002-0283-3.

Crowley, T. J.: 2000, 'Causes of climate change over the last 2000 years', *Science* **289**, 270–277.

Cubasch, U., and Voss, R.: 2000, 'The influence of total solar irradiance on climate', *Space Sci. Rev.* **94**, 185–198.

Delworth, T. L., and Knutson, T. R.: 'Simulation of early 20th century global warming', *Science* **287**, 2246–2250.

Eddy, J.: 1976, 'The late maunder minimum', *Science* **192**, 1189–1202.

Fröhlich, C.: 2006, 'Solar irradiance variability since 1978', *Space Sci. Rev.*, this volume, doi: 10.1007/s11214-006-9046-5.

Gerber, E. P., and Vallis, G. K.: 2005, 'A stochastic model for the spatial structure of annular patterns of variability and the north atlantic oscillation', *J. Climate* **18**, 2102–2118.

Hansen, J., Sato, M., and Ruedy, R.: 1997, 'Radiative forcing and climate response', *J. Geophys. Res.* **192**, 6831–6864.

Hasselmann, K.: 1976, 'Stochastic climate models. Part 1: Theory', *Tellus* **28**, 273–485.

Held, I. M.: 2005, 'The gap between simulation and understanding in climate modeling', *Bull. Amer. Meteor. Soc.* **86**, 1609–1614.

Hoyt, D. V., and Schatten, K. H.: 1993, 'A discussion of plausible solar irradiance variations, 1700–1992', *J. Geophys. Res.* **98**, 18,895–18,906.

Hurrell, J. W., Kushnir, Y., Ottersen, G., and Visbeck, M.: 2003, 'The North Atlantic Oscillation: Climate significance and environmental impact', *Geophysical Monograph Series* **134**, 279 pp.

IPCC: 2001, 'Climate change 2001: The scientific basis', *Contribution of Working Group I to the Third Assessment Report of the Intergovernmental Panel on Climate Change*, J. T. Houghton, Y. Ding, D. J. Griggs, M. Noguer, P. J. van der Linden, X. Dai, K. Maskell, and C. A. Johnson (Eds.). Cambridge University Press, Cambridge, UK, 881 pp.

Jones, P. D., and Moberg, A.: 2003, 'Hemispheric and large-scale surface air temperature variations: an extensive revision and an update to 2001', *J. Climate* **16**, 206–223.

Jungclaus, J., Botzet, M., Haak, H., Keenlyside, N., Luo, J.-J., Latif, M., Marotzke, J., Mikolajewicz, U., and Roeckner, E.: 2006, 'Ocean circulation and tropical variability in the coupled model ECHAM5/MPI-OM', *J. Climate* **06**, in press.

Lean, J., 2000: 'Evolution of the Sun's spectral Irradiance since the maunder minimum', *Geophys. Res. Lett.* **27**, 2425–2428.

Lean, J., Beer, J., and Bradley, R.: 1995, 'Reconstruction of solar irradiance since 1610: Implications for climate change', *Geophys. Res. Lett.* **22**, 3195–3198.

Lean, J. L., Wang, Y. M., and Sheeley, N. R. Jr.: 2002, 'The effect of increasing solar activity on the sun's total and open magnetic flux during multiple cycles: Implications for solar forcing of climate', *Geophys. Res. Lett.* **29**, doi:10.1029/2002GL015880.

Loptien, U., and Ruprecht, E.: 2005, 'Effects of synoptic systems on the variability of the north atlantic oscillation', *Mon. Weather Rev.* **133**, 2894–2904.

Luterbacher, J., Dietrich, D., Xoplaki, E., Grosjean, M., and Wanner, H.: 2004, 'European seasonal and annual temperature variability, trends and extremes since 1500', *Science* **303**, 1499–1503.

Manabe, S., and Stouffer, R. J.: 1996, 'Low-frequency variability of surface air temperature in a 1000-year integration of a coupled atmosphere-ocean-land surface model', *J. Climate* **9**, 376–393.

Oldenborgh, van G. J., Philip, S., and Collins, M.: 2005, 'El Niño in a changing climate: A multi-model study', *Ocean Science* **1**, 81–95.

Räisänen, J.: 2002, '$CO_2$-Induced changes in interannual temperature and precipitation variability in 19 CMIP2 experiments', *J. Climate* **15**, 2395–2411.

Rind, D.: 2002, 'The Sun's role in climate variations', *Science* **296**, 673–677.

Shindell, D. T., Schmidt, G. A., Mann, M. E., Rind, D., and Waple, A.: 2001, 'Solar forcing of regional climate change during the maunder minimum', *Science* **294**, 2149–2152.

Stephens, G. L.: 2005, 'Cloud feedbacks in the climate system: A critical review', *J. Climate* **18**, 237–273.

Thompson, D. W. J., Baldwin, M. P., and Solomon, S.: 2005, 'Stratosphere–troposphere coupling in the Southern Hemisphere', *J. Atmos. Sci.* **62**, 708–715.

Wang, Y., Lean, J. L., and Sheeley, N. R. Jr.: 2005, 'Modeling the Sun's magnetic field since 1713', *Astrophys. J.* **625**, 522–538.

Xoplaki, E, Luterbacher, J., Paeth, H., Dietrich, D., Steiner, N., Grosjean, M., and Wanner, H.: 2005, 'European spring and autumn temperature variability and change of extremes over the last half millennium', *Geophys. Res. Lett.* **32**, doi:10.1029/2005GL023424.

Zebiak, S. E. and Cane, M. A.: 1987, 'A model El Niño – Southern oscillation', *Mon. Wea. Rev.* **115**, 2262–2278.

# DETECTION AND ATTRIBUTION OF CLIMATE CHANGE, AND UNDERSTANDING SOLAR INFLUENCE ON CLIMATE

W. J. INGRAM
*Atmospheric, Oceanic and Planetary Physics, Department of Physics, University of Oxford, OX1 3PU, UK;*
*Hadley Centre for Climate Prediction and Research, Meteorological Office, Fitzroy Road, Exeter, Devon, Ex1 3PB, UK*
*(E-mail: ingram@atm.ox.ac.uk)*

(Received 29 July 2005; Accepted in final form 27 February 2006)

**Abstract.** The detection of externally-forced climate change in observations, and its attribution to specific forcings, sounds simple enough to some people, but with others it has a reputation as a complex and arcane specialism. In fact, both these impressions have some truth – in principle it is no more than regressing expected patterns of climate change (normally obtained from GCM simulations forced with observed or reconstructed past forcings) against the corresponding observations, with uncertainty estimates that try to be as rigorous as possible, but there are many technical complexities.

This survey begins with some motivating examples, and then summarizes the principles, problems and procedure without formal mathematics, before surveying results with an emphasis on possible solar effects, and why they are particularly problematic.

**Keywords:** climate change, detection and attribution, solar

## 1. Introduction and Motivation

The complexity of the D&A ("detection and attribution") formalism, with all its matrix algebra, means it is not easy to apply it, and often not easy to follow the physical meaning of the results. The GCMs (General Circulation Models) used to produce "expected" patterns of climate change are even more complex, and it is not clear how far we should trust them. All this may suggest we should find simpler ways of looking for climate change signals in observations – this section aims to show why not.

Figure 1 shows a fit of greenhouse gas and solar forcings to the 20th-century temperature record, with some smoothing.[1] The fit clearly attributes most of the warming to solar forcing, with the greenhouse gas contribution almost negligible.

[1](from `ftp://sprite.llnl.gov/pub/covey/IPCC_4AR_Forcing`).

Space Science Reviews (2006) 125: 199–211
DOI: 10.1007/s11214-006-9057-2

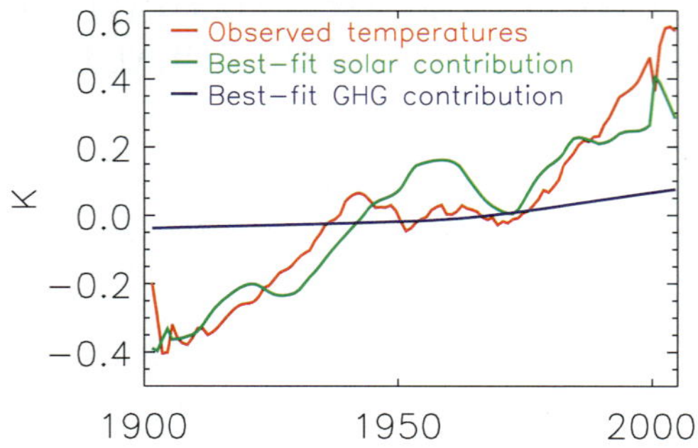


*Figure 1.* 20th-century global-mean (near-)surface temperature anomalies (red, from Jones and Moberg, 2003), with the solar (green) and greenhouse gas (blue) forcings scaled to best fit by stepwise regression. All series were smoothed with an 11-year running mean to remove short-time-scale unforced variation, "noise", such as ENSO.

But Figure 2 shows just the reverse. They differ only in that an 11-year running mean was used for Figure 1, but a 9-year one for Figure 2. (Only odd numbers are used, to preserve the original length of the timeseries, by reducing the length of the running mean towards the endpoints).

Clearly these results mean nothing. And this example was not hard to find – I looked for a fit that was not robust using the simplest possible regression on the most conveniently available estimated forcing timeseries, and varying the length of the running mean was the first thing I tried. However, this is a typical artifact of stepwise regression (fitting whichever independent variable has the highest correlation with the data, removing that apparent signal from the data, and repeating till all independent variables have been fitted), which is precisely why it should not generally be used – but papers based on it still appear.

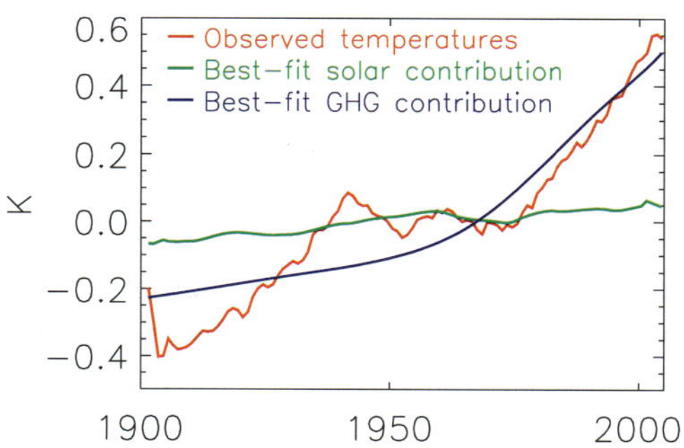


*Figure 2.* As Figure 1 but using 9-year running means.

Figure 3 resembles Figures 1 and 2, but is based on proper multilinear regressions, and also includes the potentially important forcings from volcanic and

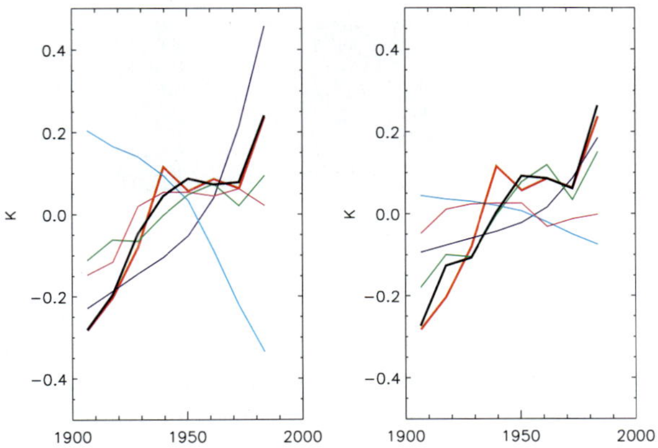


*Figure 3.* As Figures 1 and 2, but multivariate regression using 11-year means, and also allowing for volcanic (pink) and fossil-fuel (pale blue) aerosols. Overall best fit added in black. Left – volcanic timeseries is the Volcanic Explosivity Index (VEI) of Newhall and Self (1982), right – the Dust Veil Index (DVI) of Lamb (1970), both updated.

fossil-fuel aerosols. Like Figure 2, Figure 3a shows greenhouse gases dominating the 20th-century warming. But in Figure 3b changing the estimate of the volcanic forcing has made the greenhouse-gas fit no bigger than the solar. So even here, results are not as robust as we might hope. This example was harder to find, but still shows seriously non-robust results.

So what is going on? What is called *degeneracy*: the signals of different physical processes that we are fitting are not different enough in shape to be unambiguously distinguished. (Formally, a linear combination of them is close to zero). Fossil-fuel burning increased over the century as a whole, but with a hesitation in the middle – as did reconstructed solar irradiance. Thus the patterns of anthropogenic greenhouse gas forcing (roughly the integral of the burning rate), of the anthropogenic sulphate aerosol forcing (roughly the instantaneous rate) and of the solar forcing are all similar. So, we need error bars that take into account how similar our signals are, and a way to judge when they're so degenerate we shouldn't even try to fit them.

It is always well to bear in mind that the only thing we *know* about the best fit is that it has zero likelihood of being correct – without a trustworthy estimate of how close it is likely to be to reality, it is worth little.

Even when the patterns of forcing (how it varies in time or space or both) are not well known, estimating the patterns of climate response must always introduce further uncertainty. So why not fit the patterns of forcing, as in the above examples (and all too many published papers)?

Figure 1 gives a reason why not. The solar forcing looks rather like the temperatures, delayed a few years. Physically the response must of course follow the forcing, and if we shifted the forcings a few years later to reflect this (HadCM3, a well-known Hadley Centre GCM, implies about a decade would be appropriate), the solar would fit much less well, while the greenhouse gas fit would not be so

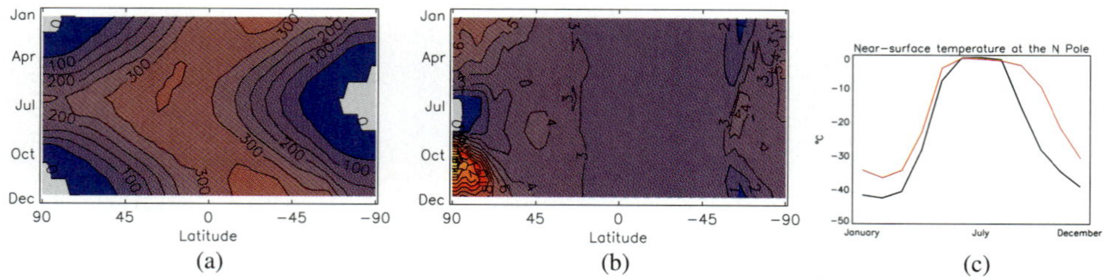


*Figure 4.* Mean seasonal cycles from the HadSM3 model of zonal means of (a) insolation absorbed by the atmosphere and surface, (b) near-surface equilibrium warming due to insolation increases, contour intervals 50 $Wm^{-2}$, 1 K respectively. (c) Mean seasonal cycle at the North Pole of near-surface temperature in the control climate (black) and warmed climate (red).

much degraded. So in this case – global-mean temperature through the 20th century – it happens that looking for the forcing pattern rather than the response pattern spuriously inflates the solar effect considerably.

Figure 4 gives a better example. 4a and 4b show mean seasonal cycles as a function of latitude and month. 4a is the insolation absorbed, which of course also gives the pattern of forcing due to TSI increases (without ozone changes). We see the expected tropical maximum, moving North and South with the sun, with decreases towards higher latitudes where the sun is lower in the sky and there is more cloud, ice and snow to reflect it, and zero in the polar night. 4b is the pattern of near-surface warming in response to TSI increases (without ozone changes) in the GCM HadSM3. Clearly, the two patterns are very different. A broad band of lower warming generally overlaps with the band of largest forcing in the tropics, while the largest warming is not only at a latitude of minimum forcing, but actually in the season of zero forcing: the polar night.

An obvious explanation is that the figure is wrongly plotted, but it isn't. Or you might think that HadSM3 is no good, but other GCMs respond similarly – and there is a physical explanation. Figure 4c shows the near-surface temperature at the North Pole. In summer, surface temperatures over the Arctic ocean cannot rise above melting point, because sea-ice is present as well as open water. Then as autumn starts, heat loss from the surface cannot cool it below freezing point till the open water has frozen over again. When this happens it insulates the surface from the comparatively warm water below the ice and local surface temperatures can rapidly drop tens of kelvins. Thus more sunlight in summer cannot raise temperatures then: rather, it melts more ice and is stored as latent heat, delaying the major cooling in autumn after the sun has set.

So we believe this effect is real. Then if we used the pattern of forcing (4a) to look for a solar response in observations we would have no chance of finding it (though we might think we had – for example, we might fit ENSO, which does give a global warming concentrated in the tropics).

Of course all GCMs are imperfect, and limited in the scales they can represent. Nevertheless, they are the most detailed and physically-based models we have, and can include a wealth of interactions – for example, rather than the constant lag of response behind forcing suggested above, they will combine a fast response over land (e.g., the Northern continents in summer) with a component strongly lagged by ocean heat capacity.

## 2. Basic Logic of D&A

Different GCMs have very different climate sensitivities, but they agree more on the patterns of response, and these patterns can be explained physically, and seem reasonable. So we don't trust GCMs to give us the right size climate change, but we hope they get the large-scale patterns of response roughly right.

Our basic procedure is then to calculate the uncertainty range (at some significance level, typically 5–95%) for the amplitude in the observations of a particular climate change signal, i.e., how much we need to scale the GCM's pattern to fit the observations ("$\beta$"). If this range is entirely positive, we have "detected" that climate change signal in the observations (at that significance level). If the uncertainty range of $\beta$ includes 1, the GCM's sensitivity is consistent with the observations ("could be correct").

While we want to know about all the uncertainty in our results, we would like it small. "Optimal detection" increases the power of the test by giving more weight to modes with higher signal/noise ratio. It would be formally optimal if certain fairly standard idealizations applied (Hasselmann, 1993): in practice it is probably usually quite close. It does *not* bias the $\beta$s found, just reduces their uncertainty – though the transformation to empirical statistical modes (see below) does make it harder to see what's going on physically, so many papers confirm results with simpler indices such as global, zonal or continental means.

Where there are choices to make which will bias the results, we always make the conservative choice (the one which means the real significance will be higher than the nominal one).

We need to stay aware of issues of circularity (climate forcings will only be included in GCM simulations and D&A work if there is already reason to expect a response by the climate system) and test multiplicity (if none of the forcings were active we should expect spurious detections at the 5% level about 5% of the time – and should take into account not only our results, but all the never-reported negative results). Wishful thinking and neglect of these points has been so common in solar/climate work (e.g., Pittock, 1978) as to lower the reputation of the whole field – hence the considerable initial scepticism about work such as Labitzke and van Loon (1988).

We should also check for as many as we can of the uncertainties that are not included in the formal error estimates by doing sensitivity studies changing details of methodology and reconstructions of the less certain forcings: only results that are robust under these tests should be believed.

For brevity this paper concentrates on "detection", but we should note that "attribution" aims to supply rigorous probabilistic answers to questions like "How much of the 20th-century warming trend was due to greenhouse gases?" (Stott *et al.*, 2006, find that 120–220% is a robust uncertainty range, depending largely on the amount of anthropogenic aerosol cooling, with natural factors playing a small role), "How much global warming can we expect over the 21st century?" (Stott and Kettleborough, 2002, use 20th-century $\beta$s to estimate this for various emissions scenarios, and Stott *et al.* (2006), confirm this is not very dependent on the GCM used) or "Was the central European heatwave of summer 2003 anthropogenic?" (in a chaotic system one cannot say that anything "caused" a particular observed event, but Stott *et al.* (2004), found it very likely that human interference had more than doubled the chance of such an event).

## 3. Methods

- To combine and scale expected signals, we assume everything combines linearly. This is true enough for GCMs' large-scale temperature changes (Boer and Yu, 2003; Gillett *et al.*, 2004), and must be true in any dynamical system with enough random variation, when the forced responses of interest are not too large compared to the internal variability.
- We test for excessive degeneracy and don't regress on any signal combinations which fail the test. (If there are any, they will of course include the complete set of signals considered, so, even we had if perfect estimates of all relevant signals, we, like Tett *et al.* (1999), might not end up with a unique best explanation).
- We take not only the "signal" (pattern of response) from GCMs, but also the "noise" (internal variability – signals must be large compared with this to be "detected"). Obviously any use of observations themselves to define a pattern to look for in those same observations introduces a circularity that would make "detection" meaningless.
- Even if we have simulated millennia to estimate the noise from, for century or half-century D&A we can only test a fairly small number of degrees of freedom. Until recently the largest spatial scales were always chosen – the most important and most likely to be well-simulated. (It was also generally assumed that regional and local detection would be impossible because the noise would be too large without the smoothing effect of large-scale averaging, but this is not actually true; IDAG, 2005). Often we project onto the model's dominant modes of unforced variation. Then the behaviour to be expected under the null hypothesis is well sampled and we can formally optimize the statistical power. But if the forced simulation does something unlike anything in the unforced one, and the observations do the same, that fact is ignored (perhaps not ideal, but certainly conservative). The modes used are typically Empirical Orthogonal Functions (von Storch and Zwiers, 1999) – formally these are eigenfunctions of covariance or correlation

matrices, but hopefully the leading ones at least have physical interpretations (e.g., global warming, hemispheric contrast, Cold Ocean Warm Land).

- The timescales considered are typically decadal, averaging out ENSO and other short-scale variability, but also, of course, removing the 11-year solar cycle, often considered too short to count as climate change.
- GCMs give us spatial patterns of response, which can resolve problems of the sort shown in the Introduction (Tett *et al.*, 1999). But of course atmospheric and oceanic flows move heat around, smoothing the input energy distributions, and then the response patterns are strongly influenced by local feedbacks which work similarly for all forcings (such as the sea-ice processes discussed above), so the spatial patterns of the modelled responses are more similar than those of their forcings (Boer and Yu, 2003) and the extra useful information is less than we might hope. Most GCM-based D&A work has derived the full space-time patterns. An alternative is to assume separability and look for a single spatial pattern, varying in time (e.g., Crooks and Gray, 2005) – assuming the time variation is given by the forcing timeseries means fitting only the instantaneous component of change, but this matters less for temperatures aloft where the lag due to the ocean's heat capacity is less important.
- Ordinary least-squares fitting assumes the signal is perfectly known, giving an unnecessary conservative bias and so making it harder to detect weaker signals in particular. "Total least squares" allows for sampling error in the signal, but gives larger, asymmetric uncertainty ranges that can extend to infinity (Allen and Stott, 2003). (Ensembles of GCM simulations are usually run to reduce this sampling error, but computational cost keeps them small).
- Both the optimization and the uncertainty estimation are thus based on using samples of the model variation. One technical issue is that they need to use different samples to avoid a liberal bias. Samples can come from a long control simulation (if the "drift" is not too bad) and/or the difference between an ensemble of runs with the same forcing.
- It is usual to assume the variable of interest has a Gaussian distribution, which again seems to be true enough for large-scale temperature. This can and should be checked in the residuals, though arguably one should expect any error introduced to be conservative.
- We omit model data where there are no observations. Observational error has usually been assumed to be negligible compared to possible modelling error (but Thorne *et al.*, 2005, argue it should not be).

Thus everything depends very much on the GCM. This is a major constraint, but GCMs are, overall, the best models we have. We do have some ways of checking each GCM:

- We can apply a consistency test – are the observations a plausible combination of the GCM's patterns and the GCM's internal variability (at some chosen significance level)? This is inevitably a weak test in that errors in different aspects could cancel, since all we have to check against are observations which combine all the forced responses and internal variability. (Still, it is not so weak that GCMs all pass easily – HadCM2, HadCM3's predecessor, did not pass it on smaller scales; Tett *et al.*, 1999).

- There are many GCMs of completely different formulation. D&A results from them are broadly consistent (Hegerl *et al.*, 2000; Stott *et al.*, 2006), implying they must be right unless we have very bad luck, important physics consistently missing (unknown – or omitted because we don't know how to include it), or consistent artifacts from lack of resolution.
- We can and should try to understand physically what's going on and check it's sensible.

But still, weaker signals, or signals whose forcing is less well known, or where the response is less accurately simulated, are less detectable. And of course all of these would work against detecting any solar signal.

In principle D&A could be applied to any observed variable, but the one most studied has been (near-)surface temperature – the most reliably observed, and with the most public interest. Temperature aloft has been used (Tett *et al.*, 1996; Crooks and Gray, 2005), but though it has promise for usefully differentiating forcings with similar patterns of response at the surface, its observational record is much disputed (Thorne *et al.*, 2005) as well as rather shorter, and the generation of GCM that most D&A work has been done with did not generally have enough vertical resolution in the stratosphere to rely on signals there. Other variables have included ocean heat content, precipitation, sea-level pressure and other measures of circulation. The forcings most studied have been anthropogenic greenhouse gases and aerosols, as most is known about them, and they are most policy-relevant (under human control). Even where variations in solar irradiance are included, this may be in combination with volcanic stratospheric aerosol as "natural" forcing, or in "all forcings" simulations.

## 4. Results

I shall continue to concentrate on work at the centres I am attached to – much excellent work has been done elsewhere, but a general survey will not suffer from focussing on what I know best. Thorough and authoritative summaries of work in the field are provided by the IPCC reports (Santer *et al.*, 1996; Mitchell *et al.*, 2001, and the 4th Assessment Report to appear in 2007).

The much-reproduced graphs in Figure 5 compare global-mean warming over the 20th century with the results of 3 HadCM3 ensembles. Plainly the ensemble with only natural forcings is deficient in the late-century warming, while the one with only anthropogenic forcings is deficient in the early-century warming – but the one with both natural and anthropogenic forcings looks just right!

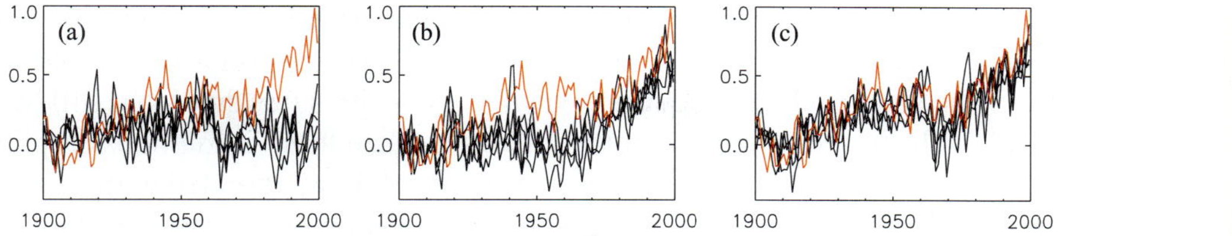


*Figure 5.* Global-mean temperature anomalies (K) through the 20th century from observations (red), and ensembles of simulations with the HadCM3 GCM (black) forced with (a) natural, (b) anthropogenic and (c) all forcings.

Since D&A is concerned with patterns rather than their size, and looks at spatial structure as well as global means, Figure 5 alone would not mean much – but in fact, what it seems to show is confirmed by detailed D&A results using HadCM3 (Tett *et al.*, 2002). We found the late 20th century dominated by greenhouse gas warming and sulphate aerosol cooling, both anthropogenic. Any solar effect was swamped, as might be expected given the smaller size of the forcing and its partial degeneracy with the anthropogenic forcings, as shown above. But in the early 20th century we detected both anthropogenic and solar signals, whichever estimate of historical solar variation we used, and attributed the warming to a combination of greenhouse-gas- and solar-forced effects, as well as internal variability.

Looking across the range of D&A studies, anthropogenic warming late in the century is always robustly detected. The warming earlier in the century is attributed to various combinations of greenhouse gases, solar, volcanoes and internal variability, with solar not always detected.

Before the 20th century, there is even more uncertainty in both forcing and the actual climate variation. The estimated solar signal, having been inconveniently degenerate with the anthropogenic signals in the 20th century, becomes degenerate with the *volcanic* signal over the previous centuries (Hegerl *et al.*, 2003) and either can be used to fit the proxy data. (Strict parsimony would then mean ascribing the effects to volcanic forcing, as that has clearly shown an effect on climate with the Pinatubo eruption).

Looking only at the last few decades, the "best-observed" period with satellite data, also has its problems, and not just those of data homogeneity (Thorne *et al.*, 2005; Trenberth *et al.*, 2002). For Europe, over the late 20th century, solar forcing correlates quite well with the North Atlantic Oscillation index if both are smoothed – but by chance: the records are long enough to show this did not happen earlier (Rodwell, 2003). Globally, there are more serious degeneracies – the two major eruptions of these recent, globally well-observed, decades, El Chichón and Pinatubo, both coincided with both solar maximum and El Niño. (Even the previous major eruption, Agung, also coincided with solar maximum). Of course the pattern of variation in time will differ, but this is not as helpful as it will seem if inadequate allowance is made for internal variability, or if other methodological errors are made. An easy error if one does not understand Fourier analysis is to think that because the e-folding time of the volcanic eruptions is much less than the period of the solar cycle, the Fourier components on the latter timescale will have negligible volcanic contribution. In fact, if you idealize these 2 eruptions erupting at the same phase of the solar cycle as an exponential decay repeated with a period 5 times the decay time, just over half their power goes into the periodic ("solar") component.

"Solar amplification" – the idea that climate is much more sensitive to solar forcing than, say, to greenhouse gases – has attracted attention over the years. It may have originated in wishful thinking given that greenhouse gas forcing is much larger than estimates of solar irradiance forcing, or in attempting to explain spurious fits such as Figure 1 and Figure 3a, but is not physically absurd. A particular GCM will usually show fairly similar climate sensitivities to different forcings – i.e., the same amount of heating produces a similar warming – but cloud feedbacks, and so climate sensitivity, can differ greatly between GCMs, and so conceivably might be quite different for different forcings in the real world. A similar end result would follow if other solar-related climate forcings were of comparable size to the direct effect of irradiance changes, though there seems to be no evidence that any are. (The one well-established example of such a forcing, ozone change induced by solar UV variation, has effects in the solar and longwave parts of the spectrum which tend to compensate (Haigh, 1994) and while it does tend to produce a specific pattern of response, this is from a generic mechanism which also applies to the direct irradiance effect (Haigh and Blackburn, 2006). Since we do not have quantitative confidence in either modelled or observational estimates of the ozone response, its omission from "solar-forced" GCM runs therefore seems an acceptable approximation, though clearly not ideal).

But for years analysis of HadCM3 had pointed to its solar response needing to be roughly doubled to match the real world best, while its anthropogenic response was about the right size. Of course this might be just random – the $\beta$s were formally consistent. But Stott *et al.* (2003) concluded that, despite artifacts of degeneracy apparently tending to exaggerate the solar signal seen and reduce the apparent greenhouse gas signal, there was evidence that HadCM3's solar response needs to be roughly doubled to match the real world – consistent with missing solar forcing or feedbacks.

More recent analysis has shown that this is in fact because HadCM3 is only half as sensitive to solar forcing as it is to other forcings (Gregory *et al.*, 2004). This model eccentricity requires full ocean dynamics, and so had been missed in our previous investigations of climate sensitivity. Thus the implied real-world climate sensitivities from a HadCM3 analysis are in fact very similar between solar and greenhouse gases – *if* the Lean *et al.* (1995) reconstruction of solar irradiance variation used in these runs is correct. If on the other hand this reconstruction exaggerates the actual long-term variation of solar irradiance by a factor of, say, 2 or

3 (Lean *et al.*, 2002), then the HadCM3 analysis still implies a "solar amplification" by a factor of about 2 or 3. It will be interesting to see results from newer GCMs with better vertical resolution in the stratosphere.

## 5. Conclusions

Looking for possible climate change signals in observations is not simple – "standard" statistical issues are complicated by degeneracy, uncertainties in many forcings, and uncertainties in the forcing-response relationship. D&A does its best to provide a rigorous methodology, but care is still needed over those uncertainties which are not naturally included in D&A's error bars (such as uncertainties in the reconstruction of solar irradiance variation).

Anthropogenic signals in the late 20th century are very robustly detected. Reconstructed solar signals are repeatedly degenerate with others, particularly anthropogenic ones for the 20th century and the volcanic one for previous centuries. Uncertainties in the reconstruction of irradiance variation and in the full physics of the response would also hinder solar detection, as, in recent decades, would the larger anthropogenic signal. HadCM3 still gives evidence for a solar signal in the early 20th century, but not all GCMs agree, so D&A cannot currently refute the possibility that there is no long-term variation in solar irradiance (Lean *et al.*, 2002). "Solar amplification" remains an open issue.

## References


Allen, M. R. and Stott, P. A.: 2003, 'Estimating signal amplitudes in optimal fingerprinting, part I: theory', *Clim. Dyn.* **21**, 477–491.

Boer, G. J. and Yu, B.: 2003, 'Climate sensitivity and response', *Clim. Dyn.* **20**, 415–429.

Crooks, S. A. and Gray, L. J.: 2005, 'Characterisation of the 11-year solar signal using a multiple regression analysis of the ERA-40 dataset', *J. Climate* **18**, 996–1015.

Gillett, N. P., Wehner, M. F., Tett, S. F. B., and Weaver, A. J.: 2004, 'Testing the linearity of the response to combined greenhouse gas and sulfate aerosol forcing', *Geophys. Res. Lett.* **31**, doi:10.1029/2004GL020111.

Gregory, J. M., Ingram, W. J., Palmer, M. A., Jones, G. S., Stott P. A., Thorpe, R. B., Lowe, J. A., Johns T. C., and Williams K. D.: 2004, 'A new method for diagnosing radiative forcing and climate sensitivity', *Geophys. Res. Lett.* **31**, doi:10.1029/2003gl018747.

Haigh, J. D.: 1994, 'The role of stratospheric ozone in modulating the solar radiative forcing of climate', *Nature* **370**, 544–546.

Haigh, J. D. and Blackburn, M.: 2006, 'Solar influences on dynamical coupling between the stratosphere and troposphere', *Space Sci. Rev.*, this volume, doi: 10.1007/s11214-006-9067-0.

Hasselmann, K.: 1993, 'Optimal fingerprints for the detection of time-dependent climate change', *J. Clim.* **6**, 1957–1971.

Hegerl, G. C., Stott, P. A., Allen, M. R., Mitchell, J. F. B., Tett, S. F. B., and Cubasch, U.: 2000, 'Optimal detection and attribution of climate change: Sensitivity of results to climate model differences', *Clim. Dyn.* **16**, 737–754.

Hegerl, G. C., Crowley, T. J., Baum, S. K., Kim, K. Y., and Hyde, W. T.: 2003, 'Detection of volcanic, solar and greenhouse gas signals in paleo-reconstructions of northern hemispheric temperature', *Geophys. Res. Lett.* **30**, doi:10.1029/2002GL016635.

IDAG (International Detection and Attribution Group), Barnett T., Zwiers, F., Hegerl, G., Allen, M., Crowley T., Gillett, N., Hasselmann, K., Jones, P., Santer, B., Schnur, R., Stott, P., Taylor, K., and Tett, S.: 2005, 'Detecting and attributing external influences on the climate system: A review of recent advances', *J. Clim.* **18**, 1291–1314.

Jones, P. and Moberg, A.: 2003, 'Hemispheric and large-scale surface air temperature variations: An extensive revision and an update to 2001', *J. Clim.* **16**, 206–223.

Labitzke, K. and van Loon, H.: 1988, 'Associations between the 11-year solar cycle, the QBO and the atmosphere. Part I: The troposphere and stratosphere in the northern hemisphere', *J. Atmos. Terr. Phys.* **50**, 197–206.

Lamb, H. H.: 1970, 'Volcanic dust in the atmosphere, with a chronology and assessment of its meteorological significance', *Phil. Trans. Roy. Soc.* **A266**, 425–533.

Lean, J., Beer, J., and Bradley, R.: 1995, 'Reconstruction of solar irradiance since 1610: Implications for climate change', *Geophys. Res. Lett.* **22**, 3195–3198.

Lean, J. L., Wang, Y. M., and Sheeley, N. Jr: 2002, 'The effect of increasing solar activity on the sun's total and open magnetic flux during multiple cycles: Implications for solar forcing of climate', *Geophys. Res. Lett.* **29**, doi:10.1029/2002GL015880.

Mitchell, J. F. B., Karoly, D. J., Hegerl, G. C., Zwiers, F. W. Allen, M. R., and Marengo, J.: 2001, 'Climate Change 2001: The Scientific Basis', Contribution of Working Group I to the *Third Assessment Report of the Intergovernmental Panel on Climate Change*, Chapter 12: Detection of Climate Change and Attribution of Causes, Houghton J. T., Ding, Y., Griggs, D. J., Noguer, M., van der Linden, P. J., Dai X., (eds.), Cambridge University Press, pp. 525–582.

Newhall, C. and Self, S. S.:1982, 'The volcanic explosivity index (VEI): An estimate of explosive magnitude for historical volcanism', *J. Geophys. Res.* **87**, 1231–1238.

Pittock, A. B.: 1978, 'A critical look at long-term Sun-weather relationships', *Revs. Geophys. and Space Phys.* **16**, 400–420.

Rodwell, M.: 2003, 'On the predictability of North Atlantic climate', in J. W. Hurrell, Y., Kushnir, G., Ottersen and M. Visbeck (eds.), *The North Atlantic Oscillation*, American Geophysical Union, pp. 173–192.

Santer, B., Wigley, T. M. L., Barnett, T. P., and Anyamba, E.: 1996, 'Detection of climate change and attribution of causes', In *Climate Change 1995: the Science of Climate Change*, J. T. Houghton, L. G. Meira Filho, Callender B. A., Harris, N., A. Kattenberg and K. Maskell (eds.), Cambridge University Press, pp. 407–443.

Stott, P. A. and Kettleborough, J. A.: 2002, 'Origins and estimates of uncertainty in twenty-first century temperature rise', *Nature* **416**, 723–726.

Stott, P. A., Jones, G. S., and Mitchell, J. F. B.: 2003, 'Do models underestimate the solar contribution to recent climate change?' *J. Clim.* **16**, 4079–4093.

Stott, P. A., Stone, D. A., and Allen, M. R.: 2004, 'Human contribution to the European heatwave of 2003', *Nature* **432**, 610–614.

Stott, P. A., Mitchell, J. F. B., Gregory, J. M., Santer, B. D., Meehl, G. A., Delworth, T. L., and Allen, M. R.: 2006, 'Observational constraints on past greenhouse warming and predictions of future global warming', *J. Clim.* **19**, 3055–3069.

Tett, S. F. B., Mitchell, J. F. B., Parker, D. E., and Allen, M. R.: 1996, 'Human influence on the atmospheric vertical temperature structure: Detection and observations', *Science* **274**, 1170–1173.

Tett, S. F. B., Stott, P. A., Allen, M. R., Ingram, W. J., and Mitchell, J. F. B.: 1999, 'Causes of twentieth century temperature change near the earth's surface', *Nature* **399**, 569–572.

Tett, S. F. B., Jones, G. S., Stott, P. A., Hill, D., Mitchell, J. F. B., Allen, M. R., Ingram, W. J., Johns, T. C., Johnson, C., Jones, A., Roberts, D. L., Sexton, D. M. H., and Woodage, M. J.: 2002,

'Estimation of natural and anthropogenic contributions to 20th century temperature change', *J. Geophys. Res.* **107**, doi:10.1029/2000JD000028.

Thorne, P. W., Parker, D. E., Christy, J. R., and Mears, C. A.: 2005, 'Uncertainties in climate trends: Lessons from upper-air temperature records', *Bull. Am. Meteorol. Soc.* **86**, 1437–1442.

Trenberth, K. E., Wielicki, B. A. Del Genio, A. D., Wong T., Chen, J., Carlson, B. E., Allan, R. P., Robertson, F., Jacobowitz, H., Slingo, A., Randall, D. A., Kiehl, J. T., Soden, B. J., Gordon, C. T., Miller, A. J., Yang, S. K., and Susskind, J.: 2002, 'Changes in tropical cloud and radiation', *Science*, **296**, 2095.

von Storch, H. and Zwiers, F. W.: 1999, 'Statistical analysis in climate research'. Cambridge University Press, Cambridge, UK, 484 pp.

# THE CLIMATE RESPONSE TO THE ASTRONOMICAL FORCING

M. CRUCIFIX*, M. F. LOUTRE and A. BERGER
*Institut d'Astronomie et de Géophysique G. Lemaître, Louvain-la-Neuve, Belgium*
*(*Author for correspondence: E-mail: michel.crucifix@uclouvain.be)*

(Received 30 August 2005; Accepted in final form 22 February 2006)

**Abstract.** Links between climate and Earth's orbit have been proposed for about 160 years. Two decisive advances towards an astronomical theory of palæoclimates were Milankovitch's theory of insolation (1941) and independent findings, in 1976, of a double precession frequency peak in marine sediment data and from celestial mechanics calculations. The present chapter reviews three essential elements of any astronomical theory of climate: (1) to calculate the orbital elements, (2) to infer insolation changes from climatic precession, obliquity and eccentricity, and (3) to estimate the impact of these variations on climate. The Louvain-la-Neuve climate-ice sheet model has been an important instrument for confirming the relevance of Milankovitch's theory, but it also evidences the critical role played by greenhouse gases during periods of low eccentricity. It is recognised today that climatic interactions at the global scale were involved in the processes of glacial inception and deglaciation. Three examples are given, related to the responses of the carbon cycle, hydrological cycle, and the terrestrial biosphere, respectively. The chapter concludes on an outlook on future research directions on this topic.

**Keywords:** astronomical theory of palæoclimates, Milankovitch, insolation, $CO_2$

## 1. Calculating Orbital Parameters

The solar energy received at the top of the atmosphere in one true solar day is a function of the total solar irradiance ($S$), the latitude ($\phi$), the tilt of the Earth's orbit on the ecliptic ($\varepsilon$) and the Earth-Sun distance ($r$). This distance is known from the length of the semi-major axis ($a$) of the Earth's orbit, its eccentricity ($e$) and the true anomaly ($\nu$) giving the position of the Earth on its orbit. Figure 1 and the *key to symbols* at the end of this chapter may be referred to for a definition of the various angles. Given that the seasons are defined by reference to the vernal equinox (V.E.), $\nu$ is expressed as the difference between the longitude of the Earth ($\lambda$) to the vernal equinox, and that of the perihelion ($\varpi$). An astronomical theory of palæoclimates is built on the hypothesis that the evolution of climate is somehow determined by the seasonal and spatial distribution of insolation. This distribution is a functional of $a$, $e$, $\varepsilon$ and $\varpi$. It therefore links the evolution of climate to the long-term (*secular*) evolution of Earth's orbital elements.

There are no secular variations of the semi-major axis of the planets at first order with respect to the masses (Laplace, 1773; Lagrange, 1781), and actually, the variations of the semi-major axis of the Earth are so small that they will not induce

Space Science Reviews (2006) 125: 213–226
DOI: 10.1007/s11214-006-9058-1

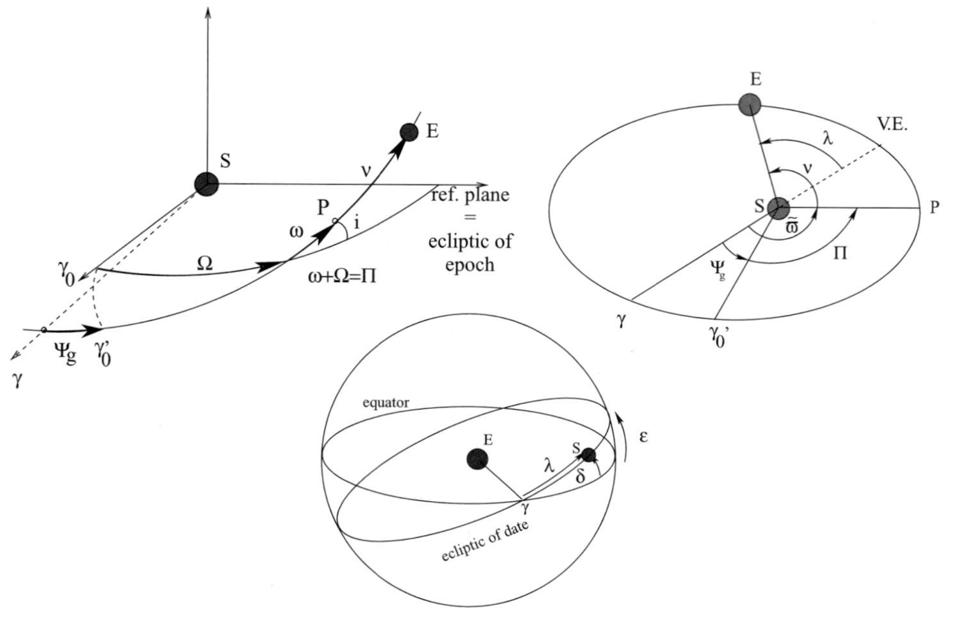


*Figure 1.* Angles relevant for the astronomical theory of climate, represented on *(left)* the ecliptic plane of epoch, *(right)* the actual orbit plane, also called ecliptic of date, and *(middle)* in geocentric coordinates. P is perihelion and V.E., vernal equinox.

any noticeable change of the mean Earth-Sun distance over the past 250 million years (Laskar *et al.*, 2004). Eccentricity is obtained from a perturbation of the two-body problem by the other planets as first formalised by Lagrange (1781). The perturbed equations come as two systems (e.g., Chapront *et al.*, 1975), determining $(e, \Pi)$ and $(i, \Omega)$, respectively.

The movement of the perihelion with respect to the moving vernal equinox results from the combination of two effects ($\tilde{\varpi} = \Psi_g + \Pi$): the first one is the general precession in which the torque of the Sun, the Moon and the planets on the Earth's equatorial bulge causes the axis of Earth rotation to wobble like a spinning top with a period of ~25,800 years; the second one is the rotation of the elliptical figure of the Earth's orbit relative to the stars, at a period of 100 kyr (thousand years). The two effects together produce the climatic precession, with a mean period of 21 kyr. Once the $(e, \Pi)$ and $(i, \Omega)$ systems are solved, the angles $\Psi_g$ and $\varepsilon$ can be obtained by resolution of the precession equation of the Earth.

Berger (1977) provides a comparison of various analytical solutions to the perturbation problem differing by the number of terms retained for the series expansion of perturbation equations. Berger (1978) reorganised the expressions of climatic precession, obliquity and eccentricity in a form appropriate for the astronomical theory of palæoclimates:

$$e \sin \varpi = \sum_i P_i \sin(\alpha_i t + \eta_i)$$

$$\varepsilon = \varepsilon^* + \sum_i A_i \cos(\gamma_i t + \zeta_i) \qquad (1)$$

$$e = e^* + \sum_i E_i \cos(\lambda_i t + \phi_i)$$

The amplitudes ($P_i$, $A_i$ and $E_i$), frequencies ($\alpha_i$, $\gamma_i$ and $\lambda_i$) and phases ($\eta_i$; $\zeta_i$ and $\phi_i$), provided by Berger (1978) on the basis of the orbital solution of Bretagnon (1974), form the BER78 solution. The main advantage of this formulation is to readily provide the spectrum of orbital parameters. It can thus be easily compared with palæoclimatic records. The essential characteristics of the evolution of the orbital elements are: (a) the existence of four dominant periods in precession: 23.7, 22.4, 18.9 and 19.1 kyr – two peaks are resolved in the marine palæoclimate record by Hays *et al.* (1976), (b) obliquity varying between 22 and 25° with a period of approximately 41 kyr, and (c) dominant periods of eccentricity of 404, 95, 124, 99 and 131 kyr – they correspond to the beatings of climatic precession. It is important to realise that the so-called 100-kyr period of eccentricity is actually a combination of several periods. The 404 kyr period began to dominate the oscillations of eccentricity about 1 Myr ago. The last cycle is particularly well defined (Figure 2).

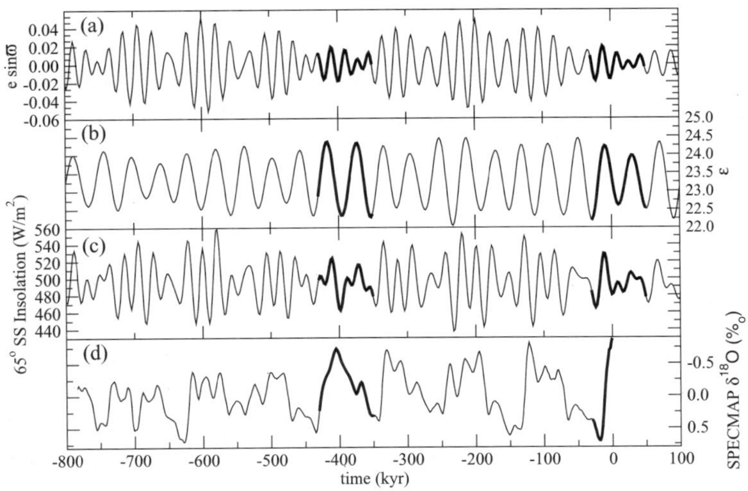


*Figure 2.* Long term variations of (a) the climatic precession parameter, (b) obliquity, (c) summer solstice insolation at 65°N, and (d) ice volume reconstructed by Imbrie *et al.* (1984) based on the oxygen isotopic composition of benthic foraminefera (the *Y*-axis is configured such that sea-level increases upwards (ice volume increases downwards). Time zero corresponds to year 1950 AD. Portions corresponding to the epochs [−430 kyr; −350 kyr] and [−30 kyr; +50 kyr] are in bold. They are considered as analogous because of low eccentricity and similar phasing of precession. However, it is seen that insolations for the two periods differ significantly.

Note finally that system (1) is not convergent, but will provide a solution of suitable precision over a finite period of time (roughly ±1 Myr with respect to the present).

Laskar (1984) indeed realises that the traditionnal perturbative method cannot be used to infer a valid astronomical solution over much longer time scales due to strong divergences appearing in the inner planets system. He therefore switches to a numerical integration in an averaged from (short period terms removed, Laskar, 1988). Berger and Loutre (1991) then provide a revised solution, called BER90, to be used for [−3 Myr; 3 Myr]. This solution combines Laskar's (1988) numerical solution of the $(e, \Pi)$ and $(i, \Omega)$ relativistic systems with an analytical solution of the luni-solar precession. The time series of orbital elements then undergo a spectral decomposition in order to be expressed in the form of Equation (1), which remains the most convenient for palæoclimate studies.

The most up-to-date solution (presented in the form of a computer code, but a spectral decomposition is also available) is provided by Laskar *et al.* (2004) (the La04 solution), which is actually an update to the previous La93 solution (Laskar *et al.*, 1993). It is obtained by direct integration (no time averaging) of a planetary model that includes the nine planets plus the Moon. The model considers tidal and other dissipative effects, but the latter have to be parameterised and uncertainties are associated to them. The eccentricity components of the La04 solution are expected to be reliable over ±65 Myr, while precession and obliquity should not be used without caution beyond 20 Myr. Laskar (1999) indeed demonstrates the cahotic nature of the planetary system, destroying the hope to obtain a precise astronomical solution over more than a few tens of millions of years. A solution to this problem may lie in using geological information on the structure of the precession beatings to constrain dissipative parameters (Lourens *et al.*, 2001; Pälike *et al.*, 2001), although this exercise remains a challenge when it is to be performed over several tens of millions of years.

## 2. Insolation

We consider the top-of-the atmosphere irradiance ($W$), that is, irradiance on a surface tangent to the Earth before sunlight is submitted to scattering and absorption. Formulæ for daily mean irradiance (mean irradiance received in a true solar day) can be established by considering the geometry of the problem (Milankovitch, 1941):

$$
\begin{aligned}
W &= \frac{S}{\pi\rho^2}(H_0 \sin\phi \sin\delta + \sin H_0 \cos\phi \cos\delta) \\
\rho &= \frac{r}{a} = \frac{1-e^2}{1+e\cos\nu} \\
\sin\delta &= \sin\varepsilon \sin\lambda
\end{aligned} \tag{2}
$$

$$H_0 = \begin{cases} 0, & \text{for } \tan\phi \tan\delta < -1 \\ \pi, & \text{for } \tan\phi \tan\delta > 1 \\ \arccos(-\tan\phi\tan\delta) & \text{otherwise} \end{cases}$$

The true anomaly ($\nu$) is expressed as a function of time by inverting Kepler's third law (e.g., Danjon, 1980). It is possible to infer a number of conclusions based on symmetry considerations:

1. Insolation integrated over the whole Earth is a function of the Earth-Sun distance ($r$), only. It is therefore independent on obliquity. The *climatic precession parameter* ($e \sin \varpi$) provides a measure of the Sun-Earth distance at northern hemisphere's winter solstice.
2. Annual mean insolation is, at any latitude, independent on precession. Increasing obliquity produces an increase in annual mean insolation poleward of $\pm 43°$, and a decrease equatorward of these latitudes. These variations are of the order of a few W/m$^2$. Loutre *et al.* (2004) discuss the consequences for palæoclimatic records analysis.
3. Kepler's third law also imposes time-integrated insolation over any particular interval bounded by fixed Earth's true longitudes to be independent on precession. Milankovitch (1941) provides a mathematical demonstration.
4. Precession dominates the secular variations of daily insolation at most latitudes, except close to the polar night where obliquity is proportionally more important. A high obliquity enhances the seasonal contrast near the poles. Daily mean insolation may vary by more than 100 W/m$^2$, depending on latitude.
5. Precession produces variations in daily-mean insolation (for a determined Earth true longitude) that are in phase with each other at different latitudes. They are out of phase by 30° (i.e., roughly 2000 years) with the variations in insolation calculated for the previous month. For example, the secular variations of July insolation lag by 2000 years those of June.

There are two ways of comparing daily insolation for two epochs. The first one is to compare insolation for the same Earth true longitude (e.g., $\lambda = 90°$ corresponds to the summer solstice). The other possibility is to compare insolation for the same day number, but this depends on a calendar convention. Most climate models impose the 21st March to correspond with the vernal equinox (Joussaume and Braconnot, 1997).

Figure 2 summarises the evolution of orbital parameters and summer solstice insolation at 65°N over the interval [−800 kyr; 100 kyr]. This insolation curve is commonly used after Milankovitch's hypothesis that summer insolation determines the mass balance of ice sheets. However, Milankovitch himself did not produce this curve; he described the secular march of insolation through the concept of *caloric seasons*. Eccentricity is presently small ($e = 0.017$) and the orbit will be very close to a circle in 27 kyr. This situation, that occurs every 400 kyr (cf. Section 1), is characterised by a long period of *weak orbital forcing*, i.e., almost constant insolation at the top of the atmosphere (for a given latitude and Earth true longitude). Northern summer solstice occurs today at aphelion. It occurred at perihelion (closest to the Sun) 11 kyr ago. The seasonal cycle of insolation was therefore strengthened in the northern hemisphere, which, for example, enhanced the dynamics of summer monsoon in the northern hemisphere (Kutzbach, 1981; Braconnot *et al.*, 2004).

## 3. Challenges Posed by Quaternary Glaciations

As briefly alluded in Section 1, geological records exhibit similar spectral signatures as those calculated by Berger (1978) for precession and obliquity. This correspondence proves a link between the secular march of insolation and palæoclimates. It has therefore been attempted to establish a one-to-one correspondence between the cycles observed in palæoclimatic records (ocean sediments: Imbrie *et al.*, 1984; ice core records: Shackleton, 2000) and those of insolation, thereby providing refined (or *orbitally tuned*) age models for the palæorecords. The record presented by Imbrie *et al.* (1984) (the SPECMAP curve, Figure 2), dated with this empirical method, reveals that interglacial periods have occurred approximately every 100 kyr over the last 800 kyr. Before 1 Myr ago, the dominant period of ice volume variations is that of obliquity: 40 kyr (Ruddiman *et al.*, 1986; Mudelsee and Schulz, 1997). Two questions are therefore posed:

1. By which non-linear mechanism(s) can the presence of a 100 kyr period in the palæorecord be explained, while this period is quasi absent from the daily insolation curves?
2. Assuming that the 100 kyr results from a beating of precession periods, why would this period begin to dominate in the palæorecord, that is, around 800 kyr ago, while the 100 kyr precession beating actually weakens?

Several authors tackled the first problem in the 1980s by means of zero- or one-dimension climate-models (i.e., prognostic variables are represented by zero- or one-dimension arrays). Imbrie and Imbrie (1980) present a conceptual model with insolation thresholds and relaxation times; Ghil and Le Treut (1981) point the dynamics of the lithosphere (isostasy) as an important source of non-linearity. Parrenin and Paillard (2003) formulate the hypothesis that a glacial climate becomes unstable once ice level exceeds a certain threshold. This hypothesis – which is justified by considerations on the carbon cycle developed in Section 5 – provides a semi-empirical explanation to the massive deglaciation underwent 420 kyr ago, while the orbital forcing was weak. Long ice-core records from the Antarctic (Petit *et al.*, 2001; EPICA community members, 2004) pose at least two additional challenges:

3. Important variations in greenhouse gases ($CO_2$ and $CH_4$) occur quasi-simultaneously to ice-volume variations. What are their origins and consequences with respect to ice-volume variations?
4. Why were oscillations of temperature, ice volume and greenhouse gases damped during the period 800 to 500 kyr (EPICA community members, 2004)?

## 4. The LLN Model to Study Ice Volume Variations

Understanding quaternary glaciations requires mechanistic, numerical climate models. Gallée *et al.* (1992) developed a model of the northern hemisphere (the Louvain-la-Neuve, or LLN model). It includes the zonally-averaged, quasi-geostrophic equations of the atmosphere on two levels coupled, on the one hand, with a dynamic model of the ocean mixed layer and, on the other hand, three ice sheet models. The radiative transfer of the atmosphere is fairly comprehensive (absorption, back-scattering, treatment of longwave emission and absorption in different spectral bands). This model is designed to understand the response of northern hemisphere ice sheets to changes in the seasonal and latitudinal distributions of insolation. However, it does not include a model of the carbon cycle. The time-evolution of the atmospheric concentration in $CO_2$ therefore needs to be prescribed. The latter is accurately known and dated over the last two glacial-interglacial cycles, which enabled to test the model, with success, for that period (Berger *et al.*, 1998). A number of sensitivity experiments were subsequently attempted, the conclusion of which can be summarised as follows:

1. Variations in ice mass balance are primarily determined by summer insolation, through its action on ice melt. This comforts Milankovitch's (1941) theory.
2. Radiative feedbacks due to the variations of areas of snow, ice and vegetation, critically determine the amplitude of the climate response (Gallée *et al.*, 1992).
3. Glacial-interglacial cycles occur if the $CO_2$ concentration is constant, provided that this concentration is low enough (typically 210 ppmv). However, the resulting ice volume has too little variance in the 100 kyr spectral band with respect to SPECMAP (Berger *et al.*, 1998).
4. No glacial-interglacial cycle occurs in absence of orbital forcing – whatever the orbit – if realistic $CO_2$ concentration variations are prescribed.
5. The spectral content of ice volume concentrations depends on the $CO_2$ concentration. Namely, the shift from 40 kyr to 100 kyr glacial/interglacial periods can be explained by a gradual decline in the mean $CO_2$ concentration (Berger *et al.*, 1999). Data support to the latter assumption remains contentious (Pearson and Palmer, 2000; de Garidel-Thoron *et al.*, 2005).

Greenhouse gas data exist for the last 800 kyr (Petit *et al.*, 2001; Siegenthaler *et al.*, 2005), but there are sizeable uncertainties on the time scale (Parrenin *et al.*, 2004).

6. The LLN model shows that the phasing between the $CO_2$ and the orbital elements is particularly crucial during periods of low eccentricity, i.e., periods of weak orbital forcing (Figure 3). This concerns *stage 11* (400 kyr ago) and the present interglacial (*Holocene*) and its future (Loutre and Berger, 2003).

For this reason, stage 11 is said to be analogous to the Holocene (Loutre and Berger, 2003). Nevertheless, insolation variations that preceded stage 11 differ substantially from those that preceded the present interglacial (in bold on Figure 2). Furthermore, the mass balance of ice sheets responds non-linearly to insolation changes. Therefore, the present interglacial will not necessarily have the same duration as stage 11, contrary to what has been hypothesised (EPICA community members, 2004). We consider that predicting the next glacial inception must be based on climate models previously tested against palæodata. The LLN model predicts a glacial inception in 50 kyr (Figure 3). In a series of recently published articles,

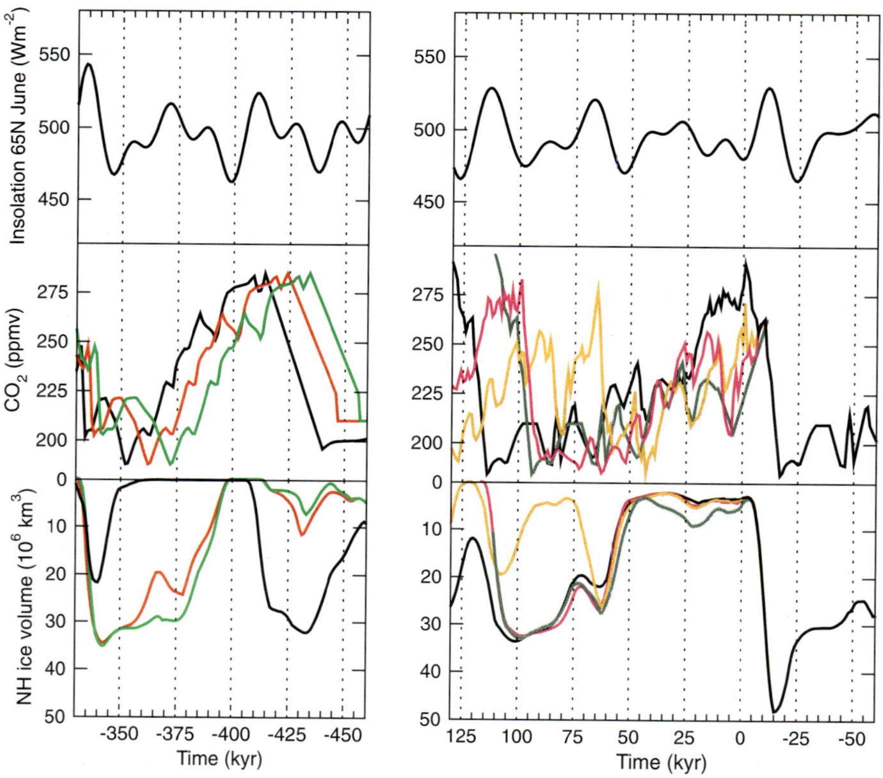


*Figure 3.* Time evolution of northern hemisphere ice volume simulated by the LLN model (lower panel) forced by insolation (top) and various scenarios the evolution of $CO_2$ concentration (middle) for marine isotopic stage 11 (around −400 kyr, left) and the present interglacial period plus the next 125 kyr (right). Geological plotting conventions are adopted (time and ice-volume axes inverted). The Figure illustrates, for these two periods, the sensitivity of simulated ice volume to the exact timing of $CO_2$ concentration changes.

Ruddiman (2003, 2005) considers that $CO_2$ and $CH_4$ concentrations would have declined during the Holocene in absence of land management and wetlands cultures that, according to him, perturbed the chemical composition of the atmosphere as early as 6000 years ago. Experiments with a general circulation model suggest that this *natural scenario* would have led to glacial inception (Ruddiman *et al.*, 2005). Experiments with climate-ice sheet models (Claussen *et al.*, 2005; Crucifix *et al.*, 2005) are less conclusive, but in the latter the suggested draw-down in methane concentration was admittedly neglected.

## 5. The Future of Palæoclimate Modelling

Climate records show connections between the low-, mid- and high latitudes of both hemispheres. Some of these connections probably determine the climate response to the astronomical forcing. Three examples are cited here.

The first one is related to carbon cycle dynamics. There is evidence of formation of dense water off Antarctica during glacial periods (Adkins *et al.*, 2002). This formation acts as a carbon sink for the atmosphere and arguably contributes to maintain low levels of atmospheric carbon dioxide concentration (Paillard and Parrenin, 2004). Deep-water formation in the Southern Ocean depends on brine released by sea-ice formation, which predominantly occurs on submerged continental platforms. Parrenin and Paillard (2003) then speculate that extreme glaciations, during which the area of submerged continental platforms is reduced, are no longer favourable to carbon sequestration to the deep ocean. The subsequent increase in $CO_2$ prepares the system for a deglaciation, the timing of which is set by the orbital forcing. An alternative view, perhaps better compatible with the earliest part of the EPICA record (Köhler and Fischer, 2006) consists in describing a series of feedbacks by which a decrease in ice volume causes an almost-immediate increase in $CO_2$ concentration (decreased insulating action of southern sea-ice, Stephens and Keeling, 2000; coral-reef build-up, Ridgwell *et al.*, 2003; decrease in southern ocean iron fertilisation, Martin *et al.*, 1990; changes in ocean alkalinity distribution related to ocean circulation changes, Broecker and Peng, 1989; plus various other mechanisms, Archer *et al.*, 2000).

The second example given here concerns the role of the hydrological cycle in the glacial inception. Vettoretti and Peltier (2003b) present evidence for an atmospheric-cryospheric pump, by which reduced summer insolation and the subsequent increased snow cover enhance the total moisture transport into the Arctic circle. The supplied moisture contributes to the growth of nucleating ice sheets. They also suggest that decreased summer insolation is partly compensated for by enhanced northward heat transport on the ridges of the stationary waves, that is, in Scandinavia and Alaska. This is how they explain the absence of ice sheet nucleation at that time in Alaska and in Northern Europe. Finally, they argue that the growth of the Laurentide ice sheet may have altered the stationary wave pattern such as to trigger ice nucleation in Fennoscandia.

The last example considers the response of the terrestrial biosphere to insolation. A number of climate models coupled to biosphere models predict a partial disparition of the boreal forest in response to reduced summer insolation (Harrison *et al.*, 1995; de Noblet *et al.*, 1996). This response begins to be documented and dated in palæoclimatic records (Sanchez Goñi *et al.*, 2005), although accurately dated data at high latitudes are still lacking. In turn, this response of vegetation impacts on the radiative balance of the northern hemisphere via the snow-vegetation feedback (Otterman *et al.*, 1984). Experiments with coarse-resolution models (Crucifix and Loutre, 2002; Meissner *et al.*, 2003; Kageyama *et al.*, 2004) suggest that this biosphere feedback strongly contributed to the last glacial inception.

This non-comprehensive list illustrates complex, but potentially relevant processes to be included in a revised astronomical theory of palæoclimates. Future numerical simulations should therefore be carried out with 3-D climate models including oceanic, biospheric and cryospheric components. However, this is not trivial. First, glacial inception and deglaciation are in essence bifurcation phenomena. An accurate representation of modern climatic features, especially at high latitudes, is essential (Vettoretti and Peltier, 2003a). Yet, even state-of-the-art general circulation models (GCM) exhibit significant errors on the distribution of precipitation. Second, GCMs are expansive. Even the fastest one would not produce a simulation of the entire glacial-interglacial cycle within less than a year of actual computer time. A try/error method (testing different parameters, switching on and off interactive components) may therefore not be an appropriate way to simulate the response of climate to the secular march of insolation. It is the reason why we advocate the development of palæodata assimilation schemes, by which model parameters are dynamically constrained by palæodata (Hargreaves and Annan, 2002). The outputs are therefore compatible with observations, but they also include information verifiable by reference to independent data. In particular, this method may help to identify which data would best constrain existing uncertainties on past climates, a valuable information for field scientists. Simulations of glacial-interglacial cycles achieved this way would then constitute a reference background material to study the feedbacks of individual components on climate.

## Key to symbols

$\phi$ latitude

$t$ time

$\nu$ true anomaly, i.e., angle between perihelion and Earth

$e$ eccentricity of the Earth orbit

$i$ angle between ecliptic of (reference) epoch and orbit plane

$\gamma_0$ (direction of) vernal point of epoch. Projects to $\gamma'_0$ on ecliptic of date
$\gamma$ (direction of) vernal point of date
$\lambda$ Earth true longitude with respect to vernal equinox (V.E)
$\Omega$ longitude of ascending node to $\gamma_0$
$\omega$ argument of perihelion relative to ascending node
$\Pi$ longitude of perihelion relative to $\gamma_0$ $(= \Omega + \omega)$
$\Psi_g$ angle $\gamma - \gamma'_0$. The general precession is the variation rate of this angle.
$\tilde{\varpi}$ longitude of perihelion relative to $\gamma$.
$\varpi$ longitude of perihelion relative to vernal equinox $(=\tilde{\varpi} - \pi)$
$\varepsilon$ obliquity, i.e., angle between equator and ecliptic
$\delta$ Sun declination
$a$ semi-major axis of Earth's orbit
$r$ Earth-Sun distance
$H_0$ hour angle at sunset

## References


Adkins, J. F., McIntyre, K., and Schrag, D. P.: 2002, 'The salinity, temperature, and delta O-18 of the glacial deep ocean', *Science* **298**, 1769–1773.

Archer, D., Winguth, A., Lea, D., and Mahowald, N.: 2000, 'What caused the glacial/interglacial atmospheric $pCO_2$ cycles?', *Rev. Geophys.* **38**, 159–189.

Berger, A.: 1977, 'Long-term variations of the Earth's orbital elements', *Celest. Mec.* **15**, 53–74.

Berger, A.: 1978, 'Long-term variations of daily insolation and Quaternary climatic changes', *J. Atmos. Sci.* **35**, 2362–2367.

Berger, A. and Loutre, M. F.: 1991, 'Insolation values for the climate of the last 10 million years', *Quat. Sci. Rev.* **10**, 297–317.

Berger, A., Loutre, M. F., and Gallée, H.: 1998, 'Sensitivity of the LLN climate model to the astronomical and $CO_2$ forcings over the last 200 ky', *Clim. Dyn.* **14**, 615–629.

Berger, A., Li, X. S., and Loutre, M. F.: 1999, 'Modelling northern hemisphere ice volume over the last 3 Ma', *Quat. Sci. Rev.* **18**, 1–11.

Braconnot, P., Harrison, S. P., Joussaume, S., Hewitt, C. D., Kitoh, A., Kutzbach, J. E., Liu, Z., Otto-Bliesner, B., Syktus, J., and Weber, N.: 2004, 'Evaluation of PMIP coupled ocean-atmosphere simulations of the mid-holocene', in R. W. Batterbee, F. Gasse, and C. E. Stickley (eds.), *Past Climate Variability Through Europe and Africa*, pp. 515–534.

Bretagnon, P.: 1974, 'Termes à longues périodes dans le système solaire', *Astron. Astroph.* **30**, 141–154.

Broecker, W. S. and Peng, T. H.: 1989, 'The cause of the glacial to interglacial atmospheric $CO_2$', *Global Biogeochem. Cycles* **3**, 215–239.

Chapront, J., Bretagnon, P., and Mehl, M.: 1975, 'Un formulaire pour le calcul des perturbations d'ordres élevés dans les problèmes planétaires', *Celes. Mech.* **34**, 165–184.

Claussen, M., Brovkin, V., Calov, R., Ganopolski, A., and Kubatzki, C.: 2005, 'Did humankind prevent a Holocene glaciation?', *Clim. Change* **69**, 409–417.

Crucifix, M. and Loutre, M. F.: 2002, 'Transient simulations over the last interglacial period (126–115 kyr BP): feedback and forcing analysis', *Clim. Dyn.* **19**, 419–433.

Crucifix, M., Loutre, M. F., and Berger, A.: 2005, 'Commentary on "the anthropogenic greenhouse era began thousands of years ago"', *Clim. Change* **69**, 419–426.

Danjon, A.: 1980, *Astronomie générale*. Libraire scientifique et technique A. Blanchard.

de Garidel-Thoron, T., Rosenthal, Y., Bassinot, F., and Beaufort, L.: 2005, 'Stable sea surface temperatures in the western Pacific warm pool over the past 1.75 million years', *Nature* **433**, 294–298.

de Noblet, N., Braconnot, P., Joussaume, S., and Masson, V.: 1996, 'Sensitivity of simulated Asian and African summer monsoons to orbitally induced variations in insolation at 126, 115 and 6 kBP', *Clim. Dyn.* **12**, 589–603.

EPICA community members: 2004, 'Eight glacial cycles from an Antarctic ice core', *Nature* **429**, 623–628.

Gallée, H., van Ypersele, J. P. Fichefet, T., Marsiat, I., Tricot, C., and Berger, A.: 1992, 'Simulation of the last glacial cycle by a coupled, sectorially averaged climate-ice sheet model. Part II: Response to insolation and $CO_2$ variation', *J. Geophys. Res.* **97**, 15, 713–15, 740.

Ghil, M. and Le Treut, H.: 1981, 'A climate model with cryodynamics and geodynamics', *J. Geophys. Res.* **86**, 5262–5270.

Hargreaves, J. C. and Annan, J. D.: 2002, 'Assimilation of paleo-data in a simple Earth system model', *Clim. Dyn.* **19**, 371–381.

Harrison, S. P., Kutzbach, J. E., Prentice, I. C., Behling, P. J. and Sykes, M. T.: 1995, 'The response of Northern Hemisphere extratropical climate and vegetation to orbitally induced changes in insolation during the last interglacial', *Quat. Res.* **43**, 174–184.

Hays, J., Imbrie, J. and Shackleton, N.: 1976, 'Variations in the Earth's orbit: Pacemaker of ice ages', *Science* **194**, 1121–1132.

Imbrie, J. and Imbrie, J. Z.: 1980, 'Modelling the climatic response to orbital variations', *Science* **207**, 943–953.

Imbrie, J. J., Hays, J. D., Martinson, D. G., McIntyre A., Mix, A. C., Morley, J. J., Pisias, N. G., Prell, W. L., and Shackleton, N. J.: 1984, 'The orbital theory of Pleistocene climate: Support from a revised chronology of the marine $\delta^{18}$Orecord', in A. Berger, J., Imbrie J., Hays, J., Kukla, and B. Saltzman (eds.), *Milankovitch and Climate, Part I*. Norwell, Mass., pp. 269–305.

Joussaume, S. and Braconnot, P.: 1997, 'Sensitivity of paleoclimate simulation results to season definition', *J. Geophys. Res.* **102**, 1943–1956.

Kageyama, M., Charbit, S., Ritz, C., Khodri, M., and Ramstein, G.: 2004, 'Quantifying ice-sheet feedbacks during the last glacial inception', *Geophysical Research Letters* **31**, L24903, doi:10.1029/2004GL021339.

Köhler, P. and Fischer, H.: 2006, 'Proposing a mechanistic understnading of changes in atmospheric $CO_2$ during the last 740 000 years', *Clim. Past Discussions* **2**, 1–42.

Kutzbach, J. E.: 1981, 'Monsoon climate of the early Holocene: Climate experiment using the Earth's orbital parameters for 9000 years ago', *Science* **214**, 59–61.

Lagrange, J. L.: 1781, 'Théorie des variations séculaires des éléments des planètes 1.', in *Nouveaux mémoires de l'Académie Royale des Sciences et Belles-Lettres*, Berlin, pp. 199–276.

Laplace, P. S.: 1773, 'Tome VIII', in *Oeuvres complètes*. Compilation published by Gauthier-Villars, in 1891, p. 199.

Laskar, J.: 1984, 'Théorie générale planétaire: Eléments orbitaux des planètes sur 1 million d'années', Ph.D. thesis, Obervatoire de Paris, Meudon, France.

Laskar, J.: 1988, 'Secular evolution of the solar system over 10 millions years', *Astron. Astroph.* **198**, 341–362.

Laskar, J.: 1999, 'The limits of Earth orbital calculations for geological time-scale use', *Phil. Trans. R. Soc. Lond. A* **357**, 1735–1759.

Laskar, J., Joutel, F., and Boudin, F.: 1993, 'Orbital, precessional, and insolation quantities for the Earth from −20 Myr to +10 Myr', *Astron. Astroph.* **270**, 522–533.

Laskar, J., Robutel, P., Joutel, F., Boudin, F., Gastineau, M., Correia, A. C. M., and Levrard, B.: 2004, 'A long-term numerical solution for the insolation quantities of the Earth', *Astron. Astroph.* **428**, 261–285.

Lourens, L. J., Wehausen, R., and Brumsack, H. J.: 2001, 'Geological constraints on tidal dissipation and dynamical ellipticity of the Earth over the past three million years', *Nature* **409**, 1029–1033.

Loutre, M. F. and Berger, A.: 2003, 'Marine Isotope Stage 11 as an analogue for the present interglacial', *Glob. Plan. Change* **36**, 209–217.

Loutre, M. F., Paillard, D., Vimeux, F., and Cortijo, E.: 2004, 'Does mean annual insolation have the potential to change the climate?', *Earth Planet. Sci. Lett.* **221**, 1–14.

Martin, J. H., Gordon, R. M., and Fitzwater, S. E.: 1990, 'Iron in Antarctic waters', *Nature* **345**, 156–158.

Meissner, K. J., Weaver, A. J., and Matthews, H. D.: 2003, 'The role of land surface dynamics in glacial inception: A study with the UVic Earth System Model', *Clim. Dyn.* **21**, 515–537.

Milankovitch, M.: 1941, *Canon of Insolation and the Ice-Age Problem*. Edited and translated by the Serbian Academy of Science and Arts, 1998, Narodna biblioteka Srbije, Beograd.

Mudelsee, M. and Schulz, M.: 1997, 'The Mid-Pleistocene climate transition: Onset of 100 ka cycle lags ice volume build-up by 280 ka', *Earth Planet. Sci. Lett.* **151**, 117–123.

Otterman, J., Chou, M.-D., and Arking, A.: 1984, 'Effects of nontropical forest cover on climate', *J. Appl. Meteor.* **23**, 762–767.

Paillard, D. and Parrenin, F.: 2004, 'The Antarctic ice sheet and the triggering of deglaciations', *Earth Planet. Sci. Lett.* **227**, 263–271.

Pälike, H., Shackleton, N. J., and Rohl, U.: 2001, 'Astronomical forcing in Late Eocene marine sediments', *Earth Planet. Sci. Lett.* **193**, 589–602.

Parrenin, F. and Paillard, D.: 2003, 'Amplitude and phase of glacial cycles from a conceptual model', *Earth Planet. Sci. Lett.* **214**, 243–250.

Parrenin, F., Remy, F., Ritz, C., Siegert, M. J., and Jouzel, J.: 2004, 'New modeling of the Vostok ice flow line and implication for the glaciological chronology of the Vostok ice core', *J. Geophys. Res.* **109**, doi:10.1029/2004JD004561.

Pearson, P. N. and Palmer, M. R.: 2000, 'Atmospheric carbon dioxide concentrations over the past 60 million years', *Nature* **406**, 695–699.

Petit, J. R., Jouzel, J., Raynaud, D., Barkov, N. I. Barnola, J.-M. Basile, I., Bender, M., Chappellaz, J., Davis, M., Delaygue, G., Delmotte, M., Kotlyakov, V. M., Legrand, M., Lipenkov, V. Y., Lorius, C., Pépin, L., Ritz, C., Saltzman, E., and Stievenard, M.: 2001, 'Climate and atmospheric history of the past 420, 000 years from the Vostok ice core, Antarctica', *Nature* **399**, 429–436.

Ridgwell, A. J., Watson, A. J., Maslin, M. A., and Kaplan, J.: 2003, 'Implications of coral reef buildup for the controls on atmospheric $CO_2$ since the Last Glacial Maximum', *Paleoceanogr.* **18**, Art. No. 1083.

Ruddiman, W. F.: 2003, 'The anthropogenic greenhouse era began thousands of years ago', *Clim. Change* **61**, 261–293.

Ruddiman, W. F.: 2005, 'The early anthropogenic hypothesis a year later – An editorial reply', *Clim. Change* **69**, 427–434.

Ruddiman, W. F., Raymo, M., and McIntyre, A.: 1986, 'Mutuyama 41, 000-year cycles: North Atlantic Ocean and northern hemisphere ice sheets', *Earth Planet. Sci. Lett.* **80**, 117–129.

Ruddiman, W. F., Vavrus, S. J., and Kutzbach, J. E.: 2005, 'A test of the overdue-glaciation hypothesis', *Quat. Sci. Rev.* **24**, 1–10.

Sanchez Goñi, M. F., Loutre, M. F., Crucifix, M., Peyron, O., Santos, L., Duprat, J., Turon, J, -L., and Peypouquet, J.-P.: 2005, 'Increasing vegetation and climate gradient in Western Europe over the Last Glacial Inception (122-110 ka): models-data comparison', *Earth Planet. Sci. Lett.* **231**, 111–130.

Shackleton, N. J.: 2000, 'The 100,000-year ice-age cycle identified and found to lag temperature, Carbon Dioxide and orbital eccentricity', *Science* **289**, 1897–1902.

Siegenthaler, U., Stocker, T. F., Monnin, E., Lüthi, D., Schwander, J., Stauffer, B., Raynaud, D., Barnola, J.-M., Ficher, H., Masson-Delmott, V., and Jouzel, J.: 2005, 'Stable carbon cycle-climate relationship during the late Pleistocene"', *Science* **310**, 1313–1317, doi:10.1126/science.1120130.

Stephens, B. B. and Keeling, R. F.: 2000, 'The influence of Antarctic sea-ice on glacial-interglacial $CO_2$ variations', *Nature* **404**, 171–174.

Vettoretti, G. and Peltier, W. R.: 2003a, 'Post-Eemian glacial inception. Part I: the impact of summer seasonal temperature bias', *J. Climate* **16**, 889–911.

Vettoretti, G. and Peltier, W. R.: 2003b, 'Post-Eemian glacial inception. Part II: Elements of a cryospheric moisture pump', *J. Climate* **16**, 912–927.

# REGIONAL RESPONSE OF THE CLIMATE SYSTEM TO SOLAR FORCING: THE ROLE OF THE OCEAN

H. GOOSSE[1,*] and H. RENSSEN[2]
[1]*Institut d'Astronomie et de Géophysique G. Lemaître, Université Catholique de Louvain, Chemin du cyclotron, 21348 Louvain-la-Neuve, Belgium*
[2]*Faculty of Earth and Life Sciences, Vrije Universiteit, Amsterdam, The Netherlands*
*(*Author for correspondence: E-mail: hgs@astr.ucl.ac.be)*

(Received 15 July 2005; Accepted in final form 20 January 2006)

**Abstract.** A coupled climate model is used to explore the regional response of the climate system to solar forcing, with emphasis on the role of the ocean. It is shown that both the transient and the equilibrium response of surface temperature to changes in total solar irradiation is smaller over ocean than over land because of the ocean's large heat capacity and the feedback involving evaporation. Furthermore, the advection of temperature anomalies and changes in ocean currents have an impact on the timing and the geographical distribution of the response. Nevertheless, at regional scales, the response to the forcing is embedded within the large internal variability of the system making the detection and analysis of the forced response difficult. Furthermore, this forced response could imply both changes in the mean state of the system as well as in its variability.

**Keywords:** climate variability, solar forcing, ocean dynamics, evaporation, thermal inertia

## 1. Introduction

The response of the climate system at hemispheric scale to a change in total solar irradiance is relatively easy to interpret: the Northern Hemisphere annual mean surface temperature tends to increase when solar irradiance is higher than the mean value and to decrease when it is lower. The magnitude of the response could generally be understood by taking into account the climate sensitivity (which is related to the equilibrium temperature response to a change in radiative forcing) and the thermal inertia of the system (for more information, see for instance Hansen *et al.*, 1985, 1997). Unfortunately, those characteristics of the climate system are not precisely known and climate models generally provide different answers (e.g., Ingram, 2006). The uncertainties in the reconstruction of past changes of total solar irradiance have also an impact on estimated temperature changes (e.g., Bertrand *et al.*, 2002). Nevertheless, relatively simple models could already provide very useful information on the temperature response at hemispheric scale (e.g., Crowley, 2000).

The problem is much more complex at regional scale since the temperature pattern of the response to the forcing is much more influenced by the internal dynamics of the climate system than by the geographical distribution of the forcing (e.g., Bengtsson, 2006). Furthermore, the internal variability of the climate system, which is present even in the absence of change in the forcing, is much larger at

Space Science Reviews (2006) 125: 227–235
DOI: 10.1007/s11214-006-9059-0

regional scale than at hemispheric scale. Because of this larger noise level, any attempt to analyze or detect the response to forcing, using observations or model results, is more difficult at regional scale than at the hemispheric one (Stott *et al.*, 2001; Goosse *et al.*, 2005).

The internal processes responsible for the geographical distribution of the temperature response involve some relatively well understood thermodynamic processes like the ones leading to the temperature-albedo feedback that could be described as follows. If surface temperatures increase (decrease) at high latitudes, the ice and snow cover tends to decrease (increase), resulting in a decrease (increase) of the surface albedo and a higher (lower) absorption of incoming solar radiation at surface and thus a subsequent warming (cooling). This feedback is for an important part responsible for the larger temperature change at high latitude compared to the global mean when a positive radiative perturbation is applied in coupled atmosphere-ocean general circulation models (AOGCMs). The effect is often called "polar amplification" (Holland and Bitz, 2003).

The modifications in the atmospheric circulation or in coupled atmosphere-ocean processes like ENSO (e.g., Mann *et al.*, 2005) and the monsoons (Meehl *et al.*, 2003) in response to external forcing have also a large influence on the spatial distribution of temperature changes. In particular, during the last few years, a lot of attention has been paid to changes in the annular modes of atmospheric circulation related to solar variations, taking into account the important changes in the ultraviolet band and the potential feedbacks related to variations in ozone concentration (e.g., Haigh and Blackburn, 2006; Kodera, 2006). For instance, Shindell *et al.* (2001) have suggested that during the Maunder Minimum (mid 1600s to early 1700s AD), because of reduced solar irradiance, the northern annular mode was preferentially in a negative phase resulting in very cold conditions over Europe in winter while the North Atlantic area was relatively mild at that time. They also emphasized in that study that this signal was embedded in large internal variability that could mask the forced signal for some periods.

On the other hand, various analyses based on observations suggest an important role of the ocean (e.g., Bond *et al.*, 2001; Waple *et al.*, 2002; Jiang *et al.*, 2005) but this has been much less studied than the one of the atmosphere. As a consequence, in this brief review, we will focus on the influence of the ocean in the spatial distribution of temperature changes in response to changes in total solar irradiance. This will be done using the information available in some recently published papers. In addition, we will present a new simulation and display new analyses in order to illustrate the points put forward in the present framework. We will first analyze the role of thermodynamic processes related to the large heat capacity of the ocean and to the potential negative feedback associated with changes in latent heat flux (Section 2). In Section 3, we will illustrate by an example the influence of the transport by the ocean of a temperature anomaly. In Section 4, we will investigate the role of the changes in ocean currents before some concluding remarks.

## 2. Thermodynamical Processes

The heat capacity of the water is much larger than the one of the air (e.g., Gill, 1982). Furthermore, because of intense mixing in the upper ocean, a layer from 50 up to several hundreds of meters is in contact with the atmosphere on seasonal time scale. In comparison, the heat conductivity of the land surface is much lower and only a few meters at best could interact with the surface on interannual timescales. As a consequence, any imbalance of the fluxes at the ocean surface will be used to warm up (or cool) a thick layer with a large heat capacity (e.g., White *et al.*, 1998), inducing thus a delayed temperature change compared to land surfaces.

This is illustrated here by performing a new 1000-year long sensitivity experiment with the coarse resolution 3-D atmosphere-ocean model of intermediate complexity ECBILT-CLIO, which results from the coupling of a general circulation model of the ocean and the sea-ice cover with a simplified quasi-geostrophic atmospheric model (Opsteegh *et al.*, 1998; Goosse and Fichefet, 1999; for more info about the model, see *http://www.knmi.nl/onderzk/CKO/ecbilt.html*). Starting from a long control experiment using constant external forcing, we have abruptly decreased the total surface irradiance by 5 W/m$^2$. We have then computed anomalies (Figure 1) by subtracting the results of the control experiment from the ones of the sensitivity experiment.

During the first 50 years of the numerical experiment, the response over the ocean is indeed much smaller than over land. On Figure 1, we also see clearly the polar amplification of the temperature response briefly described in the introduction,

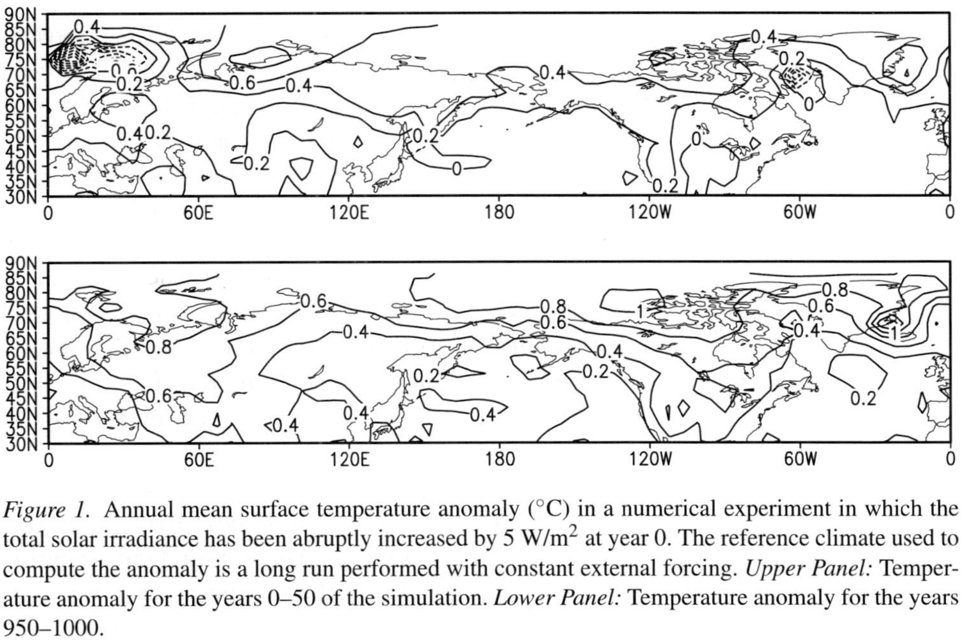


*Figure 1.* Annual mean surface temperature anomaly (°C) in a numerical experiment in which the total solar irradiance has been abruptly increased by 5 W/m$^2$ at year 0. The reference climate used to compute the anomaly is a long run performed with constant external forcing. *Upper Panel:* Temperature anomaly for the years 0–50 of the simulation. *Lower Panel:* Temperature anomaly for the years 950–1000.

which is relatively strong in ECBILT-CLIO compared to other models (e.g., Holland and Bitz, 2003).

After 1000 years of integration, the ocean is nearly in equilibrium with the forcing, at least the surface layers. The ocean has warmed up compared to the first decades of the simulation. Nevertheless, the ocean response is still much lower than over land. The ocean heat capacity should not be responsible for this behaviour since, by definition, at equilibrium heat capacity should not play anymore a role. Besides, as proposed by Manabe *et al.* (1991), the land-sea contrast at equilibrium is partly attributable to a larger increase in latent heat loss over ocean than over land. Indeed, the evaporation change from the soil is smaller than over the ocean, less water being available. In addition, Manabe *et al.* (1991) argued that changes in snow cover could also play a role in the larger response over land.

## 3. Transport of Temperature Anomalies by the Ocean

In addition to the processes described above, changes in the oceanic heat transport have also an influence on the spatial distribution of the temperature response. Those heat transport changes could be attributed to the advection of temperature anomalies, to modifications of ocean currents or to a combination of those two processes.

This Section is devoted to the formation and transport of temperature anomalies. The processes involved are illustrated by analysing the temperature evolution in the Southern Ocean in a 5000-year simulation performed with the climate model ECBILT-CLIO, using anomalies of solar irradiance following an idealised sinusoidal evolution with an amplitude of 5 $W/m^2$ and a period of 1000 years (Goosse *et al.*, 2004).

In this experiment, the global mean annual temperature averaged over the top 500 meters of the ocean displays a time delay compared to the sinusoidal forcing of about 100 years, as illustrated on Figure 2 which presents an additional diagnostic performed on the results of the experiment described in Goosse *et al.* (2004). This is firstly due to the thermal inertia of the oceanic layer as discussed in Section 2 and to the slow propagation of the temperature signal in the oceanic thermocline.

When the temperature is averaged over the Southern Ocean only, the magnitude of the changes is slightly higher and the delay reaches 200 years. This is related to the fact that relatively old intermediate and deep water upwells in the Southern Ocean. Depending on their origin, those waters have been in contact with the surface decades to centuries before they reach the top 500 meters of the Southern Ocean. As a consequence, the transport to the Southern Ocean of temperature anomalies formed at the surface by the external forcing could be particularly long, resulting in the delay observed in Figure 2. This has led Goosse *et al.* (2004) to propose the hypothesis that in the Northern Hemisphere the surface response to a stronger radiative forcing during the first centuries of the second millennium AD was more

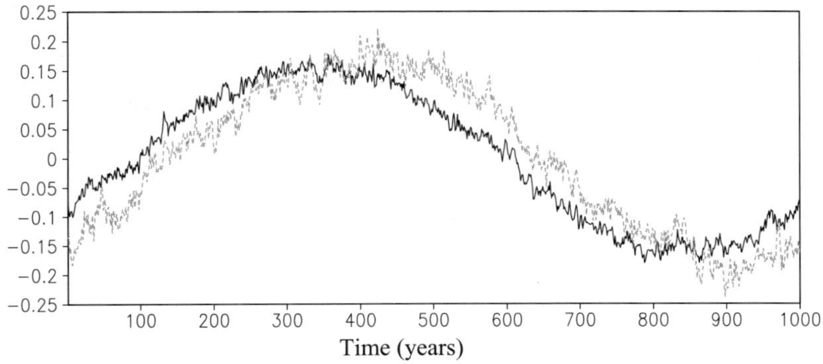


*Figure 2.* Temperature anomaly (°C) averaged over the top 500 meters of the ocean in a numerical experiment driven by total solar irradiance following an idealised sinusoidal evolution with an amplitude of 5 $W/m^2$ and a period of 1000 years. The maximum of the forcing occurs in year 250 and the minimum in year 750. The plain dark curve is the mean over the whole ocean while the dashed grey curve is the mean over the latitude band 50°S – 80°S. Reference climate to compute the anomaly is the average over the whole simulation.

or less in phase with the forcing (the so-called Medieval Warm Period), while it was delayed by about 150 years in the Southern Ocean.

## 4. Changes in Ocean Currents

The response of the climate models to an increase (decrease) in total solar irradiance generally implies a weakening (strengthening) of the Meridional Overturning Circulation (MOC) in the North Atlantic (Cubasch *et al.*, 1997; van der Schrier *et al.*, 2002; Weber *et al.*, 2004; Figure 3). The MOC is associated with a northward flow of surface water, which progressively cools by contact with the atmosphere, becomes thus denser and sinks eventually to great depths (mainly in the Greenland and Labrador Seas), inducing a southward deep flow that compensates for the surface northward mass transport. As the temperature of the deep current is much lower than the one at surface, this overturning circulation is associated with a northward heat transport that contributes to maintaining warmer surface temperatures in the Northern North Atlantic compared to the one observed at similar latitudes in the Pacific.

The decrease in the MOC intensity when solar irradiance is increased is likely due to reduced heat loss and/or higher freshwater fluxes at high latitudes that induced a reduction of the density and then of the sinking of water in the Northern North Atlantic. This behaviour could be compared to the response of models to an increase in atmospheric $CO_2$ content, which has been more thoroughly analysed. In particular, experiments have been recently performed in the framework of the Coupled Model Intercomparison Project (CMIP) in which the $CO_2$ concentration has been increased during 140 years by up to 4 times the initial level in 11 different

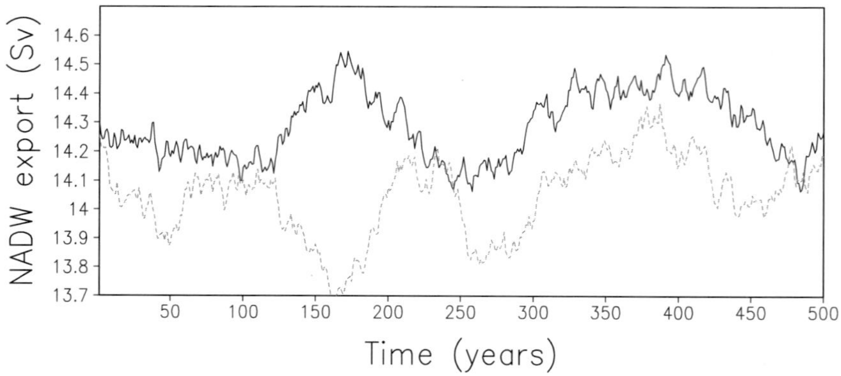


*Figure 3.* Maximum of the meridional overturning streamfunction in the Atlantic at 20°S in Sv ($10^6$ m$^3$s$^{-1}$) (which is a measure of the export of North Atlantic Deep Water out of the Atlantic and thus of the intensity of the MOC) in a numerical experiment in which the total solar irradiance has been abruptly increased by 5 W/m$^2$ at year 0 (grey dashed curve) and in the control experiment (black plain curve). A 25-year running mean has been applied to the time series to filter high frequency changes. Nevertheless, internal variability could mask the forced changes during some periods (e.g., around year 200–250).

models (Gregory *et al.*, 2005). In those experiments, the intensity of the MOC declines by between 10 and 50%. This type of response of the oceanic circulation is similar to the one simulated for an increase in solar irradiance but, as the positive radiative perturbation is much larger in the CMIP experiments, the changes of the MOC are expected to be smaller for the response to changes in solar irradiance. Indeed, the reduction of the MOC to a 5 W/m$^2$ increase in the Total Solar Irradiance reaches at most 0.5 Sv (3%) in ECBILT-CLIO (Figure 3) while using the same model a reduction of 25% was simulated in the CMIP experiments. In a 10,000 year-long simulation, Weber *et al.* (2004), using the atmospheric model ECBILT coupled to a flat-bottom ocean, also find a significant but weak response of the MOC to changes in solar irradiance that could explain 10–20% of the variability of the MOC in their model. Furthermore, they emphasised that the solar forcing at time scales 200–250 years could amplify an internal damped mode of the MOC that would result in a stronger response of the MOC at that time scale.

Despite their relatively low magnitude, the simulated changes in the MOC could have a clear regional impact, leading for instance to a cooling or to a reduced warming in the northern North Atlantic in periods when the solar irradiance is increased (e.g., Cubasch *et al.*, 1997). In the new simulation described in Section 2, this effect is large in the Greenland-Norwegian Sea and in the Labrador Sea at the beginning of the sensitivity experiment but it is weak at equilibrium (Figure 1).

In addition, anomalies of the MOC could have an impact on the salinity distribution and on the sea level as illustrated by van der Schrier *et al.* (2002). On the other hand, some simulations performed over the last millennium do not display any clear relationship between the MOC intensity and solar forcing (Zorita *et al.*, 2004;

Goosse *et al.*, 2004; Tett *et al.*, 2005) but, in those simulations, a weak forced solar signal could have been masked by internal variability. Unfortunately, the evidence from proxy records is not very clear for the last millennia (e.g., Keigwin, 1996; Bianchi and McCave, 1999; Hall *et al.*, 2004) and it is very difficult to determine if variations of the MOC have played a significant role in the temperature changes observed during this period (e.g., Jones and Mann, 2004). Proxy records could thus not help a lot in the interpretation of the MOC changes simulated in models.

In addition to changes in the MOC, surface currents could be modified in response to modifications in the atmospheric circulation. For instance, Zorita *et al.* (2004) using an AOGCM and van der Schrier and Barkmeijer (2005) using ECBILT-CLIO suggest a modified surface oceanic transport during the Maunder and Dalton Minima (more precisely the period 1790–1820 AD) respectively, that lead to cold sea surface conditions in the North Atlantic area, with a potential impact on temperature in Europe. Unfortunately, the exact separate roles of the solar and volcanic forcing and of internal variability in the simulated changes were not investigated in those studies.

Surface ocean current changes could also play a large role in the variability of the climate system at high latitudes. For instance, Goosse and Renssen (2004) show that a reduction of the solar forcing in their model has a moderate effect on the time-averaged oceanic circulation but can strongly modify the probability to have years with significantly reduced inflow of warm Atlantic waters at high latitudes. This leads to a higher probability to have very cold conditions in the Nordic Seas and over Scandinavia, resulting overall in a very strong amplification of the response to solar forcing in the Northern North Atlantic and surrounding regions during those cold periods. This lower solar forcing also significantly increased the temperature variability of the model in Polar Regions.

## 5. Conclusions

The brief overview presented above has shown that the ocean plays a significant role in the response of the system to solar irradiance changes. First, because of its thermal inertia and of the negative feedback associated with latent heat flux, the temperature changes are generally much weaker over oceanic area than over land. Furthermore, the transport of temperature anomalies created in a region to another one by the ocean could induce a delayed response to the forcing in some area and modifications of ocean currents could have an impact on heat transport and thus on the pattern of the temperature changes.

Nevertheless, the changes in oceanic circulation simulated by model are generally relatively weak so that disentangling the forced response to changes in solar irradiance from internal variability is difficult. The interpretation of the response is also complex as it could involve both changes in the mean state as well as in the probability of extreme events.

Furthermore, as analysing the ocean changes requires long simulations because of the long response timescale of the ocean compared to the one of the atmosphere, only a few model studies have been devoted to the subject and the models used in those studies have generally a coarse resolution and/or simplified physics. Their results must thus be taken with caution until they are confirmed by additional evidence.

## Acknowledgements

H. Goosse is research associate with the Belgian National Fund for Scientific Research. H. Renssen is supported by the Netherlands Organization for Scientific Research. This study was done within the scope of the Second Multiannual Scientific Support Plan for a Sustainable Development Policy (Belgian Federal Science Policy Office, contracts EV/10/7D and EV/10/9A). All support is gratefully acknowledged.

## References

Bengtsson, L.: 2006, 'On the response of the climate system to solar forcing', *Space Sci. Rev.*, this volume, doi: 10.1007/s11214-006-9056-3.

Bertrand, C., Loutre, M.-F., Crucifix M., and Berger, A.: 2002, 'Climate of the last millennium: A sensitivity study', *Tellus* **54A**, 221–244.

Bianchi, G. G. and McCave, I.: 1999, 'Holocene periodicity in North Atlantic climate and deep-ocean flow south of Iceland', *Nature* **397**, 515–517.

Bond, G., Kromer, B., Beer, J., Muscheler, R., Evans, M. N., Showers, W., Hoffmann, S., Lotti-Bond, R., Hajdas, I., and Bonani, G.: 2001, 'Persistent solar influence on North Atlantic climate during the Holocene', *Science* **294**, 2130–2133.

Crowley, T. J.: 2000, 'Causes of climate change over the past 1000 years', *Science* **289**, 270–277.

Cubasch, U., Voss, R., Hegerl, G. C., Waszkewitz, J., and Crowley, T. J.: 1997, 'Simulation of the influence of solar radiation variations on the global climate with an ocean-atmosphere general circulation model', *Clim. Dyn.* **13**, 757–767.

Gill, A. E.: 1982, 'Atmosphere-ocean dynamics', *International Geophysics Series* **30**, Academic Press, San Diego, 662 pp.

Goosse, H. and Fichefet, T.: 1999, 'Importance of ice-ocean interactions for the global ocean circulation: A model study', *J. Geophys. Res.* **104**, 23,337–23,355.

Goosse, H. and Renssen, H.: 2004, 'Exciting natural modes of variability by solar and volcanic forcing: Idealized and realistic experiments', *Clim. Dyn.* **23**, 153–163.

Goosse, H., Masson-Delmotte, V., Renssen H., Delmotte, M., Fichefet, T., Morgan, V., van Ommen, T., Khim, B. K., and Stenni, B.: 2004, 'A late medieval warm period in the Southern Hemisphere as delayed response to external forcing?', *Geophys. Res. Lett.* **31**, doi:10.1029/2003GL019140.

Goosse, H., Renssen, H., Timmermann, A., and Bradley, R. S.: 2005, 'Internal and forced climate variability during the last millennium: A model-data comparison using ensemble simulations', *Quat. Sci. Rev.* **24**, 1345–1360.

Gregory, *et al.*, 2005: 'A model intercomparison of changes in the Atlantic thermohaline circulation in response to increasing atmospheric $CO_2$ concentration', *Geophys. Res. Lett.* **32**, doi:10.1029/2005GL023209.

Haigh, J. D. and Blackburn, M.: 2006, 'Solar influences on dynamical coupling between the stratosphere and troposphere', *Space Sci. Rev.*, this volume, doi: 10.1007/s11214-006-9067-0.

Hall, I. R., Bianchi, G. G., and Evans, J. R.: 2004, 'Centennial to millennial scale Holocene climate-deep water linkage in the North Atlantic', *Quat. Sci. Rev.* **23**, 1529–1536.

Hansen, J., Russel, G., Lacis, A., Fung, I., and Rind, D.: 1985, 'Climate response times: Dependence on climate sensitivity and ocean mixing', *Science* **229**, 857–859.

Hansen, J., Sato, M., and Ruedy, R.: 1997, 'Radiative forcing and climate response', *J. Geophys. Res.* **102**, 6831–6864.

Holland, M. M. and Bitz, C. M.: 2003, 'Polar amplification of climate change in coupled models', *Clim. Dyn.* **21**, 221–232.

Ingram, W.: 2006, 'Detection and attribution of climate change, and understanding solar influence on climate', *Space Sci. Rev.*, this volume, doi: 10.1007/s11214-006-9057-2.

Jiang, H., Eiriksson, J., Schulz, M., Knudsen, K.-L., and Seidenkrantz, M.-S.: 2005, 'Evidence for solar forcing of sea-surface temperature on the North icelandic Shelf during the late Holocene', *Geology* **33**, 73–76.

Jones, P. D. and Mann, M. E.: 2004, 'Climate over past millenia', *Rev. Geophys.* **42**, doi:10.29/2003RG000143.

Keigwin, L. D.: 1996, 'The little ice age and medieval warm period in the Sargasso Sea', *Science* **274**, 1504–1508.

Kodera, K.: 2006, 'The role of dynamics in solar forcing', *Space Sci. Rev.*, this volume, doi: 10.1007/s11214-006-9066-1.

Manabe, S., Stouffer, R. J., Spelman, M. J., and Bryan, K.: 1991, 'Transient responses of a coupled atmosphere-ocean model to gradual changes of atmospheric $CO_2$. I. Annual mean response', *J. Clim.* **4**, 785–818.

Mann, M. E., Cane, M. A., Zebiak, S. E., and Clement, A.: 2005, 'Volcanic and solar forcing of the tropical Pacific over the past 1000 years', *J. Clim.* **18**, 447–456.

Meehl, G. A., Washington, W. M., Wigley, T. M., Arblaster, J. M., and Dai, A.: 2003, 'Solar and greenhouse gas forcing and climate response in the twentieth century', *J. Clim.* **16**, 426–444.

Opsteegh, J. D., Haarsma, R. J., Selten, F. M., and Kattenberg, A.: 1998, 'ECBILT: A dynamic alternative to mixed boundary conditions in ocean models', *Tellus* **50A**, 348–367.

Shindell, D., Schmidt, G. A., Mann, M. E., Rind, D., and Waple, A.: 2001, 'Solar forcing of regional climate change during the Maunder Minimum', *Science* **294**, 2149–2152.

Stott, P. A., Tett, S. F. B., Jones, G. S., Ingram, W. J., and Mitchell, J. F. B.: 2001, 'Attribution of twentieth century temperature change to natural and anthopogenic causes', *Clim. Dyn.* **17**, 1–21.

Tett, S. F. B., *et al.*: 2005, 'The impact of natural and anthropogenic forcings on climate and hydrology since 1550', submitted to *Clim. Dyn.*

van der Schrier, G. and Barkmeijer, J.: 2005, 'Bjerknes' hypothesis on the coldness during A: 1790–1820 revisited', *Clim. Dyn.* **24**, 335–371.

van der Schrier, G., Weber, S. L., and Drijfhout, S. S.: 2002, 'Sea level changes in the North Atlantic by solar forcing and internal variability', *Clim. Dyn.* **19**, 435–447.

Waple, A. M., Mann, M. E., and Bradley, R. S.: 2002, 'Long-term patterns of solar irradiance forcing in model experiments and proxy based surface temperature reconstructions', *Clim. Dyn.* **18**, 563–578.

Weber, S. L., Crowley, T. J., and van der Schrier, G.: 2004, 'Solar irradiance forcing of centennial climate variability during the Holocene', *Clim. Dyn.* **22**, 539–553.

White, W. B., Cayan, D. R., and Lean, J.: 1998, 'Global upper ocean heat storage response to radiative frocing from changing solar irradiance and increasing greenhouse gas/aerosol concentration', *J. Geophys. Res.* **103**, 21,355–21,366.

Zorita, E., von Storch, H., Gonzalez-Rouco, F. J., Cubasch U., Luterbacher, J., Legutke, S., Fischer-Bruns, I., and Schlese, U.: 2004, 'Climate evolution in the last five centuries simulated by an atmosphere-ocean model: Global temperatures, the North Atlantic Oscillation and the Late Maunder minimum', *Meteorologische Zeitschrift* **13**, 271–289.

# DISCUSSION OF THE SOLAR UV/PLANETARY WAVE MECHANISM

*Introductory Paper*

M. A. GELLER

*Stony Brook University, Stony Brook, NY 11794–5000, USA*

*(E-mail: marvin.geller@sunysb.edu)*

(Received 29 May 2006; Accepted in final form 8 June 2006)

**Abstract.** Of all the suggested mechanisms for solar activity influences on climate, the solar UV/planetary wave mechanism has been investigated most carefully in a manner that allows comparison with observations. Some historical background, the observations, and modeling investigations of this mechanism are briefly discussed in this review so that the following more detailed papers can be put into context.

**Keywords:** climate, solar influences, planetary waves

## 1. Introduction

The study of solar activity influences on climate has a long and controversial history. One reason for this is that many analyses of data have been questionable on statistical grounds (e.g., Pittock, 1978; Laut, 2003), and another reason has been the lack of a credible mechanism to produce the "observed" solar activity climate effects. Quoting Pittock (1978):

> *"... despite a massive literature of the subject, there is at present little or no convincing evidence of statistically significant or practically useful correlations between sunspot cycles and the weather or climate."*

In the following, we will examine how this situation has changed in the almost three decades since Pittock (1978).

Certainly, nobody would question that variations in solar radiation plays a large role in climate variability. The seasons themselves are produced by the annual variations of the solar energy falling upon a point on the Earth throughout the year resulting from the rotation of the Earth around the Sun, and from the obliquity of the Earth's rotation axis with respect to the ecliptic (the geometry of the angle between the planes defined by that normal to the Earth's axis of rotation, the Equatorial plane, and that defined by the Earth's orbit around the Sun, the ecliptic plane). Furthermore, much paleoclimate variability is believed to be due to small variations in the annual cycle of the solar radiation incoming to each hemisphere resulting from slow periodic changes in the eccentricity of the Earth's orbit around the Sun


DOI: 10.1007/s11214-006-9060-7

with a period of about 100,000 years, changes in the tilt of the Earth's axis of rotation with respect to the ecliptic plane with a period of about 41,000 years, and the precession of the Earth's axis of rotation with a period of about 22,000 years (e.g., Crucifix, 2006).

It is also undeniable that the amount of solar radiation reaching Earth varies with a period of about 11 years due to changing solar activity, as manifested in such visible features as sunspot activity. These variations have been directly observed from satellites since about 1978 (Fröhlich, 2006). The variation in total solar radiation (integrated over all wavelengths) has amplitude of about 0.1% (Fröhlich, 2006). On the other hand, the shorter wavelengths of solar radiation show much larger changes with solar activity (Rottman, 2006).

Starting with the seminal work of Labitzke (1987) and continuing with many papers that have followed up this research by K. Labitzke and her colleagues as well as many others, there is now quite solid evidence of significant observed solar activity effects on the troposphere-stratosphere system. These observational analyses provide many characteristics against which theory and models can be tested against observations.

Discussion of solar impacts on climate has found its way into the issue of global warming induced by increasing atmospheric concentrations of greenhouse gases. There have been claims that the recent global warming that the IPCC (2001) has identified as most likely being due to the increasing atmospheric concentration of greenhouse gases might instead be attributed to changes in solar activity (e.g., Friis-Christensen and Lassen, 1991; Svensmark and Friis-Christensen, 1997). Thus, there are both scientific and policy considerations in assessing where the field of solar activity induced climate change stands with respect to identifying viable physical mechanisms, since this helps to put solar-induced climate changes in perspective relative to other influences on the climate.

Although several mechanisms are suggested for solar activity influences on climate variability (e.g., Goosse, 2006; Arnold, 2006; Curtius, 2006), this introductory paper will concentrate on only one such mechanism, which involves the varying direct solar ultraviolet (UV) radiation producing changes in temperature and ozone, which in turn imply relatively small changes in the mean zonal wind structure in the middle atmosphere. These changes in middle atmospheric winds and temperatures affect planetary wave propagation, which in turn changes planetary wave influences on the mean zonal wind structure, and these seem to propagate downward to give rise to solar activity effects that are much larger than the direct solar UV influences that initiate this chain of events.

## 2. A Little Bit of History

The search for solar activity effects on climate in atmospheric observations is made difficult by the length of the solar cycle, 11 years in the case of the Schwabe sunspot cycle, relative to the length of the observational record. Re-analyses are thought to be less reliable before about 1958 because of a lack of tropical and Southern Hemisphere radiosonde data and fewer balloons reaching high altitudes. It should also be noted that satellite data only began to be incorporated in re-analyses after 1979. A much longer and more reliable data record is thus now available for analysis than was the case when Pittock (1978) reviewed the observational evidence for solar activity effects. The research on some suggested mechanisms for solar influences on climate has also advanced a great deal since that review paper.

Any mechanism that can successfully explain why observed solar activity effects on the climate system are so much larger than expected on purely energetic grounds must somehow modulate the large energy of the natural climate system rather than just involve direct radiative forcing. More than thirty years ago, Hines (1974) suggested that a viable mechanism for solar activity effects on climate might involve direct radiatively forced changes in the upper atmosphere (the term middle atmosphere was not yet in vogue). This would modulate planetary wave propagation, thus giving rise to changes in planetary wave structure that could extend throughout the entire atmosphere. Indeed, Hines' (1974) suggestion did involve solar effects modulating the energetics of atmospheric planetary waves. Calculations relating to this mechanism followed. Geller and Alpert (1980) did experiments in which small changes were made to the zonal mean winds at various altitudes, and they found that significant changes in tropospheric planetary wave structure resulted if changes in the mean zonal winds penetrated down into the lower stratosphere. This work was followed by the work of Callis *et al.* (1985) in which an early two-dimensional photochemical model was used to examine the predicted changes on the middle atmosphere winds and temperatures produced directly by solar cycle modulations in solar UV radiation. These solar UV produced changes in the zonally symmetric atmospheric state were then used in the planetary wave model of Geller and Alpert (1980) to see what planetary wave changes resulted. Callis *et al.* (1985) found relatively modest, but noticeable, solar UV-induced changes in the middle atmosphere planetary wave structure. At about this same time, Bates (1981) also performed calculations of the changes in planetary waves that would be expected from solar UV changes.

In 1987, Karin Labitzke published her seminal paper (Labitzke, 1987) that suggested a much larger and more coherent solar influence on north polar stratospheric temperatures could be seen when the data were separated according to the phase of the quasi-biennial oscillation (QBO). At that time, these results were somewhat controversial due to the relatively short data record (about three solar cycles), the possible biases induced by segmenting the data according to the phase of the QBO, and the possibility that a natural quasi-decadal cycle in the atmosphere might by coincidence be in phase with solar activity during the examined period. Geller (1988), in commenting on the Labitzke (1987) paper, did suggest that a radiative-dynamic feedback might give a larger response to solar UV variations than expected. Quoting Geller (1988):

> *"It seems that radiative changes can alter the meridional (Equator-to-pole) temperature gradient, which in turn changes the polar vortex. This alters the planetary-wave propagation, which further changes the polar vortex, so that the process iterates."*

The pioneering work of Haigh (1994, 1996, 1999) and Shindell *et al.* (1999) introduced an additional positive feedback, and that is the role of ozone in amplifying the direct solar UV influences on the zonal mean state of the middle atmosphere. The remainder of this introduction will discuss how the papers that follow have advanced this line of research.

## 3. Progress on the Solar UV/Planetary Wave Mechanism

Karin Labitzke and her colleague Harry van Loon pioneered the study of QBO-Solar Cycle relationships, and Labitzke (2006) updates some of her earlier results on Solar Cycle-QBO relationships, to stress the interaction between the Northern and Southern Hemispheres, and to summarize the influence of the QBO on the solar variability signal as well as the influence of solar variability on the QBO itself throughout the year. It should be noted that an additional 19 years data since Labitzke (1987) are used in her analyses.

Her work indicates that the observed solar/QBO influences on the atmosphere involve a modulation of the Brewer-Dobson circulation. Further, it is found that a great deal of her observed signal results from the presence or absence of major midwinter warmings, and that the winter hemisphere influence gives results that extend over both the winter and summer hemispheres. Also, in agreement with Salby and Callaghan (2006), she finds evidence for a solar modulation of the QBO. Labitzke's (2006) analyses of stratospheric data give many observed characteristics against which theory and models can be tested.

Marie-Lise Chanin has been studying the thermal structure of the middle and upper atmosphere for more than four decades. Different sign effects have been seen in different atmospheric regions, and this was rather puzzling for a long time. In Chanin (2006), she gathers this information together and compares her observational results to the results of a mechanistic model to demonstrate that this observational evidence is consistent with the solar UV/planetary wave mechanism. She also notes that attempts to see the solar variability at the mesopause level are greatly complicated by the fact that the response changes sign near this level.

As mentioned previously, an important factor in obtaining the correct solar UV influences on climate is to include the correct solar UV-induced ozone changes. Both Haigh (1996) and Shindell *et al.* (1999) emphasized the importance of properly including solar UV changes in ozone because that magnifies the UV effect on temperature over what would exist when no ozone changes are considered, but Haigh (1996) and Shindell *et al.* (1999) used quite different ozone variations. In particular, the latitudinal variations of their ozone changes were different, and this leads to changes in the latitudinal temperature changes, which in turn results in different directly forced changes in the mean zonal winds. Calisesi and Matthes (2006) consider the problem of determining the solar UV induced ozone variations from observations. They conclude that there remains significant uncertainty in the altitude-latitude ozone response to solar variability that should be included in models for solar influences on climate.

In the introduction, it was indicated that the simple linear planetary wave model of Geller and Alpert (1980) implied that solar-induced changes in the zonal mean atmospheric state must extend down to the lower stratosphere to significantly affect the troposphere. The paper by Haigh and Blackburn (2006) addresses the tropospheric response to heating perturbations in the lower stratosphere. The direct solar UV effects on temperature and ozone are maximum in the upper stratosphere, so Gray *et al.* (2006) explore how the effects penetrate downward into the lower stratosphere. The work by Labitzke and her colleagues have shown the very important role of the QBO in solar effects on the troposphere/stratosphere system, and Salby and Callaghan (2006) summarize several of their earlier studies to see how this might occur.

Haigh and Blackburn (2006) report results from two types of numerical experiments. They used an idealized atmospheric general circulation model to explore what effects result from an imposed temperature perturbation on the stratosphere. It should be noted that the model they use has its top level at 18.5 hPa, so their perturbations are in the lower stratosphere. They find a banded response that very much resembles the solar 11-year response derived from observations. Furthermore, they found that the resulting zonal wind changes were primarily maintained by changes in the poleward eddy momentum flux. As an aid to interpreting these results, they conducted some transient spin-up experiments where these stratospheric heatings were suddenly imposed and the evolving responses were examined. Of course, given the limited vertical extent of the models used by Haigh and Blackburn (2006), the question remains by what means does the solar input, whose UV effect is maximum near the tropical stratopause, propagate down to the lower stratosphere?

Gray (2003) has shown, rather surprisingly, that even though the QBO has maximum amplitude at around 25–30 km, it is likely that the solar cycle effect on the QBO at higher levels at 40–50 km plays the largest role in modulating stratospheric sudden warmings. Gray *et al.* (2006) pursue these ideas further by reviewing their published work that has used observational analysis, mechanistic models, and general circulation models. A particularly interesting result is that they find that the timing of their modeled stratospheric warmings is sensitive to relatively small anomalies in the equatorial, subtropical, upper stratosphere in early winter.

Salby and Callaghan (2006) show statistical evidence that the solar effect first discovered by Labitzke (1987) is related to the solar modulation of the QBO period. They distinguish between the "linear" effect of changing UV radiation on the stratosphere where small changes are seen that maximize in the tropical upper stratosphere and "nonlinear" effects. The "linear" effects are relatively small and are physically straightforward to understand through solar UV heating, both directly and indirectly, by photochemically enhancing stratospheric ozone at these altitudes and latitudes, which increases the absorbed UV radiation. The larger effects that are seen in lower altitudes and higher latitudes are nonlinear and include two distinct physical mechanisms. These are planetary wave interactions with tropical winds that are affected by the direct solar effect and by solar modulations in the QBO period. Both of these latter two effects are imperfectly understood at the present time.

For instance, Gray *et al.* (2006) find that the QBO winds in the upper stratosphere are key in modulating the lower stratospheric planetary wave activity and the strength of the winter polar vortex. Statistical analysis of observations, mechanistic modeling, and general circulation modeling all indicate this to be the case. Yet, it is somewhat surprising to see that the upper stratosphere tropical winds seem to be more important than the lower and middle stratospheric winds in modulating stratospheric sudden warmings, with the accompanying large changes in the temperatures at high latitudes and the strength of the winter vortex. Perhaps, this just follows from the fact that the planetary wave Eliassen-Palm vectors are normally directed upwards and equatorward (see Andrews *et al.*, 1987). It is also not clear how solar activity modulates the QBO period although some mechanistic model results by Cordero and Nathan (2005) and Mayr *et al.* (2006) give this result, albeit for different reasons.

Kodera (2006) discusses two different mechanisms by which the solar UV effects on the stratosphere might influence climate in the troposphere. Both start with the solar UV-planetary wave mechanism that has been referred to in previous Sections, which acts to reduce the Brewer-Dobson circulation at solar maximum.

This scenario can produce solar-climate effects in the troposphere in two ways. One is by the downward extension of the planetary waves themselves. Some evidence for this is that the strongest correlation between surface temperature and the F10.7 cm solar flux is found in two regions – east of Japan and north of the Eurasian mountains. It is interesting that these two regions are also the two main regions from which planetary waves propagate upwards from the troposphere (see Plumb, 1985, Figure 5).

The other mechanism involves the solar modulation of the tropical upwelling of the Brewer-Dobson circulation. It is suggested that this might modulate equatorial convective activity. Kodera (2006) goes on to show statistical evidence for a solar modulation of the Walker circulation. Kodera (2006) closes with the remark that given that the total solar irradiance (TSI) shows so much less variation with solar activity than does the solar UV radiation, one might expect much less solar-induced variation in globally averaged temperature than in the regional changes that result from solar-induced changes in the circulation.

A new generation of climate models is appearing that extend up from the troposphere to the thermosphere. Schmidt and Brasseur (2006) show results from one of these models, the HAMMONIA general circulation model whose top is at 250 km. The HAMMONIA model is based on the ECHAM-5 climate model and includes fully interactive chemistry and radiation using the MOZART3 chemistry scheme. They point out that this type of a model extends in altitude where the atmosphere absorbs those portions of the solar spectrum that show large variations with the solar cycle, whereas most climate models do not. Also, the HAMMOMIA model includes physics appropriate to the upper atmosphere that traditional climate models lack. Their modeling results for the solar UV resemble observations, but there are also significant differences between their modeled ozone and temperature responses from those derived from observations.

It should be noted that Giorgetta *et al.* (2002) were successful in modeling the QBO with the ECHAM-5 model, but at the present time, the HAMMONIA model lacks the vertical resolution needed to do this. Given the apparent importance of the QBO in solar influences on the atmosphere, it will be interesting to see how results might change once there is a QBO in HAMMONIA. Also, if the HAMMONIA model were formulated in a similar manner to what was done to successfully model the QBO in Giorgetta *et al.* (2002), this would better resolve the planetary wave interactions that are essential for the solar UV/planetary wave mechanism.

An important point made by Schmidt and Brasseur (2006) is that their modeled solar response has a great deal of longitudinal structure, which might not be treated properly in analysis of satellite observations. It will be very interesting to analyze the HAMMONIA model, and other similar models, for transient simulations of varying solar flux, including volcanoes, the QBO, and changing atmospheric composition, in the same way as satellite data analyses have been carried out to see if the statistical analyses that have been used properly separate these modes of variation for the length of the data record that has been analyzed.

## 4. Some Closing Remarks

With the improved modeling of the troposphere-stratosphere system, model results are showing increasing consistency with the observational analyses of Labitzke, van Loon, Chanin, and others. The relatively recent work of Lesley Gray and her colleagues suggests that solar UV influences on the lower stratosphere are most sensitive to alterations in atmospheric structure in the tropical upper stratosphere, and Matthes *et al.* (2004) find that proper inclusion of this does indeed improve their general circulation model results. Another interesting line of research is the recent work originally due to Murry Salby that suggests a solar effect on the quasi-biennial oscillation in the stratospheric mean zonal wind in the tropics. If this is a mechanism by which solar activity affects the lower stratosphere and troposphere, there are several implications. One is that to treat this mechanism properly, one must use an atmospheric model that extends through the stratosphere and goes sufficiently high for the stratosphere to be well modeled. To account for the indirect effect of solar ultraviolet heating of the stratosphere, the ozone variations that result from the solar UV variations should also be in the model. Another point relates to comments made in the Kodera (2006) and Schmidt and Brasseur (2006) papers. This is that the solar UV ⇒ stratosphere ⇒ lower atmosphere effect on climate mainly influences regional climate whereas variations in the solar constant affect globally averaged temperatures and other climate parameters.

Finally, there is the interesting suggestion in Kodera (2006) of a completely new mechanism by which solar-induced effects in the stratosphere can affect the troposphere, and that is the effect of modulations in the Brewer-Dobson circulation on tropical convection, and its possible subsequent possible effect on the global circulation. In this regard, it should be noted that Collimore *et al.* (2003) have found evidence of a QBO effect on tropical convection, and they conclude that this is most likely related to QBO modulations in tropical tropopause heights.

This summary paper has discussed several of the papers that follow in this book, but it is certainly not a comprehensive presentation of all mechanisms for solar influences on climate. One such mechanism is discussed elsewhere in this volume, and that is the mechanism that involves cosmic rays on cloud amounts. The Curtius (2006) and Arnold (2006) papers address this issue.

Other topics relating to mechanisms for solar influences on the climate are also not covered here. One is the suggestion of Meehl *et al.* (2003) that variations in total solar insolation lead to more evaporation in the cloud-free subtropics, which in turn affects moisture transports and latent heat release. To treat the solar UV/QBO/planetary wave mechanism properly in a climate model, the model (i) should extend at least into the mesosphere; (ii) include solar modulation of ozone, (iii) ideally should account for the QBO; and (iv) include good resolution of the solar variations in the UV portion of the solar spectrum. The Meehl *et al.* (2003) mechanism does not require all these things. It also is not yet being researched as intensively as the solar UV/planetary waves mechanism.

It is clear that the study of solar influences on climate is now in a new phase where quantitative treatments of mechanisms are being studied. Furthermore, in the case of solar UV effects on climate, these studies contribute to the more general study of stratospheric influences on climate, where the stratospheric changes may occur from natural or anthropogenic influences. The introduction started with a rather pessimistic quote from Pittock (1978). Five years later, Pittock (1983) said the following:

> *"Advances have been made in the modeling of variable ultraviolet influences on middle stratosphere temperatures and ozone content, and of the modulation of reflection and absorption of tropospheric planetary wave energy by solar-induced variations in stratospheric winds and temperature profiles."*

He went on to say that this solar UV/planetary wave mechanism was

> *". . . a complex phenomenon which as yet has only been modeled in a very simplified manner."*

How right he was, and how far we have come.

## References


Andrews, D. G., Holton, J. R., and Leovy, C. B.: 1987, *Middle Atmosphere Dynamics*, Academic Press, San Diego, 489 pp.

Arnold, F.: 2006, 'Atmospheric aerosol and cloud condensation nuclei formation: A possible influence of cosmic rays', *Space Sci. Rev.*, this volume, doi: 10.1007/s11214-006-9055-4.

Bates, J. R.: 1981, 'A dynamical mechanism through which variations in solar ultra violet radiation can affect tropospheric climate', *J. Geophys. Res.* **104**, 27,321–27,339.

Calisesi, Y., and Matthes, K.: 2006, 'The middle atmospheric ozone response to the 11-year solar cycle', *Space Sci. Rev.*, this volume, doi: 10.1007/s11214-006-9063-4.

Callis, L. B., Alpert, J. C., and Geller, M. A.: 1985, 'An assessment of thermal, wind, and planetary wave changes in the middle and lower atmosphere due to 11-year UV flux variations', *J. Geophys. Res.* **90**, 2273–2282.

Chanin, M.-L.: 2006, 'Signature of the 11-year cycle in the upper atmosphere', *Space Sci. Rev.*, this volume, doi: 10.1007/s11214-006-9062-5.

Collimore, C. C., Martin, D. W., Hitchman, M. H., Huesmann, A., and Waliser, D. E.: 2003, 'On the relationship between the QBO and tropical deep convection', *J. Clim.* **16**, 2552–2568.

Cordero, E. C., and Nathan, T. R.: 2005, 'A new pathway for communicating the 11-year solar cycle signal to the QBO', *J. Geophys. Res.* **32**, doi:10.1029/2005GL023696.

Crucifix, M.: 2006, 'The climate response to the astronomical forcing', *Space Sci. Rev.*, this volume, doi: 10.1007/s11214-006-9058-1.

Curtius, J.: 2006, 'Atmospheric ion-induced aerosol nucleation', *Space Sci. Rev.*, this volume, doi: 10.1007/s11214-006-9054-5.

Friis-Christensen, E., and Lassen, K.: 1991, 'Length of the solar cycle: An indicator of solar activity closely associated with climate', *Science* **254**, 698–700.

Fröhlich, C.: 2006, 'Solar irradiance variability since 1978', *Space Sci. Rev.*, this volume, doi: 10.1007/s11214-006-9046-5.

Geller, M. A.: 1988, 'Solar cycles and the atmosphere', *Nature* **332**, 584–585.

Geller, M. A., and Alpert, J. V.: 1980, 'Planetary wave coupling between the troposphere and middle atmosphere as a possible sun-weather mechanism', *J. Atmos. Sci.* **37**, 1197–1215.

Giorgetta, M. A., Manzini, E., and Roeckner, E.: 2002, 'Forcing of the quasi-biennial oscillation from a broad spectrum of atmospheric waves', *Geophys. Res. Lett.* **29**, doi:10.1029/2002GL014756.

Goosse, H.: 2006, 'Regional response of the climate system to solar forcing: The role of the ocean', *Space Sci. Rev.*, this volume, doi: 10.1007/s11214-006-9059-0.

Gray, L. J., 2003: The influence of the equatorial upper stratosphere on stratospheric sudden warmings', *Geophys. Res. Lett.*, **30**, doi:10.1029/2002GL016430.

Gray, L., Crooks, S., Palmer, M., Pascoe, C., and Sparrow, S.: 2006, 'A possible transfer mechanism for the 11-year solar cycle to the lower stratosphere', *Space Sci. Rev.*, this volume, doi: 10.1007/s11214-006-9069-y.

Haigh, J. D.: 1994, 'The role of stratospheric ozone in modulating the solar radiative forcing of climate', *Nature* **370**, 544–546.

Haigh, J. D.: 1996, 'The impact of solar variability on climate', *Science* **272**, 981–984.

Haigh, J. D.: 1999, 'A GCM study of climate change in response to the 11-year solar cycle', *Quart. J. Roy. Meteorol. Soc.* **125**, 871–892.

Haigh, J. D., and Blackburn, M.: 2006, 'Solar influences on dynamical coupling between the stratosphere and troposphere', *Space Sci. Rev.*, this volume, doi: 10.1007/s11214-006-9067-0.

Hines, C. O.: 1974, 'A possible mechanism for the production of sun-weather correlations', *J. Atmos. Sci.* **31**, 589–591.

IPCC (Intergovernmental Panel on Climate Change): 2001, *Climate Change 2001: The Scientific Basis: Contribution of Working Group I to the Third Assessment Report of the Intergovernmental*

*Panel on Climate Change, 2001*, J. T., Houghton, Y., Ding, D. J., Griggs, M., Noguer, van der P. J., Linden, Dai, X., *et al.* (Eds.), Cambridge University Press, Cambridge, U.K., 881 pp.

Kodera, K.: 2006, 'The role of dynamics in solar forcing', *Space Sci. Rev.*, this volume, doi: 10.1007/s11214-006-9066-1.

Labitzke, K.: 1987, 'Sunspots, the QBO, and the stratospheric temperature in the north polar region', *Geophys. Res. Lett.* **14**, 535–537.

Labitzke, K.: 2006, 'Solar variation and stratospheric response', *Space Sci. Rev.*, this volume, doi: 10.1007/s11214-006-9061-6.

Laut, P.: 2003, 'Solar activity and terrestrial climate: An analysis of some purported correlations', *J. Atmos. Solar-Terr. Phys.* **65**, 801–812.

Matthes, K., Langematz, U., Gray, L. J., Kodera, K., and Labitzke, K.: 2004, 'Improved 11-year solar signal in the Frei Universität Berlin Climate middle Atmosphere Model (FUB-CHAM)', *J. Geophys. Res.* **109**, doi:10.1029./2003JD004012.

Mayr, H. G., Mengel, J. G., Wolff, C. L., Porter, H. S.: 2006, 'QBO as potential amplifier of solar cycle influence', *Geophys. Res. Lett.* **33**, doi:10.1029/2005GL025650.

Meehl, G. A., Washington, W. M., Wigley, T. M. L., Arblaster, J. M., and Dai, A.: 2003, 'Solar and greenhouse gas forcing and climate response in the twentieth century', *J. Clim.* **16**, 426–444.

Pittock, A. B.: 1978, 'A critical look at long-term Sun-weather relationships', *Rev. Geophys. Sp. Phys.* **16**, 400–420.

Pittock, A. B.: 1983, 'Solar variability, weather, and climate: An update', *Q. J. Roy. Met. Soc.* **109**, 23–55.

Plumb, A.: 1985, 'On the three-dimensional propagation of stationary waves', *J. Atmos. Sci.* **42**, 217–229.

Rottman, G.: 2006, 'Measurement of total and spectral solar irradiance', *Space Sci. Rev.*, this volume, doi: 10.1007/s11214-006-9045-6.

Salby, M., and Callaghan, P. F.: 2006, 'Influence of the solar cycle on the general circulation of the stratosphere and upper troposphere', *Space Sci. Rev.*, this volume, doi: 10.1007/s11214-006-9064-3.

Schmidt, H., and Brasseur, G. P.: 2006, 'The response of the middle atmosphere to solar cycle forcing in the Hamburg Model of the Neutral and Ionized Atmosphere', *Space Sci. Rev.*, this volume, doi: 10.1007/s11214-006-9068-z.

Shindell, D., Rind, D., Balachandran, N., Lean, J., and Lonergan, P.: 1999, 'Solar cycle variability, ozone, and climate', *Science* **284**, 305–308.

Svensmark, H., and Friis-Christensen, E.: 1997, 'Variation of cosmic ray flux and global cloud coverage – a missing link in solar-climate relationships', *J. Atmos. Solar-Terr. Phys.* **59**, 1225–1232.

# SOLAR VARIATION AND STRATOSPHERIC RESPONSE

K. LABITZKE
*Institute for Meteorology, Free University of Berlin, Carl-Heinrich-Becker-Weg 6–10, D-12165 Berlin, Germany*
*(E-mail: labitzke@strat01.met.fu-berlin.de)*

(Received 27 June 2005; Accepted in final form 30 January 2006)

**Abstract.** We have shown in several recent publications that it is necessary to group the meteorological data according to the phase of the Quasi-Biennial Oscillation (QBO) *throughout the year*, in order to find a clear signal of the 11-year sunspot cycle (SSC). This work is summarized here. It is the purpose of this paper (1) to update earlier results of the solar cycle – QBO relationship for the northern winter, (2) to stress the interaction between the hemispheres and (3) to summarize the influence of the QBO on the solar variability signal, as well as the influence of the solar variability signal on the QBO throughout the year. For this, the *constructed annual mean of the solar cycle – QBO relationship* is introduced.

**Keywords:** 11-year solar cycle, solar signal, stratosphere-troposphere interaction, hemispheric interaction, Quasi-Biennial Oscillation, mean meridional circulation systems

## 1. Introduction

Until recently it was generally doubted that the solar variability in the 11-year sunspot cycle (SSC), as measured by satellites, has a significant influence on weather and climate variations (see, e.g., Pittock's review, 1983, and Hoyt and Schatten, 1997). But several studies, both empirical and modeling, have in recent years pointed to probable and certain influences. For instance, Labitzke (1982) suggested that the Sun influences the intensity of the north polar vortex (i.e., the Arctic Oscillation, AO) in the stratosphere in winter, and that the Quasi-Biennial Oscillation (QBO) is needed to identify the solar signal (e.g., Labitzke, 1987; Labitzke and van Loon, 1988, 2000).

Different observations indicate that the mean meridional circulation systems, like the Brewer-Dobson Circulation (BDC) and the Hadley Circulation (HC) are influenced by the 11-year solar cycle (Kodera and Kuroda, 2002; Hood, 2003; Labitzke, 2003, 2004a,b; Salby and Callaghan, 2004; van Loon *et al.*, 2004).

Today, there is general agreement that the *direct* influence of the changes in the UV part of the spectrum (6 to 8% between solar maxima and minima) leads to more ozone and warming in the upper stratosphere (around 50 km) in solar maxima (Haigh, 1994; Hood *et al.*, 1993; Hood, 2004; Crooks and Gray, 2005). This leads to changes in the thermal gradients and thus in the wind systems, which in turn

Space Science Reviews (2006) 125: 247–260
DOI: 10.1007/s11214-006-9061-6 

lead to changes in the vertical propagation of the planetary waves that drive the global circulation. Therefore, the relatively weak, *direct* radiative forcing of the solar cycle in the upper stratosphere can lead to a large *indirect* dynamical response in the lower atmosphere through a modulation of the polar night jet (PNJ) as well as through a change in the Brewer Dobson Circulation (BDC) (Kodera and Kuroda, 2002; Matthes *et al.*, 2004).

## 2. Data and Methods

The *global* re-analyses by NCEP/NCAR (Kalnay *et al.*, 1996) are used throughout this paper. The re-analyses are less reliable for earlier periods (before 1958), mainly because of the lack of radiosonde stations over the tropics and over the Southern Hemisphere, the lack of high reaching balloons in the early years and the scarce satellite information before 1979. However, we note that the inclusion of the early data nevertheless yields similar results.

It is difficult to determine the statistical significance of the correlations which depends on the number of solar cycles available (Salby and Callaghan, 2004). With the analyses of the winter data starting in 1958, 5 solar cycles are available and therefore the significance of the correlations is the following: $r = 0.5$ yields about 90%; $r = 0.65$ about 95%; and $r = 0.7$ about 99%. Recently, more data became available which go back to 1942, and over the Arctic the correlations (now with 6.5 solar cycles) remained the same as for the data starting in 1958. This shows the stability of the data and high significance of the solar signal, especially in the west phase of the QBO (Labitzke *et al.*, 2006). The summer data (starting in 1968) cover a shorter period, but the correlations are higher and the interannual variability is much lower during the northern summer (Labitzke and van Loon, 1999). Therefore, the significance of the correlations is the following: $r = 0.6$ yields about 90%; $r = 0.7$ about 95%; and $r = 0.8$ about 99%.

The monthly mean values of the 10.7 cm solar flux are used as a proxy for variations through the SSC. The flux values are expressed in solar flux units: 1 s.f.u. = $10^{-22}$ $\mathrm{Wm^{-2}Hz^{-1}}$. This is an objectively measured radio wave intensity, highly and positively correlated with the 11-year SSC and particularly with the UV part of the solar spectrum (Hood, 2003). For the *range* of the SSC, the mean difference of the 10.7 cm solar flux between solar minima (about 70 units) and solar maxima (about 200 units) is used, i.e., 130 units (Labitzke, 2003).

More information on the QBO can be found in Labitzke (2004b, 2005).

## 3. Variability in the Stratosphere

The arctic stratosphere reaches its highest variability in winter. Figure 1 gives an example of the interannual variability of the stratosphere during the northern

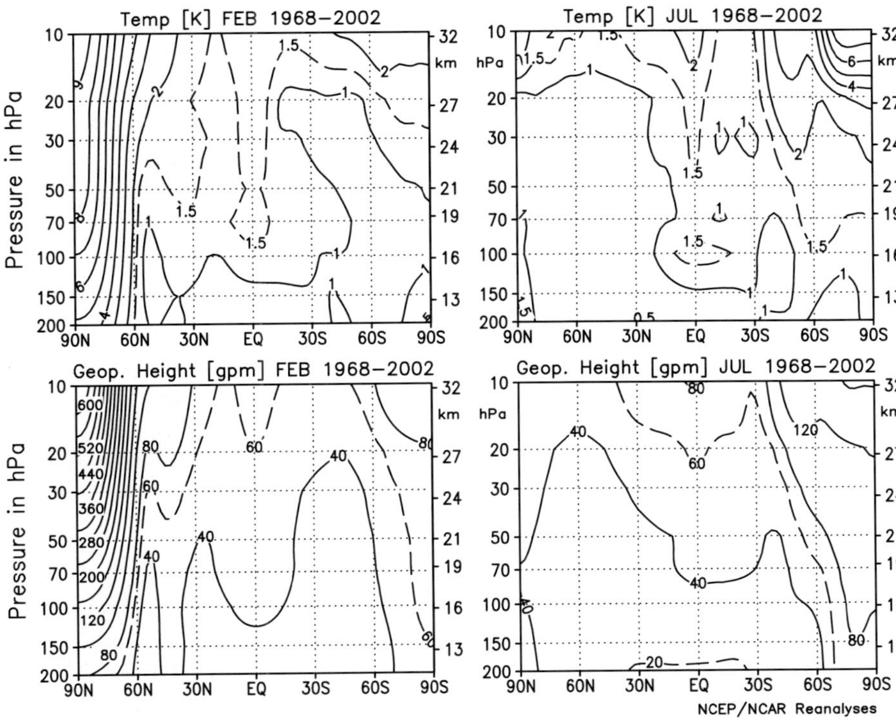


*Figure 1.* Vertical meridional sections of the standard deviations for February (left) and July (right) of the zonal mean monthly mean temperatures (K), upper panels, and of the zonal mean monthly mean geopotential heights (geopot. m), lower panels, for the period 1968–2002 (NCEP/NCAR re-analyses).

winter (February) and summer (July). It is remarkable that in the lower and middle stratosphere the standard deviations in the arctic winter are three to four times larger than those in the antarctic winter; this is due to the fact that the Major Mid-Winter Warmings (MMWs) which create the large variability of the Arctic, do usually not penetrate to the lower stratosphere over the Antarctic. But the variability is large in the upper stratosphere over the Antarctic where so-called Minor Mid-Winter Warmings occur frequently (Labitzke and van Loon, 1972).

When the antarctic westerly vortex breaks down in spring (September–November) the middle stratosphere varies so much from one spring to another that the standard deviation at the South Pole in October (Labitzke and van Loon, 1999, their Figure 2.11) approaches that at the North Pole in January and February. In summer (July) the variability is low in both hemispheres, because the southern winter is still relatively undisturbed. A minor maximum of variability is observed on the equator due to the QBO.

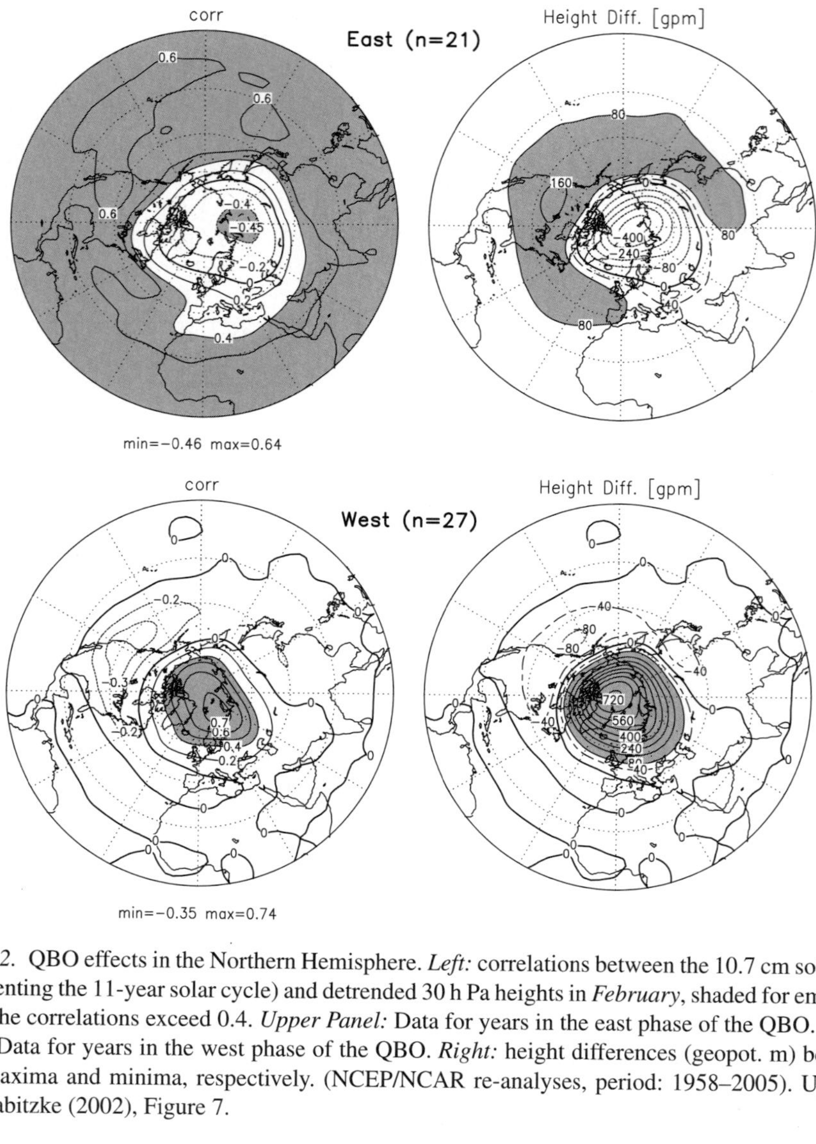


*Figure 2.* QBO effects in the Northern Hemisphere. *Left:* correlations between the 10.7 cm solar flux (representing the 11-year solar cycle) and detrended 30 h Pa heights in *February*, shaded for emphasis where the correlations exceed 0.4. *Upper Panel:* Data for years in the east phase of the QBO. *Lower Panel:* Data for years in the west phase of the QBO. *Right:* height differences (geopot. m) between solar maxima and minima, respectively. (NCEP/NCAR re-analyses, period: 1958–2005). Updated from Labitzke (2002), Figure 7.

## 4. Influences of the 11-Year Sunspot Cycle on the Stratosphere

### 4.1. The Stratosphere During the Northern Winter

Based on results published in 1982, Labitzke (1987) found that a signal of the 11-year SSC emerged when the arctic stratospheric temperatures and geopotential heights were grouped into two categories determined by the direction of the equatorial wind in the stratosphere (QBO). The reality and significance of using this approach have been confirmed by Naito and Hirota (1997) and Salby and

Callaghan (2004). Meanwhile, 19 more years of data have become available and the correlations remained stable. This phenomenon exists during the whole winter but maximizes during late winter (February) when the planetary wave activity is largest (e.g., van Loon and Labitzke, 2000). An example is given in Figure 2 which shows on the left hand side for the Northern Hemisphere in *February* maps of the correlations between the solar 10.7 cm flux and the 30 hPa heights, with the winters (Februarys) in the east phase of the QBO in the upper part of the Figure, and the winters in the west phase of the QBO in the lower part. The pattern of correlations is clearly very different in the two groups, with negative correlations over the Arctic in the east phase and large positive correlations there in the west phase. Outside of the Arctic the correlations are positive and strong in the east phase, but very weak in the west phase.

The respective height differences between solar maxima and minima are given on the right hand side of Figure 2. In the east phase of the QBO the stratosphere is colder in solar maxima than in solar minima, and the heights tend to be below normal over the Arctic (about one standard deviation, c.f. Figure 1); they are well above normal towards the equator (up to three standard deviations over Canada). In the west phase, the arctic heights tend to be well above normal (about two standard deviations, c.f. Figure 1) in solar maxima; there are only small anomalies outside the Arctic.

Figure 3 shows for *February vertical meridional sections* of correlations between the solar 10.7 cm flux and zonally averaged temperatures, as well as the corresponding temperature differences between solar maxima and minima. When all years are used in February, the correlations (top left) and the corresponding temperature differences (top right) are small. But, in the *east phase of the QBO*, the correlations of the zonally averaged temperatures in the lower stratosphere with the solar data are strongly positive from 60°N to the South Pole in the summer hemisphere, and negative north of 60°N, in the winter hemisphere. These negative correlations are connected with seven Major Midwinter Warmings (MMWs) during East/Min (solar flux below 110, Labitzke *et al.*, 2006); but because also three such MMWs took place in solar maxima, the correlations are weaker than in the west phase of the QBO, see below.

On the right hand side in the middle panel are the zonally averaged temperature differences between solar maxima and minima in the east phase of the QBO which correspond to the correlations on the left side; the shading is the same as that in the correlations where it denotes correlations above 0.4. As the standard deviations in the tropics and subtropics are very small (Figure 1), the temperature differences of 1 to 2 K reflect a solar signal of the order of 1 to 2 standard deviations.

In the *west phase of the QBO* (Figure 3, bottom), the pattern of the correlations with the 10.7 cm solar flux is completely different from the east phase winters: highly positive correlations are found over the Arctic while they are near zero or weakly negative elsewhere. The large positive correlations over the Arctic are associated with the frequent MMWs which occur at solar maxima when the QBO

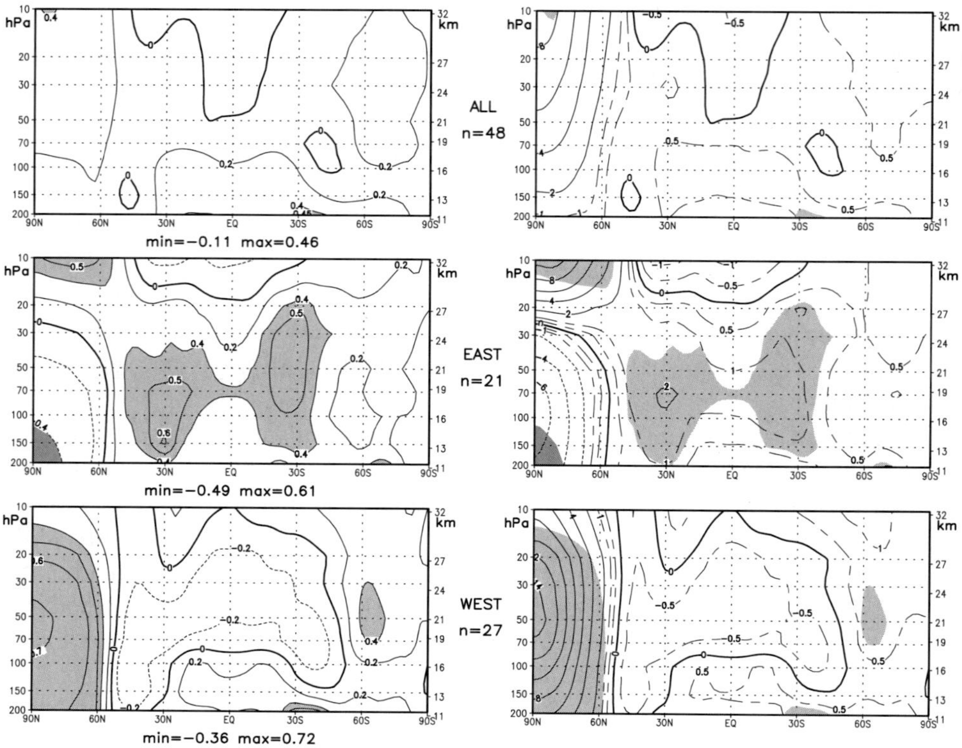

*Figure 3.* Vertical meridional sections between 200 and 10 hPa (about 11 and 32 km) of, on the left, the correlations between the detrended zonally averaged temperatures for *February* and the 10.7 cm solar flux (shaded for emphasis where the correlations are larger than 0.4), and, on the right, the respective temperature differences (K) between solar maxima and minima, shaded where the corresponding correlations on the left hand side are above 0.4. The upper panels show all years, the middle panels only years in the east phase of the QBO, and the lower panels only years in the west phase of the QBO (NCEP/NCAR re-analyses, 1958–2005). Updated from Labitzke (2002).

is in the west phase. Seven MMWs occurred during the 9 winters in the group West/Max, with the 10.7 cm solar flux larger than 150 units. But it is also important to note that *no* MMWs took place during West/Min (12 winters with the solar flux below 110 units).

The height and temperature changes shown in Figures 2 and 3 indicate that the solar cycle influences the "Mean Meridional Circulation (MMC)", also called the "Brewer-Dobson Circulation (BDC)". Forced by planetary waves the MMC regulates wintertime *polar temperatures* through downwelling and adiabatic warming (Kodera and Kuroda, 2002; Kuroda and Kodera, 2002; Hood and Soukharev, 2003; Labitzke, 2003; Hood, 2004; Salby and Callaghan, 2004).

During the *west phase of the QBO the MMC is intensified during solar maxima* (and vice versa during solar minima), with large positive anomalies over the Arctic (intensified downwelling and warming), and concurrent weak (and even negative) anomalies (anomalous upwelling/adiabatic cooling) over the tropics and subtropics, as shown in the lower maps of Figure 2 and in the lowest panels of Figure 3. During the *east phase the MMC is weakened in solar maxima* (and vice versa during solar minima), with reduced downwelling (anomalous upwelling/cooling) and negative anomalies over the Arctic in solar maxima, and concurrent anomalous downwelling with positive anomalies over the tropics and subtropics (upper maps in Figure 2 and middle panels in Figure 3).

## 4.2. The Stratosphere During the Northern Summer

The stratosphere is least disturbed during the northern summer (Labitzke and van Loon, 1999) and results for July and August are very similar to each other (Labitzke, 2003, 2004b). The correlations on the left of Figure 4 show for *July+August* the

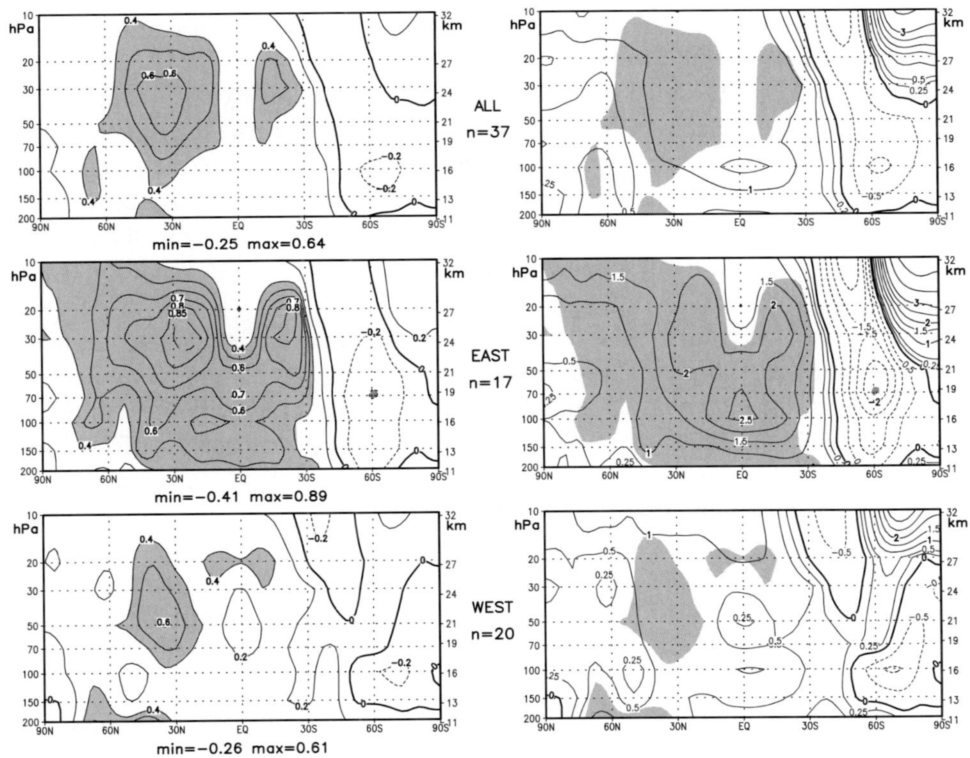


*Figure 4.* Vertical meridional sections between 200 and 10 hPa (11 to 32 km) of (on the left) the correlations between the detrended zonally averaged temperatures for *(July+August)/2* and the 10.7 cm solar flux (shaded for emphasis where the correlations are larger than 0.4) and, on the right, the respective temperature differences (K) between solar maxima and minima, shaded where the corresponding correlations on the left hand side are above 0.4. The upper panels show all years, the middle panels only years in the east phase of the QBO, and the lower panels only years in the west phase of the QBO (NCEP/NCAR re-analyses, 1968–2004).

vertical distribution of the relationship between the solar variability and the temperature, from the upper troposphere to the middle stratosphere. Again, the data are grouped in all years and in the east and west phases of the QBO; the corresponding temperature differences are on the right in the diagram. Originally we made this division according to the phase of the QBO only in the northern winter data; but it turns out that it is also a valid approach throughout the year (Labitzke, 2003, 2004a,b, 2005).

The results for the east phase (middle panels) are most striking: two centers with correlations above 0.8 are found over the subtropics between 20 and 30 hPa; further down, the double maximum in the east phase changes into one maximum, centered on the equatorial tropopause (Labitzke, 2003). The temperature differences (right hand side) between solar maxima and minima are large, more than two standard deviations in some regions, Figure 1. This warming, i.e., positive anomalies, can only be explained by anomalous downwelling (i.e., reduced upwelling) over the subtropics and tropics (Kodera and Kuroda, 2002; Shepherd, 2002) which – in other words – means a weakening of the BDC for solar maxima/east phase of the QBO, as discussed above already for the northern winter.

The solar variability signal is much weaker in the west phase years. It hints at an intensification of the Hadley Circulation (HC) over the Northern Hemisphere, with stronger rising motions over the equator in the troposphere, i.e., weak warming due to latent heat release in the upper troposphere, accompanied by relative cooling in the lower stratosphere (50 hPa) and some anomalous heating (downwelling) over the subtropics of the northern summer hemisphere (Shepherd, 2002; van Loon *et al.*, 2004).

Figure 5 shows maps of the height differences (between solar maxima and minima) over the subtropics and tropics in *July* for the 30 hPa level (about 24 km): again, the east phase dominates the solar signal (top of figure). In addition, the anomalous zonal (west-east) wind in the equatorial belt is affected by the solar variability on the decadal scale. At the top of the Figure an anomalously high value (more than two standard deviations, Figure 1) is centered over the equator. It means that an anomalous anticyclonic circulation is centered on the equator in the solar maximum east years, connected with anomalous winds from the west. Therefore, during solar maxima in QBO/east years the low-latitude east wind is weakened, and conversely in the solar minimum years.

In the west phase of the QBO (bottom of Figure 5), the pattern of the anomalies of the geopotential heights is completely opposite, with the lowest values on the equator in the solar maxima. The anomalous winds are from the east, around the anomalous low on the equator, and conversely in the solar minima and west years. *The QBO thus not only modulates the solar signal on the decadal scale, but is itself modulated by the solar variability* (Salby and Callaghan, 2000; Gray *et al.*, 2001; Soukharev and Hood, 2001; Labitzke, 2003, 2004b).

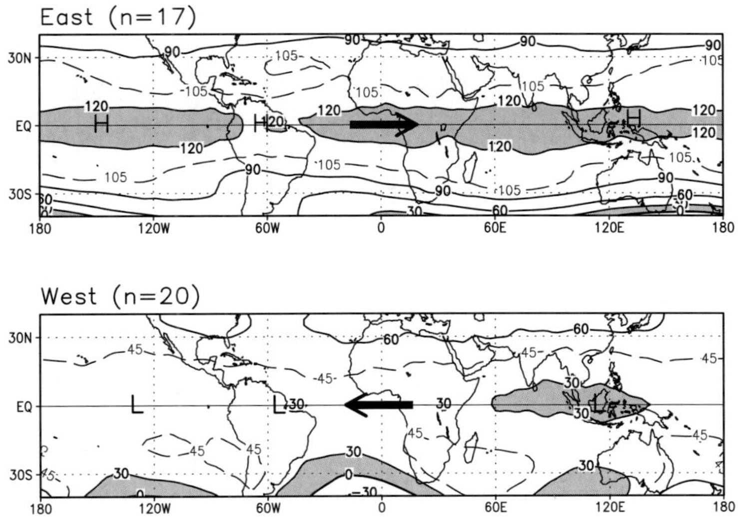


*Figure 5.* Maps of the 30 hPa height differences (geopot. m) in *July* between solar maxima and minima. The upper map is for years in the east phase of the QBO, and the lower map for years in the west phase of the QBO. Arrows indicate the direction of the anomalous winds (NCEP/NCAR re-analyses, 1968–2004). Adapted from Labitzke (2003), Figure 5.

## 5. The QBO – Solar Cycle Relationship Throughout the Year

*During the east phase of the QBO* the structure shown in Figure 5 for the 30 hPa level, with the largest height differences between solar maxima and minima directly over the equator, exists from March through October, that is for 8 months. Only during the northern winter does a very weak anomalous height gradient exist over the tropics. Therefore *one finds an influence of the solar cycle on the tropical QBO for the whole year*, as discussed by Soukharev and Hood (2001) and Labitzke (2004b).

*During the west phase of the QBO*, most of the year the solar signal is much weaker than during the east phase. During the northern spring and summer (April through August) a clear signal exists from the northern subtropics to the Arctic, with a secondary maximum over the Southern Hemisphere. At the same time the solar variability signal during the southern summer (i.e., northern winter) is much reduced compared with the results obtained during northern summer. This is due to the dynamical interactions between high and low latitudes, as discussed above.

Over the equatorial region the height differences show a structure completely different from the other phase of the QBO. A minimum of the height differences is found over the equator for most of the year, implying anomalous winds from the east, that is a weaker QBO-west wind in solar maximum.

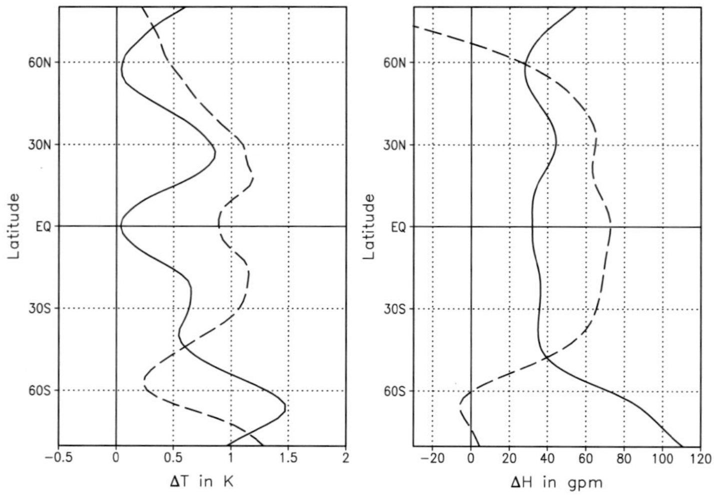


*Figure 6. Left Panel:* Meridional profiles (80°N to 80°S) of the *Constructed Annual Mean* 30 hPa temperature differences (K). *Right panel:* the same, but for the 30 hPa height differences (geopot. m). These are both arranged for the QBO west phase (solid lines) and QBO east phase (dashed lines), respectively. Updated from Labitzke (2004b) (NCEP/NCAR re-analyses, 1968–2004).

Figure 6 summarizes the differences discussed above. As it is practically impossible to derive an annual mean for the QBO data, the data shown in Figure 6 display a *constructed annual mean* of the differences between solar maxima and minima at the 30 hPa level. The anomalies were derived for each month separately and then linearly averaged into an annual mean.

The main features are clearly identified from Figure 6. In the middle stratosphere (here at 30 hPa) the signal of the 11-year SSC (differences between solar maxima and minima) is much stronger during the east phase of the QBO, from 60°N to 50°S. This is the case for the 30 hPa temperatures and heights. The height differences (Figure 6, right) have a clear maximum over the equator during the east phase, and a clear minimum during the west phase. These differences are connected with the weaker QBO winds (in both phases) during solar maxima, as described above.

The summarized solar variability signal in the 30 hPa temperatures shows two maxima in the subtropics in both phases of the QBO. It is of importance to note that there are large differences between the two phases of the QBO in the tropics, with almost no solar signal directly over the equator in the west phase and a strong signal (almost 1 K) in the east phase (Figure 6, left).

Figure 7 shows the vertical meridional sections of the *constructed annual mean* of the temperature and height differences between solar maxima and minima, for all years (upper panels ) and for the two different phases of the QBO. *The temperature differences for all years (upper left) are practically positive from the ground up to 32 km, i.e., as high as the data are available. Both the troposphere and the*

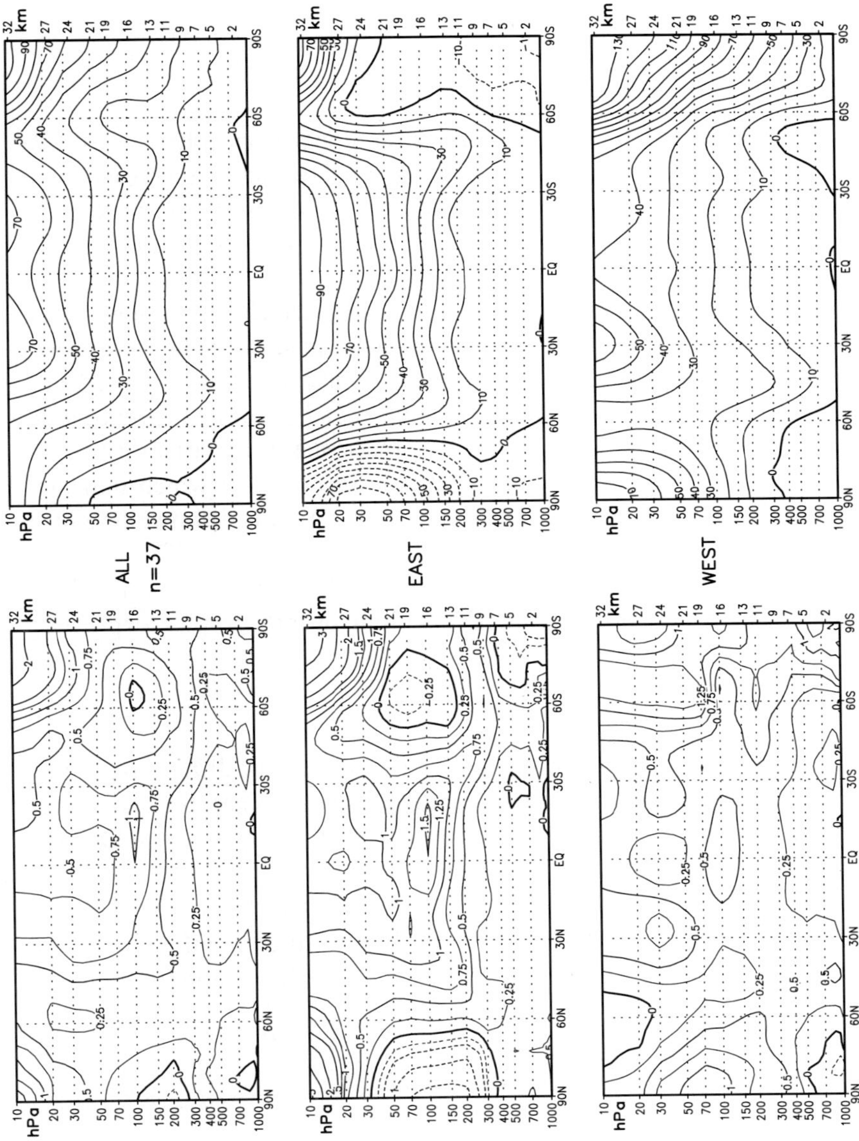


*Figure 7.* Vertical meridional sections of the *Constructed Annual Means. Left Panels:* Annual means of the temperature differences (K) between solar maxima and minima. *Right Panels:* Annual means of the height differences (geopot. m). The upper panels show all years, the middle panels only years in the east phase of the QBO, and the lower panels only years in the west phase. Updated from Labitzke (2005) (NCEP/NCAR re-analyses, 1968–2004).

*stratosphere are warmer during solar maxima than during solar minima* (Labitzke, 2005).

In the middle and lower panels, the solar variability signal is convincingly summarized for the two different phases of the QBO, and all the details discussed before for the respective winters and summers stand out clearly. They can be summarized as follows:

1. Temperature differences:

(a) Over the latitude range 30°N to 30°S, from the upper troposphere to 10 hPa the solar variability signal over the tropics and subtropics is much larger in the east phase. This must be explained by anomalous downwelling (or reduced upwelling, i.e., less adiabatic cooling), which *indicates a weakened BDC during solar maxima/QBO-east phase.*

(b) The annual mean over the polar regions is much influenced by the winter events. Therefore, over both polar regions the solar signal is positive in the west phase, reflecting the warmer polar stratosphere during solar maxima in winter (i.e., *an enhanced BDC with enhanced downwelling during solar maxima/QBO-westphase*); see discussion above. In the east phase the solar variability signal over the polar regions is negative in the middle stratosphere which indicates a weakened BDC during solar maxima/east phase, i.e., less downwelling over the poles in winter.

(c) The warmer polar stratosphere during solar maxima/QBO-west phase in winter is connected with dynamically enhanced upwelling over the tropics, explaining the smaller solar signal there.

2. Height differences:

(a) The height differences relate to the integral of the temperature differences below (hydrostatic relationship). Accordingly, in the QBO/east phase they are larger than in the QBO/west phase in the stratosphere over the tropics and subtropics and weaker (negative) over the polar regions.

(b) There is a maximum of the height differences over the equator in the QBO/east phase and a minimum in the QBO/west phase, reflecting the influence of the solar cycle on the QBO throughout the year, see above discussion for Figure 5.

## Acknowledgements

I thank the members of the Stratospheric Research Group, FUB for professional support and Dipl. Met. Markus Kunze for doing the computations and graphics. Special thanks go to Harry van Loon for long discussions and to Michael Rycroft for reviewing the manuscript. The 10.7 cm solar flux data are from the World Data Center A, Boulder, Colorado.

## References


Crooks, S. A. and Gray, L. J.: 2005, 'Characterization of the 11-year solar signal using a multiple regression analysis of the ERA-40 dataset', *J. Clim.* **18**, 996–1015.

Gray, L. J., Phipps, J., Dunkerton, T., Baldwin, M., Drysdale, E., and Allen, M.: 2001, 'A data study of the influence of the equatorial upper stratosphere on northern hemisphere stratospheric sudden warmings', *Q. J. R. Meteor. Soc.* **127**, 1985–2003.

Haigh, J. D.: 1994, 'The role of stratospheric ozone in modulating the solar radiative forcing of climate?', *Nature* **370**, 544–546.

Hood, L. L.: 2003, 'Thermal response of the tropical tropopause region to solar ultraviolet variations', *Geophys. Res. Lett.* **30**, doi:10.1029/2003GL018364.

Hood, L. L.: 2004, 'Effects of solar UV variability on the stratosphere', in: J. Pap *et al.* (eds.), *Solar Variability and its Effect on the Earth's Atmosphere and Climate System*. AGU Monograph Series, Washington D.C., pp. 283–304.

Hood, L. L. and Soukharev, B.: 2003, 'Quasi-decadal variability of the tropical lower stratosphere: The role of extratropical wave forcing', *J. Atmos. Sci.* **60**, 2389–2403.

Hood, L. L., Jirikowic, J. L., and McCormack, J. P.: 1993, 'Quasi-decadal variability of the stratosphere: Influence of long-term solar ultraviolet variations', *J. Atmos. Sci.* **50**, 3941–3958.

Hoyt, D. V. and Schatten, K. H.: 1997, 'The role of the Sun in climate change', Oxford University Press, New York, 279 pp.

Kalnay, E. R., *et al.*: 1996, 'The NCEP/NCAR 40-year re-analysis project', *Bull. Am. Meteor. Soc.* **77**, 437–471.

Kodera, K. and Kuroda, Y.: 2002, 'Dynamical response to the solar cycle', *J. Geophys. Res.* **107**, doi:10.1029/2002JD002224.

Kuroda, Y. and Kodera, K.: 2002, 'Effect of solar activity on the polar-night jet oscillation in the northern and southern hemisphere winter', *J. Met. Soc. Japan* **80**, 973–984.

Labitzke, K.: 1982, 'On the interannual variability of the middle stratosphere during the northern winters', *J. Met. Soc. Japan* **60**, 124–139.

Labitzke, K.: 1987, 'Sunspots, the QBO, and the stratospheric temperature in the north polar region', *Geophys. Res. Lett.* **14**, 535–537.

Labitzke, K.: 2002, 'The solar signal of the 11-year sunspot cycle in the stratosphere: Differences between the northern and southern summers', *J. Met. Soc. Japan* **80**, 963–971.

Labitzke, K.: 2003, 'The global signal of the 11-year solar cycle in the atmosphere: When do we need the QBO?', *Meteorolog. Z.* **12**, 209–216.

Labitzke, K.: 2004a, 'On the signal of the 11-year sunspot cycle in the stratosphere over the Antarctic and its modulation by the Quasi-Biennial Oscillation (QBO)', *Meteorolog. Z.* **13**, 263–270.

Labitzke, K.: 2004b, 'On the signal of the 11-year sunspot cycle in the stratosphere and its modulation by the Quasi-Biennial Oscillation (QBO)', *J. Atmos. Sol.-Terr. Phys.* **66**, 1151–1157.

Labitzke, K.: 2005, 'On the solar cycle – QBO relationship: A summary', *J. Atmos. Sol.-Terr. Phys.* **67**, 45–54.

Labitzke, K. and van Loon, H.: 1972, 'The stratosphere in the Southern Hemisphere', in: C. W. Newton (ed.), *Meteorology of the Southern Hemisphere. Met. Monogr.* **13**, No. 35, 113–138.

Labitzke, K. and van Loon, H.: 1988, 'Associations between the 11-year solar cycle, the QBO and the atmosphere. Part I: The troposphere and stratosphere in the northern hemisphere winter', *J. Atmos. Terr. Phys.* **50**, 197–206.

Labitzke, K. and van Loon, H.: 1999, 'The Stratosphere (Phenomena, History, and Relevance)', Springer Verlag, Berlin Heidelberg New York, 179 pp.

Labitzke, K. and van Loon, H.: 2000, 'The QBO effect on the global stratosphere in northern winter', *J. Atmos. Sol.-Terr. Phys.* **62**, 621–628.

Labitzke, K., Kunze, M., and Brönnimann, S.: 2006, 'Sunspots, the QBO, and the stratosphere in the north polar region – 20 years later', sumbitted to *Meteorol. Z.* **15**.

Matthes, K., Langematz, U., Gray, L. J., Kodera, K., and Labitzke, K.: 2004, 'Improved 11-year solar signal in the Freie Universität Berlin Climate Middle Atmosphere Model (FU-CMAM)', *J. Geophys. Res.* **109**, doi:10.1029/2003JD004012.

Naito, Y. and Hirota, I.: 1997, 'Interannual variability of the northern winter stratospheric circulation related to the QBO and the solar cycle', *J. Met. Soc. Japan* **75**, 925–937.

Pittock, A. B.: 1983, 'Solar variability, weather and climate: An update', *Q. J. Roy. Met. Soc.* **109**, 23–55.

Salby, M. and Callaghan, P.: 2000, 'Connection between the solar cycle and the QBO: The missing link', *J. Clim.* **13**, 2652–2662.

Salby, M. and Callaghan, P.: 2004, 'Evidence of the solar cycle in the general circulation of the stratosphere', *J. Clim.* **17**, 34–46.

Shepherd, T. G.: 2002, 'Issues in stratosphere – troposphere coupling', *J. Met. Soc. Japan* **80**, 769–792.

Soukharev, B. and Hood, L. L.: 2001, 'Possible solar modulation of the equatorial quasi-biennial oscillation: Additional statistical evidence', *J. Geophys. Res.* **106**, 14,855–14,868.

van Loon, H. and Labitzke, K.: 2000, 'The influence of the 11-year solar cycle on the stratosphere below 30 km: A review', *Space Sci. Rev.* **94**, 259–278.

van Loon, H., Meehl, G. E., and Arblaster, J. M.: 2004, 'A decadal solar effect in the tropics in July–August', *J. Atmos. Sol.-Terr. Phys.* **66**, 1767–1778.

# SIGNATURE OF THE 11-YEAR CYCLE IN THE UPPER ATMOSPHERE

M.-L. CHANIN
*Service d'Aéronomie du CNRS, I.P.S.L., BP 3, 91370 Verrières-le-Buisson, France*
*(E-mail: chanin@aerov.jussieu.fr)*

(Received 28 June 2005; Accepted in final form 22 November 2005)

**Abstract.** In the last 45 years I have studied the thermal structure of the atmosphere from the thermosphere down to the stratosphere, and found evidence of its variability in relationship with the change of solar irradiation during the 11-year solar cycle. I would review, in the light of recent model results, the measurements which I had made since the 1960s and which, for some of them, did not find any explanation at the time of their publication. The data were obtained by two different techniques, rockets and lidars and correspond to different regions of the atmosphere from the upper thermosphere to the stratosphere. The expectation was until recently that the atmosphere should be warmed by an increase of solar flux in the course of the solar cycle due to the increase of UV flux. It has been shown to be the case in the tropical stratosphere and at all latitudes in the upper thermosphere. But, at high and mid latitudes and at other altitudes, the reverse situation was found to exist and, until recently, this cooling observed in parts of the atmosphere with increasing solar flux had never been simulated by models. In addition to reviewing our own data, the paper will present recent results using other dataset which support our observations. It is only recently that we succeeded with a model able to tune the forcing by planetary waves at the tropopause level to reproduce such behaviour.

**Keywords:** solar cycle, temperature, high atmosphere

## 1. Introduction

I have taken the opportunity of this workshop to review more than 40 years of my own research, during which the topic of the solar influence on the atmospheric temperature was always present in my mind, and to show how the results I obtained can be interpreted in the light of present models. In the 1960s, when I started my research, the atmospheric layer I was concerned with was the thermosphere. As a matter of fact very few scientists worked on the stratosphere at that time (except up to the level reached by radiosondes). Satellites were just starting to be used and it is only in the early 1970s that OGO 6 provided the first temperature measurements in the thermosphere from 150 to 500 km by the Doppler width of the atomic Oxygen line and later SAMS and SME in the 1980s did the same for longer periods of time. In the 1960s most of the results came from rockets. At CNRS/SA we used since 1959 the sodium cloud technique which, by measuring the Doppler width of the Na D line, provided temperature profiles from 90 to 500 km. This method first experimented at Hammaguir, Algeria (31°N, 2°W), was extensively used in Heyss Islands (80.5°N, 58°E) until the mid 1970s. At that time, in parallel, we had developed, to describe the middle atmosphere, the Rayleigh lidar technique which


DOI: 10.1007/s11214-006-9062-5 

became operational at the Observatoire de Haute Provence (OHP) (44°N, 6°E) in 1979. It provided the temperature profiles from 30 to 80 km and the focus was then shifted to the upper stratosphere-lower mesosphere. A summary of these results will be presented in Section 1. Those past results will be then compared to a recent reanalysis of all dataset covering the upper stratosphere and the low mesosphere and that will be the topic of Section 2. All those observations led us to develop a mechanistic model to simulate the observations, with the idea that the forcing by planetary waves was a major parameter. The summary of these results will be the subject of Section 3. The conclusions presented in Section 4, together with the ideas which are developed in other papers in this chapter of the book, may help understanding the mechanism by which the Sun influences the climate.

## 2. Our Observational Results

### 2.1. In the Upper Thermosphere

The Earth's thermosphere is the layer of the atmosphere first exposed to the Sun's radiation and so is first heated by the Sun. The ratio of atomic to molecular oxygen is very sensitive to a change of UV flux and therefore the composition and temperature should be very sensitive to changes in solar activity. But in the early 1960s, the temperature was just starting to be measured.

On two following years (1960–1961) during the descending phase of solar cycle 19, the temperature we measured by rockets at Wallops Islands exhibited a difference of 500 $\pm$ 150 K around 400 km for a change in the solar flux of about 50 units of the 10.7 cm solar flux $F_{10.7}$ (Blamont and Lory, 1963). These measurements showed that the atmosphere was isothermal above 200 km and that its temperature increased in phase with the solar activity. Figure 1 presents those measurements above 200 km at Wallops in 1960–1961 together with results obtained at Hammaguir below 200 km between 1960–1963 (Blamont and Lory-Chanin, 1965). It gave indication that the warming under the influence of solar flux is clearly in phase with it above 150 km. The influence of particle precipitations was also observed both at Fort Churchill (58°N) and at Heyss Islands (80°N) (Blamont and Lory, 1964), but this is not the subject of this paper.

Our results were confirmed by other rocket measurements and contributed to the first thermospheric models in development then, as the Jacchia Reference Atmosphere, 1977. Later, temperature data were obtained from satellites: SAMS, SME, UARS but most of the time, the life time of the satellites was too short to study the 11-year cycle, and the studies of solar influence were mainly performed at the solar rotation time scale (27-day cycle). Today the upper thermosphere temperature is measured on a regular basis from the density of the air by satellites drag and this region is now well documented.

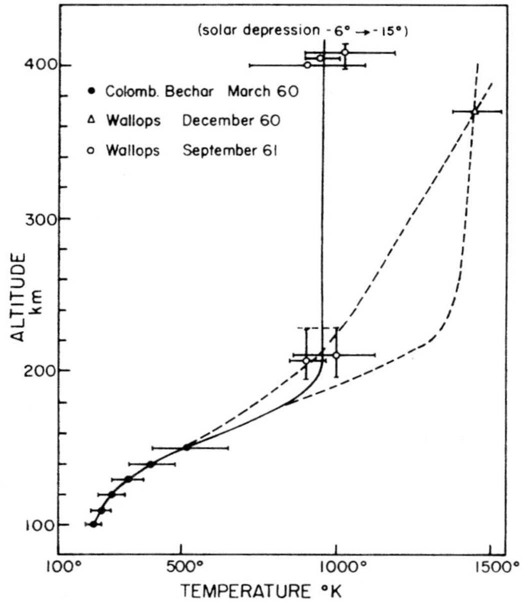


*Figure 1.* Temperature of the neutral atmosphere measured at Wallops Islands (38°N) in December 1960 and September 1961 for a difference of solar flux of 50 units of $F_{10.7}$. The values used to represent the temperature of the lowest thermosphere were obtained in 1960 at Hammaguir (31°N). From Blamont and Lory (1963).

## 2.2. In the Lower Thermosphere

This is the region of transition region between the mesopause and the isothermal thermosphere. This region until recently has not been the subject of long series of satellites measurements. In the 1960s and 1970s it was only accessible by rockets. The longer series of results came from 50 rocket launches with emission of sodium clouds, which we performed at Heyss Islands (80.5°N) between 1968 and 1976. The data showed an inverse correlation between the temperature and the solar flux (here given as sunspot number) (Chanin and Toulinov, 1979). They also seem to indicate a displacement of the mesopause level. The difference of temperature corresponding to a decrease of 75 units of solar flux was 500 K at a constant altitude of 165 km (Figure 2). A similar effect was observed from the series of measurements at Hammaguir (31°N) between 1959 and 1966, but with smaller amplitude: the temperature at 165 km increased by 100 K for a decrease of $F_{10.7}$ of 60 units, as already visible on Figure 1 between 1960 and 1963. We interpreted this anticorrelation as due to a displacement of the height of the mesopause, which moved down with increasing solar flux as already seen by other authors. In the light of recent modelling, we may relate this change of height to a solar induced change of planetary wave activity in the middle atmosphere.

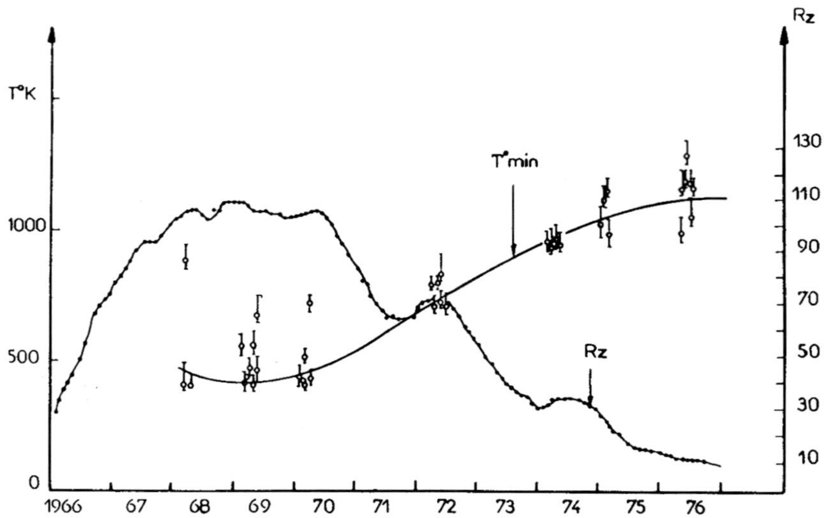


*Figure 2.* The set of temperature measurements represented by a circle with error bars were obtained at 165 km between 1968 and 1976. The smooth curve is a fit with the values corresponding to small geomagnetic activity ($Kp < 3$). The monthly mean values of the sunspot numbers, $R_z$, for the same period are plotted on the same figure (dots connected by curve). From Chanin and Toulinov (1979).

Another result relevant to our review is that the shift from a positive solar signature in the mesosphere to a negative one above the mesopause was also seen at the mid latitude site of Saint-Santin, France (45°N, 2°E) from the incoherent scatter radar data (Chanin *et al.*, 1989). This complexity, and the fact that the observations mostly from airglow emissions are too few to be conclusive, may explain why the conclusion about both the sign and the amplitude of the solar response at the mesopause are still uncertain today (Beig *et al.*, 2004).

### 2.3. In the Low Mesosphere – High Stratosphere

The survey of temperature by Rayleigh lidar in that region started at OHP in 1979 has been pursued since then and is being performed now in all the NDSC (Network for Detection of Stratospheric Changes) stations. After one solar cycle we had observed that the solar response was not in phase with the solar cycle in the high stratosphere and that its amplitude was larger than expected from 2D models (Chanin *et al.*, 1989; Chanin and Keckhut, 1991). The results showed a significant solar negative response in the upper stratosphere, whereas the response was larger and positive in the middle mesosphere. This anticorrelation has been confirmed since then, with more than 2 solar cycles, but the updated results indicate smaller amplitudes and show that the correlation is only significant in winter (Figure 3).

The role of the sign of the QBO was also shown to influence the stratosphere-mesosphere response in winter, both at high and middle latitudes. This result was

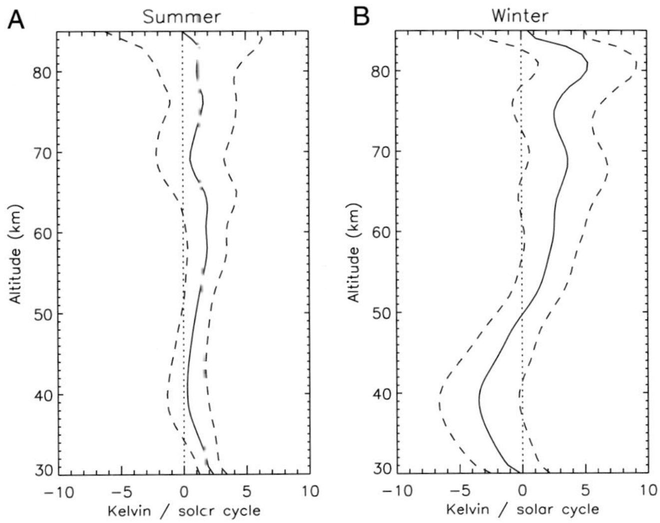


*Figure 3.* Temperature solar response as a function of altitude deduced from Lidar data obtained since 1979 up to 2001. Response is calculated for 6 months labeled as: (A) Summer (April–September) and (B) Winter (October–Mars). From Keckhut *et al.* (2005).

published in a paper by Labitzke and Chanin (1988) using rocket data from Heyss Islands and lidar data from OHP. At high latitude, the stratospheric responses were of opposite sign for the two phases of the QBO. The effect of the QBO was also shown to be slightly visible in the mesopause and in the lower thermosphere in Chanin *et al.* (1989). The influence of the QBO in the mesosphere, even though visible, did not change the fact that the temperature responsed positively to the increased solar flux. Therefore the separation of the data according to the sign of the QBO was considered as not critical for the mesosphere and above, and because of the sparcity of data, and the cost in term of significence of the results, it was not taken into account; thus the phase of the QBO was not taken into account in the following data analysis.

One can summarize our observations by saying that the results showed an alternance of positive and negative responses to the change of solar flux within the 11-year cycle with amplitudes increasing with height; the negative response occurring mostly in the upper stratosphere and the lower mesosphere.

## 3. Comparison with Other Datasets

Recently a reanalysis of all the available datasets in the stratosphere and low mesosphere was performed as a follow-up of the assessment on long-term stratospheric temperature trend (Ramaswamy *et al.*, 2001). This analysis used an identical multi

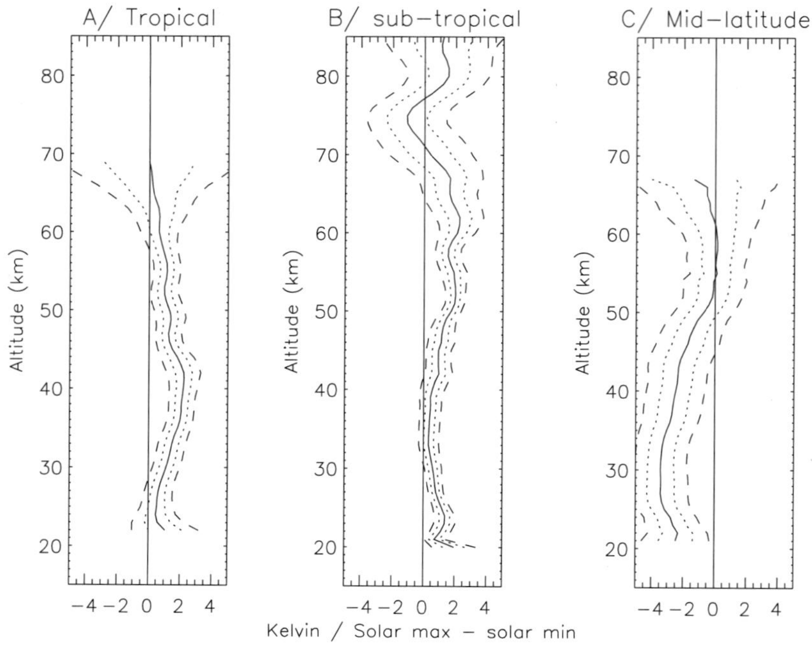


*Figure 4.* Average temperature response to solar activity changes as a function of altitude for three latitude bands: (A) Tropics (Ascension Island, 8°S; and Kwajalein, 9°N), (B) Northern subtropics (Barking Sands, 22°N; Cape Kennedy, 28°N; Point Mugu, 34°N) and (C) mid and high-latitudes (Shemya, 53°N; and Primrose Lake, 55°N). Mean error bars (2 standard deviations) have been derived from the multi-regression analysis statistical error. Solar response is given for a full solar cycle having a mean amplitude of the solar forcing estimated from the last 3 solar cycles. From Keckhut *et al.* (2005).

parameters regression analysis for all the data sets: in this process the US rocket sondes (Figure 4), the OHP lidar and the SSU dataset were reanalysed to identify the solar signature (Keckhut *et al.*, 2005). The results led to the following conclusions: as expected from 2D models, a positive and weak response to the solar cycle at low latitude (1–2 K) and a succession of alternatively positive and negative responses at higher latitudes, with amplitude increasing with latitude and with a very large seasonal variability, the maximum occurring during winter. The only available satellite dataset available for 2 solar cycles, SSU, confirms this result, but furthermore it provides the latitudinal dependence of the solar response. Figure 5 shows clearly the response from 60°S to 60°N with a strong high latitude signal during winter-spring periods in both hemisphere in the mid-stratosphere.

Other rocket stations in Japan and ex-USSR were reanalysed recently and they allow a comparison of solar responses as a function of longitude for 6 mid-latitude sites (Keckhut and Kodera, 1999; Kubicki *et al.*, 2006). They clearly

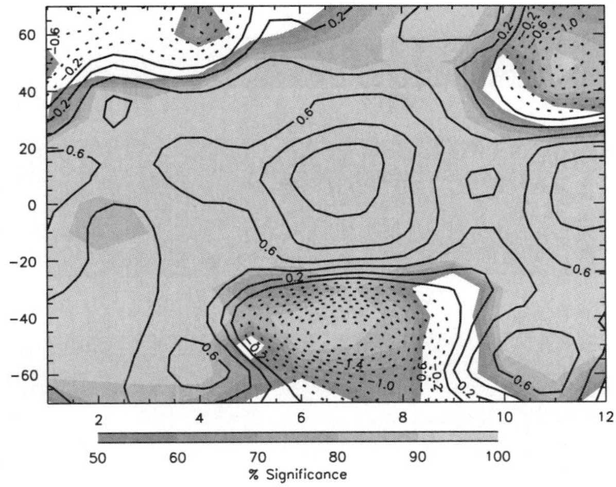


*Figure 5.* Zonal temperature response to solar activity based on SSU/MSU data from 1979 up to 1998 presented as monthly solar responses as a function of latitudes and months for a given height level corresponding to 6 hPa (35 km). The gray shaded regions indicate statistically significant signal. From Keckhut *et al.* (2005).

show the influence of longitude on the amplitude and the sign of the response, indicating clearly that zonal mean, as often done on satellite data, could mask the effect observed locally (bottom part of Figure 8) (Hampson *et al.*, 2006). This could be the reason why a high quality dataset as the one from HALOE does not show any significant solar variation in the stratosphere and only a weak one in the mesosphere (1–1.7 K/solar cycle) as shown in Remsberg *et al.* (2002). This longitudinal dependence should be taken into consideration when comparing observations.

All these results (alternate sign of the response, seasonal dependence, influence of QBO, longitudinal variation) indicate a strong role played by planetary waves in the response of the atmosphere to solar variability on the time scale of the 11-year cycle.

## 4. Comparison with a Mechanistic Model

Therefore we decided to develop a mechanistic model to examine the effect of the solar cycle on the atmospheric temperature under the influence of planetary wave activity. The model is dedicated to the stratosphere and mesosphere with detailed chemical, radiative and dynamical schemes. In this model, planetary waves are initiated at the lowest boundary level of the model that corresponds to the tropopause height. Model runs have been carried out in pairs, with one run using UV solar forcing corresponding to solar minimum and the other to solar maximum. Details of the models can be found in Hampson *et al.* (2006).

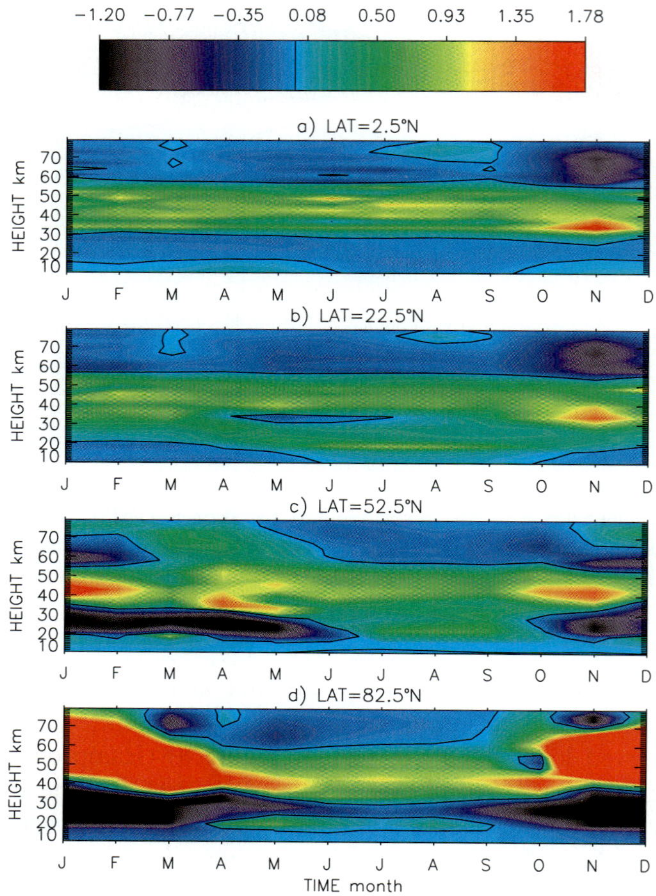


*Figure 6.* Climatology of temperature difference (in Kelvin) between model runs with solar maximum forcing and with solar minimum forcing, for a pair of model runs with the planetary wave forcing multiplied by a factor $C = 1.8$, at four different latitudes: (a) 2.5°N, (b) 22.5°N, (c) 52.5°N, (d) 82.5°N. From Hampson *et al.* (2005).

The level of lower boundary planetary wave forcing is varied between pairs of model runs. Figure 6 presents the difference in temperature signal between the pairs of runs for an amplification of the planetary wave amplitude of 1.8, value which gave the largest signal. These results illustrate the crucial role played by the planetary wave forcing in the solar cycle temperature response at mid and high latitudes. The solar cycle temperature signal in the tropics and subtropics is about 1 K for all values of wave forcing. However, in the extra-tropics the solar signal varies critically with wave forcing, giving a solar signal as strong as 16 K for intermediate values of wave forcing. Despite some spatial differences, the simulations with a specific wave forcing show good qualitative agreements with observational results described before. Above a critical level of wave activity, the non-linear interaction

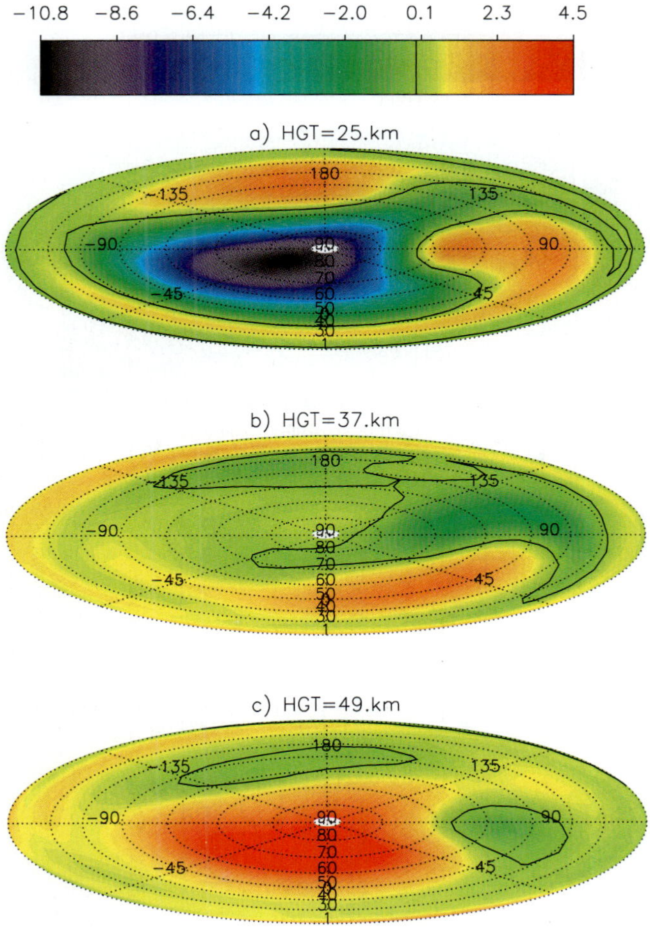


*Figure 7.* Solar signal (i.e., difference between solar minimum forcing and solar maximum forcing) in temperature (K) at three heights: (a) 25 km, (b) 37 km, (c) 49 km. From Hampson *et al.* (2006).

with the mean flow induces a stratospheric warming and a correlated mesospheric cooling and therefore a strong temperature change in both the stratosphere and the mesosphere (Figure 7).

The critical wave-forcing amplitude necessary to produce such event is very sensitive to the initial state of the atmosphere and a small change of the mean wind, due for example to an enhancement of the solar forcing, can generate a large difference in temperature according to the level of the wave activity. The dependence as a function of longitude is clearly due to the asymmetry of the stratospheric warming and reproduces quite well the observations at mid latitude as shown in Hampson *et al.* (2006) (Figure 8).

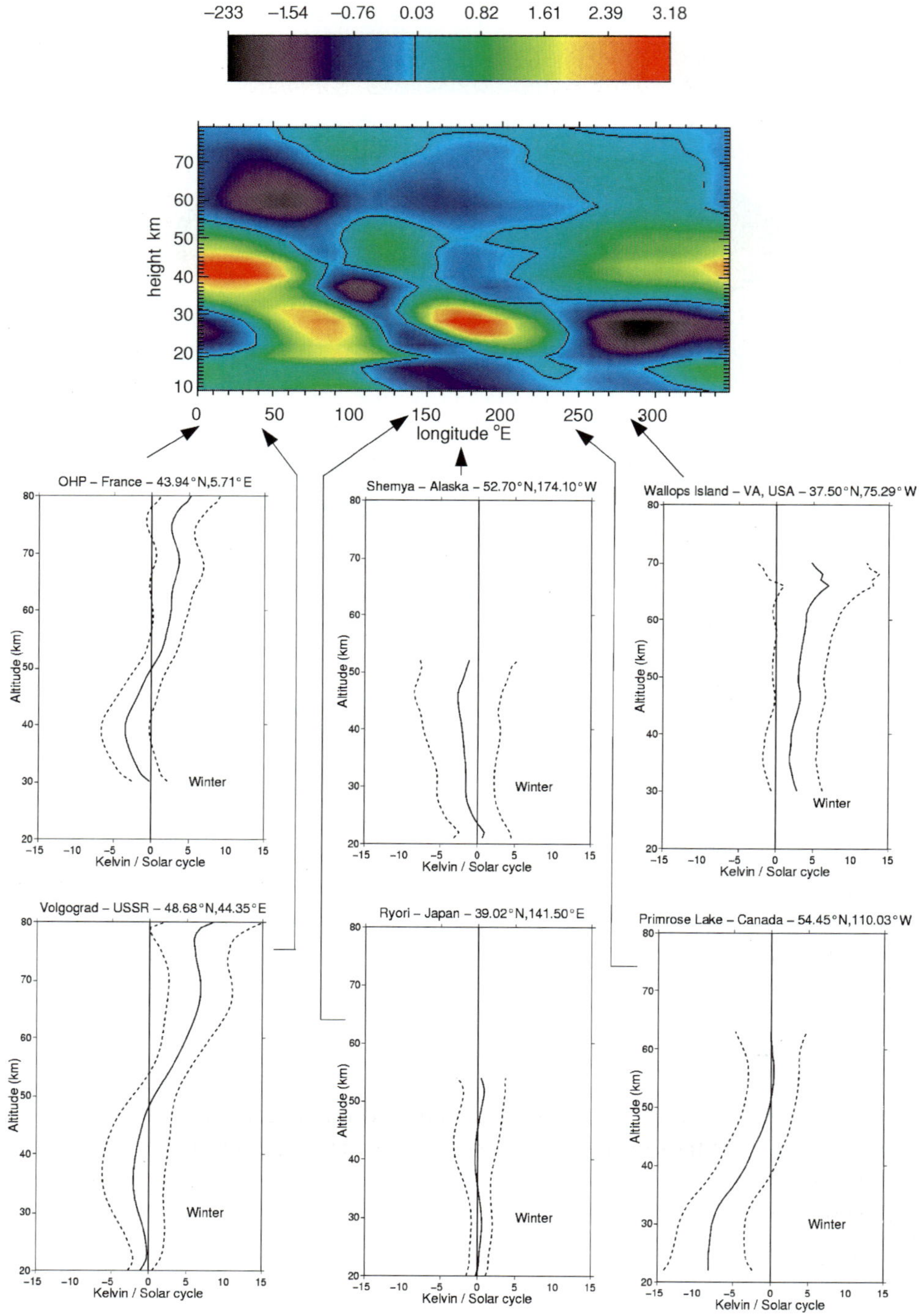


*Figure 8.* Comparison between the solar signal in temperature for the model simulations at mid-latitudes (averaged between 35°N and 55°N) and six observational rocket-sonde and lidar vertical profiles. From Hampson *et al.* (2006).

## 5. Conclusions

Using our own observations, we have showed the following points:

- The temperature in the high thermosphere is strongly influenced by solar activity and varies in phase with it. However this situation is not observed in the lower thermosphere, due to a change of level of the mesopause with solar activity, whether or not induced by change of the planetary wave activity.
- In the upper stratosphere-lower mesosphere, the equatorial solar response is close to the prediction of photochemical calculation (1–2 K). Elsewhere the response is strongly related to planetary wave activity. At mid and high latitude, the response is highly variable (±0–25 K) with latitude, altitude, season, sign of QBO, and longitude.

Numerical simulations taking into account the change of the UV flux and its influence of ozone distribution and using a variable planetary wave level have been able to reproduce the observed patterns in the upper stratosphere-low mesosphere and suggest a mechanism by which a small change induced by the solar forcing can generate a large stratospheric response.

As it had been predicted by several authors and as we already mentioned in a precedent publication (Chanin and Keckhut, 1991), the way the solar variability influences the upper layers of the atmosphere is strongly influenced by dynamics, essentially by the propagation of planetary waves which redistributed the solar energy deposited in a non uniformed manner around the Earth. We have showed that the way stratospheric warmings are influenced by solar activity contribute to the signal detected when comparing the state of the atmosphere for different solar activity levels.

The influence of solar activity on the occurrence of stratospheric warming is adding to the complexity of the detection of solar signature. The difficulty in interpreting the apparently discordant observations can now be understood by taking into account the change of the solar signature with altitude, latitude and longitude. Furthermore, due to the large longitudinal asymmetry of the solar response, the signal average over a zonal mean may be different from local results and could be inexistent or much weaker than observed locally and therefore, and this can explain why most satellite data do not detect the solar influence.

The UV influence of ozone distribution modifies the temperature structure and thus the general circulation of the stratosphere and influences the propagation of planetary waves. As those waves are shown in this paper to have a critical role in the response of the stratosphere, the solar signal may be amplified and have much larger signature on the troposphere and the climate. This will need to be checked with climate GCMs.

## Acknowledgements

All along these four decades, I benefited from the support of the National Center of Scientific Research (CNRS) and of the National Center of Space Studies (CNES).

The recent work was also supported by a grant from the European Commission (SOLICE Project, Contract EVK2–CT–1999). During the sixties and seventies, J.E. Blamont and C.F. Toulinov contributed to this work and for the recent work, I would like to acknowledge the contribution of my present collaborators: Philippe Keckhut, John Hampson and Alain Hauchecorne.

## References

Beig, G., Keckhut, P., Lowe, R. P., Roble, R. G., Mlynczak, M. G., Scheer, J., Fomichev, V. I., *et al.*: 2004, 'Review of mesospheric temperature trends', *Rev. Geophys.* **41**, doi:10.1029/2002RG000121.

Blamont, J., and Lory, M. L.: 1963, 'New direct measurements of ionospheric temperatures', in *Proc. of the First International Symposium on Rocket and Satellite Meteorology*, Washington, USA, 23–25 April 1962, pp. 71–75.

Blamont, J., and Lory, M. L.: 1964, 'Sudden heating of the ionosphere in the auroral zone', *Nature* **201**, 593–595.

Blamont, J., and Lory-Chanin, M. L.: 1965, 'Temperature measurements in the ionosphere from 100 to 400 km between 1960 and 1964', *Space Res.* **V**, 1137–1139.

Chanin, M. L., and Toulinov, G. F.: 1979, 'The polar thermospheric temperature behaviour during the 11 year solar cycle', *J. Geophys. Res.* **84**, 406–410.

Chanin, M. L., and Keckhut, P.: 1991, 'Influence on the middle atmosphere of the 27-day and 11-year solar cycles: Radiative and/or dynamical forcing?', *J. Geomag. Geoelectr.* **43**, 647–655.

Chanin, M. L., Keckhut, P., Hauchecorne, A., and Labitzke, K.: 1989, 'The solar activity – QBO effect in the lower thermosphere', *Ann. Geophys.* **7**, 463–470.

Hampson, J., Keckhut, P., Hauchecorne, A., and Chanin, M. L.: 2005, 'The 11-year solar-cycle in the temperature in the upper-stratosphere and mesosphere: Part II, numerical simulation and role of planetary waves', *J. Atmos. Sol. Terr. Phys.* **67**, 948–958.

Hampson, J., Keckhut, P., Hauchecorne, A., and Chanin, M. L.: 2006, 'The effect of the 11-year solar-cycle on the temperature in the upper stratosphere and mesosphere: Part III investigations of zonal asymmetry', *J. Atmos. Sol. Terr. Phys.* **68**, 1591–1599.

Keckhut, P., and Kodera, K.: 1999, 'Long-term changes of the upper stratosphere as seen by rocket-sondes at Ryori (39°N, 141°E)', *Ann. Geophys.* **17**, 1210–1217.

Keckhut, P., Cagnazzo, C., Chanin, M. L., Claud, C., and Hauchecorne, A.: 2005, 'The 11-year solar-cycle in the temperature in the upper-stratosphere and mesosphere: Part I assessment of observations', *J. Atmos. Sol. Terr. Phys.* **67**, 940–947.

Kubicki, A., Keckhut, P., Chanin, M. L., Hauchecorne, A., Lysenko, E., and Golitsyn, G.: 2006, 'Temperature trends in the middle atmosphere of the mid-latitude as seen by systematic rocket launches above Volgograd', *J. Atmos. Sol. Terr. Phys.* **68**, 1075–1086.

Labitzke, K., and Chanin, M. L.: 1988, 'Changes in the middle atmosphere in winter related to the 11-year solar cycle', *Ann. Geophys.* **6**, 643–644.

Ramaswamy, V., Chanin, M. L., Angell, J., Barnett, J., Gaffen, D., Gelman, M. E., Keckhut, P., *et al.*: 2001, 'Stratospheric temperature trends: Observations and model simulations', *Rev. Geophys.* **39**, 71–122.

Remsberg, E. E., Bhatt, P. P., and Deaver, L. E.: 2002, 'Seasonal and longer-term variations in middle atmosphere temperature from HALOE on UARS', *J. Geophys. Res.* **107**, ACL 18, 1–13.

# THE MIDDLE ATMOSPHERIC OZONE RESPONSE TO THE 11-YEAR SOLAR CYCLE

Y. CALISESI[1,*] and K. MATTHES[2,3]
[1]*International Space Science Institute, Hallerstrasse 6, CH – 3012 Bern, Switzerland*
[2]*Institut für Meteorologie, Freie Universität Berlin, Germany*
[3]*Atmospheric Chemistry Division, National Center for Atmospheric Research, Boulder, CO, USA*
*(*Author for correspondence: E-mail: yasmine@issibern.ch)*

(Received 10 January 2006; Accepted in final form 3 April 2006)

**Abstract.** Because of its chemical and radiative properties, atmospheric ozone constitutes a key element of the Earth's climate system. Absorption of sunlight by ozone in the ultraviolet wavelength range is responsible for stratospheric heating, and determines the temperature structure of the middle atmosphere. Changes in middle atmospheric ozone concentrations result in an altered radiative input to the troposphere and to the Earth's surface, with implications on the energy balance and the chemical composition of the lower atmosphere. Although a wide range of ground- and satellite-based measurements of its integrated content and of its vertical distribution have been performed since several decades, a number of uncertainties still remain as to the response of middle atmospheric ozone to changes in solar irradiance over decadal time scales. This paper presents an overview of achieved findings, including a discussion of commonly applied data analysis methods and of their implication for the obtained results. We suggest that because it does not imply least-squares fitting of prescribed periodic or proxy data functions into the considered times series, time-domain analysis provides a more reliable method than multiple regression analysis for extracting decadal-scale signals from observational ozone datasets. Applied to decadal ground-based observations, time-domain analysis indicates an average middle atmospheric ozone increase of the order of 2% from solar minimum to solar maximum, which is in reasonable agreement with model results.

**Keywords:** ozone-climate relationships, stratospheric ozone, solar signal, 11-year solar cycle

## 1. Introduction

Ozone absorbs ultraviolet (UV) radiation from the Sun, and modulates outgoing infrared (IR) radiation from the Earth. Because of these properties, and because it is sensitive to both natural and anthropogenic perturbations, middle atmospheric ozone constitutes a key element in the Earth's climate system (Figure 1). Although natural ozone variability had become an object of study since the pioneering work of G. M. B. Dobson and S. Chapman (Dobson and Harrison, 1927; Chapman, 1930), it is only recently that the need emerged for a precise evaluation of the amplitude and spectral characteristics of the various cycles affecting stratospheric and mesospheric ozone. After discovery of the springtime Antarctic ozone hole (Farman *et al.*, 1985; Chubachi, 1985), the urge for assessing the impact of anthropogenic emissions on

 

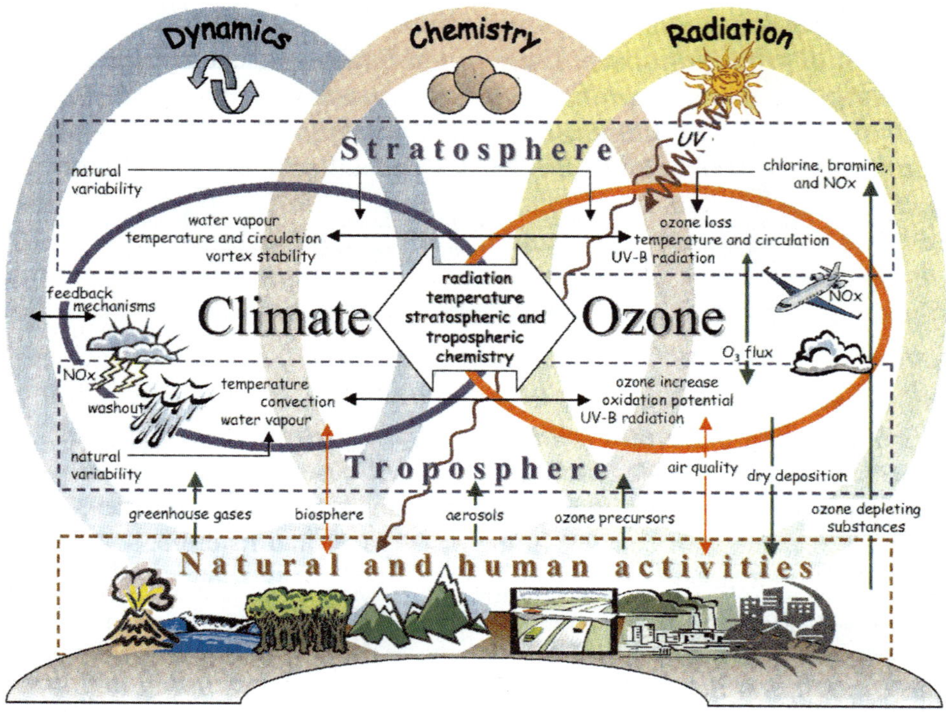


*Figure 1.* Processes determining the ozone – climate interaction in the stratosphere and the troposphere. From European Commission (2003).

stratospheric ozone triggered the development of numerous ground- and space-based experiments, and, simultaneously, of elaborated simulation and regression models to account for the long-term change observed also in midlatitude ozone (Stolarski *et al.*, 1992). Besides confirmation of the threat exerted on the ozone layer by the release to the atmosphere of man-made chlorofluoromethanes (Crutzen, 1974; Molina and Rowland, 1974), changes in solar irradiance, and in particular the 11-year, or Schwabe cycle of solar activity, was soon identified as another source of middle atmospheric ozone variability on a decadal time scale. Today, more than 15 years after the regulation of their emissions by the Montreal Protocol and its amendments, concentrations of ozone depleting substances in the stratosphere are observed to level off (WMO, 2003) and discussions are open as to the detection of the first signs of ozone recovery (Newchurch *et al.*, 2003; Steinbrecht *et al.*, 2004; Cunnold *et al.*, 2004). In this context, precise knowledge of the ozone response to solar variability is of renewed interest, due to the similar and potentially confusable character of the solar and turnaround ozone signals over the considered time scales.

In the frame of climate studies, assessment of the stratospheric and mesospheric ozone response to solar variability is of relevance to the understanding of the propagation of the solar signal to the Earth's lower atmosphere. Changes in incoming solar UV radiation affect photochemical ozone production, resulting in altered ozone concentrations and stratospheric thermal properties (e.g., Brasseur, 1993; Haigh, 1994; Fleming *et al.*, 1995). Modification of the stratospheric vertical and meridional temperature gradients affects the propagation of planetary waves, and influences thus the global atmospheric circulation pattern (Hines, 1974; Bates, 1981). These thermal and dynamical changes in turn affect the concentration and global distribution of long-lived chemical tracers (European Commission, 2003). With regard to the small amplitude of the involved initial perturbation (e.g., Rottman, 2006; Fröhlich, 2006), radiative forcing associated to solar UV changes could thus nevertheless trigger a significant dynamical response of the middle atmosphere (Kodera and Kuroda, 2002; Hood, 2004; see also Gray *et al.*, 2006; Haigh, 2006; Kodera, 2006; Labitzke, 2006; Salby and Callaghan, 2006) and could result in large climatic effects as shown by several observational and modeling studies (e.g., Haigh, 1996, 1999; Shindell *et al.*, 1999, 2001; Gleisner and Thejll, 2003; Matthes *et al.*, 2003, 2004, 2006; Tourpali *et al.*, 2003, 2005; Kodera, 2004; Egorova, 2004; Rozanov *et al.*, 2004; van Loon *et al.*, 2004).

The present contribution provides an overview of the middle atmospheric ozone response to decadal-scale solar variability, as derived from ground and satellite-based observations and from model simulations. The presentation of the results is preceded by a discussion of the methods applied to the extraction of the solar signal from long-term data sets, and of their implication for the obtained results. For recent measurements and interpretations of the mesospheric and upper stratospheric ozone response to shorter-term variations, the readers are referred to Jackman *et al.* (2006) and Langen (2006).

## 2. Extraction of the 11-Year Solar Signal from Observational Data Sets

In numerical simulations, the solar signal in any relevant atmospheric quantity is most simply investigated by comparing model results obtained under different solar input conditions, the difficulty lying then in the choice of the solar input parameters and, of course, in the representation of the involved physical and chemical processes. When considering observational datasets, on the other hand, extraction of the solar signal is hampered by the limited length of available times series, that rarely encompass more than three solar cycles (Figure 2), and is also complicated by the faint amplitude of the searched signal with respect to noise and to more dominant modes of variability. Results extracted from both ground- and satellite-based datasets indicate thus a maximal amplitude of the solar signal in middle and upper stratospheric ozone of the order of a few percents, which contrasts with a seasonal cycle amplitude of the order of 30% at midlatitudes (Staehelin *et al.*, 2001), or with peak Antarctic ozone hole depletion levels of up to 70% in record years (WMO, 2003).

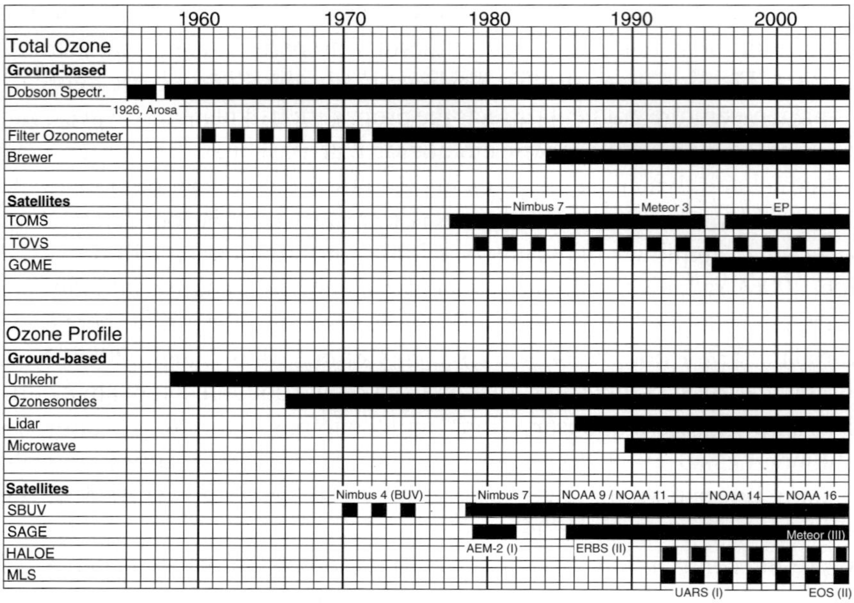


*Figure 2.* Length of archived ozone measurements series suitable for long-term data analysis. Data sets of restricted quality are marked with broken bars. Updated from Staehelin *et al.* (2001) with permission of American Geophysical Union.

In this context, several methods are employed to extract the solar signal from long-term ozone observations. In each case, a linear relationship between solar signal and ozone response is assumed, thus excluding the investigation of non-linear feedbacks.

The first method of identifying the solar signal in ozone measurements consists in the multiple linear regression analysis of the achieved times series. A model of the form (Staehelin *et al.*, 2001):

$$Y(t) = \mu_i + \sum_{n=1}^{k} \xi_n \Phi_n(t) + \beta_i R(t) + N(t) \tag{1}$$

is applied to monthly-mean averaged ozone data $Y(t)$, where $t$ is a linear time index, $i$ is the periodic calendar month index applying to the time $t$, $\mu_i$ is the average component of ozone seasonal variation at month $i$, $\Phi_n(t)$ is an explanatory variable, e.g., a monthly-averaged proxy value at time $t$, with corresponding coefficient $\xi_n$, $R(t)$ is a steady ramp function accounting for long-term continuous changes, $\beta_i$ is the trend, or long-term change amplitude on month $i$, and $N(t)$ is an autocorrelated noise term. A system of linear equations is solved to determine the amplitude of the searched trend and proxy data coefficients, $\beta_i$ and $\xi_n$. The set of explanatory variables commonly taken into account includes the 11-year Schwabe cycle of solar activity, represented by the solar radio flux at 10.7 cm (F10.7 cm in the following) or the MgII UV index, and the Quasi-Biennial Oscillation (QBO), typically represented by the 45 hPa zonal wind amplitude at Singapore. Optionally, other parameters such as volcanic activity (represented by the aerosol optical depth), the El Niño Southern Oscillation (ENSO), or the North Atlantic Oscillation (NAO) (represented by their respective indexes), are sometimes also considered (see, for instance, Zerefos *et al.*, 1997).

Originally, multiple regression analysis was applied to the assessment of long-term trends from ozone measurement data (e.g., Staehelin *et al.*, 2001, and references therein). In this case, the explanatory variables account for the data natural variability, and the unexplained remaining trend is attributed to anthropogenic ozone depletion. In the context of climate studies, however, this technique is applied to the "reverse" problem, which consists in determining the exact contribution of the solar signal to long-term ozone changes (e.g., McCormack and Hood, 1996; Miller, 1996; Hood, 1997; Zerefos *et al.*, 1997, 2005; Reinsel *et al.*, 2002). This brings to the fore the two major caveats associated with the use of this technique, namely, the necessity of defining in advance the processes or associated proxy signals that determine natural ozone variability, and the dependency of the obtained results on the considered set of explanatory variables (signal degeneracy). In the search for the ozone solar signal in particular, concurring long-term trends, QBO, or volcanic eruptions signals might contaminate the achieved regression results (Lee and Smith, 2003).

Alternatively, frequency-domain signal analysis (via Fourier Transformation of the original data set, and digital filtering in the Fourier domain, see, e.g., Press *et al.*, 1992) might provide a useful tool for extracting the solar signal from observations. In the present case, however, frequency-domain signal analysis is hampered by the limited length of the measurements times series, and by the relatively faint amplitude of the searched signal compared to random and systematic noise. In particular, special care should then be devoted to the identification and proper handling of aliasing or sidelobe effects (Press *et al.*, 1992). In addition, and similarly to the above described multiple regression technique, frequency-domain signal analysis involves least-squares fitting of a prescribed infinite set of periodic functions into the considered times series. As above, casually matching events such as quasi-periodic volcanic eruptions in part of the time series will thus be accounted for by periodic signals, that can be misinterpreted if only part of the fitted power spectrum is retained in the final results analysis.

Finally, a somewhat less widespread technique of extracting the solar signal from long-term ozone measurements is represented by time-domain (as opposed to frequency-domain) signal analysis (Press *et al.*, 1992). Fixed-width moving window averages are computed from the considered data sets, in analogy to frequency-domain lowpass filtering. The assumed window width is then assimilated to a lowpass filter cut-off frequency. Pseudo bandpass filtering is obtained when subtracting from each other two lowpass-filtered datasets with different cut-off frequencies. The amplitude of the searched solar signal can then be evaluated from adequately bandpass-filtered data series. In contrast to multiple regression and frequency-domain signal analysis, time-domain signal analysis presents the advantage of avoiding aliasing or signal degeneracy problems, as it does not imply least-squares fitting of a prescribed periodic or proxy data function into the considered data set. However, and as illustrated by Ingram (2006), special care should then be devoted to the choice of the employed window width, in order to ensure robust analysis results.

## 3. Observed and Modeled Solar Signal in Total Ozone

Analyses of ground- and satellite based observations of total ozone extending over two to three decades reveal the existence of a decadal scale variation in column ozone, that is approximately in phase with the solar cycle (Angell, 1989; Zerefos *et al.*, 1997, 2001; Hood, 1997, 2004; Reinsel *et al.*, 2002; WMO, 2003). The latitudinal structure of the solar cycle extracted from TOMS total ozone data using multiple regression analysis for one solar cycle from 1979–1992 is shown in Figure 3. The results indicate an asymmetric total ozone response to the 11-year solar cycle as a function of latitude, with maxima around 30°N and 60°S, and minima in the equatorial and sub-tropical zone and at the northern hemisphere high latitudes. The amplitude of the observed total ozone changes ranges from 0 to 6 Dobson Units (DU) per 100 units change of the F10.7 cm solar flux. Assuming an average amplitude of the solar cycle of 130 F10.7 cm units, and ozone background levels

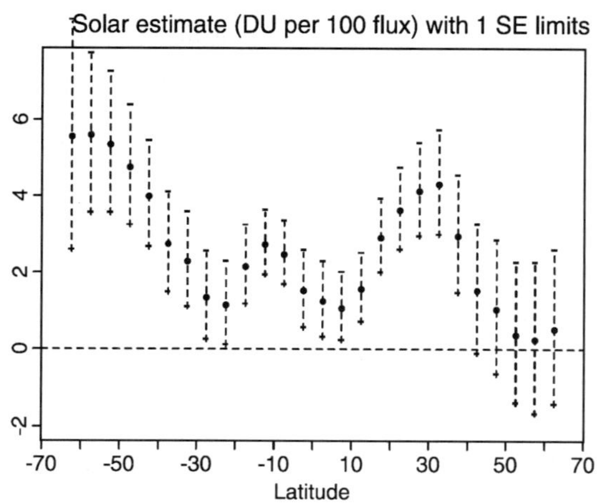


*Figure 3.* Latitudinal structure of the solar cycle regression coefficient calculated from TOMS zonal (5°) total ozone data for the period 1979–1992. Vertical dashed lines indicate 1 standard error (SE) limits. From Reinsel *et al.* (2002).

of 260 to 330 DU depending on latitude, this corresponds to an overall amplitude of the total ozone solar cycle of 1–2%.

Three mechanisms have been proposed to explain the observed features. The first mechanism invokes photochemical control of total ozone, which would however imply a maximum response amplitude at the equator and could only account for a limited share of the observed total ozone changes (Hood, 2004). The second hypothesis involves changes in mean lower stratospheric dynamics between solar minimum and solar maximum. This hypothesis is supported by model results, and by isolated observations of a phase-shift between ozone and solar variations, with ozone appearing to lead the solar flux changes by 2 seasons (Angell, 1989). Finally, observational and numerical evidence was found recently (Sinnhuber *et al.*, 2005; Langematz *et al.*, 2005; Rozanov *et al.*, 2005) for the influence of solar cycle-related energetic electrons precipitation effects on stratospheric ozone, that could induce enhanced total ozone values at high latitudes during solar maximum.

## 4. Vertically-Resolved Observed and Modeled Solar Signal

The main characteristic of the 11-year solar signal retrieved using multiple regression analysis from vertically-resolved stratospheric ozone measurements is that of a dipole structure, with values peaking at the upper stratospheric mid-latitudes (Figure 4). As suggested by McCormack and Hood (1996), this "double-lobed" structure might be related to the superposition of a seasonally varying solar signal in both hemispheres. A secondary maximum in the ozone response amplitude

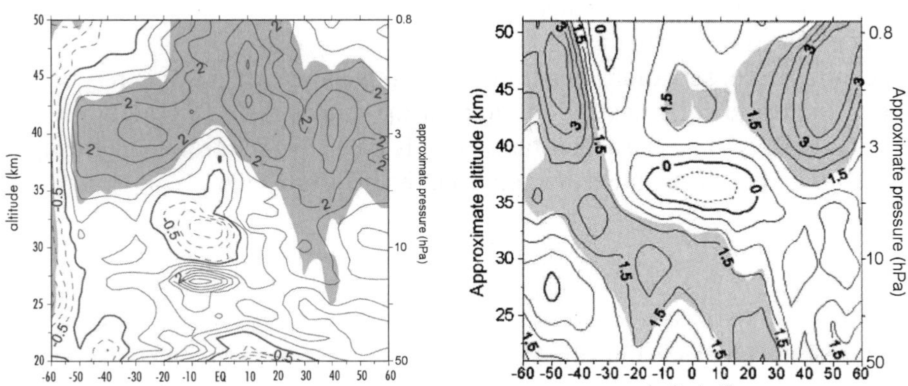


*Figure 4.* Annual mean ozone response to solar activity in units of percentage change per 100 F10.7 cm units (recent solar cycle amplitude ~130 F10.7 cm units). *Left Panel:* Zonal mean response from SAGE II data for the period October 1984–June 2002. From Haigh *et al.* (2004). *Right Panel:* zonal mean response from SBUV data for the period 1978–2004. From Zerefos *et al.* (2005). Contour intervals 0.5%. Shading indicates regions of 95% statistical significance.

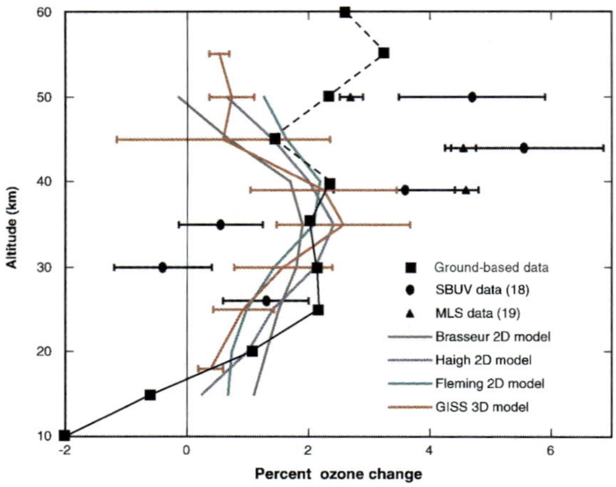


*Figure 5.* Annual percentage ozone differences between solar maximum and solar minimum (global average from 60°S to 60°N) from satellite observations (data points) and model results (solid lines). The data points are from satellite observations covering 15 years of Solar Backscatter Ultraviolet (SBUV) and 3 years of Microwave Limb Sounder (MLS) observations. Model error bars are for the GISS 3D Model. Median results obtained from the data sets presented in Figure 6 are indicated with squares and black line. Dashed line indicates regions with larger uncertainties due to restricted number of considered data sets. Updated from Shindell *et al.* (1999) with permission of AAAS.

is found in the equatorial lower stratosphere, just below a region of unsignificant or negative ozone response to increased solar irradiance. Analysis of both ground-based and satellite-borne measurements (Miller, 1996; McCormack and Hood, 1996; Newchurch *et al.*, 2003; Haigh *et al.*, 2004; Zerefos *et al.*, 2005) yield solar cycle amplitudes of 2 to 3.5% per 100 F10.7 cm units in upper stratospheric ozone, which corresponds to an amplitude of 3–5% assuming an average solar cycle amplitude of 130 F10.7 cm units.

In contrast to observations, most model results indicate a positive response of ozone to increased solar activity throughout the stratosphere (Figure 5, see also Schmidt and Brasseur, 2006), and succeed only partially in reproducing the upper stratospheric dipole structure extracted from the measurements.

## 5. Discussion

Considering the above results, the question arises of the causes of the discrepancy still existing between modeled and observed solar signal in middle atmospheric ozone. As proposed by Lee and Smith (2003), the middle to lower stratospheric response structure could be induced by aliasing of the solar cycle by the QBO and/or volcanic signals in multiple regression analysis. On the other hand, Langematz *et al.* (2005) and Rozanov *et al.* (2005) suggest that the noted structure is possibly induced by energetic particle precipitation.

As observed by Lee and Smith (2003), the similarity between signatures in ozone of different controling factors such as the equatorial QBO, volcanic effects, and the 11-year solar cycle, associated to the limited length of the available measurements series, could lead to signal degeneracy and aliasing of the QBO and volcanic signals by the fitted solar cycle in regression analysis results. Lee and Smith (2003) verified this hypothesis by showing that multiple regression analysis applied to model data simulations including both effects of the equatorial QBO and volcanic eruptions led to the extraction of a dipole structure as well as to a negative signal in the middle stratosphere and a secondary maximum in the lower stratosphere similar to the pattern shown in Figure 4. When the 11-year solar flux variation is considered as the only external forcing of ozone change, the dipole response is replaced by a uniformly positive stratospheric response.

As a consequence, the pattern observed in Figure 4 might thus simply be explained by an inadequate rendering of the above mentioned signals through multiple regression analysis. Alternatively, the discrepancy might also be solved by a more accurate rendering of the QBO and volcanic eruption effects in chemistry climate models.

In order to obtain an additional reference as to the influence of the solar cycle on middle atmospheric ozone, time-domain analysis was applied to 11 ozonesondes, lidar, and ground-based microwave remote sensing data sets acquired within the frame of the Network for the Detection of Atmospheric Composition Change (NDACC) (formerly Network for the Detection of Stratospheric Change, NDSC). Three- and 10-year running-windows averaging was successively applied to each data set, the difference between the successive times series yielding the 3–10 years bandpass-filtered ozone times series represented in Figure 6. For the shorter-term data sets, only the 3-year running average was applied and the average background signal was removed from each times serie. In the lower stratosphere and the troposphere, the obtained results suggest a negative ozone response to the solar signal in three of the four considered ozonesonde data sets (see also the corresponding profile represented with black squares in Figure 5). Detailed inspection of the obtained times series however indicates a non-uniform anticorrelation over time, thus ruling out a direct influence of the solar signal on tropospheric ozone. A large positive anomaly with respect to the averaged background signal appears in the 1980s in all three data sets below 20 km, possibly in connection with the 1982 El Chichón eruption. According to the principles of least-squares fitting, this anomaly would most probably be accounted for by a significant negative tropospheric ozone/solar variability relationship in multiple regression analysis results, if no specific proxy for volcanic activity was taken into account in the considered regression model.

In agreement with model results (see also Figure 5), a positive ozone response to the solar activity cycle is in addition observed at all observation sites throughout the 25 to 40 km altitude range. This result also agrees with a more intuitive expectation, based upon consideration of the large seasonal cycles observed at

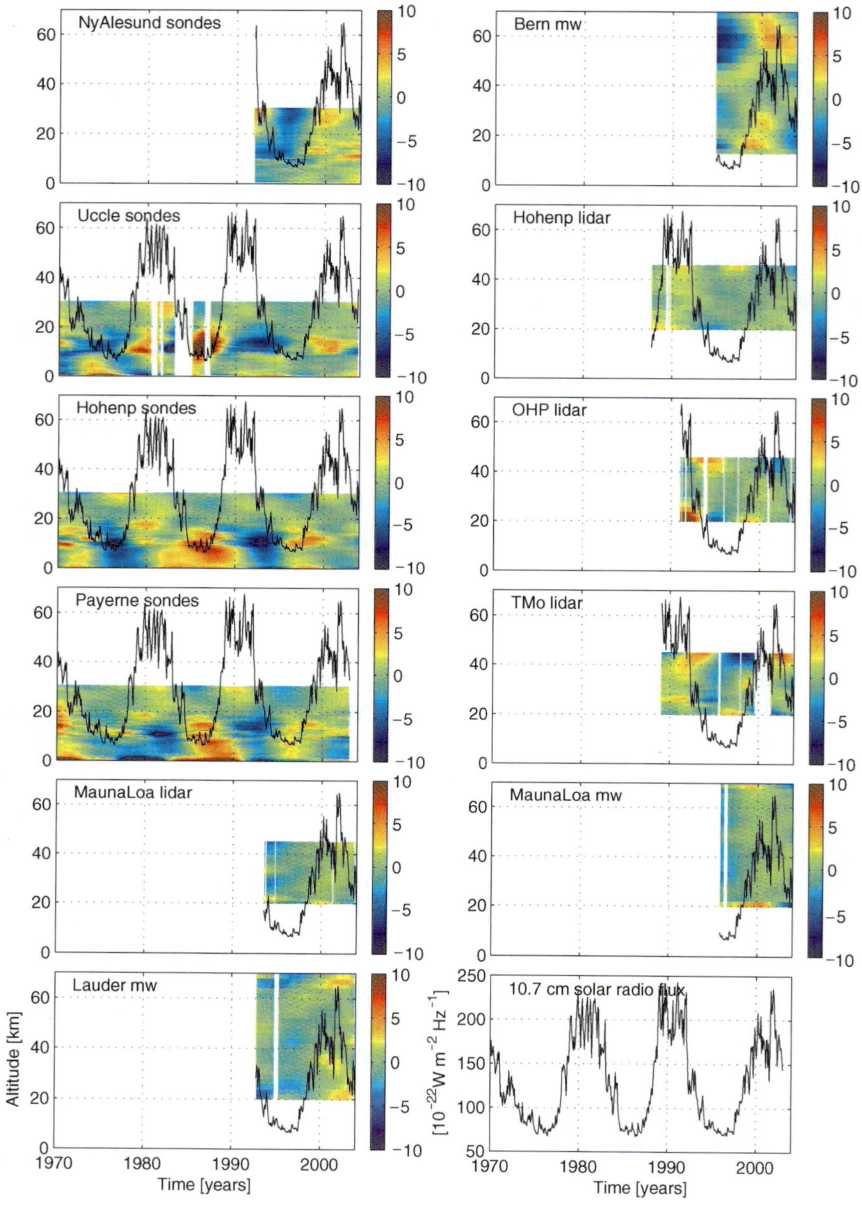


*Figure 6.* Three to 10-year bandpass filtered (short datasets: 3-year lowpass filtered) ozone residuals (%) as extracted from ozonesonde, lidar, and microwave (mw) observations performed at NyAlesund, Uccle, Hohenpeissenberg, Payerne, Mauna Loa, Lauder, Bern, Observatoire de Haute Provence, and Table Mountain within the frame of the Network for the Detection of Atmospheric Composition Change (NDACC).

the middle and high latitudes in the photochemically-controlled regions of the stratosphere.

## Acknowledgements

The authors wish to thank all cited authors for their contribution to the present article. Contributors of ground-based ozone data to the Network for the Detection of Atmospheric Composition Change (NDACC) and to the World Ozone and Ultraviolet Radiation Data Centre (WOUDC) databases are also fully acknowledged. The considered data sets are publicly available from the WOUDC and NDACC databases at `http://www.woudx.org` and `http://www.ndacc.org`. Katja Matthes is supported by a Marie Curie International Fellowship within the 6th European Community Framework Programme.

## References

Angell, J. K.: 1989, 'On the relation between atmospheric ozone and sunspot number', *J. Clim.* **2**, 1404–1416.

Bates, J. R.: 1981, 'A dynamical mechanism through which variations in solar ultra violet radiation can influence tropospheric climate', *J. Geophys. Res.* **104**, 27,321–27,339.

Brasseur, G.: 1993, 'The response of the middle atmosphere to long-term and short-term solar variability: A two-dimensional model', *J. Geophys. Res.* **98**, 20,079–23,090.

Chapman, S.: 1930, 'On ozone and atomic oxygen in the upper atmosphere', *Phil. Mag. S.* **110**, 369–383.

Chubachi, S.: 1985, 'A special ozone observation at Syowa Station, Antarctica from February 1982 to January 1983', in *Atmospheric Ozone, Proceedings of the Quadrennial Ozone symposium*, Halkidiki, Greece, September 3–7, 1984, 285–289.

Crutzen, P. J.: 1974, 'Estimates of possible future ozone reductions from continued use of fluoro-chloro-methanes ($CF_2Cl_2$, $CFCl_3$)', *Geophys. Res. Lett* **1**, 205–208.

Cunnold, D. M., Yang, E.-S., Newchurch, M. J., Reinsel, G. C., Zawodny, J. M., and Russell III, J. M.: 2004, 'Comment on "Enhanced upper stratospheric ozone: Sign of recovery or solar cycle effect?" by W. Steinbrecht *et al.*', *J. Geophys. Res.* **109**, doi:10.1029/2004JD004826.

Dobson, G. M. B. and Harrison, D. N.: 1927, 'Measurements of the amount of ozone in the Earth's atmosphere and its relation to other geophysical conditions. – Part II', *Proc. R. Soc. Lond. A* **114**, 521–541.

European Commission: 2003, 'Ozone Climate Interactions', Air pollution research report No. 81, EUR 20623, 143 pp., Luxembourg.

Egorova, T., Rozanov, E., Manzini, E., Haberreiter, M., Schmutz, W., Zubov, V., and Peter, T.: 2004, 'Chemical and dynamical response to the 11-year variability of the solar irradiance simulated with a chemistry-climate model', *Geophys. Res. Lett.* **31**, doi:10.1029/2003GL019294.

Farman, J. C., Gardiner, B. G., and Shanklin, J. D.: 1985, 'Large losses of total ozone in Antarctica reveal seasonal $ClO_x/NO_x$ interaction', *Nature* **315**, 207–210.

Fleming, E. L., Chandra, S., Jackman, C. H., Considine, D. B., and Douglass, A. R.: 1995, 'The middle atmospheric response to short and long-term solar UV varaitions: Analysis of observations and 2D model results', *J. Atmos. Terr. Phys.* **57**, 333–365.

Fröhlich, C.: 2006, 'Solar irradiance variability since 1978', *Space Sci. Rev.*, this volume, doi: 10.1007/s11214-006-9046-5.

Gleisner, H. and Thejll, P.: 2003, 'Patterns of tropospheric response to solar variability', *Geophys. Res. Lett.* **30**, doi:10.1029/2003GL017129.

Gray, L., Crooks, S., Palmer, M., Pascoe, C., and Sparrow, S.: 2006, 'A possible transfer mechanism for the 11-year solar cycle to the lower stratosphere', *Space Sci. Rev.*, this volume, doi: 10.1007/s11214-006-9069-y.

Haigh, J. D.: 1994, 'The role of stratospheric ozone in modulating the solar radiative forcing of climate', *Nature* **370**, 544–546.

Haigh, J. D.: 1996, 'The impact of solar variability on climate', *Science* **272**, 981–984.

Haigh, J. D.: 1999, 'A GCM study of climate change in response to the 11-year solar cycle', *Q. J. Roy. Meteorol. Soc.* **125**, 871–892.

Haigh, J. D.: 2006, 'Solar influences on dynamical coupling between the stratosphere and troposphere', *Space Sci. Rev.*, this volume, doi:10.1007/s11214-006-9067-0.

Haigh, J. D., Austin, J., Butchart, N., Chanin, M.-L., Crooks, S., Gray, L. J., Halenka, T., Hampson, J., Hood, L. L., Isaksen, I. S. A., Keckhut, P., Labitzke, K., Langematz, U., Matthes, K., Palmer, M., Rognerud, B., Tourpali, K., and Zerefos, C.: 2004, 'Solar variability and climate: Selected results from the SOLICE project', *SPARC Newsletter* **23**, 19–29.

Hines, C. O.: 1974, 'A possible mechanism for the production of Sun-weather correlations', *J. Atmos. Sci.* **31**, 589–591.

Hood, L. L.: 1997, 'The solar cycle variation of total ozone: Dynamical forcing in the lower stratosphere', *J. Geophys. Res.* **102**, 1355–1370.

Hood, L. L.: 2004, 'Effects of solar UV variability on the stratosphere', in J. Pap *et al.* (eds.), *Solar Variability and its Effect on the Earth's Atmosphere and Climate System*, AGU Monograph Series, American Geophysical Union, Washington D.C., pp. 283–303.

Ingram, W. J.: 2006, 'Detection and attribution of climate change, and understanding solar influence on climate', *Space Sci. Rev.*, this volume, doi:10.1007/s11214-006-9057-2.

Jackman, C. H., DeLand, M. T., Labow, G. J., Fleming, E. L., Lopez-Puertas, M.: 2006, 'Satellite measurements of middle atmospheric impacts by solar proton events in Solar Cycle 23', *Space Sci. Rev.*, this volume, doi:10.1007/s11214-006-9071-4.

Kodera, K.: 2004, 'Solar influence on the Indian Ocean Monsoon through dynamical processes', *Geophys. Res. Lett.* **31**, doi:10.1029/2004GL020928.

Kodera, K.: 2006, 'The role of dynamics in solar forcing', *Space Sci. Rev.*, this volume, doi: 10.1007/s11214-006-9066-1.

Kodera, K. and Kuroda, Y.: 2002, 'Dynamical response to the solar cycle', *J. Geophys. Res.* **107**, doi:10.1029/2002JD002224.

Labitzke, K.: 2006, 'Solar variation and stratospheric response', *Space Sci. Rev.*, this volume, doi: 10.1007/s11214-006-9061-6.

Langematz, U., Grenfell, J. L., Matthes, K., Mieth, P., Kunze, M., Steil, B., and Brühl, C.: 2005, 'Chemical effects in 11-year solar cycle simulations with the Freie Universität Berline Climate Middle Atmosphere Model with online chemistry (FUB-CMAM-CHEM)', *J. Geophys. Res.* **32**, doi:10.1029/2005GL022686.

Langen, J.: 2006, 'Recent space data – introductory paper part V', *Space Sci. Rev.*, this volume, doi: 10.1007/s11214-006-9070-5.

Lee, H. and Smith, A. K.: 2003, 'Simulation of the combined effects of solar cycle, quasi-biennial oscillation, and volcanic forcing on stratospheric ozone changes in recent decades', *J. Geophys. Res.* **108**, doi:10.1029/2001JD001503.

Matthes, K., Kodera, K., Haigh, J. D., Shindell, D. T., Shibata, K., Langematz, U., Rozanov, E., and Kuroda, Y.: 2003, 'GRIPS solar experiments intercomparison project: Initial results', *Pap. Meteorol. Geophys.*, **54**, 71–90.

Matthes, K., Langematz, U., Gray, L. J. Kodera, K., and Labitzke, K.: 2004, 'Improved 11-year solar signal in the Freie Universität Berlin Climate Middle Atmosphere Model (FUB-CHAM)', *J. Geophys. Res.* **109**, doi:10.1029/2003JD004012.

Matthes, K., Kuroda, Y., Kodera, K., and Langematz, U.: 2006, 'The transfer of the solar signal from the stratosphere to the troposphere: Northern Winter', *J. Geophys. Res.* **111**, doi:10.1029/2005JD006283.

McCormack, J. P. and Hood, L. L.: 1996, 'Apparent solar cycle variations of uppers stratospheric ozone and temperature: Latitude and seasonal dependences', *J. Geophys. Res.* **101**, 20,933–20,944.

Miller, A. J., Hollandsworth, S. M., Flynn, L. E., Tiao, G. C., Reinsel, G. C., Bishop, L., McPeters, R. D., Planet, W. G., DeLuisi, J. J., Mateer, C. L., Wuebbles, D., Kerr, J., and Nagatani, R. M.: 1996, 'Comparisons of observed ozone trends and solar effects in the stratosphere through examination of ground-based Umkehr and combined solar backscattered ultraviolet (SBUV) and SBUV 2 satellite data', *J. Geophys. Res.* **101**, 9017–9021.

Molina, M. J. and Rowland, F. S.: 1974, 'Stratospheric sink for chlorofluoromethanes: Chlorine atom-catalysed destruction of ozone', *Nature* **249**, 810–812.

Newchurch, M. J., Yang, E.-S., Cunnold, D. M., Reinsel, G. C., Zawodny, J. M., and Russell III, J. M.: 2003, 'Evidence for slowdown in stratospheric ozone loss: First stage of ozone recovery', *J. Geophys. Res.* **108**, doi:10.1029/2003JD003471.

Press, W. H., Teukolsky, S. A. Vetterling, W. T., and Flannery, B. P.: 1992, 'Numerical recipes in C - the art of scientific computing (2nd Edition)', Cambridge University Press, Cambridge, 994 pp.

Reinsel, G. C., Weatherhead, E. C., Tiao, G. C., Miller, A. J., Nagatani, R. M., Wuebbles, D. J., and Flynn, L. E.: 2002, 'On detection of turnaround and recovery in trend for ozone', *J. Geophys. Res.* **107**, doi:10.1029/2001JD000500.

Rottman, G.: 2006, 'Measurement of total and spectral solar irradiance', *Space Sci. Rev.*, this volume, doi:10.1007/s11214-006-9045-6.

Rozanov, E. V., Schlesinger, M. E., Egorova, T. A., Li, B., Andronova, N., and Zubov, V. A.: 2004, 'Atmospheric response to the observed increase of solar UV radiation from solar minimum to solar maximum simulated by the University of Illinois at Urbana-Champaign climate-chemistry model', *J. Geophys. Res.* **109**, doi:10.1029/2003JD003796.

Rozanov, E., Callis, L., Schlesinger, M., Yang, F., Andronova, N., and Zubov, V.: 2005, 'Atmospheric response to $NO_y$ source due to energetic electron precipitation', *J. Geophys. Res.* **32**, doi:10.1029/2005GL023041.

Salby, M. L. and Callaghan, P. F.: 2006, 'Influence of the solar cycle on the general circulation of the stratosphere and upper troposphere', *Space Sci. Rev.*, this volume, doi:10.1007/s11214-006-9064-3.

Schmidt, H. and Brasseur, G. P.: 2006, 'The response of the middle atmosphere to solar cycle forcing in the Hamburg model of the neutral and ionized atmosphere', *Space Sci. Rev.*, this volume, doi: 10.1007/s11214-006-9068-z.

Shindell, D., Rind, D., Balachandran, N., Lean, J., and Lonergan, P.: 1999, 'Solar cycle variability, ozone, and climate', *Science* **284**, 305–308.

Shindell, D. T., Schmidt, G. A., Miller, R. L., and Rind, D.: 2001, 'Northern hemisphere winter climate response to greenhouse gas, ozone, solar, and volcanic forcing', *J. Geophys. Res.* **106**, 7193–7210.

Sinnhuber, B.-M., von der Gathen P., Sinnhuber, M., Rex, M., König-Langlo G., and Oltmans, S. J.: 2005, 'Large decadal scale changes of polar ozone suggest solar influence', *Atmos. Chem. Phys. Discuss.* **5**, 12,103–12,117.

Staehelin, J., Harris, N. R. P., Appenzeller, C., and Eberhard, J.: 2001, 'Ozone trends: A review', *Rev. Geophys.* **39**(2), 231–290.

Steinbrecht, W., Claude, H., and Winkler, P.: 2004, 'Enhanced upper stratospheric ozone: Sign of recovery or solar cycle effect?', *J. Geophys. Res.* **109**, doi:10.1029/2003JD004284.

Stolarski, R., Bojkov, R., Bishop, L., Zerefos, C., Staehelin, J., and Zawodny, J.: 1992, 'Measured trends in stratospheric ozone', *Science* **256**, 342–349.

Tourpali, K., Schuurmans, C. J. E., van Dorland, R., Steil, B., and Bruehl, C.: 2003, 'Stratospheric and tropospheric response to enhanced solar UV radiation: A model study', *Geophys. Res. Lett.* **30**, doi:10.1029/2002GL016650.

Tourpali, K., Schuurmans, C. J. E., van Dorland, R., Steil, B., Brühl, C., and Manzini, E.: 2005, 'Solar cycle modulation of the Arctic oscillation in a chemistry-climate model', *Geophys. Res. Lett.* **32**, doi:10.1029/2005GL023509.

van Loon, H., Meehl, G.A., and Arblaster, J.M.: 2004, 'A decadal solar effect in the tropics in July-August', *J. Atm. Sol.-Terr. Phys.* **66**, 1767–1778.

WMO (World Meteorological Organization): 2003, 'Scientific Assessment of Ozone Depletion: 2002', Global Ozone Research and Monitoring Project – Report No. 47, Geneva, 498 pp.

Zerefos, C. S., Tourpali, K., Bojkov, B. R., Balis, D. S., Rognerud, B., and Isaksen, I. S. A.: 1997, 'Solar activity-total column ozone relationships: Observations and model studies with heterogeneous chemistry', *J. Geophys. Res.* **102**, 1561–1569.

Zerefos, C. S., Tourpali, K., Isaksen, I. S. A., and Schuurmans, C. J. E.: 2001, 'Long term solar induced variations in total ozone, stratospheric temperatures and the tropopause', *Adv. Sp. Res.* **27**, 1943–1948.

Zerefos, C. S., Tourpali, K., and Balis, D.: 2005, 'Solar activity-ozone relationships in the vertical distribution of ozone', *Int. J. Rem. Sensing* **26**, 3449–3454.

# INFLUENCE OF THE SOLAR CYCLE ON THE GENERAL CIRCULATION OF THE STRATOSPHERE AND UPPER TROPOSPHERE

M. L. SALBY[1,*] and P. F. CALLAGHAN[2]
[1]*University of Colorado, Boulder, CO, USA*
[2]*Atmospheric Systems and Analysis, Broomfield, CO, USA*
*(*Author for correspondence, E-mail: gratrix@icarus.colorado.edu)*

(Received 24 August 2005; Accepted in final form 31 January 2006)

**Abstract.** The record of dynamical structure reveals a systematic variation that operates coherently with the 11-yr variation of UV irradiance. Involving periods shorter than 5 years, the systematic variation reflects the influence of the QBO on the polar-night vortex. It has the same basic structure as interannual changes associated with the residual mean circulation of the stratosphere. A signature of the solar cycle also appears in the direct correlation to solar flux, as recovered through regression of the entire monthly record. That signature, however, is sharply enhanced around solstice, when the residual circulation is active, and during extremal phases of the QBO. In the tropics, the solar signature follows, throughout the year, from a decadal modulation in the frequency of the QBO. The modulation is manifested to either side of the QBO's mean frequency, in two spectral peaks where the QBO dwells: one at (24 months)$^{-1}$, reflecting a Biennial Oscillation (BO), and another at (36 months)$^{-1}$. Intrinsic to the QBO, those peaks are separated from its mean frequency by $\sim$11 years$^{-1}$. Through the QBO's residual circulation, the decadal modulation introduces anomalies in the subtropics, with symmetry about the equator. Accompanying anomalous temperature in the subtropics is a stronger signature over the winter pole. Discovered by Labitzke and van Loon (1988), that solar signature reflects anomalous downwelling of the Brewer-Dobson circulation. It is shown to follow through the BO, which is intrinsic to the QBO and its modulation of the polar-night vortex.

**Keywords:** solar cycle, stratosphere, quasi-biennial oscillation, uv

## 1. Introduction

The 11-year variation of solar irradiance becomes important in the upper stratosphere, where ozone heating involves wavelengths shorter than 200 nm. Unlike longer wavelengths, which prevail at lower altitude, UV irradiance at short wavelengths changes significantly between solar min and solar max (e.g., WMO, 1987; Lean, 2000). Implied are analogous changes of ozone heating, which shapes thermal structure and the global circulation. A signature of such changes is, in fact, visible in wintertime polar temperature, when it is stratified according to the phase of the equatorial QBO (Labitzke and van Loon, 1988, 1996).

Such changes are also visible in QBO itself. The duration of equatorial westerlies and easterlies varies systematically between solar min and solar max (Quiroz, 1981; Salby and Callaghan, 2000; Soukharev and Hood, 2001; McCormack, 2003). The

Space Science Reviews (2006) 125: 287–303
DOI: 10.1007/s11214-006-9064-3 

lengthening and shortening of individual cycles represents a modulation in the frequency of the QBO, which varies about its mean value by ~20%.

A signature of the solar cycle also emerges in traditional treatments, wherein the entire (monthly) atmospheric record is regressed onto the record of 10.7-cm solar flux, $F_s$ (van Loon and Labitzke, 1998, 1999; Crooks and Gray, 2005; Labitzke and van Loon, 1997). Associated changes represent the direct (linear) response to the 11-yr variation of UV irradiance, one that gradually drifts with $F_s$. Those changes are relatively weak, typically less than 0.5 K. This contrasts with wintertime polar temperature when it is stratified according to phase of the QBO. The anomaly that tracks $F_s$ is then of order 10 K – comparable to the overall interannual variance of polar temperature (Labitzke and van Loon, 1988; cf Figure 8a).

Anomalous structure that results when records are sampled during an individual month resembles structure that is recovered when all months are included in the regression onto $F_s$ (Crooks and Gray, 2005). Both exhibit hemispheric symmetry. However, correlations to $F_s$ intensify and become more symmetric about the equator when stratospheric records have been sampled during an individual month and phase of the QBO (Labitzke, 2003; Salby and Callaghan, 2006). This features suggests that correlations to $F_s$ derived from the entire monthly record may reflect the residual of a more complex signal in which the QBO figures centrally.

Here, we discuss a key mechanism through which solar changes influence the stratosphere and, in turn, can be conveyed to the troposphere. The mechanism is first illustrated in Section 2 through interannual variations. Anomalous dynamical structure that varies coherently with solar activity is then considered in Section 3 in light of that mechanism and in Section 4 in relation to the QBO.

## 2. Interannual Variations of Dynamical Structure

Changes in the stratosphere are linked to the residual-mean circulation, the so-called Brewer-Dobson circulation. Illustrated in Figure 1, the residual circulation is driven by planetary waves that transmit momentum upward from the troposphere, as measured in the upward EP flux at the tropopause ($F_z$). When absorbed, that momentum drives a poleward drift ($\overline{v}^* > 0$) that converges to produce downwelling ($\overline{w}^* < 0$) and adiabatic warming over the winter pole. Compensating it at lower latitudes is upwelling ($\overline{w}^* > 0$) and adiabatic cooling. Interannual changes of planetary wave activity lead to an intensification or weakening of the residual circulation, with commensurate changes of temperature.

A similar influence is exerted by the QBO of equatorial wind. The QBO controls the critical line ($\overline{u} = 0$), indicated in green in Figure 1, where planetary waves suffer strong absorption that forces residual mean motion. During QBO easterlies (red), the critical line is advanced into the winter hemisphere, along with downwelling. This favors a polar-night vortex that is anomalously warm and weak (Holton and Tan, 1980; Labitzke, 1982). During QBO westerlies (not shown), the critical line

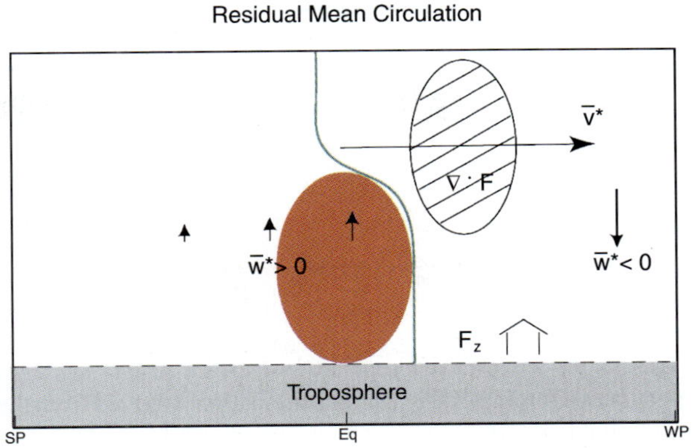


*Figure 1.* Schematic representation of the residual mean circulation of the stratosphere. Known also as the Brewer-Dobson circulation, it is driven by planetary waves that transmit momentum upward from the troposphere, as measured in the upward EP flux $F_z$. When absorbed ($\nabla \cdot \mathbf{F}$), that momentum drives a poleward drift ($\overline{v}^*$), which converges over the winter hemisphere to produce downwelling ($\overline{w}^* < 0$). Compensating it at lower latitudes is upwelling ($\overline{w}^* > 0$). Superposed is the critical line (green), where $\overline{u} = 0$ and planetary waves suffer strong absorption that drives residual motion. During QBO easterlies (red), the critical line is advanced into the winter hemisphere, along with downwelling and adiabatic warming over the winter pole. During QBO westerlies (not shown), those influences are removed into the summer hemisphere.

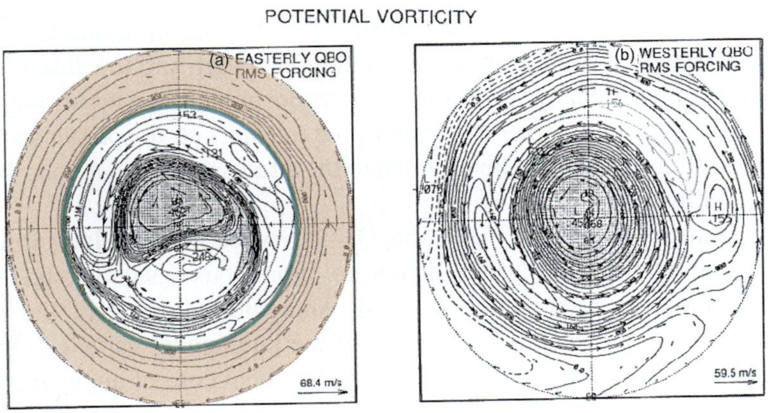


*Figure 2.* Potential vorticity (contoured) and horizontal motion (vectors) in a numerical simulation under conditions of (a) QBO easterlies and (b) QBO westerlies. The critical line (green) divides zonal-mean westerlies from zonal-mean easterlies. Adapted from O'Sullivan and Salby (1990).

is removed into the summer hemisphere, along with downwelling. It then favors a vortex that is anomalously cold and strong (e.g., Naito and Hirota, 1997; Gray *et al.*, 2004).

Figure 2 compares potential vorticity (a marker of individual bodies of air) from simulations under QBO easterlies and QBO westerlies. North of the critical line (green), planetary waves steepen and overturn the potential vorticity distribution,

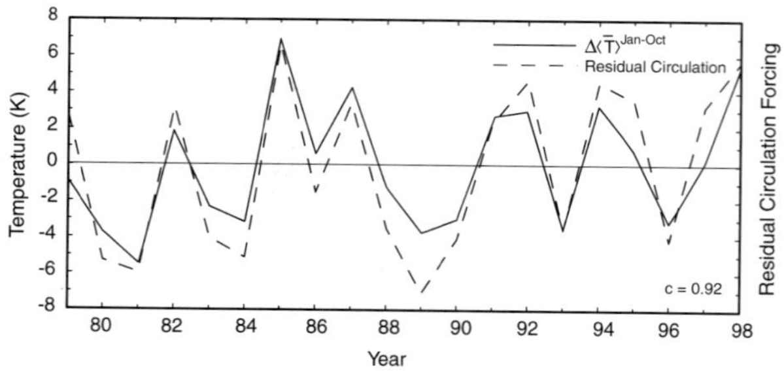


*Figure 3.* Record of anomalous wintertime temperature over the Arctic, averaged from 100 mb to 10 mb (solid), and of anomalous forcing of the residual circulation. Derived from the NCEP reanalysis record. After Salby and Callaghan (2002).

producing nonlinear mixing that, in tandem with mean downwelling and adiabatic warming, limits the size and strength of the vortex. QBO easterlies (Figure 2a) advance the critical line into the winter hemisphere, yielding a vortex that is anomalously weak and disturbed from polar symmetry. Arctic mean temperature is then anomalously warm. In contrast, QBO westerlies (Figure 2b) remove the critical line into the summer hemisphere, leaving the vortex anomalously strong and close to polar symmetry. Arctic mean temperature is then anomalously cold. This dependence on equatorial wind operates only during winter, when midlatitude westerlies support planetary waves that couple the circulation at high and low latitudes.

These stratospheric changes are coupled to tropospheric changes, through transfers of mass. Downwelling over the winter pole transfers stratospheric air into the troposphere (Figure 1). It must be returned to the stratosphere at subpolar latitudes through upwelling.

Jointly, changes of EP flux and the QBO represent anomalous forcing of the residual circulation. The latter drives an anomalous residual circulation, a perturbation about its climatological mean. Anomalous forcing of the residual circulation, it turns out, accounts for much of the interannual variance of wintertime temperature and ozone. Plotted in Figure 3 is the record of anomalous Arctic temperature over the last 2 decades (solid). Superposed is anomalous forcing of the residual circulation (dashed). The two records track one another closely, achieving a correlation of 0.92.

Figure 4 plots the structure of anomalous wintertime temperature (contoured) that varies coherently with anomalous temperature at 70 mb and 90°N (open circle), a proxy for anomalous downwelling. Shading marks regions where anomalous temperature and downwelling over the Arctic are highly coherent. Anomalous temperature is strong and positive over the Arctic. A signature of anomalous downwelling, the warm anomaly extends downward coherently into the troposphere – to as low as 500 mb. Positive anomalous temperature over the Arctic reflects a polar-night vortex that is anomalously warm and disturbed from zonal symme-

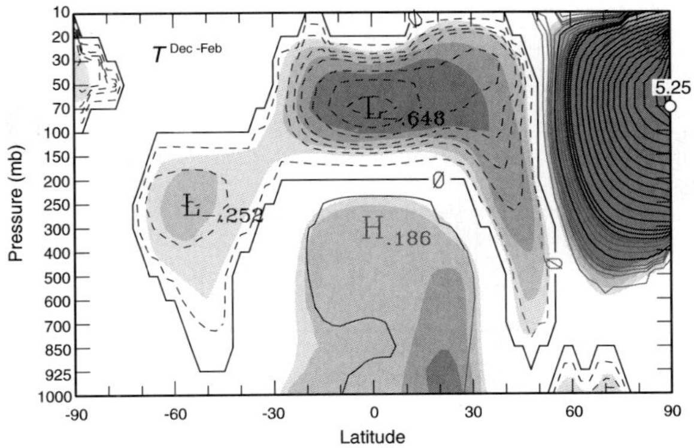


*Figure 4.* Anomalous wintertime (DJF) temperature (contoured) that varies coherently with that at 90°N and 70 mb (open circle), which is a proxy for anomalous downwelling over the Arctic. Analogous to regression, values represent the anomalous temperature (K) associated with a 1-std deviation increase of downwelling. Shading marks regions where the two are coherent at the 95%, 99% , and 99.9% significance levels. Derived from the record of NCEP reanalyses over 1955 – 2000. After Salby and Callaghan (2005).

try (cf Figure 2a). Compensating it at subpolar latitudes is negative anomalous temperature, a signature of anomalous upwelling. Visible in each hemisphere, cold anomalies in the midlatitude troposphere join a stronger signature of anomalous upwelling just above the tropical tropopause. Just below, inside the tropical troposphere, anomalous temperature reverses. A warm anomaly appears below 200 mb and equatorward of 30°, where thermal structure is shaped by convection and latent heating. It too is highly coherent with anomalous downwelling over the Arctic. Strongest in the subtropics of the winter hemisphere, it reflects a change of the Hadley circulation (Salby and Callaghan, 2005).[1] Jointly, the features in Figure 4 imply a tropical tropopause that is anomalously cold and high during those winters when downwelling over the Arctic is anomalously strong.

## 3. Systematic Changes Coherent with Solar Activity

The interannual changes described above are coupled to the QBO, which contributes to anomalous forcing of the residual circulation (e.g., Gray and Pyle, 1989; Tung and Yang, 1994). Included in the QBO is a component that varies systematically with solar activity. Plotted in Figure 5 is the anomalous frequency of the equatorial QBO (deviation from its climatological mean). The QBO's frequency increases near solar max in each of the last 4 decades: near 1960, 1970, 1980, and again near

[1] Also coherent with anomalous downwelling over the Arctic is divergence in the tropical troposphere, which is intrinsic to convection and the Hadley circulation (ibid).

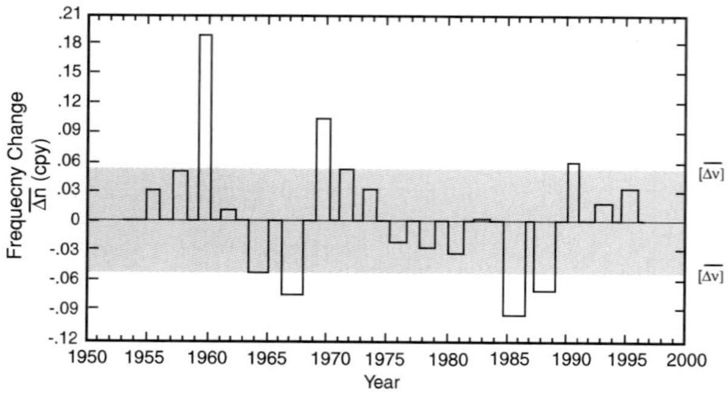


*Figure 5.* Anomalous frequency of the equatorial QBO, as a function of year. Derived from the radiosonde record. After Salby and Callaghan (2000).

1990. Conversely, it decreases in the middle of each of the decades. The behavior in Figure 5 represents a frequency modulation of the QBO. In contrast, the QBO's amplitude does not change.

Anomalous frequency varies coherently and nearly in phase with $F_s$, achieving a correlation of $\sim$0.7 – highly significant. (The duration of westerlies varies less systematically, achieving a correlation with $F_s$ of only 0.46 (Hamilton, 2002; see also Salby and Callaghan, 2002). It should be noted that, although systematic over the record shown, the relationship implied by Figure 5 need not hold in every decade, as $F_s$ represents but one of several sources of interannual variance.

Through the mechanisms described in Section 2, the quasi-biennial reversal of equatorial wind is manifested over the Arctic in the strength and symmetry of the polar-night vortex (Figure 2). Implied are changes of polar temperature $T_{NP}$ that vary *out of phase* with equatorial wind $u_{EQ}$: Equatorial easterlies ($u_{EQ} < 0$) are attended by anomalously-warm polar temperature ($T_{NP} > 0$), implying a correlation between $u_{EQ}$ and $T_{NP}$ of $-1$.

Figure 6 presents the running correlation between $T_{NP}$ and $u_{EQ}$ (solid). In the middle of each decade, the correlation is indeed close to $-1$ ($T_{NP}$ varying *out of phase* with $u_{EQ}$). However, their correlation swings systematically towards $+1$ ($T_{NP}$ varying *in phase* with $u_{EQ}$) near solar max – likewise in each of the 4 decades. Superimposed is the record of $F_s$, lagged by one year (dashed). The swing in the relationship between $T_{NP}$ and $u_{EQ}$, evident in Figure 6, is plainly visible in the raw time series (Salby and Callaghan, 2004).

Plotted in Figure 7 is the anomalous wintertime tendency of temperature that varies coherently with $u_{EQ}$ and $F_s$ (contoured), along with its significance (shaded). It represents temperature fluctuations that, locally in time, vary coherently with $u_{EQ}$, but with relative phase that gradually drifts with $F_s$ (Figure 6). The anomalous temperature in Figure 7 exhibits the same basic structure as that which varies coherently with anomalous downwelling over the winter pole (Figure 4). The latter, in turn, varies coherently with anomalous forcing of the residual circulation (Figure 3).

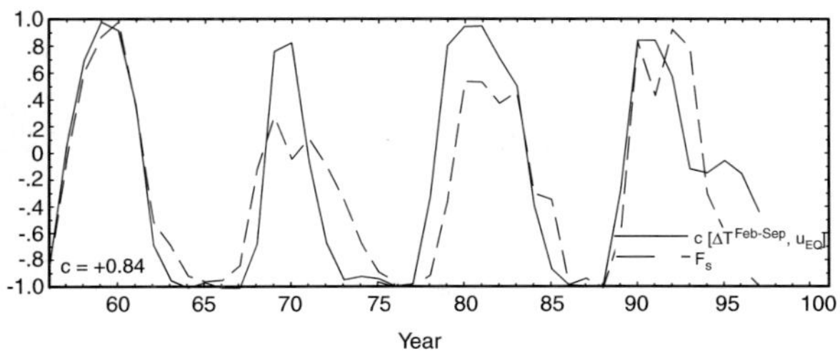


*Figure 6.* Running correlation between equatorial wind at 50 mb and wintertime temperature at 90°N and 70 mb, derived from the NCEP record. Superimposed is the record of $F_s$, lagged by 1 year. After Salby and Callaghan (2004).

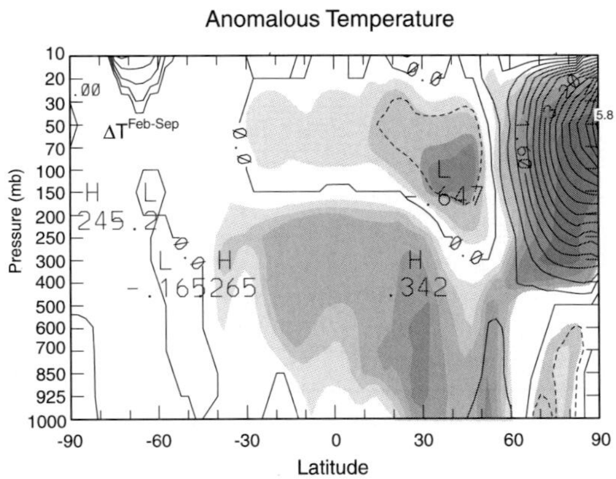


*Figure 7.* Anomalous wintertime tendency of temperature, $\Delta T^{\text{Feb-Sep}} = T^{\text{Feb}} - T^{\text{Sep}}$ (contoured), that varies coherently with 50 mb $u_{\text{EQ}}$ and $F_s$. Anomalous temperature during individual winters (K) follows by scaling values by the product of anomalous $F_s$ and anomalous $u_{\text{EQ}}$ during those winters (each normalized).

The signature of anomalous downwelling (warm anomaly) is visible well into the troposphere. Compensating it at subpolar latitudes is a signature of anomalous upwelling (cold anomaly). As in Figure 4, anomalous temperature reverses sign inside the tropical troposphere (warm anomaly), where temperature is shaped by the Hadley circulation. It too is strongest in the subtropics of the winter hemisphere.[2]

Anomalous temperature during individual years is recovered by scaling values in Figure 7 by the instantaneous product of anomalous $u_{\text{EQ}}$ and $F_s$ (normalized). Implied is a polar-night vortex that is anomalously warm during equatorial easterlies ($u_{\text{EQ}} < 0$) and near solar min ($F_s < 0$), consistent with the behavior in Figure 2.

[2]The implied interaction with the Hadley circulation shares features with the *low-frequency* (linear) signature of the solar cycle that has been inferred from regression (Haigh *et al.*, 2005). However, Figure 7 describes *high-frequency* variability, namely, that associated with the QBO, which undergoes a decadal modulation (Figure 5) and, as will be seen in Section 4, accounts for significantly greater variance.

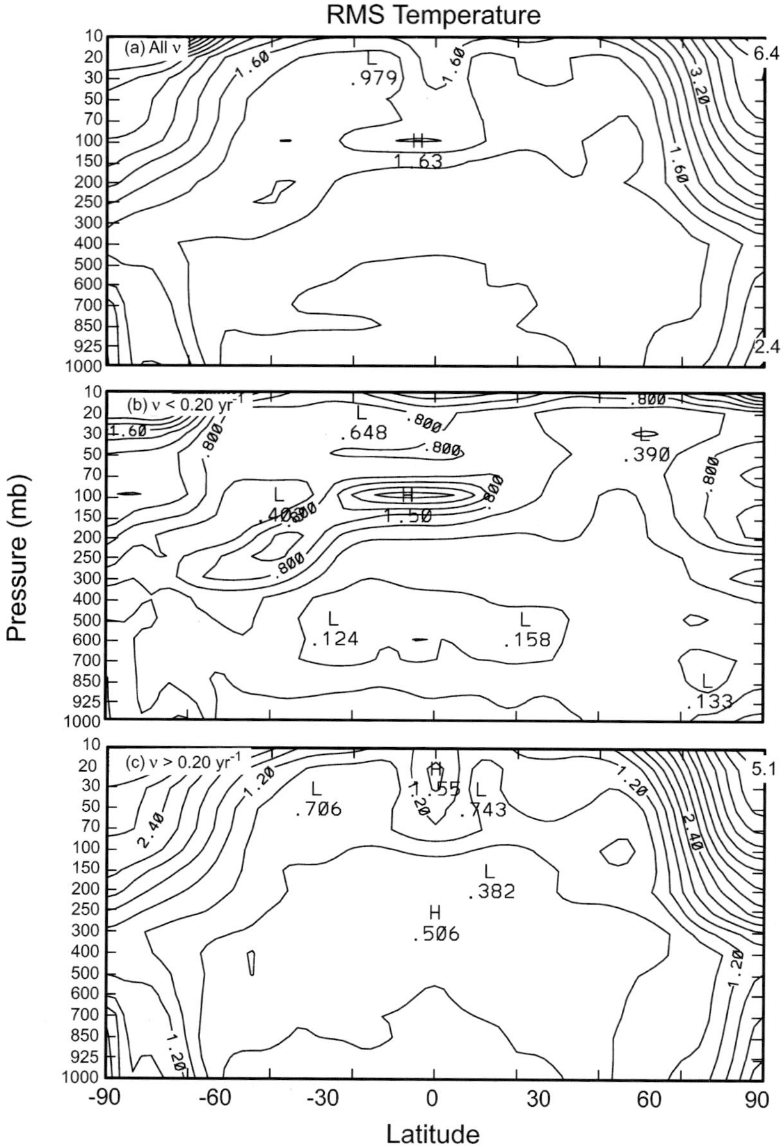


*Figure 8.* Root mean square variation of temperature in the entire monthly record: (a) inclusive of all periods, (b) for periods longer than 5 years, (c) for periods shorter than 5 years. Derived from the NCEP record. After Salby and Callaghan (2006).

However, the same behavior is also implied during equatorial westerlies ($u_{EQ} > 0$) and near solar max ($F_s > 0$). The relationship between $T_{NP}$ and $u_{EQ}$ must therefore reverse during the solar cycle, as was discovered by Labitzke and van Loon (1988).

The temperature changes implied by Figure 7 are not small. In fact, they are almost as large as the entire (rms) interannual anomaly (cf Figure 8a). Yet, in the record of wintertime $T_{NP}$, those changes are virtually invisible. This paradoxical feature stems from anomalous temperature reflecting, not a simple drift with the solar cycle (ie, a linear response), but something more complex: a systematic shift of the QBO's phase with respect to winter months, when the polar vortex is sensitive to equatorial wind. Anomalous wintertime $T_{NP}$ then drifts through the full range of interannual variance (cf Figure 2), tracking the decadal drift in QBO's frequency (Figure 5).

## 4. Origin of the Solar Correlation

Plotted in Figure 8a is the rms interannual anomaly of temperature, based on the entire monthly record. Anomalous temperature is dominated by variance at the poles, approaching 7 K over the Arctic. Near the tropical tropopause is a much weaker maximum, of about 1 K.

Plotted in Figure 8b is the same information, but lowpass filtered to periods longer than 5 years. The low-frequency component contains the direct correlation to $F_s$, namely, that recovered by regression of the entire monthly record. It reflects a linear response to changes of solar activity, one that simply drifts with the 11-yr variation of $F_s$. The low-frequency component accounts for much of the anomaly near the tropical tropopause, but virtually none of the dominant anomaly over the poles.

Plotted in Figure 8c is the same information, but now highpass filtered to periods shorter than 5 years. The high-frequency component contains the indirect correlation to $F_s$, namely, that associated with the QBO (Figure 6). It reflects a nonlinear response to changes of solar activity. The high-frequency component accounts for nearly all of the dominant anomaly over the poles. Also accounted for is anomalous temperature in the equatorial stratosphere of $\sim$1 K.

Figure 9 plots the power spectrum of $T_{NP}$ at 30 mb. A peak appears near 0.41 cycles per year (cpy). Corresponding to a period of 29 months, it reflects the QBO's influence on $T_{NP}$ (cf Figures 1 and 2). Another peak appears near 0.59 cpy. That peak follows through interaction between the QBO and the annual cycle of midlatitude westerlies, which limits the QBO's influence to winter months (Salby *et al.*, 1997).

Between the two side lobes is a central peak at 0.50 cpy. Corresponding to a period of 24 months, it represents a Biennial Oscillation (BO). Harmonics appear at higher frequency, near 8 months and 6 months (not shown). Notice: The BO accounts for almost as much temperature variance as the QBO. It is separated from the two side lobes by $\sim$11 years$^{-1}$.

Contoured in Figure 10 is anomalous temperature regressed onto $F_s$, along with the corresponding correlation (shaded). Each is derived from temperature during February, when the residual circulation is strong, and during extremal phases of the

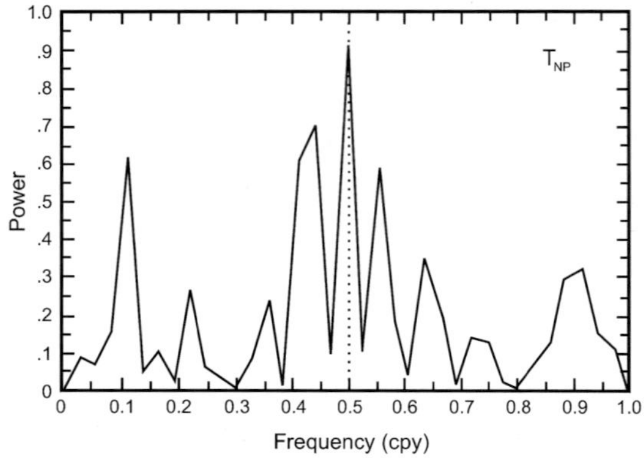


*Figure 9.* Power spectrum of North Polar temperature at 30 mb, derived from the monthly record. Side lobes that straddle 0.5 cpy follow from the QBO and its interaction with the annual cycle. Adapted from Salby *et al.* (1997).

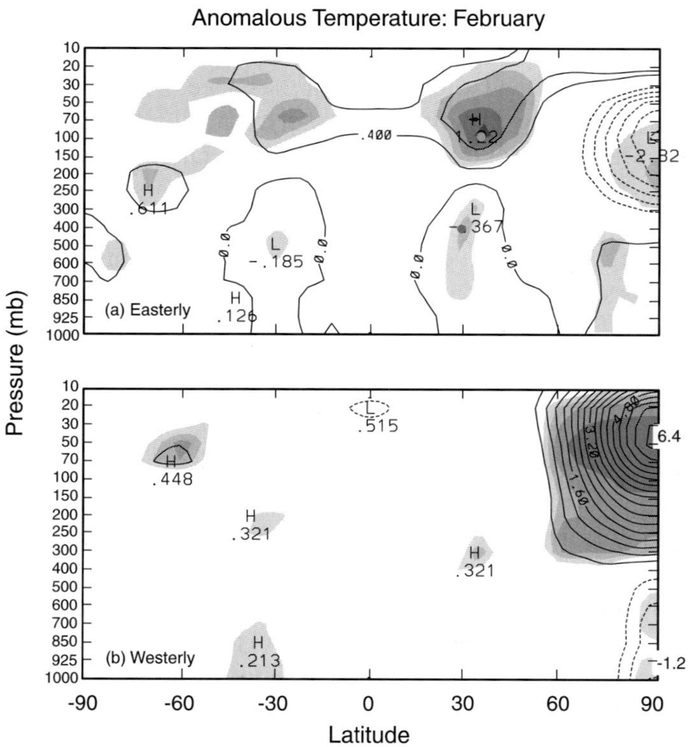


*Figure 10.* Anomalous temperature (K) that varies coherently with $F_s$ (contoured) and corresponding correlation (shaded) during (a) Februaries when the QBO is easterly and (b) Februaries when the QBO is westerly. Correlation shaded at ±0.40, ±0.50, etc. Blue and green dots mark sites referred to in Figures 11 and 12. After Salby and Callaghan (2006).

QBO.[3] During easterly Februaries (Figure 10a), the correlation to $F_s$ is intensified to either side of the equator, where it exceeds 0.80. There, the correlation to $F_s$ is highly significant, even by a strict standard (Labitzke, 2003; Salby and Callaghan, 2006). The strong latitudinal symmetry reflects the residual circulation of the QBO, which induces downwelling and adiabatic warming to either side of the equator (e.g., Baldwin *et al.*, 2001). Inside the tropical troposphere, the correlation to $F_s$ also maximizes to either side of the equator. Positioned near the edge of the Hadley circulation, the correlation there approaches −0.60, marginally significant even by a strict standard.

Over the Arctic is a stronger anomaly of opposite sign (cf Figure 7). It reflects a change of the Brewer-Dobson circulation. Implied is adiabatic cooling and anomalous upwelling during those Februaries when $u_{EQ}$ is easterly and near solar max. During westerly Februaries (Figure 10b) is similar structure, but reversed. The opposing anomalies over the Arctic are consistent with the reversal in the running correlation between $T_{NP}$ and $u_{EQ}$ (Figure 6). As before, those signatures of anomalous vertical motion are almost as strong as the entire rms interannual anomaly (Figure 8a).

The strong correlation in the subtropical stratosphere reflects a close correspondence between anomalous temperature during easterly winters and contemporaneous changes of $F_s$. Plotted in Figure 11 is the record of anomalous temperature near 32°N and 100 mb. Of 14 Februaries, all but 2 track contemporaneous changes of $F_s$. They achieve a correlation with $F_s$ of 0.82.

Analogous information is plotted in Figure 12, but at 30°N and 400 mb, inside the subtropical troposphere (blue). Of 14 Februaries, all but 3 track opposite to contemporaneous changes of $F_s$. They achieve a correlation to $F_s$ of −0.67.

Some insight into the correlation with $F_s$ comes from a running spectrum analysis. It reveals how spectral features evolve. Plotted in Figure 13a is the running spectrum of temperature in the subtropical stratosphere, where the correlation to $F_s$ is large (Figure 10). Displayed along the horizontal axis is carrier frequency, that of the instantaneous spectrum. The vertical axis displays year. Temperature variance in the subtropical stratosphere is concentrated at periods of the QBO. However, the central period actually meanders, between 24 and 36 months.[4]

Plotted in Figure 13b is the spectrum of the running spectrum: The vertical axis now displays the periods at which individual carrier frequencies are modulated. In the bivariate spectrum, variance of the QBO is concentrated in two peaks, at which the QBO's instantaneous period dwells: one near 24 months and another near 36 months. The peak at 24 months evident at low latitude is the same one seen earlier at high latitude, in the univariate spectrum of $T_{NP}$ (Figure 9), associated with

[3]Similar structure is recovered if the temperature record is first highpass filtered to periods shorter than 5 years (cf Figures 8a and 8c).

[4]The weakened temperature signal after 1979 reflects the introduction of nadir-viewing satellite data, which, because of its deep vertical weighting functions, tends to suppress the temperature anomaly of the QBO (Huesmann and Hitchman, 2003).

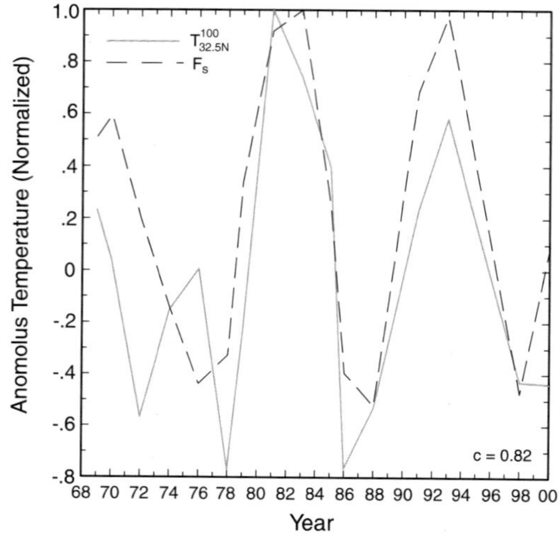


*Figure 11.* Record of anomalous temperature at 32.5°N and 100 mb during easterly Februaries (cf Figure 10a). Correlation with $F_s$ equals 0.82. After Salby and Callaghan (2006).

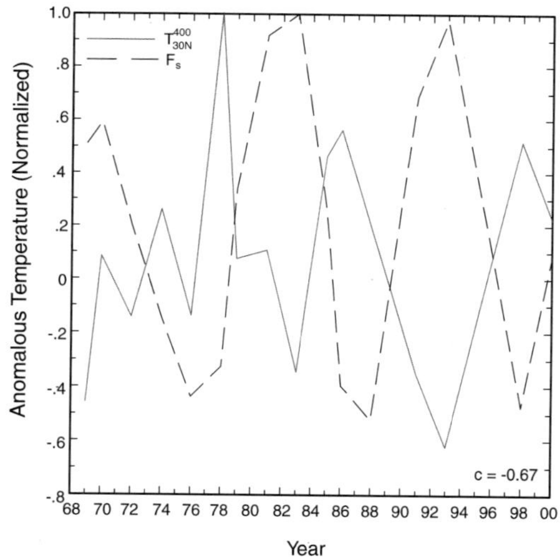


*Figure 12.* As in Figure 11, but at 32°N and 400 mb, after the monthly record has been filtered to periods shorter than 5 years (cf Figure 10a). Correlation with $F_s$ equals −0.67. After Salby and Callaghan (2006).

the BO. In this light, the BO is intrinsic to the equatorial QBO. The two peaks in Figure 13b are separated from the QBO's mean frequency by 11 years$^{-1}$. Both are magnified about a modulation period of ~11 years.

With increasing latitude in the winter hemisphere, the peak associated with the BO intensifies. It reflects an interaction between the QBO and the annual cycle,

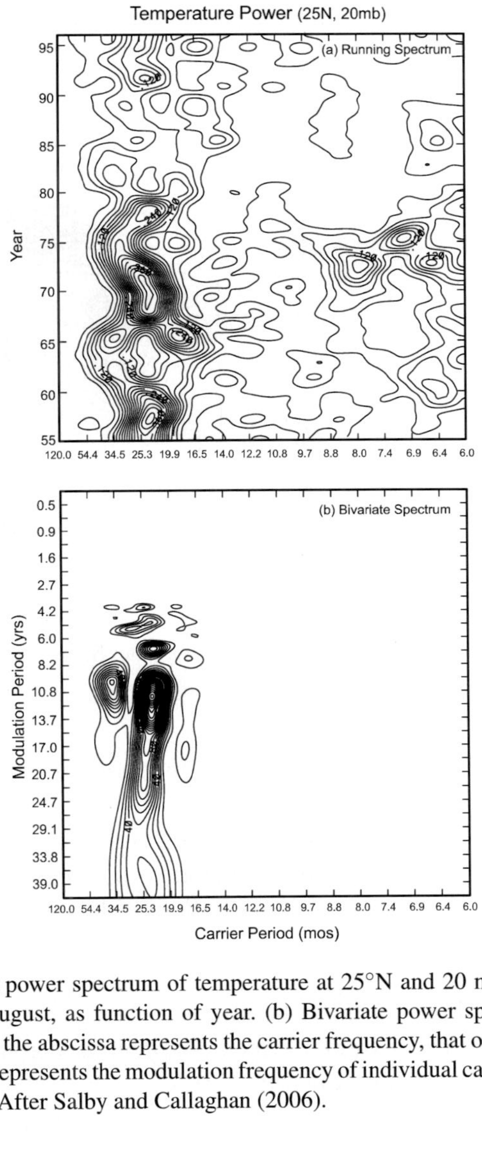


*Figure 13.* (a) Running power spectrum of temperature at 25°N and 20 mb, where the correlation to $F_s$ is large during August, as function of year. (b) Bivariate power spectrum derived from the running spectrum in (a): the abscissa represents the carrier frequency, that of the instantaneous power spectrum. The ordinate represents the modulation frequency of individual carrier frequencies. Derived from the NCEP record. After Salby and Callaghan (2006).

corresponding to a decadal modulation of the QBO's phase with respect to winter months (Gray and Dunkerton, 1990). Figure 14 plots the bivariate spectrum at 90°N and 70 mb. The peak associated with the BO, found at a carrier period of ~24 months, is now the dominant member of the pair. Accompanying it is the next harmonic, at a carrier period of ~8 months.

Those peaks are intrinsic to the strong correlation between wintertime $T_{NP}$ and $F_s$ (Figure 10). Bandpass filtering the monthly record of $T_{NP}$ to the BO and its harmonics and then sampling during February yields the record in Figure 15a. Anomalous temperature alternates between consecutive Februaries, virtually un-

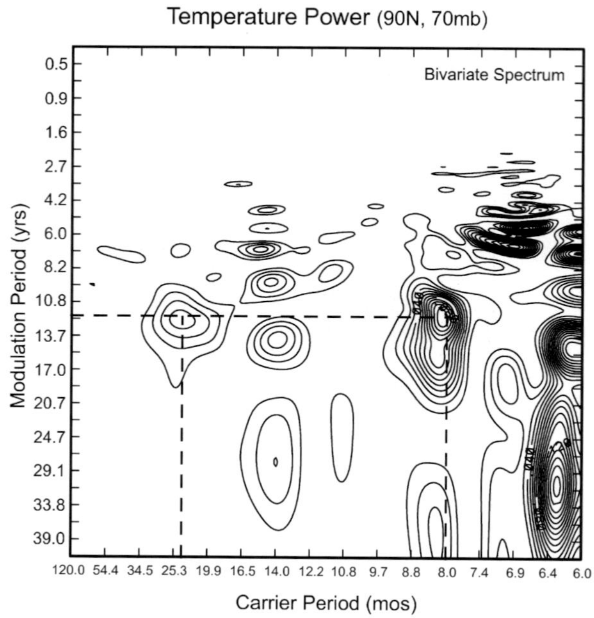


*Figure 14.* As in Figure 13b, but at 90°N and 70 mb. After Salby and Callaghan (2006).

correlated to $F_s$. It is accompanied, however, by an amplitude modulation – one that is synchronized with $F_s$.

Sampling the record during those Februaries when $u_{EQ}$ is easterly yields the record in Figure 15b. By discriminating to an individual phase of the QBO, easterly sampling rectifies the alternating carrier oscillation, leaving its decadal modulation. Anomalous $T_{NP}$ then varies *out of phase* with $F_s$, achieving a correlation of nearly −0.80.

Sampling the record during those Februaries when $u_{EQ}$ is westerly yields the record in Figure 15c. It has a similar effect. Anomalous $T_{NP}$ then varies *in phase* with $F_s$, achieving a correlation of nearly 0.70.

Figure 15 is the behavior that was discovered by Labitzke and van Loon (1988). It derives from the BO and its harmonics, which in turn are intrinsic to the QBO and its influence on the polar-night vortex.

## 5. Closing

The results collected here have the following implications:

(i) In addition to a low-frequency (linear) response, the 11-yr variation of UV introduces a high-frequency (nonlinear) response that is considerably stronger.

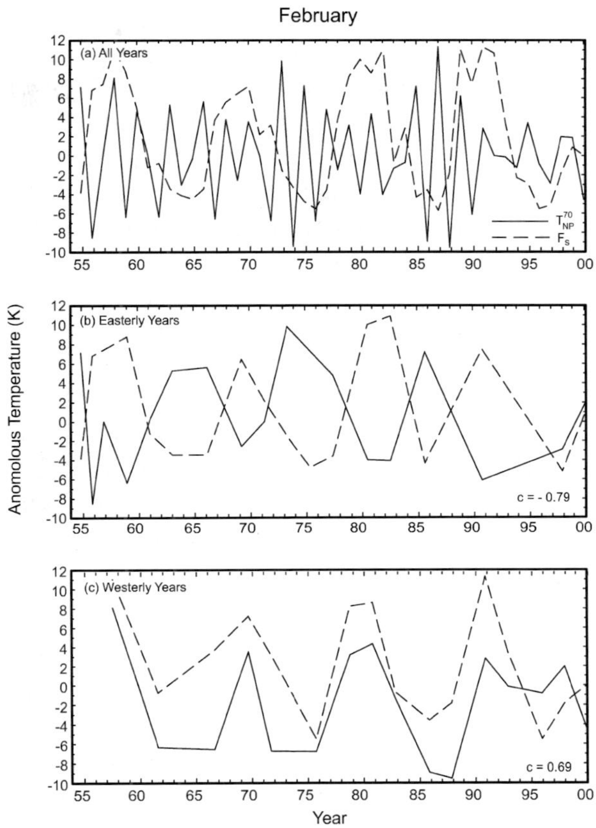


*Figure 15.* Record of anomalous temperature at 90°N and 70 mb (solid), after bandpass filtering the entire monthly record about the BO and its harmonics and then sampling during February, (a) during all years, (b) during easterly years, (c) during westerly years. Superimposed is the corresponding record of $F_S$ (dashed). After Salby and Callaghan (2006).

Involving interannual fluctuations, the high-frequency response is associated with the QBO and its influence on the Brewer-Dobson circulation.

(ii) The high-frequency response is as large as the interannual variance – even though, in the raw record, it is virtually invisible. It reflects the 11-yr modulation of the QBO's phase with respect to winter months, when the polar vortex is sensitive to equatorial wind.

(iii) Intrinsic to the QBO and its modulation is the BO, which is magnified at extratropical latitudes through interaction with the annual cycle. The BO and its harmonics account for the 11-yr modulation of polar temperature that was discovered by Labitzke and van Loon (1988).

(iv) Although weaker, analogous behavior is visible in the troposphere, where it is associated with interannual fluctuations of the Hadley circulation.

How the solar signature apparent in the QBO is actually conveyed to the Brewer-Dobson circulation remains to be determined. Almost surely involved are planetary waves, which, through interaction with the annual cycle of zonal wind, provide the bridge between the tropics and polar vortex. It is noteworthy that salient features of the observed behavior are reproduced by a simple model of the QBO and its influence on the extratropical circulation (Salby and Callaghan, 2006). Accounted for are the BO, its amplification over the winter pole, and the reversal of anomalous polar temperature during the solar cycle.

In reality, the QBO's influence on polar temperature is conveyed nonlinearly, by planetary waves and their forcing of residual mean motion. Their involvement is, in fact, indicated by anomalous horizontal structure that varies coherently with equatorial wind and $F_s$, in which the Aleutian high figures centrally (van Loon and Labitzke, 1994; Salby and Callaghan, 2004). Understanding how these features of the wintertime circulation are influenced by the 11-yr variation of UV irradiance will require 3D simulations, but in a model that accounts for the observed modulation of the QBO and its interaction with the extratropical circulation.

## Acknowledgements

The authors are grateful for constructive remarks provided during the workshop and during review. Figures were produced by Jackie Gratrix. This work was supported by NSF Grant ATM-0127671.

## References

Baldwin, M., Gray, L., Dunkerton, L., Hamilton, K., Haynes, P., Randel, W., Holton, J., Alexander, J., Hirota, I., Horinouchi, T., Jones, D., Kinnersley, C., Marquardt, K., Sato, T., and Takahashi, M.: 2001, 'The quasi-biennial oscillation', *Rev. Geophys.* **39**, 179–229.

Crooks, S. and Gray, L.: 2005, 'Characterisation of the 11-year solar signal using a multiple regression analysis of the ERA-40 dataset', *J. Clim.* **18**, 996–1015.

Gray, L. and Pyle, J.: 1989, 'A two-dimensional model of the quasi-biennial oscillation of ozone', *J. Atmos. Sci.* **46**, 203–220.

Gray, L. and Dunkerton, T.: 1990, 'The role of the seasonal cycle in the quasi-biennial oscillation of ozone', *J. Atmos. Sci.* **47**, 2429–2451.

Gray, L., Crooks, S., Pascoe, C., and Sparrow, S.: 2004, 'Solar and QBO influences on the timings of stratospheric sudden warmings', *J. Atmos. Sci.* **61**, 2777–2796.

Haigh, J., Blackburn, M., and Day, R.: 2005, 'Response of tropospheric circulation to perturbations in lower-stratospheric temperature', *J. Climate* **18**, 3672–3685.

Hamilton, K.: 2002, 'On the quasi-decadal modulation of the stratospheric QBO period', *J. Climate* **15**, 2562–2565.

Holton, J. R. and H-C. Tan: 1980, 'The influence of the equatorial quasi-biennial oscillation on the global circulation at 50 mb', *J. Atmos. Sci.* **37**, 2200.

Huesmann, A. and Hitchman, M.: 2003, 'The 1978 shift in the NCEP reanalysis stratospheric quasi-biennial oscillation', *Geophys. Res. Lett.* **30**, doi:10.1029/2002GL016323.

Labitzke, K.: 1982, 'On the interannual variability of the middle stratosphere during the northern winters', *J. Meteorol. Soc. Jpn.* **60**, 124–139.

Labitzke, K.: 2003, 'The global signal of the 11-year sunspot cycle in the atmosphere: When do we need the QBO?', *Meteorol. Zeit.* **12**, 209–216.

Labitzke, K. and van Loon, H.: 1988, ' Association between the 11-year solar cycle, the QBO, and the atmosphere, I, The troposphere and stratosphere on the northern hemisphere winter', *J. Atmos. Terr. Phys.* **50**, 197–206.

Labitzke, K. and van Loon, H.: 1996, 'On the stratosphere, the QBO, and the sun', *Meteor. Z.* **5**, 166–169.

Labitzke, K. and van Loon, H.: 1997, 'The signal of the 11-yr sunspot cycle in the upper troposphere and lower stratosphere', *Space Sc. Rev.* **80**, 393–410.

Lean, J.: 2000, 'Evolution of the sun's spectral irradiance since the Maunder Minimum', *Geophys. Res. Lett.* **27**, 2425–2428.

McCormack, J.: 2003, ' The influence of the 11-year solar cycle on the quasi-biennial oscillation', *Geophys. Res. Lett.* **30**, doi:10.1029/2003GL018314.

Naito, Y. and Hirota, I.: 1997, 'Interannual variability of the northern winter stratosphere circulation related to the QBO and solar cycle', *J. Meteor. Soc. Japan* **75**, 925–937.

O'Sullivan, D. and Salby, M. L.: 1990, ' Coupling of the quasi-biennial oscillation and the extratropical circulation in the stratosphere through planetary wave transport', *J. Atmos. Sci.* **47**, 650.

Quiroz, R.: 1981, 'Periodic modulation of the stratospheric quasi-biennial oscillation', *Mon. Wea. Rev.* **109**, 665–674.

Salby, M., Callaghan, P., and Shea, D.: 1997, ' On the interdependence of the tropical and extratropical QBO and its relationship to the solar cycle', *J. Geophys. Res.* **102**, 29,789–29,798.

Salby, M. and Callaghan, P.: 2000, 'Connection between the solar cycle and the QBO: The missing link', *J. Climate* **13**, 2652–2662.

Salby, M. and Callaghan, P.: 2002, 'Interannual changes of the stratospheric circulation: Relationship to ozone and tropospheric structure', *J. Climate* **15**, 3673–3685.

Salby, M. and Callaghan, P.: 2004, 'Evidence of the solar cycle in the general circulation of the stratosphere', *J. Climate* **17**, 31–46.

Salby, M. and Callaghan, P.: 2005, 'Interaction between the Brewer-Dobson circulation and the Hadley circulation', *J. Climate* (in press).

Salby, M. and Callaghan, P: 2006, 'Relationship of the QBO to the stratospheric signature of the solar cycle', *J. Geophys. Res.* (in press).

Soukharev, B. and Hood, L.: 2001, 'Possible solar modulation of the quasi-biennial oscillation: Additional statistical evidence', *J. Geophys. Res.* **106**, 14,855–14,868.

Tung, K. and Yang, H.: 1994, 'Global QBO in circulation and ozone, II: A simple mechanistic model', *J. Atmos. Sci.* **51**, 2708.

van Loon, H. and Labitzke, K.: 1994, 'The 10–12 year atmospheric oscillation', *Meteor. Zeitschrift* **3**, 259–266.

van Loon, H. and Labitzke, K.: 1998, 'The global range of the stratospheric decadal wave. Part I: Its association with the sunspot cycle in summer and in the annual mean, and with the troposphere', *J. Clim.* **11**, 1529–1537.

van Loon, H. and Labitzke, K.: 1999, 'The signal of the 11-yr solar cycle in the global stratosphere', *J. Atmos. Solar-Terrest. Phys.* **61**, 53–61.

van Loon, H. and Labitzke, K.: 2000, 'The influence of the 11-yr solar cycle on the stratosphere below 30 km: A review', *Space Sci. Rev.* **94**, 259–278.

WMO, 1987: World Meteorological Organization, *Atmospheric Ozone: Assessment of Our Understanding of the Processes Controlling its Present Distribution and Change*, Report No. 16. Available from NASA, Office of Mission to Planet Earth, Two Independence Square, 300 E Street SW, Washington DC.

# MULTIDECADAL SIGNAL OF SOLAR VARIABILITY IN THE UPPER TROPOSPHERE DURING THE 20TH CENTURY

S. BRÖNNIMANN[1,*], T. EWEN[1], T. GRIESSER[1] and R. JENNE[2]

[1]*Institute for Atmospheric and Climate Science, ETH Zürich, Universitätstrasse 16, CH-8092 Zürich, Switzerland*

[2]*National Center for Atmospheric Research (NCAR), Boulder, CO, USA*

*(*Author for Correspondence, E-mail: stefan.broennimann@env.ethz.ch)*

(Received 31 August 2005; Accepted in final form 21 December 2005)

**Abstract.** Studies based on data from the past 25–45 years show that irradiance changes related to the 11-yr solar cycle affect the circulation of the upper troposphere in the subtropics and midlatitudes. The signal has been interpreted as a northward displacement of the subtropical jet and the Ferrel cell with increasing solar irradiance. In model studies on the 11-yr solar signal this could be related to a weakening and at the same time broadening of the Hadley circulation initiated by stratospheric ozone anomalies. Other studies, focusing on the direct thermal effect at the Earth's surface on multidecadal scales, suggest a strengthening of the Hadley circulation induced by an increased equator-to-pole temperature gradient. In this paper we analyse the solar signal in the upper troposphere since 1922, using statistical reconstructions based on historical upper-air data. This allows us to address the multidecadal variability of solar irradiance, which was supposedly large in the first part of the 20th century. Using a simple regression model we find a consistent signal on the 11-yr time scale which fits well with studies based on later data. We also find a significant multidecadal signal that is similar to the 11-yr signal, but somewhat stronger. We interpret this signal as a poleward shift of the subtropical jet and the Ferrel cell. Comparing the magnitude of the two signals could provide important information on the feedback mechanisms involved in the solar climate relationship with respect to the Hadley and Ferrel circulations. However, in view of the uncertainty in the solar irradiance reconstructions, such interpretations are not currently possible.

**Keywords:** solar variability, Hadley cell, Ferrel cell, multidecadal variability

## 1. Introduction

Solar irradiance changes have been found to influence global surface climate on various time scales (e.g., Bond *et al.*, 2001; Haigh, 2003; Lohmann *et al.*, 2004; see also Rind, 2002), including multidecadal variability. One of the key periods in this respect is the first half of the 20th century. According to reconstructions, total solar irradiance (TSI) increased by around 0.8 to 1.5 W/m$^2$ during this period (see Figure 1) and global surface air temperatures increased by about 0.5 °C, which is comparable in magnitude to the warming since the 1970s (Jones *et al.*, 1999). To what extent the warming was caused by solar variability is debated. Model simulations suggest that both anthropogenic and natural influences contributed (Schlesinger and Andronova, 2004; Stott *et al.*, 2001; Ingram, 2006), but

Space Science Reviews (2006) 125: 305–317
DOI: 10.1007/s11214-006-9065-2

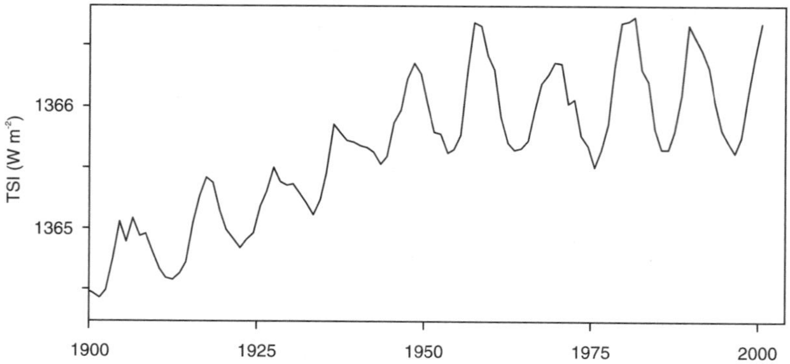


*Figure 1.* Reconstruction of solar radiation by Lean (2004).

the warming is still underestimated by many simulations and the question is not yet solved (e.g., Meehl *et al.*, 2003).

While the solar impact on climate is increasingly well accepted in the scientific community, the mechanisms that produce it are not. Some of the detected influences energetically do not match well with the supposed forcing, feedbacks have to be involved, and the dynamics of the whole ocean-atmosphere system needs to be considered (see Rind, 2002; Haigh, 2003). Stratospheric ozone seems to play an important role for the stratospheric signal, which then can be communicated to the troposphere, but the details of this mechanism are not clear (e.g., Balachandran *et al.*, 1999; Shindell *et al.*, 1999; see also contributions by Gray *et al.*, and Haigh and Blackburn in this issue). Solar variability has also been suggested to influence climate via a cloud feedback (Svensmark and Friis-Christensen, 1997; Tinsley and Yu, 2004), but again the mechanisms are debated.

One of the key questions relates to the response of the Hadley and Ferrel circulations. An 11-yr solar signal has been found in zonal mean geopotential height (GPH) and temperature at midlatitudes in the free troposphere at around 30°–60°N (e.g., Labitzke and van Loon, 1997; van Loon and Shea, 1999; Labitzke, 2003; Haigh, 2003; Gleisner and Thejll, 2003; Crooks and Gray, 2005), with higher temperatures and GPH accompanying higher irradiance. The signal has been interpreted as a northward displacement of the subtropical jet and the Ferrel cell. In a series of model studies on the 11-yr solar signal considering both irradiance changes and ozone changes (Haigh, 1999; Larkin *et al.*, 2000), this signal could be related to a weakening and at the same time broadening of the Hadley circulation initiated by stratospheric ozone anomalies. Other studies focusing on the direct thermal effect at the Earth's surface on multidecadal scales (using long integrations with models that do not include ozone changes), suggest a strengthening of the Hadley circulation due to a solar-induced change in tropical temperature gradients (Meehl *et al.*, 2003, 2004; van Loon *et al.*, 2004). According to this mechanism, the thermal solar signal is strongest in subtropical cloud free areas and affects the Hadley cell via moisture transport, precipitation, and latent heat release (and a feedback on subtropical cloud cover). This is consistent with observations that point to an increase in strength of the Hadley circulation during the past century, but the uncertainty in the data is very large (Evans and Kaplan, 2004).

It would be desirable to find a framework that allows for an overarching discussion of solar-climate effects on the Hadley and Ferrel circulations in the 20th century. However, many problems remain to be solved. This concerns not only the poorly known feedbacks of the climate system, but also the actual forcing mechanisms, which might be different at the two time scales (e.g., due to different spectral characteristics of the irradiance change).

One important gap is that the studies on the solar signal in the upper troposphere are based on the last two to four solar cycles only, which were characterised by a large amplitude of the 11-yr cycle, but no or little low-frequency variability (Figure 1). It would thus be very interesting to repeat the studies for the first part of the 20th century, when TSI (according to Lean, 2004) increased substantially. Up to now, data from the upper troposphere that would be needed for such a study have not yet been available. During the past few years we have compiled, digitised, and re-evaluated a large amount of global upper-air data reaching back to around 1920. The data originate from radiosonde, pilot balloon, aircraft, and kite ascents performed in many parts of the world. These data are used in a statistical reconstruction approach to derive zonal mean GPH and temperature up to the 200 hPa level. We focus on the latitude band 30°–60°N, where we expect the signal to be strongest. Then we extract the solar signal from the zonal mean series using a simple regression approach and try to answer the following question: Is there a multidecadal signal in the upper troposphere during the 20th century, and if so, how does it compare with the 11-yr signal?

## 2. Data and Methods

We used monthly historical upper-level data that have been re-evaluated in recent years (Brönnimann *et al.*, 2005b). These data sets include the UA39_44 upper-air data for the extratropical northern hemisphere, 1939–1944 (Brönnimann, 2003), with recent updates (Version 1.1, S_DCRM), consisting of radiosonde and aircraft observations of temperature and GPH on pressure levels. In addition, earlier U.S. data from kites and aircraft were taken from the *Monthly Weather Review*, 1922–1938, consisting mainly of temperature on geometric altitude levels up to 5 km from kites and aircraft. Radiosonde data on pressure levels from the IGRA data set (1945–1947, Durre *et al.*, 2005) and from the Finnish station Ilmala, 1942–1947, were also used (Brönnimann *et al.*, 2005a) as well as upper level wind (on geometric altitude levels) from selected stations, 1922–1947 (NCAR, TD52 an TD53 data sets, see *http://dss.ucar.edu/docs/papers-scanned/papers.html*, documents RJ0167, RJ0168).

In total, 253 upper-air stations were used (Figure 2). Note, however, that at each station the record covers only part of the period. The bottom part of Figure 2 shows the number of available predictor values per month during the historical period (1922–1947). The data availability is best for the War time, but some upper-air data are available back to 1922. The data clearly do not suffice to directly calculate reliable zonal averages for different levels, which is also made impossible by the different vertical coordinates (pressure, altitude). Hence, we used a statistical reconstruction approach to obtain the zonal means.

The upper-level data were supplemented with data from the Earth's surface, which also contribute information to the statistical reconstruction approach. For this purpose we used a set of 387 surface air temperature records from the NASA-GISS database (Hansen *et al.*, 1999). The stations are chosen to cover the northern extratropics. If possible, mountain sites were chosen (see Figure 2, top). In addition, we used gridded sea-level pressure (SLP) (HadSLP2, Allan and Ansell, 2005) and sea-surface tempertaure (SST) (ERSST V2, Smith and Reynolds, 2004) data. These data were included in the form of latitude-weighted principal component (PC) time series of the monthly anomaly fields north of 20°N (calculated for 1900–2003.) For SLP and SST, we retained 20 and 32 PCs, respectively, corresponding to approximately 95 and 90% of the total variance.

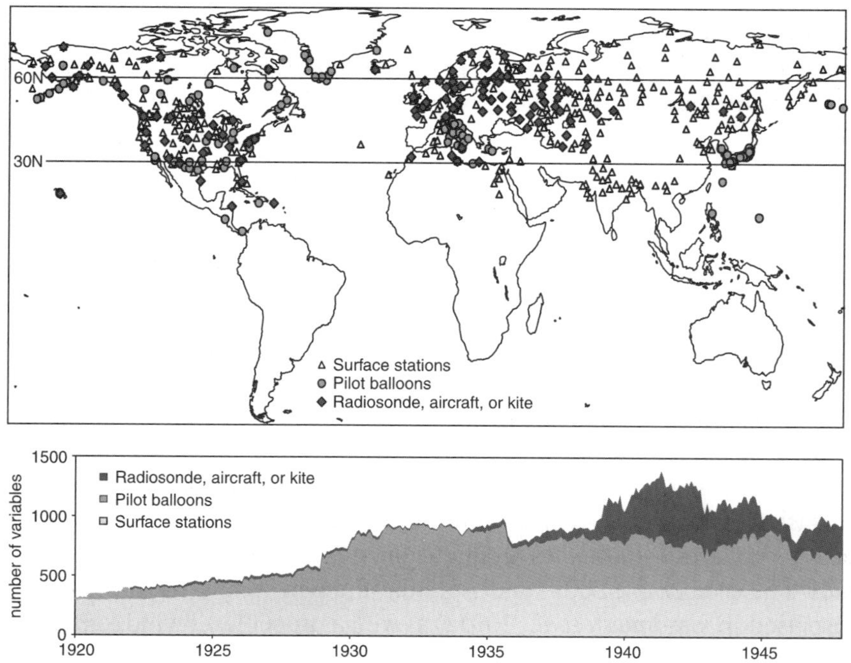


*Figure 2. Top:* Map of upper-air and surface stations used for the reconstruction. *Bottom:* Number of available variables for the reconstruction model as a function of time.

In order to reconstruct zonal mean temperature and geopotential height from the historical upper-level station data, statistical models need to be calibrated in a period for which both the station data and the zonal means are available and have no or only few gaps. We used the 1948–2003 period, where the zonal means can be calculated from NCEP/NCAR reanalysis data (Kistler *et al.*, 2001). In many cases, upper-level station data are not available for the same locations, or they have long gaps. Therefore, we decided to use NCEP/NCAR reanalysis data, interpolated to the station location and levels, to supplement the historical upper-level data in all cases. However, reanalysis data are a model product which is somewhat different from observations. Additional errors arise from the fact that historical data have a lower quality than more recent observations and from the interpolation procedure. To account for these errors, we randomly perturbed the supplemented predictors by a normally distributed noise. The error (standard deviation of the noise) was estimated based on our quality assessment (see references above), which itself was based on NCEP/NCAR reanalysis as a reference.

Gaps in the surface station data are short. They were filled with corresponding NCEP/NCAR reanalysis data at 925 hPa after the standardising procedure (see below). The PC time series for the SLP and SST fields have no gaps. In this way, a calibration data set can be constructed that has no missing values in the 1948–2003 period. All predictor data series were then expressed as anomalies from the 1961–1990 mean annual cycle and were standardised.

The statistical model used is a simple principal component regression. Due to missing values prior to 1948, each month in the historical period has a different combination of available variables and therefore reconstructions were performed independently for each month from 1922 to 1947. A three-calendar-months moving window was used for calibration, e.g., to form a reconstruction model for January 1922 we used all Decembers, Januaries, and Februaries in the calibration period. First, for each time step the available variables were weighted as follows. We attributed 25% of the weight to the surface and the rest to the upper-level data. Within the surface data, half of the weight was attributed to the SLP PCs and one fourth to station temperature and one fourth to the SST PCs. Within the SLP PCs and SST PCs, the weight was split according to their correlation with SLP or surface (land and marine) air temperature averaged from 30°–60°N, respectively. Within the upper-level data, half of the weight was attributed to the pilot balloon data, half of the weight to the upper-level temperature or geopotential height data. This weighting scheme gave much better results than using no weighting at all.

After the weighting, a principal component analysis was performed on all available predictor variables in the calibration period in order to reduce the number of variables. Thereby the amount of variance retained was varied between 70 and 98% and the optimum fraction (according to the validation experiments, see below) was chosen for each month.

The reconstructions were validated using two split-sample validation experiments, i.e., the calibration period was split and one part was used for calibrating the model, the other for validating the model. In the first experiment we used 1948–1966 for validation, in the second 1985–2003. Out of the many different reconstruction models (varying amount of retained variance), we chose for each time step (one month) the one that had the highest reduction of error (RE, see Cook *et al.*, 1994), averaged from both validation experiments. Values of RE can be between $-\infty$ and 1 (perfect reconstructions). RE $> 0$ determined in an independent period is normally considered an indication that the model has predictive skill.

## 3. Results

### 3.1. Reconstructions

Figure 3 shows the reconstructed 300 hPa GPH anomalies as an example. The reconstructions capture variability from the month-to-month to the interannual scale. The anomalously high value in January 1932 is robust with respect to the amount of variance retained and the weighting used. The strongly negative values in 1940–1941 coincide with a global climate anomaly related to an El Niño event (Brönnimann *et al.*, 2004). There is a slight positive bias of 3.8 gpm which, however, does not vary much with the amount of retained variance and only slightly with season. The RE statistics (lower panel) shows a clear annual cycle. Reconstructions are good in late winter, whereas they are less reliable in May. Also, the RE changes with time due to the change in the predictor variables. During the War, a large amount of upper-air data is available and the reconstructions are much better. Over all, the reconstructions can be considered as relatively good, with RE values mostly in the

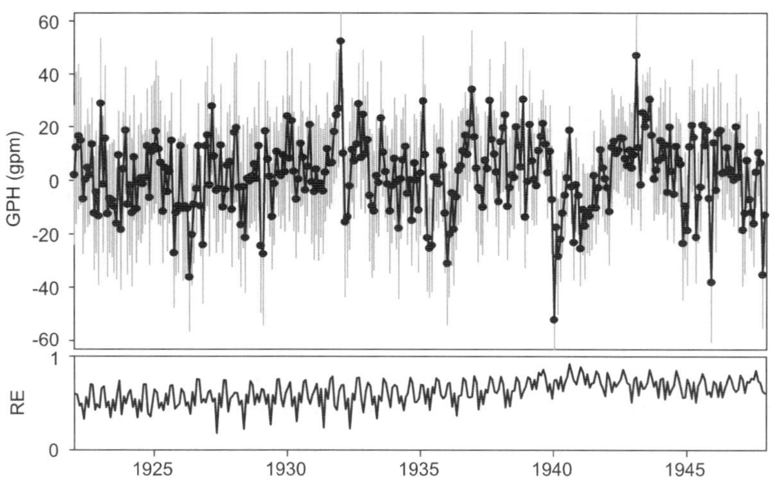


*Figure 3. Top:* Reconstructed monthly 300 hPa GPH from 1922 to 1947 with 95% confidence intervals calculated from the two split-sample validation experiments. *Bottom:* RE from the two split-sample validation experiments.

range of 0.6 to 0.8 (higher than 0.5 in 83% of the cases). This is sufficient for the following analyses. To account for the possible bias, the regression analyses presented in the next section includes a step function in 1948.

### 3.2. Regression Model

The reconstructed zonal means were merged with NCEP/NCAR data after 1948, yielding 82-year long time series. All analyses were performed for both temperature and GPH, for summer (Jun – Aug) means and for the annual means. In the following we focus on annual mean values of GPH for which we expect a more robust signal and better reconstruction quality (results were similar for the other analyses). The time series were related to solar variability in a regression approach, similar as in Haigh (2003), Crooks and Gray (2005), and other studies. As a measure for solar variability we used the TSI reconstructions by Lean (2004), shown in Figure 1, which we updated to 2003 with satellite data from PMOD/WRC (Davos, Switzerland). In addition, we accounted for stratospheric aerosols in the Northern Hemisphere (Sato *et al.*, 1993) as well as the El Niño index NINO3.4 (assuming that El Niño/Southern Oscillation is independent from the solar signal). Both variables were used with a 6 month lead time, which is considered a typical response time of extratropical circulation to these forcings (results are not sensitive to the choice of the lead). Other studies (e.g., Crooks and Gray, 2005) also include the North Atlantic Oscillation (NAO) or other circulation variability modes as explanatory variables. However, several authors (e.g., Shindell *et al.*, 2001) suggest that the mechanism linking solar variability to tropospheric circulation might include the NAO or Northern Annular Mode. Therefore, we repeated all analyses with and without using the NAO (SLP difference between 37.73°N, 23.7°W and 64.13°N, 21.9°W from HadSLP2 data) and PNA (reconstructed exactly as described above using a subset of the data, manuscript in preparation) as explanatory variables in our analysis.

In a first step we focused on the 11-yr time scale and subtracted from each variable the low frequency variability in the form of a 4th order polynomial function in time. We then performed a multi-linear regression in three time periods: 1922–1957, 1958–2003, and 1922–2003. The coefficients are shown in Figure 4 (top, note that for display purposes, 90% confidence intervals are used). The coefficients for the 1958–2003 agree well with previous studies. No signal at all is found at the surface. With increasing altitude, the signal becomes strongly positive and highly significant. In the 1922–1957 period, confidence intervals are much larger, especially in the upper troposphere. This is expected to some extent. The signal (amplitude of the 11-yr cycle) is only about half as strong in the earlier period. At the same time the uncertainty of the reconstructions, especially at higher levels, also contributes. When analysing the whole period (1922–2003, Figure 4 top left), we find again significantly positive coefficients. The coefficients do not change much

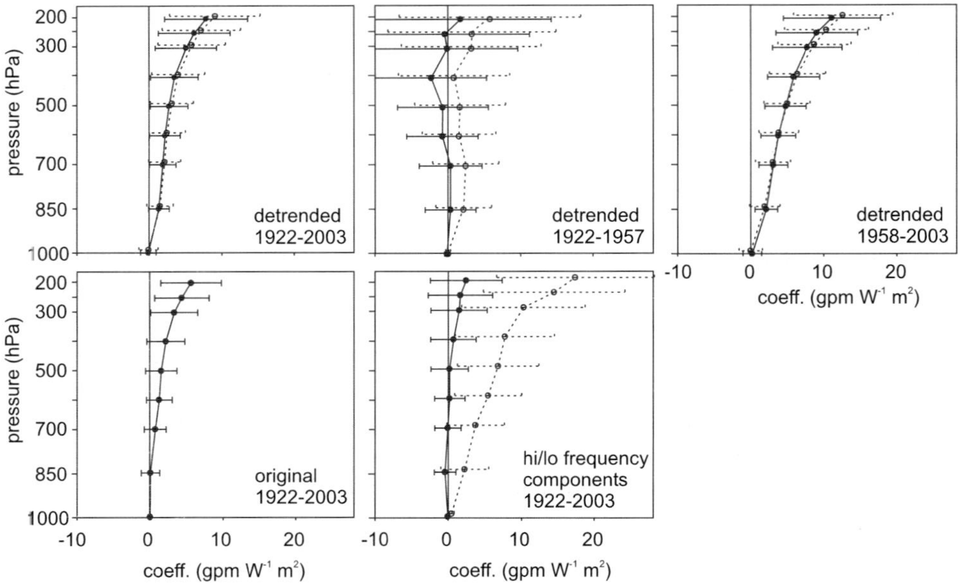


*Figure 4. Top:* Coefficient (90% confidence intervals) for TSI from multiple regression models of the annual mean GPH (30°–60°N average) using detrended data for 1922–2003 (left), 1922–1957 (middle) and 1958–2003 (right) with only TSI, NINO3.4 and volcanic aerosols as explanatory variables (dashed lines), or TSI, NINO3.4, volcanic aerosols, NAO, and PNA (solid lines). *Bottom:* Same for models which are based on the non-detrended data (1922–2003), but which include further explanatory variables in addition to TSI, NINO3.4, volcanic aerosols, NAO and PNA (left). In the middle figure TSI is split into an 11-yr (solid) and a background component (dashed).

when including NAO and PNA as explanatory variables. The confidence intervals decrease slightly. However, the overall model statistics (not shown) improve substantially. In all, we have no statistical evidence to assume that the relation between solar variability on the 11-yr scale and atmospheric circulation is non-stationary, but the signal is clearly less obvious in the earlier period.

The first part of the record is mainly interesting with respect to multidecadal variability. Therefore, in a second analysis we used the whole period but did not detrend the data. Instead, we included a number of other terms in the regression model that could possibly explain low-frequency variability in our target variables. In addition to the volcanic forcing and El Niño used above, we included greenhouse gas and tropospheric aerosol forcing from Crowley (2000). The two series were updated to 2003 using a linear relation to observed annual mean $CO_2$ concentrations (provided by CDIAC via their website *http://cdiac.esd.ornl.gov/*) or a second order polynomial in time, respectively, calibrated in the 1959–1998 period. In addition, we introduced two step functions to account for possible inhomogeneities in the data: in 1948 (merging of reconstructions with NCEP/NCAR reanalysis) and in 1979 (main start of satellite data used in NCEP/NCAR). Again, including NAO and PNA has a relatively small effect on the results, but general model statistics improve greatly, with explained variances between 54 and 68% (this is surprisingly high given the fact that a large part of the record is reconstructed). Figure 4 (bottom left) shows the coefficients found for TSI in a model that includes NAO/PNA. They are similar, though somewhat weaker, to those found on the 11-yr time scale in the top left panel. To further analyse the effect of the time scale, we used the same regression model, but additively split TSI into an 11-year and a low-frequency component (as given in Lean, 2004). Coefficients for the two components are given in Figure 4 (bottom right). For the 11-yr component we expect similar coefficients as in the other analyses, but find smaller, insignificant coefficients. On the other hand, the coefficients for the multidecadal scale are larger, and they are significantly different from zero at higher levels.

The different sets of coefficients are just within each others 90% confidence intervals. This means that from our analysis we have some weak indications, but no firm evidence to assume that the solar signal in the upper troposheric circulation is different in strength on the 11-yr scale than on the multidecadal scale.

## 4. Discussion

Up to now the effect of solar variability on upper tropospheric GPH was only studied effectively for the 11-yr cycle. Our analysis reveals a significant solar signal also on the multidecadal scale which is similar, though somewhat stronger, than the signal found at the 11-yr scale. Based on the similarity of the vertical profile of the response (considering both temperature and GPH in the summer and the annual analyses) with the results from model studies such as Haigh (1999, 2003) and Larkin *et al.* (2000), we interpret this signal as a northward displacement of the subtropical jet and the Ferrel cell. The relation to the strength of the Hadley cell remains open. (Note, however, that when using zonal means the distinction between the polar extreme of the Hadley cell and the equatorial extreme of the Ferrell cell is necessarily unsharp.)

Different mechanisms of how solar variability affects the Hadley and Ferrel cells have been suggested. In a simplifying manner we can refer to the Haigh (1999) and Meehl *et al.* (2004) mechanisms as an influence on the Hadley circulation "from above" or "from below", respectively. Changes in the Hadley circulation have also been studied by other authors. Rind and Perlwitz (2004), using a set of model simulations for very different conditions, show that the strength of the Hadley circulation is strongly related to the equator-to-pole gradient in sea surface temperature. An increase in strength with increasing TSI is consistent with the observed thermal effect at the Earth's surface on the 11-yr scale, although this effect is relatively weak (White *et al.*, 1998; North *et al.*, 2004). However, Rind and Perlwitz (2004) also find that the extent of the Hadley circulation is not related to this gradient. The two interpretations therefore do not contradict each other as much as might seem at first sight. For instance, Gleisner and Thejll (2003) find no change in strength of the Hadley circulation with solar variability (but a longitudinally heterogeneous pattern) in their analysis of NCEP/NCAR reanalysis data since 1958. They do find, however, indications for a broadening of the Hadley cell when centred around the ITCZ. It is this broadening to which our upper-tropospheric signal might be related, and not necessarily the strength of the Hadley circulation. To resolve this issue, it would be desirable for further studies to address also the longitudinal and seasonal variations in upper-tropospheric GPH as well as more specific variables such as vertical velocity, upper-level divergence, latent heat flux, and eddy heat and momentum fluxes. Gridded 3-dimensional data to perform such analyses for the entire 20th century will hopefully be available in the future (Brönnimann *et al.*, 2005b). Also, the relation with the Quasi-Biennial Oscillation should be addressed (Balachandran *et al.*, 1999; Labitzke, 2003).

Apart from the finding that there exists a multidecadal solar signal in the upper troposphere, the question remains to be discussed whether it is larger than for the 11-yr cycle and, consequently, whether there are feedback or damping mechanisms. The influence "from above" is supposed to be primarily a direct effect, initiated by stratospheric ozone anomalies. The response time is assumed to be much shorter than the averaging interval of 1 year (in our analysis we find smaller coefficients for lags of one or two years; not shown) and so we expect, to a first approximation, similar coefficients at different time scales. On the other hand, the thermal signal at the Earth's surface is damped by the global oceans (with lag times of 0–2 years, see White *et al.*, 1998; North *et al.*, 2004). Therefore, one would expect for the influence "from below" a slight damping at the 11-yr scale and a relatively stronger signal at multidecadal scales, which might be additionally amplified by interaction with the effect of increasing greenhouse gases (Meehl *et al.*, 2003, 2004). This leads to a strengthening, but not necessarily to a broadening of the Hadley circulation. Hence, another feedback mechanism must be found for the strengthening of the solar signal in the upper-troposphere (if real) at low frequencies. Note, however, that the coefficients for the background signal are very sensitive to the TSI reconstructions, which are uncertain. If the Lean (2004) reconstructions are underestimating the low-frequency variability, the coefficients for the background signal are overestimated by the same factor and no feedback is required. If Lean (2004) overestimates the low-frequency variability, as is suggested, e.g., by Wang *et al.* (2005), the coefficients for the background might easily become significantly larger than on the 11-yr scale and a feedback mechanism must be involved. Without better TSI reconstructions, no firm conclusion can be made.

## 5. Conclusions

Our analysis of the relation between solar irradiance variability and zonal mean GPH at midlatitudes during the past 82 years reveals an 11-yr signal (increasing GPH with increasing solar variability) that is consistent with previous studies based on much shorter periods. We find a quantitatively similar (statistically significant) signal also for the multidecadal variations, which we interpret as a poleward displacement of the subtropical jet and the Ferrel cell. The relation to the Hadley circulation remains unclear. Resolving this problem would be an important step towards understanding the mechanisms of solar-climate links. However, at the current time interpretations are limited by the availability and quality of reconstructed upper-level circulation and, perhaps more importantly, the large uncertainty in TSI reconstructions.

## Acknowledgements

Funding by the Swiss National Science Foundation is gratefully acknowledged. We would like to thank the Climatic Research Unit (Norwich, UK), the Hadley Centre of the Met Office (Exeter, UK), NOAA-CIRES/CDC (Boulder, USA), and NASA/GISS (New York, USA) for providing data for this study.

## References

Allan, R. J. and Ansell, T. J.: 2006, 'A new globally complete monthly historical gridded mean sea level pressure data set (HadSLP2): 1850–2004', *J. Climate*, in press.

Balachandran, N. K., Rind, D., Lonergan, P., and Shindell, D. T.: 1999, 'Effects of solar cycle variability on the lower stratosphere and the troposphere', *J. Geophys. Res.* **104**, 27,321–27,339.

Bond, G., Kromer, B., Beer, J., Muscheler, R., Evans, M. N., Showers, W., Hoffmann S., Lotti-Bond R., Hajdas, I., and Bonani, G.: 2001, 'Persistent solar influence on north Atlantic climate during the Holocene', *Science* **294**, 2130–2136.

Brönnimann, S.: 2003, 'A historical upper-air data set for the 1939–1944 period', *Int. J. Climatol.* **23**, 769–791.

Brönnimann, S., Luterbacher, J., Staehelin, J., Svendby, T. M., Svenoe, T., and Hansen, G.: 2004, 'Extreme climate of the global troposphere and stratosphere in 1940–42 related to El Niño', *Nature* **431**, 971–974.

Brönimann, S., Mohr, C., Ewen, T., and Grant, A.: 2005a, 'The Finnish radiosonde records from Ilmala, Rovaniemi, and Äänislinna, 1942–1947', Technical Report, ETH Zürich, 13 pp.

Brönimann, S., Compo, G. P., Sardeshmukh, P. D., Jenne, R., and Sterin, A.: 2005b, 'New approaches for extending the 20th century climate record', *Eos* **86**, 2–7.

Cook, E. R., Briffa, K. R., and Jones, P. D.: 1994, 'Spatial regression methods in dendroclimatology – a review and comparison of two techniques', *Int. J. Climatol.* **14**, 379–402.

Crooks, S. A. and Gray, L. J.: 2005, 'Characterization of the 11-year solar signal using a multiple regression analysis of the ERA-40 dataset'. *J. Clim.* **18**, 996–1015.

Crowley, T. J.: 2000, 'Causes of climate change over the past 1000 years', *IGBP PAGES/World Data Center for Paleoclimatology Data Contribution Series*, 2000–045, NOAA/NGDC Paleoclimatology Program, Boulder CO, USA.

Durre, I., Vose, R. S., and Wurtz, D. B.: 2005, 'Overview of the Integrated Global Radiosonde Archive', *J. Clim.* **10**, 53–68.

Evans, M. N. and Kaplan, A.: 2004, 'The Pacific sector Hadley and Walker Circulations in historical marine wind analyses', in: H. Diaz and R. Bradley (eds.) *The Hadley Circulation: Past, Present, and Future*, Advances in Global Change Research 21, Kluwer Academic Publishers, pp. 239–258.

Gleisner, H. and Thejll, P.: 2003, 'Patterns of tropospheric response to solar variabiity', *Geophys. Res. Lett.* **30**, 1711, doi:10-1029/2003GL017129.

Gray, L. J., Crooks, S. A., Palmer, M. A., Pascoe, C. L., and Sparrow, S.: 2006, 'A possible transfer mechanism for the 11-year solar cycle to the lower stratosphere', Space Sci. Rev., this volume, doi:10.1007/s11214-006-9069-y.

Haigh, J. D.: 1999, 'A GCM study of climate change in response to the 11-year solar cycle', *Q. J. Roy. Meteorol. Soc.* **125**, 871–892.

Haigh, J. D.: 2003, 'The effects of solar variability on climate', *Phil. Trans. R. Soc. Lond. A*, **361**, 95–111.

Haigh, J. D. and Blackburn, M.: 2006, 'Solar influences on dynamical coupling between the stratosphere and troposphere', Space Sci. Rev., this volume, doi:10.1007/s11214-006-9067-0.

Hansen, J., Ruedy, R., Glascoe, J., and Sato, M.: 1999, 'GISS analysis of surface temperature change', *J. Geophys. Res.* **104**, 30,997–31,022.

Ingram, W. J.: 2006, 'Detection and attribution of climate change, and understanding solar influence on climate', *Space Sci. Rev.*, this volume, doi:10.1007/s11214-006-9057-2.

Jones, P. D., New, M., Parker D. E., Martin, S., and Rigor, I. G.: 1999, 'Surface air temperature and its changes over the past 150 years', *Rev. Geophys.* **37**, 173–199.

Kistler, R., *et al.*: 2001, 'The NCEP-NCAR 50-year reanalysis: Monthly means CD-ROM and documentation', *B. Am. Meteorol. Soc.* **82**, 247–267.

Labitzke, K.: 2003, 'The global signal of the 11-year sunspot cycle in the atmosphere: When do we need the QBO?', *Met. Z.* **12**, 209–216.

Labitzke, K. and van Loon, H.: 1997, 'The signal of the 11-year sunspot cycle in the upper troposphere-lower stratosphere', *Space Sc. Rev.* **80**, 393–410.

Larkin, A., Haigh, J. D., and Djavidnia, S.: 2000, 'The effect of solar UV irradiance variations on the Earth's atmosphere', *Space Sci. Rev.* **94**, 199–214.

Lean, J.: 2004, 'Solar Irradiance Reconstruction', *IGBP PAGES/World Data Center for Paleoclimatology Data Contribution Series*, 2004–035, NOAA/NGDC Paleoclimatology Program, Boulder CO, USA.

Lohmann, G., Rimbu, N., and Dima, M.: 2004, 'Climate signature of solar irradiance variations: Analysis of long-term instrumental, historical, and proxy data', *Int. J. Clim.* **24**, 1045–1056.

Meehl, G. A., Washington, W. M., Wigley, T. M. L., Arblaster, J. M., and Dai, A.: 2003, 'Solar and greenhouse gas forcing and climate response in the twentieth century', *J. Clim.* **16**, 426–444.

Meehl, G. A., Washington, W. M., Wigley, T. M. L., Arblaster, J. M., and Dai, A.: 2004, 'Mechanisms of an intensified Hadley circulation in response to solar forcing in the 20th century', in: H. Diaz and R. Bradley (eds.), *The Hadley Circulation: Past, Present, and Future*, Advances in Global Change Research 21, Kluwer Academic Publishers.

North, G. R., Wu, Q., and Stevens, J.: 2004, 'Detecting the 11-year solar cycle in the surface temperature field', in: J. M. Pap and P. Fox (eds.), *Solar Variability and its Effects on Climate*, Geophysical Monograph 141, American Geophysical Union, pp. 251–260.

Rind, D.: 2002, 'Climatology –The sun's role in climate variations', *Science* **296**, 673–677.

Rind, D., and Perlwitz, J.: 2004, 'The response of the Hadley circulation to climate changes, past and future', In: H. Diaz and R. Bradley (eds.), *The Hadley Circulation: Past, Present, and Future*, Advances in Global Change Research 21, Kluwer Academic Publishers, pp. 399–435.

Sato, M., Hansen, J., McCormick, M. P., and Pollack, J. B.: 1993, 'Stratospheric aerosol optical depths, 1850–1990', *J. Geophys. Res.* **98**, 22,987–22,994.

Schlesinger, M. A. and Andronova, N. G.: 2004, 'Has the sun changed climate? Modelling the effect of solar variability on climate', In: J. M. Pap and P. Fox (eds.), *Solar Variability and its Effects on Climate*, Geophysical Monograph 141, American Geophysical Union, pp. 262–282.

Shindell, D. T., Rind, D., Balachandran, N., Lean, J., and Lonergan, P.: 1999, 'Solar cycle variability, ozone and climate', *Science* **284**, 305–308.

Shindell, D. T., Schmidt, G. A., Mann, M. E., Rind, D., and Waple, A.: 2001, 'Solar forcing of regional climate change during the Maunder Minimum', *Science* **294**, 2149–2152.

Smith, T. M. and Reynolds, R. W.: 2004, 'Improved extended reconstruction of SST (1854–1997)', *J. Clim.* **17**, 2466–2477.

Stott, P. A., Tett, S. F. B., Jones, G. S., Allen, M. R., Ingram, W. J., and Mitchell, J. F. B.: 2001, 'Attribution of twentieth century temperature change to natural and anthropogenic causes', *Clim. Dyn.* **17**, 1–22.

Svensmark, H. and Friis-Christensen, E.: 1997, 'Variation of cosmic ray flux and global cloud coverage – a missing link in solar-climate relationships', *J. Atmos. Sol.-Terr. Phys.* **59**, 1225–1232.

Tinsley, B. A. and Yu, F.: 2004, 'Atmospheric ionization and clouds as links between solar activity and climate', in: J. M. Pap and P. Fox (eds.), *Solar Variability and its Effects on Climate*, Geophysical Monograph 141, American Geophysical Union, pp. 321–339.

van Loon, H., and Shea, D. J.: 1999, 'A probable signal of the 11-yr solar cycle in the troposphere of the northern hemisphere', *Geophys. Res. Lett.* **16**, 2893–2896.

van Loon, H., Meehl, G. A., and Arblaster, J. M.: 2004, 'A decadal solar effect in the tropics in July–August', *J. Atmos. Sol.-Terr. Phys.* **66**, 1767–1778.

Wang, Y. M., Lean, J. L., and Sheeley, N. R.: 2005, 'Modeling the sun's magnetic field and irradiance since 1713', *Astrophys. J.* **625**, 522–538.

White, W. B., Cayan, D. R., and Lean, J.: 1998, 'Global upper ocean heat storage response to radiative forcing from changing solar irradiance and increasing greenhouse gas/aerosol concentrations', *J. Geophys. Res.* **103**, 21,355–21,366.

# THE ROLE OF DYNAMICS IN SOLAR FORCING

K. KODERA
*Meteorological Research Institute, Nagamine 1-1, Tsukuba, Japan*
*(E-mail: kodera@mri-jma.go.jp)*

(Received 31 August 2005; Accepted in final form 24 December 2005)

**Abstract.** This paper reviews the solar influence on climate through stratospheric dynamical processes. There are two possible processes, both being a consequence of the wave-mean flow interaction in the upper stratosphere. One involves changes in the vertical propagation of planetary waves and the resultant tropospheric circulation change in the extratropics of the winter hemisphere. The other involves change in the global meridional circulation in the stratosphere and associated convective activity change in the tropics. These processes have been discussed on an 11-year solar cycle, but they are also applicable for centennial-scale solar influence on climate.

**Keywords:** solar influence, climate change

## 1. Introduction

The solar influence on climate has been a subject of debate for a long time (e.g., Pittock, 1978). The simplest idea is that the Earth's surface temperature becomes higher in response to increased solar energy, or the total solar irradiance (TSI). However, no global structure related to the 11-year solar cycle has been found in the surface temperature. Northern Hemisphere (NH) land surface air temperatures exhibit an 11-year cycle only in the center of the Eurasian continent (Figure 1a). Extratropical sea-surface temperature (SST) is correlated with the 11-year solar cycle only in the western North Pacific (Figure 1b). The global SST of the tropical ocean has a variation in phase with the 11-year solar cycle (White *et al.*, 1997). However, recent analysis (White *et al.*, 2003) suggests that the ocean is not radiatively heated, but that the variation is mainly due to the variation of heat flux (sensible + latent) between the atmosphere and ocean.

This does not necessarily mean that the 11-year cycle signal in the Earth's surface temperature is not of solar origin, but it may not be direct radiative forcing of the Earth's surface. In fact, the observed TSI change associated with the 11-year solar cycle is too small, only around 0.1% (Fröhlich, 2006), to produce meaningful climate impact. However, the ultraviolet (UV) range of solar irradiance, which is responsible for the stratospheric ozone heating, possesses more variability (2 to 8%) during a solar cycle (Rottman, 2006), and the air temperature of the stratopause region varies about 1 to 2 K from solar minimum (min) to solar maximum (max) (Chanin, 2006; Gray, 2006; Labitzke, 2006). The solar irradiance

Space Science Reviews (2006) 125: 319–330
DOI: 10.1007/s11214-006-9066-1

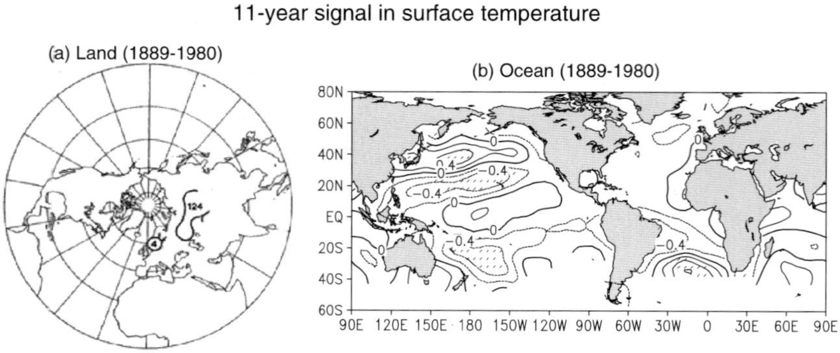


*Figure 1.* (a) 11-year cycle in land-surface temperature for 1899 to 1980. Contours indicate the area where the 11-year spectral peak is significant in seasonal temperature. 1. Annual 2. Winter 3. Spring 4. Summer 5. Autumn. From Kelly and Jones (1983). (b) Correlation coefficients between the solar irradiance and SST at each grid point for 1866 to 1991. Data were detrended and band-passed (9 to 15 years) prior to the calculation. From Lohmann *et al.* (2004).

variation and its direct chemical and radiative influence in the middle atmosphere is addressed elsewhere by other papers. Here, we concentrate on the problem of dynamical transmission of the solar signal from the upper stratosphere to the Earth's surface.

This paper is organized as follows. Section 2 outlines a mechanism for how the solar signal created in the stratopause region can be transmitted to the lower atmosphere. Sections 3 and 4 illustrate how this mechanism works using examples of the solar influence on the surface air temperature in the NH winter, and the solar influence on the monsoon and El Niño Southern Oscillation (ENSO). Section 5 provides the discussion and concluding remarks.

## 2. Downward Transmission of the Solar Signal

Because the density of air in the upper stratosphere is only a few percent of that at the surface, the process of transmitting the solar signal to the lower atmosphere implicitly includes an amplification mechanism of the solar signal. This has been demonstrated through numerical simulations of general circulation models extending up to the lower mesosphere (Shindell *et al.*, 2001; Matthes *et al.*, 2004). Figure 2 schematically illustrates a possible mechanism for the transmission of the solar signal from the upper stratosphere to the troposphere, proposed by Kodera and Kuroda (2002). The temperature in the equatorial stratopause region increases during the solar max. Accordingly, the temperature gradient between the equatorial and polar regions increases in the winter hemisphere during solar max. The subtropical lower mesosphere-stratopause jet becomes stronger and balances the increased meridional

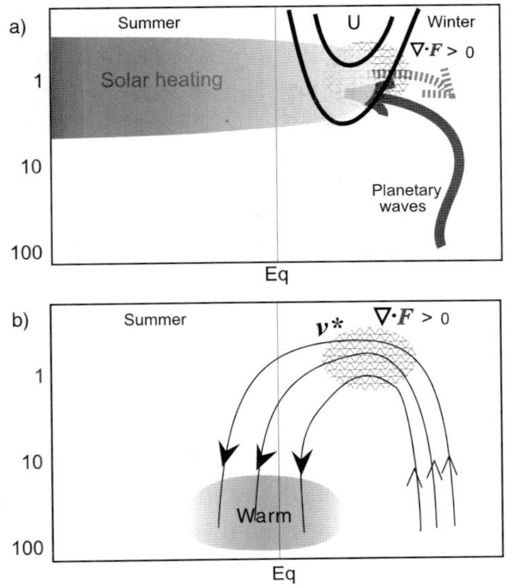


*Figure 2.* Schematic illustration of the solar influence on the lower stratosphere. (a) A stronger subtropical jet due to increased solar forcing during the maximum phase deflects planetary waves from the subtropics (dashed arrow), which creates anomalous divergence of the E-P flux (stippled). (b) A decrease in wave forcing results in reduced BD circulation (arrows) and warming in the tropical lower stratosphere (shading). From Kodera and Kuroda (2002).

temperature gradient. This initial solar effect on the subtropical jet develops by interacting with planetary waves propagating from the winter troposphere. A stronger jet deflects planetary waves. Less wave forcing, or larger Eliassen-Palm (E-P) flux divergence, makes the westerly jet even stronger. The zone of anomalous E-P flux divergence is situated poleward and downward of the center of the jet core, so that anomalous winds gradually shift poleward and downward, similar to the polar night jet oscillation (PJO), an internal mode of variation of the winter stratosphere (Kuroda and Kodera, 2001).

This process applies to the extratropics of the winter hemisphere. In addition, another downward extension of the solar influence is also possible through changes in meridional circulation. The Brewer-Dobson (BD) circulation is driven by extratropical wave forcing in the winter hemisphere (e.g., Garcia, 1987; Holton *et al.*, 1995). Thus, less wave forcing in the subtropical upper stratosphere-stratopause region makes the BD circulation weaker during solar max. Weaker BD circulation means weaker upwelling in the equatorial summer hemisphere; hence, less cooling and higher temperature in the equatorial lower stratosphere.

Thus, solar influence induced in the upper stratosphere-stratopause region can be transmitted to the lower atmosphere by two processes: one in the extratropics of the winter hemisphere, and the other in the tropics of the summer hemisphere. Because the wave-mean flow interaction is stronger in the NH, the westerly jet becomes weaker and rapidly shifts to higher latitudes in early winter. Therefore, in the NH winter, connection by change in propagation of planetary waves is more prominent. Conversely, in the Southern Hemisphere (SH), the stratopause jet is stronger and stays longer in the subtropics during the winter. To influence the equatorial vertical velocity, wave forcing should be applied sufficiently near the equatorial region (Plumb and Eluszkiewic, 1999), so that the connection through change in meridional circulation works more efficiently during the SH winter, i.e., in the NH summer.

## 3. Extratropical Winter

This Section demonstrates how the spatial structure of the extratropical solar signal, such as in Figure 1, is explained from a dynamical perspective. According to the mechanism described above, the solar signal manifests as a change in the zonal-mean flow and planetary waves in the NH winter.

Figure 3a depicts the correlation coefficient between the solar index (10.7 cm radio flux, F10.7) and December mean zonal-mean zonal wind (contour) and E-P flux (arrow) based on the NCAR reanalysis data for 1968 to 1998. The solar signal in zonal wind first appears in November at the top level of the analysis (10 hPa) and extends poleward and downward in December. A stronger westerly jet and a suppression of upward propagation of waves are found in the stratosphere (Fig-

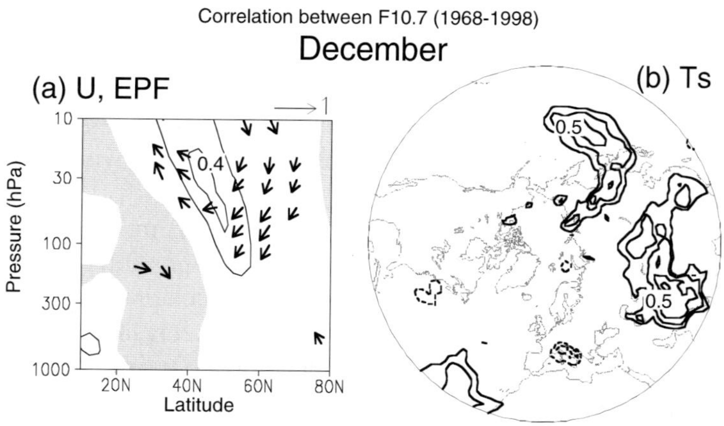


*Figure 3.* Correlation coefficient between winter mean F10.7 and December mean meteorological variables for 1968 to 1998. (a) Correlation with zonal-mean zonal wind (contour) and E-P flux (arrow). Shading indicates negative value. Contour interval is 0.1, and absolute values less than 0.3 are suppressed. For E-P flux, the correlation coefficient corresponding to 1 is indicated by an arrow in the upper right of the panel. (b) Same as in (a) but for the surface air temperature in the NH extratropics. Negative values are indicated by a dashed line.

ure 3a). Simultaneously, a correlation in the surface temperature appears in two regions: one east of Japan, the other north of the Eurasian mountainous region (Figure 3b), similar to those in Figure 1. Note that the planetary waves propagate from the troposphere mainly from the eastern Eurasia-western North Pacific region (see Figure 5 of Plumb, 1985). Therefore this surface temperature change might be associated with a modified planetary wave structure due to a changed stratospheric state.

## 4. Tropical Summer

This Section gives examples of how the solar influence appears in the tropical troposphere during the NH summer through a change in the meridional circulation.

### 4.1. Indian Monsoon

A close relationship between the Indian monsoon rainfall and the solar cycles of decadal to millennial scale is known from paleoclimate data ($\delta^{18}O$ in stalagmites) (Neff *et al.*, 2001; Burns *et al.*, 2002; Fleitmann *et al.*, 2003). The direct cause of the variation of $\delta^{18}O$ is a change in northward surface winds south of the caves in Oman, where the stalagmites were sampled (star in Figure 4a). A similar relationship is still seen today (Kodera, 2004).

Standardized July–August (Jul–Aug) mean northward component of the surface winds averaged over the area south of Oman (box in Figure 4a) is shown together with the F10.7 solar index after having been lightly smoothed with a low-pass filter (Figure 4b). The two curves vary in phase with the 11-year solar cycle conforming to the paleoclimate result. To illustrate how the monsoon and solar activity can be related, the correlation coefficient between the F10.7 index and Jul–Aug mean low-pass filtered meteorological variables was calculated. The correlation with northward surface wind at each grid point demonstrates a clear relationship between the solar activity and the monsoon circulation across the Equator (Figure 4c). This cross-equatorial wind is due to a seesaw of vertical velocity (or convective activity) over the equatorial Indian Ocean and the continent (Figure 4d). A seesaw of upward velocity between the equator and subtropics is also seen in the zonally averaged field, although in the zonal mean field the equatorial vertical velocity is more strongly correlated with the solar cycle (Figure 4f). This suggests that the variation of vertical velocity in the equatorial region is more fundamental than that of the monsoon region over the continent.

Figure 4e displays the relationship between the F10.7 index and the zonal mean air temperature. It should be noted that the tropical tropospheric temperature is not strongly related with the solar cycle, but the equatorial stratosphere exhibits a clear relationship. A pair of negative and positive correlations south and northward of

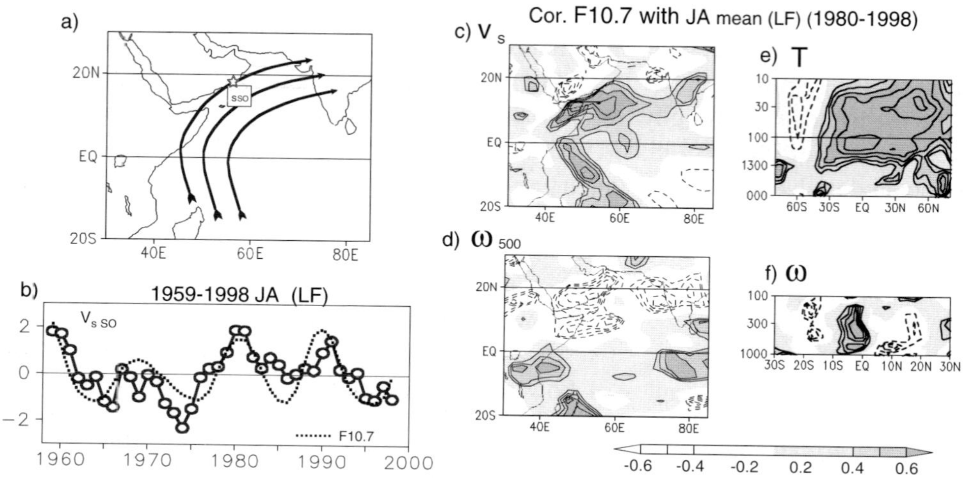


*Figure 4.* (a) Location of the site where stalagmites were sampled (star) and the climatological surface-wind pattern during the summer monsoon (solid lines with arrows). (b) Standardized low-pass-filtered July–August mean northward component of the surface winds averaged over the area south of Oman (box in Figure 4a) (solid line with circle) and F10.7 solar index (dotted line). (c) to (f) Correlation coefficient between the F10.7 index and July–August mean low-pass-filtered meteorological variables at each grid point. (c) Northward surface wind. (d) Pressure coordinate vertical velocity (negative values means upwelling). (e) Zonal mean air temperature. (f) Zonal mean pressure coordinate vertical velocity in tropics. (c)–(f) Contour interval is 0.1, and absolute values less than 0.3 are suppressed. Dashed contours denote negative values. From Kodera (2004).

40°S in the stratosphere can be attributed to anomalous cooling due to anomalous upwelling in high latitudes of the SH, and anomalous warming due to anomalously weak upwelling in the equatorial region and the summer NH. Warming in the equatorial region can be observed down to 200 hPa. A numerical model experiment (Thuburn and Craig, 2000) suggests that a change of stability near the tropopause region can impact the convective heating rate. A solar influence on equatorial convective activity may therefore be possible through changes in upward motion in the equatorial stratosphere.

## 4.2. ENSO

So far, only a linear relationship between solar activity and meteorological parameters has been considered. However, there is a more complicated relationship, such as solar-cycle dependence of the extratropical effect of the equatorial quasi-biennial oscillation (QBO) found by Labitzke and van Loon (1988). Motivated by their work, Barnett (1989) analyzed the tropospheric biennial oscillation (TBO) in the SST related to the ENSO cycle. He found that the amplitude of the TBO is modulated by the 11-year solar cycle: the amplitude is larger during the solar min

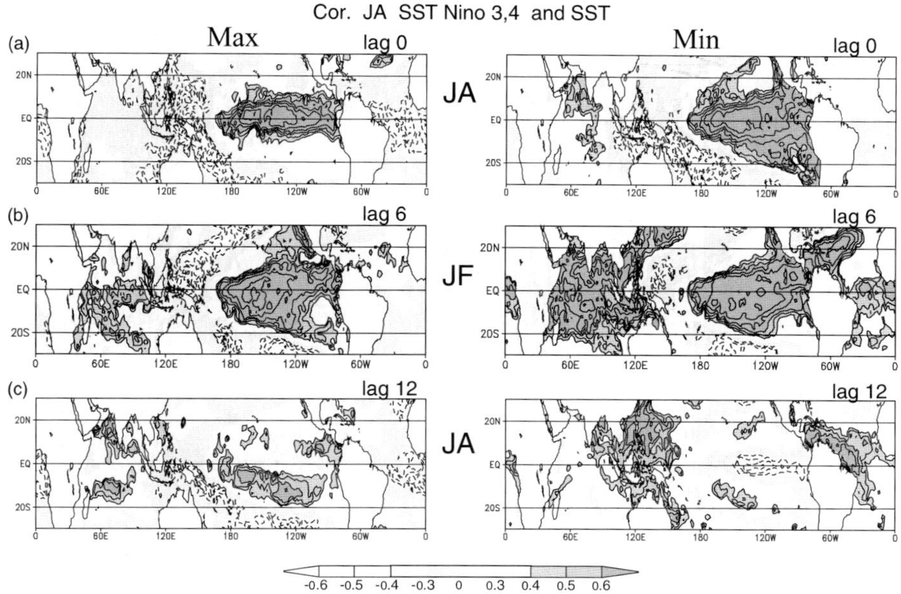


*Figure 5.* Lagged correlation coefficient between Jul–Aug mean Niño 3, 4 SST and SST at grid points of (a) Jul–Aug, (b) Jan–Feb, or (c) Jul–Aug of next year. The left-hand pane is for the solar max, and the right-hand pane is for the solar min. Correlation is calculated for 1958 to 1998. The contour interval is 0.1, and absolute values less than 0.4 are suppressed. Dashed lines indicate negative values. From Kodera (2005).

than during the solar max. Recently it was suggested (Kodera, 2005) that this is also produced through a stratospheric process.

The ENSO phenomenon, in general, develops from summer to autumn, and the influence of the ENSO extends from the Pacific to the Indian Ocean. However, as discussed above, the convective activity over the Indian Ocean is suppressed in summer during the solar max. Therefore, the influence of ENSO on the Indian Ocean is likely to differ between solar max and min. Figure 5 depicts the lagged correlation coefficient between Jul–Aug mean SST of the Niño 3, 4 region (170°E–120°W, 5°S–5°N) and SST at each grid point of (a) Jul–Aug, (b) Jan–Feb, and (c) Jul–Aug of the following year calculated separately for solar max (left) and min (right) (Kodera, 2005). In both cases, ENSO develops from summer to winter, but in the solar min case, its influence is stronger over the Indian Ocean. In the following summer, the positive correlation persists over the equatorial eastern Pacific during the solar max, but the correlation changes sign to negative for the solar min case. This means that the TBO tends to occur more frequently during the solar min, conforming to Barnett's result.

The difference of the evolution of the SSTs may be understood by the difference of the Walker circulation. Figure 6 exhibits the same correlation coefficient

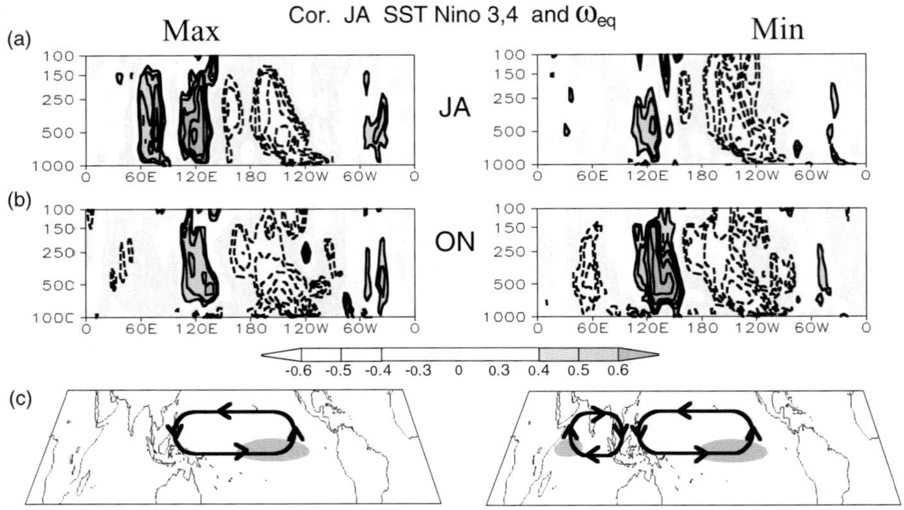


*Figure 6.* Same as in Figure 5, but for the correlation coefficient between Jul–Aug mean Niño 3, 4 SST and (a) Jul–Aug or (b) Oct–Nov mean pressure coordinate vertical velocity over equator at each longitude-height grid point. From Kodera (2005). (c) Schematic presentation of (left) single and (right) double-cell regimes of anomalous Walker circulation (arrows). Shading indicates higher SST regions.

between the Jul–Aug mean SST of Niño 3, 4, but with (a) Jul–Aug or (b) Oct–Nov mean pressure-coordinate vertical velocity ($\omega$) over the equator (5°S–5°N) at each longitude and height grid point. In Jul–Aug, anomalous upwelling over the warm SST of the East Pacific and downwelling over the Maritime continent region are observed in both cases. The anomalous upwelling over the Eastern Pacific extends up to the stratosphere for solar min, but it stops at 150 hPa for max. Accordingly, anomalous downwelling is stronger over the Indian Ocean sector for solar max, whereas no anomalous downwelling, or even upwelling, is found over the Indian Ocean for solar min.

In autumn, anomalous upwelling is formed over the western Indian Ocean in the minimum case. This means that anomalous Walker circulation changes from the single-cell type to the double-cell type for the solar minimum, but the single-cell regime persists for solar max, as schematically depicted in Figure 6c. The development of the double-cell regime is an important condition for the transition of the ENSO cycle in the following year (for more details see Kawamura *et al.*, 2003). We suggest that the ENSO tends to change phase every other year, i.e., biennial oscillation tends to occur during the solar min.

The reason why the double-cell regime does not develop during the solar max may be that the convective activity over the Indian Ocean is suppressed during solar max as can be seen in Figure 4.

## 5. Discussion and Concluding Remarks

This paper illustrates that diverse aspects of the solar influence on Earth's surface can be understood as a dynamical response to the solar forcing transmitted from the stratosphere, rather than that created in the lower atmosphere by a direct radiative effect. The present paper focused mainly on the 11-year solar cycle. However, the same mechanism is also applicable for centennial time scale phenomena, such as the solar influence on the Indian Ocean monsoon through change in stratospheric meridional circulation, as shown above. Similarly, the other processes that involve the change of the structure of planetary waves should also work for a centennial scale.

The correlation coefficient between the centennial-scale solar variation around the Maunder minimum and surface temperature estimated from proxy data (Figure 2A of Waple *et al.*, 2002), reveals two regions of high correlation in the extra-tropics of the NH: one north of the mountainous region of Eurasia and the other around Japan, similar to those in Figures 1 or 3. The cold air advected from Siberia toward Japan during the winter is a direct cause of the temperature change around Japan. This effect persists through interaction with the ocean surrounding Japan. Therefore, a centennial solar influence on surface temperature is also expected in this region.

If the solar influence on the Earth's surface is due to the change in radiative forcing, we may expect to find the clearest signal in global mean temperature, rather than a local one which contains large temperature variations due to climate noise. The global mean temperature for the last 1000 years reconstructed from proxy data by Mann and Jones (2003) is displayed in Figure 7a. Except for a large increase in the 20th century, the global temperature exhibits only small (0.1°C) fluctuations. However, the March temperature in Kyoto, estimated from a full flower day of cherry blossoms (Figure 7b), exhibits a centennial scale variation of 2 to 3°C. Note that a half of the warming after 1900 (1°C) is due to the urbanization effect (Aono, 2005, private communication). Past solar activity estimated from cosmogenic nuclei is depicted in Figure 7c. Inspection reveals that a cold period at the end of the 17th century corresponds to the Maunder minimum (Mm), and that there is also a quasi 200-year variation in phase with the Suess solar cycle. Thus, centennial-scale temperature variation may also be of dynamical origin. This is consistent with a recent remark that the long-term variation of the TSI is negligibly small (Foukal *et al.*, 2004). The global mean temperature is considered as a representative variable of the climate change, but it may not be so relevant if the changes are dynamically produced in the troposphere, although a variation of the global mean temperature can be induced by the Sun through circulation change.

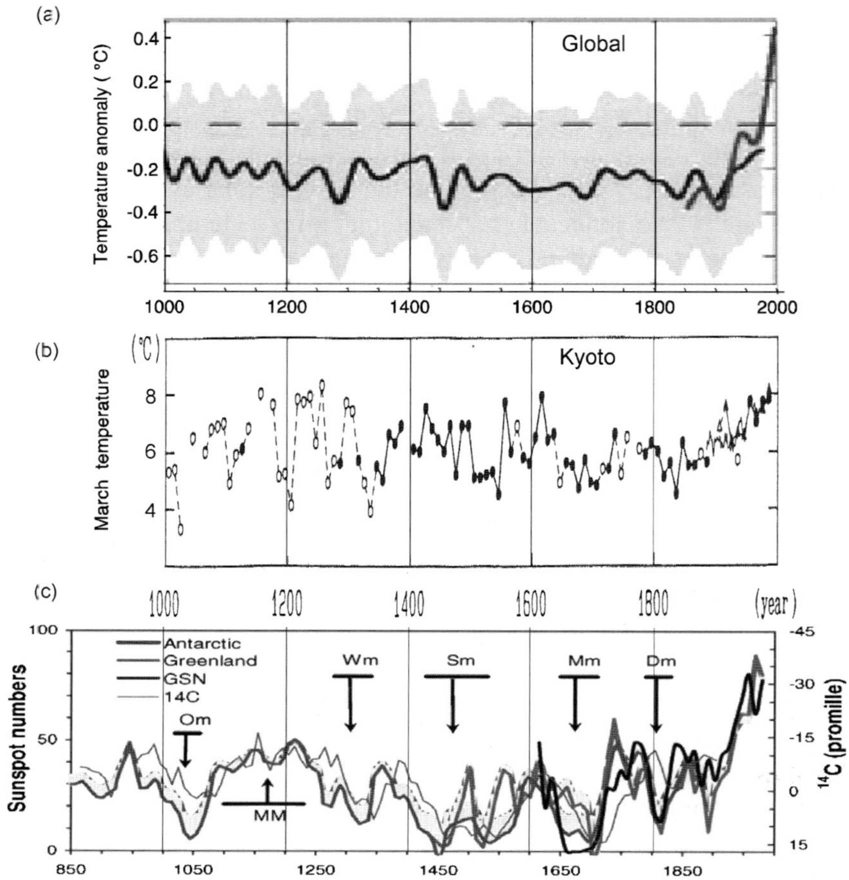


*Figure 7.* (a) Global mean temperature reconstructed from proxy data and observation (Mann and Jones, 2003). (b) March mean temperature in Kyoto, Japan, estimated from flowering data of cherry trees in old documents (Aono and Omoto, 1994). (c) Solar activity indices derived from cosmogenic nuclei (Usoskin *et al.*, 2003). Vertical lines in each 200-year increment are added to the original figures for the convenience of comparison.

## Acknowledgements

The author thanks Y. Kuroda and S. Yukimoto for useful discussions. This study is partly supported by a Grant-in-Aid for Scientific Research of the Japanese Ministry of Education, Culture, Sports, Science, and Technology (17340143).

## References

Aono, Y. and Omoto, Y.: 1994, 'Estimation of temperature at Kyoto since the 11th century using flowering data of cherry trees in old documents' (in Japanese), *J. Agricult. Meteorol.* **49**, 263–272.

Barnett, T. P.: 1989, 'A solar-ocean relation: Fact or fiction', *Geophys. Res. Lett.* **16**, 803–806.

Burns, S. J., Fleitmann, D., Mudelsee, M., Neff, U., Matter, A., and Mangini, A.: 2002, 'A 780-year annually resolved record of Indian Ocean monsoon precipitation from a speleothem from south Oman', *J. Geophys. Res.* **107**, doi:10.1029/2001JD001281.

Chanin, M.-L.: 2006, 'Signature of the 11-yr cycle in the upper atmosphere', *Space Sci. Rev.*, this volume, doi: 10.1007/s11214-006-9062-5.

Fröhlich, C.: 2006, 'Solar irradiance variability since 1978', *Space Sci. Rev.*, this volume, doi: 10.1007/s11214-006-9046-5.

Fleitmann, D., Burns, S. J., Mudelsee, M., Neff, U., Kramers, J., Mangini, A., and Matter, A.: 2003, 'Holocene forcing of the Indian monsoon recorded in a stalagmite from southern Oman', *Science* **300**, 1737–1739.

Foukal, P., North, G., and Wigley, T.: 2004, 'A Stellar view on solar variations and climate', *Science* **306**, 68–69.

Garcia, R. R.: 1987, 'On the meridional circulation of the middle atmosphere', *J. Atmos. Sci.* **44**, 3599–3609.

Gray, L.: 2006, 'A possible Transfer Mechanism for the 11-year Solar Cycle to the Lower Stratosphere', *Space Sci. Rev.*, this volume, doi: 10.1007/s11214-006-9069-y.

Holton, J. R., Haynes, P. H., McIntyre M. E., Douglass, A. R., Rood, R. B., and Pfister, L.: 1995, 'Stratospheric-troposphere exchange', *Rev. Geophys.* **33**, 403–439.

Kawamura, R., Matsuura, T., and Iizuka, S.: 2003, 'Equatorially symmetric impact of El Niño-Southern Oscillation on the South Asian summer monsoon system', *J. Meteor. Soc. Japan* **81**, 1329–1352.

Kelly, P. M. and Jones, P. D.: 1983, 'The evidence for cycles on solar time scales in climate data', in: McCormac, B. N. (ed.), *Weather and Climate Responses to Solar Variations*, Colorado Associated University Press, Boulder, Colorado, pp. 581–590.

Kodera, K.: 2004, 'Solar influence on the Indian Ocean Monsoon through dynamical processes', *Geophys. Res. Lett.* **31**, doi:10.1029/2004GL020928.

Kodera., K.: 2005, 'Possible solar modulation of the ENSO cycle', *Pap. Meteorol. Geophys.* **55**, 35–54.

Kodera, K. and Kuroda, Y.: 2002, 'Dynamical response to the solar cycle', *J. Geophys. Res.* **107**, doi:10.1029/2002JD002224.

Kuroda, Y. and Kodera, K.: 2001, 'Variability of the polar-night jet in the northern and southern hemispheres', *J. Geophys. Res.* **106**, 20,703-20,713.

Kuroda, Y. and Kodera, K.: 2004, 'Role of the polar-night jet oscillation on the formation of the Arctic Oscillation in the Northern Hemisphere winter', *J. Geophys. Res.* **109**, doi:10.1029/2003JD00412.

Labitzke, K.: 2006, 'Solar variation and stratospheric response', *Space Sci. Rev.*, this volume.

Labitzke, K., and van Loon, H.: 1988, 'Association between the 11-year solar cycle, the QBO and the atmosphere. Part I: The troposphere and stratosphere in the northern hemisphere in winter', *J. Atmos. Terr. Phys.* **50**, 197–206.

Lohmann, G., Rimbu, N., and Dima, M.: 2004, 'Climate signature of solar irradiance variations: Analysis of long-term instrumental, historical, and proxy data', *Int. J. Climatology* **24**, doi: 10.1002/joc.1054.

Mann, M. E. and Jones, P. D.: 2003, 'Global surface temperatures over the past two millennia', *Geophys. Res. Lett* **30**, doi:10.1029/2003GL017814.

Matthes, K., Langematz, U., Gray, L. L., Kodera, K., and Labitzke, K.: 2004, 'Improved 11-year solar signal in the Freie Universität Berlin Climate Middle Atmosphere Model (FUB-CMAM)', *J. Geophys. Res.* **109**, doi:10.1029/2003JD004012.

Neff, U., Burns, S. J., Mangini, A., Mudelsee, M., Fleitmann, D., and Matter A.: 2001, 'Strong coherence between solar variability and the monsoon in Oman between 9 and 6 kyr ago', *Nature* **411**, 290–293.

Pittock, A. B. A.: 1978, 'Critical look at long-term sun-weather relationships', *Rev. Geophys. Space Phys.* **16**, 400–420.

Plumb, A.: 1985, 'On the three-dimensional propagation of stationary waves', *J. Atmos. Sci.* **42**, 217–229.

Plumb, A. and Eluszkiewic, J.: 1999, 'The Brewer–Dobson circulation: Dynamics of the tropical upwelling', *J. Atmos. Sci.* **56**, 868–890.

Rottman, G.: 2006, 'Measurements of total and spectral solar irradiance', *Space Sci. Rev.*, this volume.

Shindell, D. T., Schmidt, G. A., Miller, R. L., and Rind, D.: 2001, 'Northern hemisphere winter climate response to greenhouse gas, ozone, solar, and volcanic forcing', *J. Geophys. Res.* **106**, 7193–7210.

Thuburn, J., and Craig, G. C.: 2000, 'Stratospheric influence on tropopause height: The radiative constant', *J. Atmos. Sci.* **57**, 17–28.

Usoskin, I. G., Solanki, S. K., Schüssler, M., Mursula, K., and Alanko, K.: 2003, 'A millennium scale sunspot reconstruction: Evidence for an unusually active Sun since the 1940's', *Phys. Rev. Lett.* **91**, 211,101-1–211,101-4.

Waple, A. M., Mann, M. E., and Bradley, R. S.: 2002, 'Long-term patterns of solar irradiance forcing in model experiments and proxy based surface temperature reconstructions', *Climate Dyn.* **18**, 563–578.

White, W. B., Lean, J., Cayan, D. R., and Dettinger, M. D.: 1997, 'Response of global upper ocean temperature to changing solar irradiance', *J. Geophys. Res.* **102**, 3255–3266.

White, W. B., Dettinger, M. D., and Cayan, D. R.: 2003, 'Sources of global warming of the upper ocean on decadal period scales', *J. Geophys. Res.* **108**, doi:10.1029/ 2002JC001

# SOLAR INFLUENCES ON DYNAMICAL COUPLING BETWEEN THE STRATOSPHERE AND TROPOSPHERE

J. D. HAIGH[1,*] and M. BLACKBURN[2]
[1]*Blackett Laboratory, Imperial College of Science, Technology and Medicine, London SW7 2BW, UK*
[2]*Centre for Global Atmospheric Modelling, University of Reading, UK*
*(*Author for correspondence: E-mail: j.haigh@imperial.ac.uk)*

(Received 31 August 2005; Accepted in final form 8 February 2006)

**Abstract.** We use a simplified atmospheric general circulation model (AGCM) to investigate the response of the lower atmosphere to thermal perturbations in the lower stratosphere. The results show that generic heating of the lower stratosphere tends to weaken the sub-tropical jets and the tropospheric mean meridional circulations. The positions of the jets, and the extent of the Hadley cells, respond to the distribution of the stratospheric heating, with low latitude heating displacing them poleward, and uniform heating displacing them equatorward. The patterns of response to the low latitude heating are similar to those found to be associated with solar variability in previous observational data analysis, and to the effects of varying solar UV radiation in sophisticated AGCMs. In order to investigate the chain of causality involved in converting the stratospheric thermal forcing to a tropospheric climate signal we conduct an experiment which uses an ensemble of model spin-ups to analyse the time development of the response to an applied stratospheric perturbation. We find that the initial effect of the change in static stability at the tropopause is to reduce the eddy momentum flux convergence in this region. This is followed by a vertical transfer of the momentum forcing anomaly by an anomalous mean circulation to the surface, where it is partly balanced by surface stress anomalies. The unbalanced part drives the evolution of the vertically integrated zonal flow. We conclude that solar heating of the stratosphere may produce changes in the circulation of the troposphere even without any direct forcing below the tropopause. We suggest that the impact of the stratospheric changes on wave propagation is key to the mechanisms involved.

**Keywords:** solar variability, climate variability, stratosphere-troposphere coupling, storm tracks

## 1. Introduction

The existence of signals of the 11-year solar cycle in meteorological fields of the lower atmosphere has been confirmed in analyses of observational data (see, e.g., van Loon and Shea, 2000; Haigh, 2003; Gleisner and Theijll, 2003; Crooks and Gray, 2005). The responses found are latitudinally non-uniform and locally larger than would be predicted from radiative forcing considerations alone (Haigh, 2001). Atmospheric general circulation models (AGCMs) are beginning to be able to reproduce some of these effects when forced by changes in solar ultraviolet radiation in the stratosphere (Haigh, 1996, 1999; Shindell *et al.*, 1999; Matthes *et al.*, 2004) suggesting that changes in the stratosphere may exert some dynamical control on the troposphere. It is well known that planetary waves propagating

Space Science Reviews (2006) 125: 331–344
DOI: 10.1007/s11214-006-9067-0

upwards from the troposphere can influence the state of the stratosphere but the mechanisms whereby the stratosphere can influence the troposphere are not well established and form the subject of active current research (see Haynes, 2005 for a review).

In this paper we use a simplified AGCM to investigate how thermal perturbations to the stratosphere affect the climate of the troposphere. Our work is motivated by the desire to understand how solar variability influences climate but our results are applicable more generally to understanding stratosphere-troposphere coupling.

## 2. Response of Tropospheric Zonal Wind to Solar Variability

In an earlier paper (Haigh *et al.*, 2005) we demonstrated, using a multiple regression analysis, the existence of a statistically significant signal of solar 11-year cycle variability in the lower atmosphere annual mean zonal winds of the National Centers for Environmental Prediction/National Center for Atmospheric Research (NCEP/NCAR) reanalysis dataset, provided by the Climate Diagnostics Center, Boulder, USA (`http://www.cdc.noaa.gov/`). A similar result, for northern hemisphere winter, is presented in Figure 1d: the solar signal is seen to be composed of banded structures in mid-latitudes with the largest change, a reduction of over 2 $ms^{-1}$, present on the equatorward sides of the mid-latitude jet cores, and an increase, of over 1 $ms^{-1}$, on the poleward sides. By comparison with the mean zonal wind (Figure 1c) it can be seen that this represents, when the Sun is more active, a weakening of the winter upper-tropospheric sub-tropical jet, and a poleward shift of the mid-latitude jets in both hemispheres.

Both the shape and the amplitude of this pattern are remarkably similar to those found in the GCM simulations of Haigh (1996, 1999), reproduced in Figure 1a

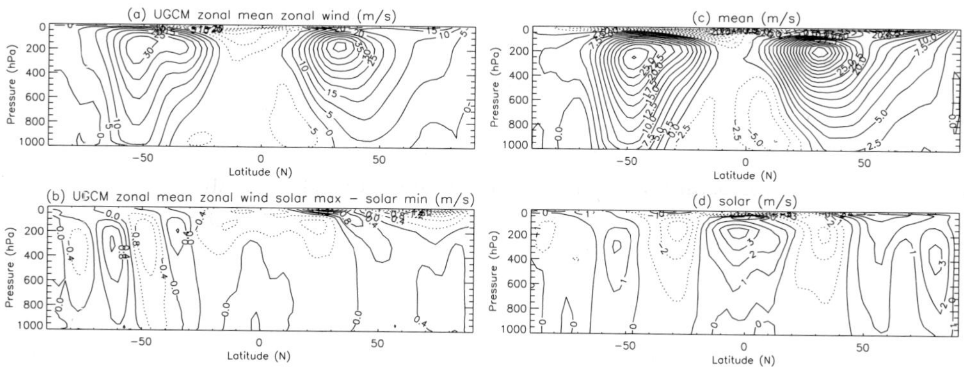


*Figure 1.* Zonal mean zonal wind (a) January mean from GCM; (b) Solar signal in January from GCM experiments; (c) DJF mean of NCEP reanalysis data 1979–2002; (d) Solar signal in DJF from multiple regression analysis of NCEP data.

and 1b. Note, however, this GCM has some problems in the southern hemisphere – producing a strong sub-tropical jet feature that is not found in observations for this season. Using entirely different models, Larkin *et al.* (2000) and Matthes *et al.* (2004) find a similar tropospheric response. These model studies included the effects of changes in solar spectral irradiance where the largest direct radiative impact, due to enhanced solar UV, is found in the stratosphere. The GCMs also predicted a response to higher levels of solar UV in the tropospheric mean meridional circulation consisting of a weakening and expansion of the Hadley cells and poleward shift of the Ferrel cells.

The similarity of the signals found in the observational data and model runs is intriguing and, by the nature of the model experiments, suggests that changes in the stratosphere, introduced by modulation of solar ultraviolet radiation and ozone, are key. It does not, however, explain the mechanisms whereby such a change in tropospheric circulation is brought about by thermal perturbations to the stratosphere. In Section 3 we describe some experiments with a simplified GCM designed to elucidate some of these mechanisms. An initial discussion of the results of these runs was included in Haigh *et al.* (2005); some of these are reviewed here, together with further analysis. In Section 4 we present results from a new experiment in which an ensemble of model spin-ups is used to investigate the mechanisms involved in producing the effects found in the perturbation experiments discussed in Section 3.

## 3. Impact of Stratospheric Heating on the Troposphere of a Simplified GCM

The model uses the spectral dynamical core described by Hoskins and Simmons (1975), modified to include the angular-momentum conserving vertical discretisation of Simmons and Burridge (1981), but retaining the original sigma coordinate. All experiments use a T42L15 resolution, denoting triangular spectral truncation at total wavenumber 42, equivalent to an equatorial grid spacing close to 475 km for the equivalent linear transform grid, and 15 levels between the surface and about 18.5 hPa. The model's climate is maintained using the linear forcing and drag scheme of Held and Suarez (1994), in place of the moist parameterisation of a sophisticated GCM. Orography is omitted. Temperature is relaxed towards the zonally symmetric equilibrium distribution shown in Figure 2a on a timescale which is 40 days above the boundary layer (pressure greater than about 700 hPa), but which reduces to 4 days at the equatorial surface. Rayleigh friction is also included in the boundary layer with a timescale of 1 day at the surface.

The dynamical response to the equilibrium structure is such as to cause adiabatic warming at mid-high latitudes and cooling at low latitudes so that, although the equilibrium temperature is uniform and constant above the tropopause, the model's temperature field (Figure 2b) shows a positive latitudinal temperature gradient there while in the troposphere the negative temperature gradients are reduced. The

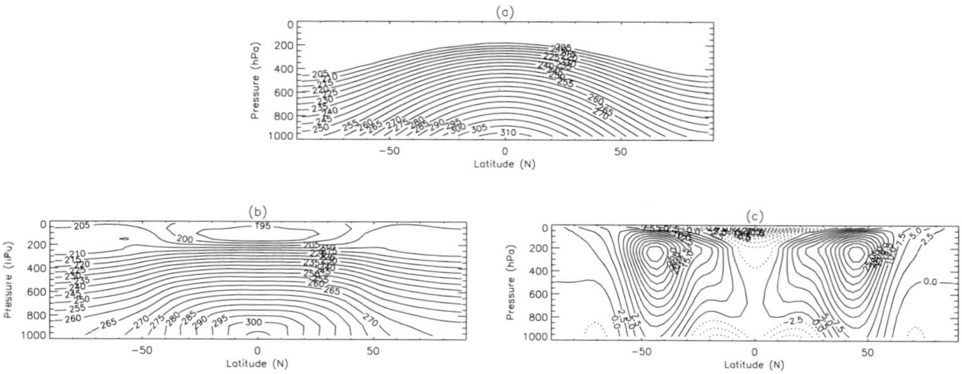

*Figure 2.* Simple GCM fields: (a) the prescribed equilibrium temperature (K); (b) the zonal mean temperature (K) for the control run C and (c) the zonal mean zonal wind ($ms^{-1}$) for run C.

model's zonal wind structure is shown in Figure 2c to aid assessment of the results of the model experiments discussed below.

The model climatology consists of a single jet which extends through the depth of the troposphere in the mid-latitudes of each hemisphere, maintained by fluxes of (heat and) momentum associated with the transient weather systems (shown later). The observed climate (Figure 1c) comprises an eddy-maintained jet in each hemisphere, and also a subtropical jet which is confined to the upper-troposphere of the winter hemisphere, and which is only partially separated from the eddy-maintained jet. The existence of, and variations in, the subtropical jet involve moist tropical processes which have been idealised as a linear relaxation in the simplified model. The model results are therefore relevant only to stratospheric influences on the eddy-driven mid-latitude jets and not to other possible mechanisms such as changes in convective activity or the vertical propagation of stationary planetary waves, such as discussed by, e.g., Rind and Balachandran (1995).

Experiments with the model have been designed to investigate the effects of perturbations to the temperature structure of the lower stratosphere. These are implemented by modifications of the equilibrium temperature distribution in the stratosphere as summarized in Table I. Experiments U1 and U5 prescribe a uniform heating perturbation of 1 K and 5 K respectively, throughout the stratosphere (defined as the region of uniform temperature at the top of Figure 2a). In experiments E1 and E5 the equatorial stratosphere is warmed by 1 K and 5 K respectively but this is gradually reduced to zero increase at the poles. In no case is any perturbation applied within the troposphere. The structure of the perturbations is very simple, and the model runs are certainly not intended as direct simulations of solar effects, but we find that the results may be useful in interpreting the effects of solar perturbations to the real atmosphere.

The U and E experiments were each carried out with the two different amplitudes, of 1 K and 5 K, in order to assess whether the patterns of response were

TABLE I

Summary of experiments with the simplified GCM.

| Run ID | Length of run (days) | Increase in stratospheric equilibrium temperature (K) |
|---|---|---|
| C | 2 × 10,000 | None |
| U1 | 2 × 10,000 | 1 |
| U5 | 2 × 1,000 | 5 |
| E1 | 2 × 10,000 | $\cos^2$(latitude) |
| E5 | 2 × 1,000 | 5 $\cos^2$(latitude) |

independent of the magnitude of the forcing and to what extent the response scaled with the forcing. The statistical significance of the responses is estimated independently at each grid-point using a Student t-test taking into account the reduction in the number of degrees of freedom due to autocorrelations in the time series. The signals produced by the 5 K perturbations are much larger than the long-term variability in the fields so that it was necessary to continue these runs for only 1,000 days, after an initial spin-up period of 200 days. For the 1 K experiments, and the control, runs of 10,000 days were carried out. In all cases the effective number of data points has been doubled by using the results from the two hemispheres as independent simulations. This is possible because an analysis of the temporal correlations between equivalent points in the two hemispheres shows very low values (of magnitude ~0.2) across most of the hemisphere.

The perturbations in zonal mean zonal wind found in the simplified GCM experiments are presented in Figure 3. Banded structures appear in the troposphere in each case, despite the forcing having only been applied in the stratosphere. In runs U1 and U5 (Figure 3a and 3b) these appear as a band of weaker winds near 55° and a band of stronger winds near 35°. This represents a weakening of the jet and an equatorward shift. This response is qualitatively similar (apart from a change in sign of both the forcing and response) to the results of Polvani and Kushner (2002) who presented results from a similar model used to study the impacts of polar lower stratospheric cooling. The response of the E experiments (Figure 3c and 3d) is different from the U runs, with a negative band near 40° flanked by positive bands near 25° and 60° respectively. These correspond to a weakening and latitudinal expansion, but mainly poleward shift, of the jets. This pattern is qualitatively similar to the solar signal found in the sophisticated GCM experiments and observational data analysis presented in Figure 1b and 1d.

Clearly the responses seen in the simple model experiments U5 and E5 are much larger than those found to be due to solar influences in the data analysis. Experiments U1 and E1, using forcings which were a factor 5 smaller than U5 and E5, respectively, might be more appropriate for interpretation in this regard. However, as the patterns of response seen in the 1 K experiments are essentially

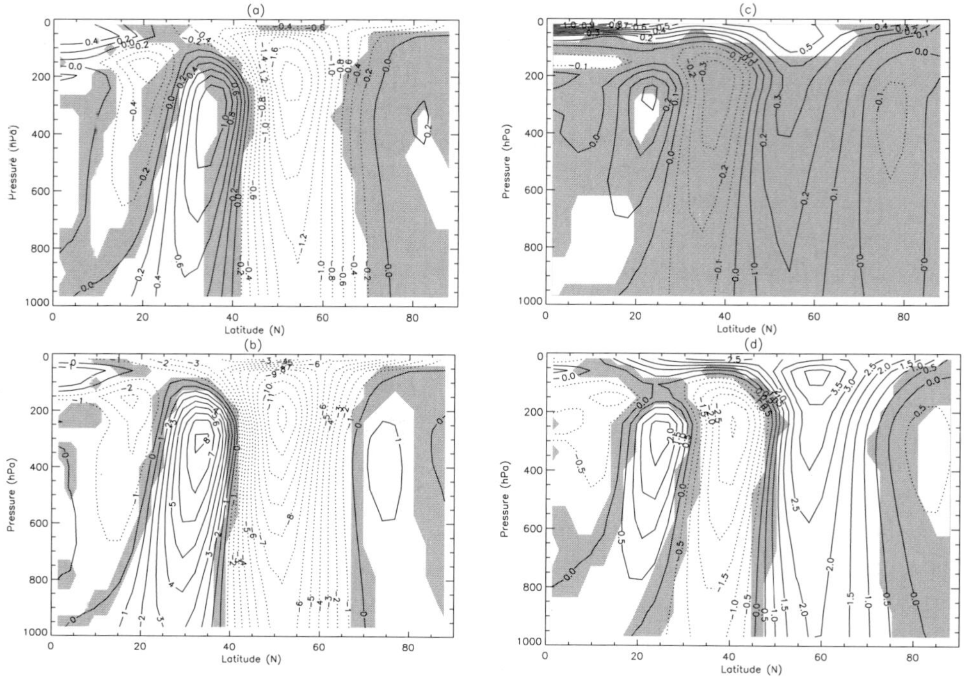


*Figure 3.* Difference from control run of zonal mean zonal winds in run (a) U1 (contour interval 0.2 ms$^{-1}$), (b) U5 (1 ms$^{-1}$), (c) E1 (0.1 ms$^{-1}$) and (d) E5 (0.5 ms$^{-1}$). Average of 2 hemispheres. Regions in which the signal does not reach the 95% confidence level are shaded.

identical to those of the corresponding 5 K runs it seems clear that the same processes are operating. On this basis, and for reasons of clarity and brevity, the analysis and interpretation of the data, below, is based on the 5 K experiments alone.

The field of poleward eddy momentum flux for control run C, and the anomalies, relative to run C, for runs U5 and E5 are shown in Figure 4. In the control run C there is poleward eddy momentum flux from the tropics into mid-latitudes, with maximum flux at 35°. The (vertically integrated) flux convergence in mid-latitudes acts as a forcing of westerly flow, maintaining the jet in Figure 2c against low-level drag, while the flux divergence in the subtropics acts as a forcing of easterly flow there, balancing the Coriolis force associated with the poleward outflow of the tropical Hadley circulation near the tropopause (shown in Figure 5a). The effect of stratospheric heating is to increase the static stability of the tropopause region and to lower the tropopause so that the wave flux convergence, and the poleward branch of the Hadley cell, are moved downwards. For run U5 the momentum flux is significantly weakened, particularly on the poleward side of the mid-latitude jet. The latitudinal gradient of the anomaly implies anomalous easterly forcing by the eddies poleward of the peak anomaly at 45° and westerly forcing equatorward of the peak. Comparison with the zonal wind anomalies in Figure 3b reveals that these

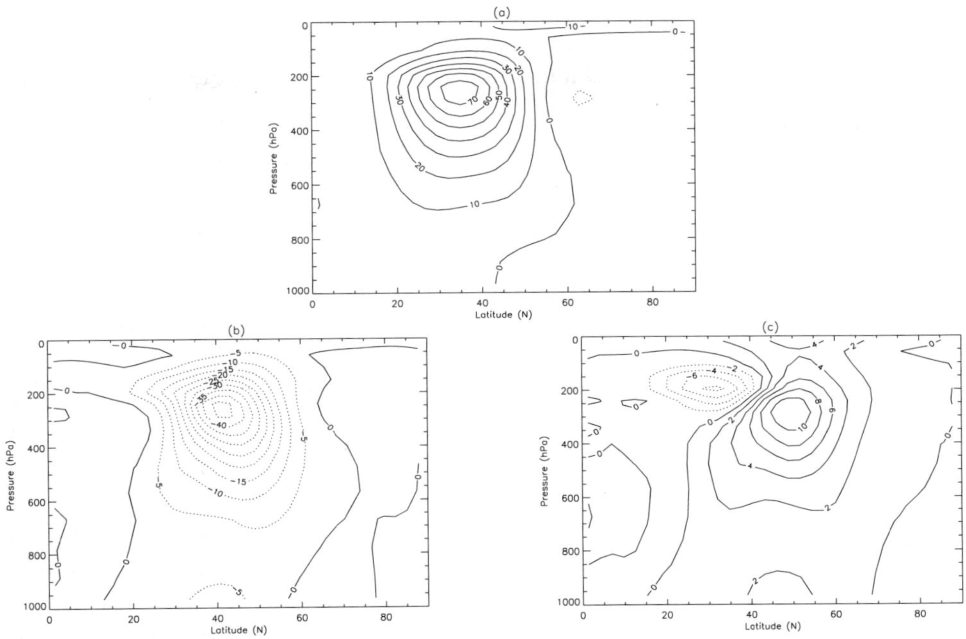


*Figure 4.* Poleward flux of westerly momentum by the zonally asymmetric eddies $[u'v']$ ($m^2s^{-2}$) for (a) the control run C; (b) the U5 – C anomaly and (c) the E5 – C anomaly. Average of 2 hemispheres.

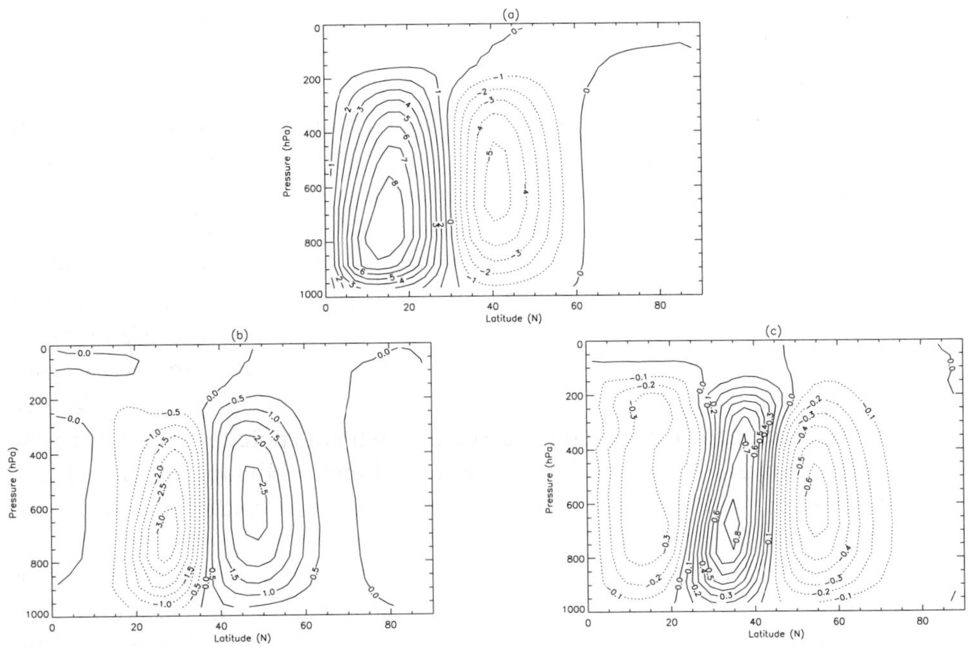


*Figure 5.* Stream-function of the mean meridional circulation (in units of $10^{10}$ kg $s^{-1}$) for (a) the control run C; (b) the U5 – C anomaly and (c) the E5 – C anomaly. Average of 2 hemispheres.

forcing regions coincide with wind anomalies of the same sign. The eddy forcing peaks in the upper troposphere, driving the zonal mean state away from thermal wind balance there. Balance is maintained by an anomalous zonally symmetric overturning circulation, whose Coriolis force balances the eddy forcing locally in the upper troposphere, and which yields a low level anomaly-balance between drag and a Coriolis force of the opposite sign to that in the upper troposphere. The streamfunction of the anomalous meridional circulation for run U5 (Figure 5b) coincides with the anomalous eddy forcing and zonal wind in precisely this manner.

The anomalous meridional circulation also affects the zonal mean temperature, by vertical advection of potential temperature in its vertical branches, in such a way as to maintain thermal wind balance in the presence of the eddy momentum forcing. This acts to increase the meridional temperature gradient below upper-level westerly eddy forcing and reduce it below easterly eddy forcing. The temperature field (not shown) reveals, for run U5, a warm anomaly poleward of 45° coincident with the anomalous mean descent shown in Figure 5b and a cold anomaly equatorward of this latitude coincident with anomalous mean ascent. Thus a thermal perturbation applied only in the stratosphere produces vertical bands of warming and cooling in the troposphere.

To complete the picture, thermal forcing by the anomalous meridional circulation is balanced locally by a combination of anomalous convergence of the northward eddy heat flux and diabatic heating. The former comprises an equatorward shift and weakening of the poleward heat flux. The latter is inevitable in the sGCM experiments, in which all thermal anomalies in the troposphere are linearly damped by the Newtonian relaxation, since the tropospheric reference state is unchanged in the integrations.

A similar relationship between the various anomaly fields holds for experiment E5, although the anomalies are of smaller magnitude. The northward eddy momentum flux anomaly (Figure 4c) now has a dipolar structure, comprising a strengthening of the flux on the poleward flank of the jet and a weakening on its equatorward flank. This corresponds to a tripolar pattern of anomalous momentum flux convergence that coincides with the zonal wind anomalies in Figure 3d. The low latitude change is larger, however, so that the net effect is to weaken the jet and displace it poleward. The anomalous meridional circulation (Figure 5c) locally counteracts the anomalous momentum forcing with its vertical branches again producing a response in tropospheric temperatures. In this case these consist of bands of cooling around 30° and 65° and a band of warming in the 40–55° latitude region. This pattern is qualitatively very similar to the solar response found by Haigh (2003) in an analysis of NCEP Reanalysis temperatures.

For interpretation of the model experiments we find a useful diagnostic to be the vertically-integrated budget of zonal momentum. The rate of change of total westerly momentum integrated through a vertical column of atmosphere can be partitioned into components due to the zonally averaged flow, $C_{\text{zonal}}$, and the zonally asymmetric flow, $C_{\text{eddy}}$, with dissipation due to surface drag. Since the model's

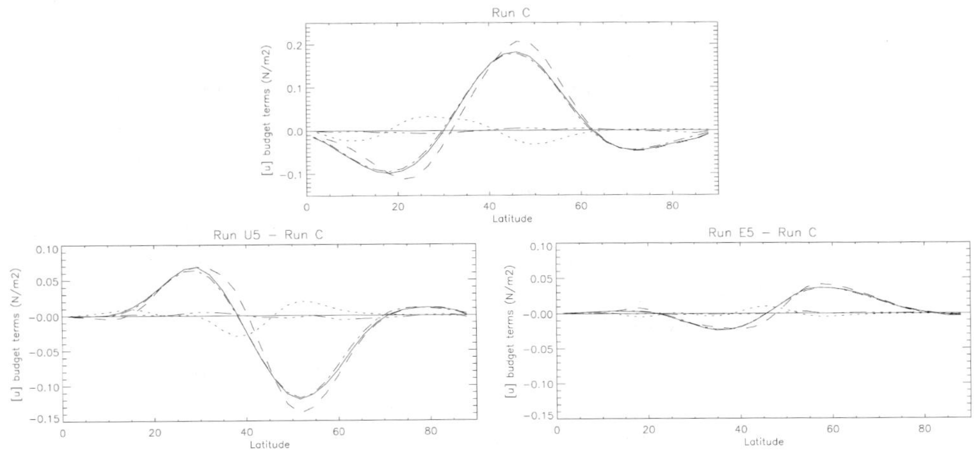


*Figure 6.* Terms in the vertically integrated budget of westerly momentum for (a) the control run C; (b) the U5 – C anomaly and (c) the E5 – C anomaly. Each panel shows the convergence of the flux of westerly momentum due to the zonally asymmetric eddies (dashed line), the zonally averaged flow (dotted) and their sum (solid line); the surface stress (dash-dotted) and the budget residual (dash-triple dotted). All quantities are in units of $Nm^{-2}$.

low-level flow is damped by the linear drag parameterization, latitudinal displacements in the westerly jet must be maintained by convergence of the poleward momentum flux anomaly $C_{total}$ $(= C_{eddy} + C_{zonal})$. Terms in the vertically integrated momentum budget for the control experiment, shown in Figure 6a, confirm that the surface stress is accurately balanced at each latitude by the convergence of the poleward flux of westerly momentum, $C_{total}$, and that this is dominated by its eddy component. Both the mid-latitude surface westerlies and the easterly trade-winds are maintained by the poleward momentum flux. The divergence of the mean circulation term, $C_{zonal}$, is generally small and reflects the latitudinal structure of the upper-tropospheric branches of the Hadley and Ferrel cells. It opposes the eddy flux convergence in mid-latitudes and leads to a small displacement of the eddy flux convergence and surface stress in the sub-tropics.

The momentum budget anomalies in experiments U5 and E5, relative to the control, are shown in Figure 6b and 6c. It is clear that the eddy flux convergence dominates the anomalies, balancing the surface stress dipole associated with the equatorward jet shift in U5 and the tripole associated with the poleward shift in E5. The mean circulation term is small in comparison, but accounts for a small displacement between the peak stress anomaly and peak eddy flux convergence anomaly on the equatorward side of the jet in each experiment. We conclude that the jet displacements in the model experiments are primarily maintained by the changes in poleward eddy momentum flux. The anomalous eddy flux convergence is concentrated in the upper troposphere, driving an anomalous mean circulation (displacement of the Ferrel cell) which effectively transfers the anomalous momentum forcing downwards, where it is balanced by the surface stress anomaly.

## 4. Investigation of Mechanisms Using an Ensemble of Model Spin-ups

We now ask to what extent the diagnostic picture presented in the preceding section can be interpreted causally, beyond the statement that the thermal perturbations imposed in the stratosphere lead to the modelled changes in the troposphere. All the experiments involve long-term averaged steady-states, in which the anomalous mean flow and eddy forcing terms in the budget equations must balance. The vertically integrated momentum balance immediately implies that latitudinal displacements of the jets necessarily involve up-gradient anomalous momentum fluxes, which we have found to be dominated by the eddy component, while thermal wind balance implies that the baroclinic component of such displacements will be associated with anomalous temperature gradients.

To investigate the chain of causality in the establishment of the equilibrium response an experiment has been conducted in which, starting from fields produced by the control run C, the thermal perturbation used in run U5 is suddenly switched on and the time-evolving state monitored. The results from a single such spin-up experiment would be sensitive to the initial conditions and subject to the effects of model internal variability so could not be statistically representative. To counteract this problem an ensemble of 200 runs has been carried out. In each of these the same perturbation is applied but the initial conditions are set from points 50 days apart in the control run. Each member of the ensemble was run for 50 days and output produced daily. In the figures that follow all the plots show results which are averaged over the 200 members of the ensemble and over the two hemispheres (with hemispherically asymmetric fields scaled appropriately).

Figure 7 shows how the zonal mean zonal wind, horizontal eddy momentum flux and mean meridional circulation evolve. Comparison with Figures 3b, 4b and 5b respectively reveals that by days 40–49 the spin-up fields have only progressed a small fraction of the way towards the equilibrium state but are already showing some of the features of the equivalent equilibrium responses: the zonal wind has the characteristic weakening and equatorward shift of the jet while the eddy momentum fluxes and mean meridional circulation are both weakened. There are also some clear differences between the patterns of the time-evolving and equilibrium responses, however, which give an indication of how the state develops. In the spin-up the signal is initially concentrated in the stratosphere with effects in the troposphere lagging behind. The peak (negative) anomaly in zonal wind by days 40–49 (approximately $-1.75$ ms$^{-1}$) occurs near 45°, 100 hPa while that in the equilibrium response (12 ms$^{-1}$) is at 50°, 250 hPa. The positive anomaly at lower latitudes (representing the equatorward shift of the jet) and the signals extending down through the troposphere are barely established.

This effect is seen, perhaps even more clearly, in the eddy momentum fluxes. By days 40–49 of the spin-up there is a clear negative anomaly but it is almost exclusively confined to the tropopause region. There is some evidence, however, of a signal propagating downwards into the troposphere, strengthening with time.

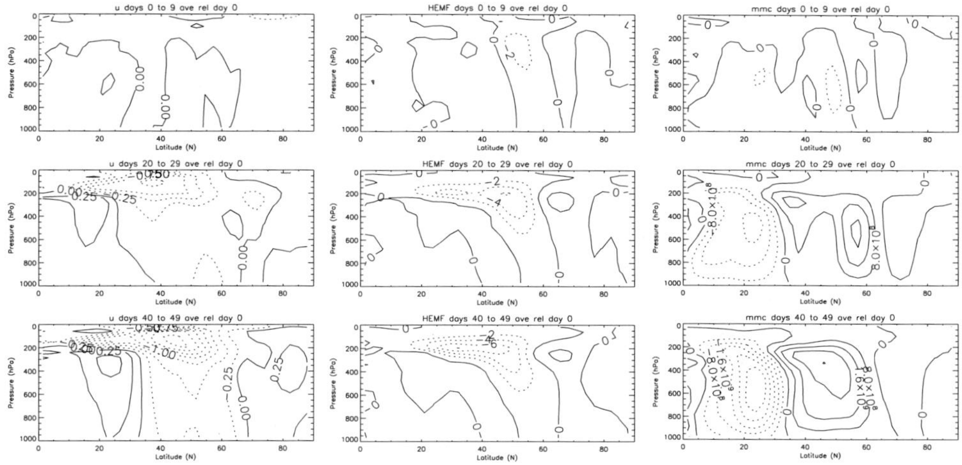


*Figure 7.* 10-day averages of fields (presented as the anomaly from the initial state) evolving after a switch-on of the U5 perturbation at day 0. *Top:* days 0–9; *Middle:* days 20–29; *Bottom:* days 40–49. *Left:* Zonal mean zonal wind ($ms^{-1}$); *Centre:* Poleward flux of westerly momentum by the zonally asymmetric eddies ($m^2s^{-2}$); *Right:* Stream-function of the mean meridional circulation (kg $s^{-1}$).

The mean meridional circulation anomaly extends through the troposphere but is more weighted towards the upper troposphere than in the equilibrium case.

Kushner and Pulvani (2004) have used a similar methodology to investigate high latitude stratosphere-troposphere interaction associated with annular variability and showed that the zonal wind anomaly takes about 500 days to equilibrate in the troposphere. They also used a zonally symmetric version of their simplified GCM in attempt to diagnose the roles of tropospheric and stratospheric eddies on the equilibrium response. They found that stratospheric eddy-driving induces a response in the troposphere but without feedback from the tropospheric eddies the full equilibrium response in the troposphere could not be reproduced.

Our results show that heating the lower stratosphere reduces the wave fluxes in the tropopause region by lowering the tropopause, and leads to coherent changes developing in the troposphere. In Section 3 we found, by analyzing the vertically integrated budget of zonal momentum for the equilibrium experiments, that the displacements of the jets were primarily maintained by the changes in poleward eddy momentum flux. Now we investigate if the time evolution of the vertically integrated momentum budget from the spin-up experiment can provide an indication of the chain of causality in the observed effects. Figure 8 shows 10-day averages of the momentum budget components as deviations from the initial state. These can be compared with Figure 6b which presented the values for the U5 equilibrium run.

The eddy term in the spin-up shows a developing negative anomaly around 50–65° similar (though smaller in magnitude) to that in the equilibrium case. The positive anomaly in the subtropics is less well developed, however and the anomalies in the zonally averaged term are shifted towards the equator. Consistent with the

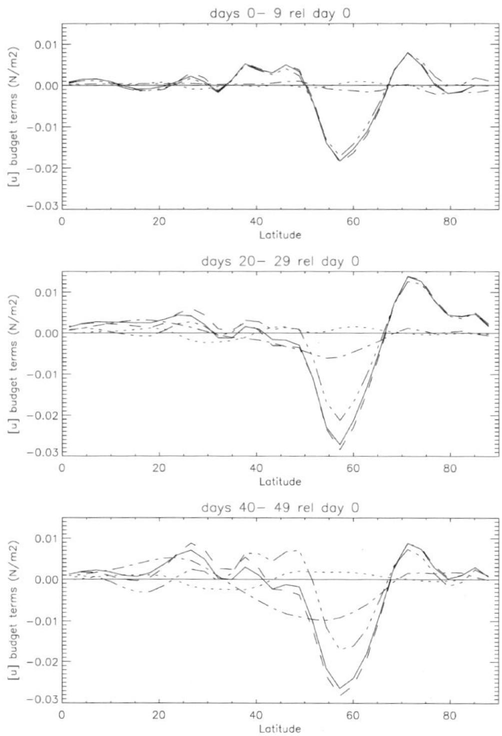


*Figure 8.* 10-day averages of terms in the vertically integrated budget of westerly momentum (presented as the anomaly from the initial state) evolving after a switch-on of the U5 perturbation at day 0. Each panel shows the convergence of the flux of westerly momentum due to the zonally asymmetric eddies (dashed line), the zonally averaged flow (dotted) and their sum (solid line); the surface stress (dash-dotted) and the budget residual (dash-triple dotted). All quantities are in units of $Nm^{-2}$.

latter, in the spin-up case the surface stress term is not balancing the net momentum flux but is lagging behind suggesting that the anomalous Ferrel cell driven by the eddies has not become sufficiently well established to transport momentum anomalies down to the surface. This provides a clear indication that the chain of causality starts with modifications to the wave flux convergence at the tropopause.

## 5. Conclusions

We conclude from our equilibrium experiments that imposed changes in the lower stratospheric temperature forcing lead to coherent changes in the latitudinal location and width of the mid-latitude jetstream and its associated storm-track, and that eddy/mean-flow feedbacks are crucial to these changes. Imposed stratospheric warming and an associated lowering of the tropopause tends to weaken the jet and storm-track eddies; equatorial stratospheric warming displaces the jet polewards while uniform warming displaces it markedly equatorwards.

We suggest that the observed climate response to solar variability is brought about by a dynamical response in the troposphere to heating predominantly in the stratosphere. The effect is small, and frequently masked by other factors, but not negligible in the context of the detection and attribution of climate change. The results also suggest that, at the Earth's surface, the climatic effects of solar variability will be most easily detected in the sub-tropics and mid-latitudes.

Details of the mechanisms involved in the transfer of the effects of the stratospheric forcing to the troposphere are still not clear. However, our analysis of the vertically integrated momentum budget in the ensemble of spin-up runs shows that changes in the static stability of the tropopause region initially cause a weakening of eddy momentum flux convergence and that the transfer of the momentum anomaly by the mean circulation to the surface, where it is dissipated, lags behind. This suggests that understanding the effects of the stratospheric perturbations on wave propagation will provide the key to unraveling the mechanisms of how a solar (or indeed any other) influence in the stratosphere may influence tropospheric climate.

## References


Crooks, S., and Gray, L. J.: 2005, 'Characterisation of the 11-year solar cycle using a multiple regression analysis of the ERA-40 dataset', *J. Clim.* **18**, 996–1015.

Gleisner, H., and Thejll, P.: 2003, 'Patterns of tropospheric response to solar variability', *Geophys. Res. Lett.* **30**, doi:10.1029/2003GL017129.

Haigh, J. D.: 1996, 'The impact of solar variability on climate', *Science* **272**, 981–984.

Haigh, J. D.: 1999, 'A GCM study of climate change in response to the 11-year solar cycle', *Quart. J. Roy. Meteorol. Soc.* **125**, 871–892.

Haigh, J. D.: 2001, 'Climate variability and the role of the Sun', *Science* **294**, 2109–2111.

Haigh, J. D.: 2003, 'The effects of solar variability on the Earth's climate', *Phil. Trans. Roy. Soc. A.* **361**, 95–111.

Haigh, J. D., Blackburn, M., and Day, R.: 2005, 'The response of tropospheric circulation to perturbations in lower stratospheric temperature', *J. Clim.* **18**, 3672–3691.

Haynes, P. H.: 2005, 'Stratospheric dynamics', *Ann. Rev. Fluid Mech.* **37**, 263–293.

Held, I. M., and Suarez, M. J.: 1994, 'A proposal for the intercomparison of the dynamical cores of atmospheric general circulation models', *Bull. Amer. Meteor. Soc.* **75**, 1825–1830.

Hoskins, B. J., and Simmons, A. J.: 1975, 'A multi-layer spectral model and the semi-implicit method', *Quart. J. Roy. Meteorol. Soc.* **101**, 637–655.

Kushner, P. J., and Polvani, L. M.: 2004, 'Stratosphere-troposphere coupling in a relatively simple AGCM: the role of eddies', *J. Climate* **17**, 629–639.

Larkin, A., Haigh, J. D., and Djavidnia, S.: 2000, 'The effect of solar UV irradiance variations on the Earth's atmosphere', *Space Sci. Rev.* **94**, 199–214.

Matthes K., Langematz, U., Gray, L. J., Kodera, K., and Labitzke, K.: 2004, 'Improved 11-year solar signal in the freie universitaet Berlin climate middle atmosphere model (FUB-CMAM)', *J. Geophys. Res.* **109**, doi:10.1029/2003JD004012.

Polvani, L. M., and Kushner, P. J.: 2002, 'Tropospheric response to stratospheric perturbations in a relatively simple general circulation model', *Geophys. Res. Lett.* **29**, 10.1029/2001GL014284.

Rind, D., and Balachandran, N. K.: 1995, 'Modeling the effects of UV variability and the QBO on the troposphere-stratosphere system, Part II: the troposphere', *J. Clim.* **8**, 2080–2095.

Shindell, D., Rind, D., Balachandran, N., Lean, J., and Lonergan, P.: 1999, 'Solar cycle variability, ozone, and climate', *Science* **284**, 305–308.

Simmons, A. J., and Burridge, D. M.: 1981, 'An energy and angular-momentum conserving vertical finite-difference scheme and hybrid vertical coordinates', *Mon. Wea. Rev.* **109**, 758–766.

van Loon, H., and Shea, D.: 2000, 'The global 11-year signal in July-August', *Geophys. Res. Lett.* **27**, 2965–2968.

# THE RESPONSE OF THE MIDDLE ATMOSPHERE TO SOLAR CYCLE FORCING IN THE HAMBURG MODEL OF THE NEUTRAL AND IONIZED ATMOSPHERE

H. SCHMIDT[1,*] and G. P. BRASSEUR[1,2]
[1]*Max Planck Institute for Meteorology, Bundesstr. 53, D-20146 Hamburg, Germany*
[2]*Now at: National Center for Atmospheric Research, Boulder, Colorado, USA*
*(*Author for correspondence: E-mail: hauke.schmidt@zmaw.de)*

(Received 7 September 2005; Accepted in final form 30 January 2006)

**Abstract.** This paper studies the response of the middle atmosphere to the 11-year solar cycle. The study is based on numerical simulations with the Hamburg Model of the Neutral and Ionized Atmosphere (HAMMONIA), a chemistry climate model that resolves the atmosphere from the Earth's surface up to about 250 km. Results presented here are obtained in two multi-year time-slice runs for solar maximum and minimum conditions, respectively. The magnitude of the simulated annual and zonal mean stratospheric response in temperature and ozone corresponds well to observations. The dynamical model response is studied for northern hemisphere winter. Here, the zonal mean wind change differs substantially from observations. The statistical significance of the model's dynamical response is, however, poor for most regions of the atmosphere. Finally, we discuss several issues that render the evaluation of model results with available analyses of observational data of the stratosphere and mesosphere difficult. This includes the possibility that the atmospheric response to solar variability may depend strongly on longitude.

**Keywords:** middle atmosphere, solar cycle, chemistry climate modeling

## 1. Introduction

Classical general circulation models (GCMs) used for climate simulations generally cover the atmosphere from the surface to the mid-stratosphere. In some cases, extensions of these models to the mesosphere exist. On the other hand, three-dimensional numerical models used to study the mesosphere and lower thermosphere usually do not include the troposphere. However, it is well known that mesospheric dynamics are largely determined by upward propagating waves of different kinds that have their origin, in general, in the troposphere. To overcome this separation, recently, models have been developed that are sometimes referred to as "whole atmosphere models". They include a detailed dynamical description of the atmosphere including the troposphere, have their upper lid in the thermosphere, and can be coupled to comprehensive chemistry modules (GCMs with interactive chemistry are referred to as chemistry climate models, CCMs). Models of this type are the *Extended Canadian Middle Atmosphere Model* (EXCMAM, Fomichev *et al.*, 2002), the *Whole Atmosphere Community Climate Model* (WACCM, e.g.,

Space Science Reviews (2006) 125: 345–356
DOI: 10.1007/s11214-006-9068-z 

Beres *et al.*, 2005) and the *Hamburg Model of the Neutral and Ionized Atmosphere* (HAMMONIA, Schmidt *et al.*, 2006) that is used in this study.

These models are particularly suited for studying the atmospheric response to the variability of solar irradiance. While total solar irradiance changes only about 0.1% within the 11-year solar cycle (e.g., Froehlich, 2004), the changes become in general larger for short wavelengths. The mean increase of the part of the UV irradiance for the wavelength band from 200 to 250 nm, e.g., is of about 3% for solar maximum conditions (hereafter called solar max) with respect to solar minimum (solar min) (Lean *et al.*, 1997). The irradiance in the Lyman-$\alpha$ line at 121.6 nm changes by more than 50% (e.g., Rottman, 2000). Since almost all the irradiance with wavelengths shorter than 200 nm is absorbed above the stratopause, classical GCMs do not cover the height range where the largest influence of solar variability can be expected. In the literature, mechanisms are discussed by which the solar signal can be transported downward in the atmosphere: e.g., the downward transport of nitrogen oxide (of which the thermospheric production depends, among other factors, on the strength of solar irradiance) in the polar night vortex (Siskind, 2000), and dynamical processes in the stratosphere (Kodera and Kuroda, 2002). Therefore, models which are able to represent vertical coupling processes in different heights of the atmosphere are useful tools to study solar variability effects.

Steady state simulations with HAMMONIA for solar max and min of a typical 11-year solar cycle have been discussed elsewhere (Schmidt *et al.*, 2006) with respect to the zonal mean response in trace gases, temperature and winds in the mesosphere and lower thermosphere. This paper is intended to discuss additional results obtained from these simulations. We will evaluate in more detail (a) the zonal mean stratospheric response to solar variability, and (b) the dependence of the response in the middle atmosphere on longitude. Our model results will be compared to earlier numerical studies and observational results. Some problems of the comparison between simulations and observations will be discussed.

## 2. Model Description and Simulation Setup

HAMMONIA consists of the vertical extension to the thermosphere (about 250 km altitude) of the MAECHAM5 model (Manzini *et al.*, 1997; Giorgetta *et al.*, 2002), which is itself a vertical extension to the lower mesosphere of the ECHAM5 atmospheric GCM (Roeckner *et al.*, 2003, 2006). ECHAM5 is the most recent version in a series of ECHAM models evolving originally from the spectral weather prediction model of the European Centre for Medium Range Weather Forecasts (ECMWF; Simmons *et al.*, 1989).

HAMMONIA also comprises a full dynamic and radiative two-way coupling with the MOZART3 chemical module (Kinnison *et al.*, 2005; 48 compounds, 148 gas phase reactions in the version used here). It is a spectral model with triangular truncation at wavenumber 31 (T31) and with 67 vertical levels ranging from the surface to 1.7 × $10^{-7}$ hPa (~250 km). The dynamical and radiative processes that have been specifically implemented for HAMMONIA include solar heating in the UV and EUV wavelength regime (down to 5 nm, Richards *et al.*, 1994), a non-local thermodynamic equilibrium long-wave radiative scheme (Fomichev and Blanchet, 1995; Fomichev *et al.*, 1998), heating and mixing due to parameterized gravity waves (Hines, 1997a,b), vertical molecular diffusion and heat conduction (parameterized following the basic equations from Banks and Kockarts, 1973), and a simple parameterization of electromagnetic forces in the thermosphere (ion drag and Lorentz forces, Hong and Lindzen, 1976). Two 20-year time-slice experiments for solar max and solar min conditions as observed in November 1989 and September 1986, respectively, have been performed. The spectral solar irradiance data used for these periods were produced by Lean (2000). A more detailed description of the model and the experiments analyzed in the present paper is given by Schmidt *et al.* (2006).

## 3. The Zonal Mean Atmospheric Response to Solar Variability

### 3.1. Annual Mean Ozone and Temperature

Figure 1 shows the relative difference between the ozone volume mixing ratios calculated by HAMMONIA for solar max and solar min in the annual mean. The response above the stratopause has been discussed by Schmidt *et al.* (2006). The

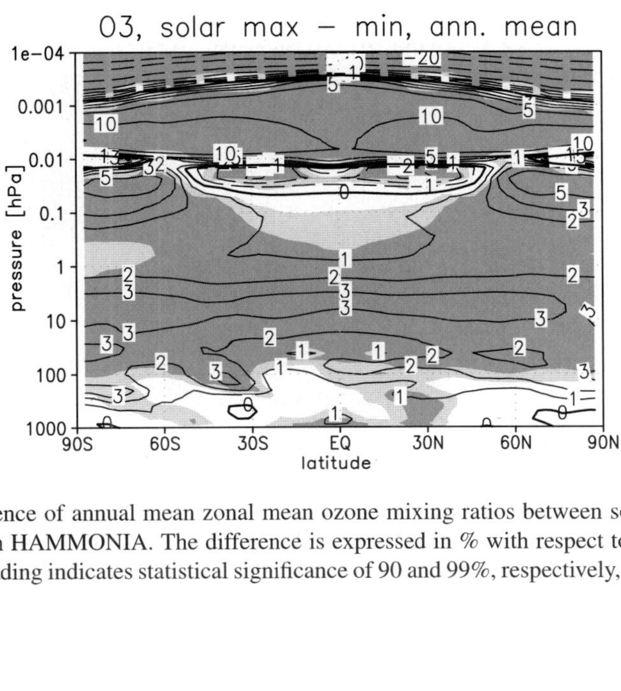


*Figure 1.* Difference of annual mean zonal mean ozone mixing ratios between solar max and min as simulated with HAMMONIA. The difference is expressed in % with respect to solar min. Light and dark grey shading indicates statistical significance of 90 and 99%, respectively, as computed with Student's *T*-test.

largest stratospheric response of slightly more than 3% occurs at pressure levels between 5 and 10 hPa (~ between 35 and 30 km) and seems to be uniform for almost all latitudes. The reason for the ozone increase is the higher rate of $O_2$ photodissociation by the increased UV irradiance. Analyses of data provided by the *Solar Backscattered Ultraviolet* (SBUV, McCormack and Hood, 1996) and the *Stratospheric Aerosol and Gas Experiment* (SAGE, Wang *et al.*, 1996) satellite instruments show maximum responses of similar magnitude but a different latitudinal pattern. Both analyses suggest a double peak structure with distinct maxima at mid-latitudes in the upper stratosphere (at slightly different altitudes for SAGE and SBUV). Additionally, a negative response occurs in the equatorial middle stratosphere. However, 2D-simulations by Lee and Smith (2003) suggest, that the ozone response pattern derived from observations may be influenced by the fact that the regression technique used fails in properly separating effects of solar variability and the quasi-biennial oscillation (QBO) of equatorial stratospheric winds. The pure solar effect may well be homogeneous in latitude and non-negative everywhere in the stratosphere.

The temperature response simulated by HAMMONIA (Figure 2) shows a stratospheric maximum of about 1 K close to the stratopause. The temperature increase is due to both the direct effect of stronger solar irradiance at solar max and the increased availability of ozone which is the main absorber at these altitudes. The absolute stratospheric maximum of slightly more than 1 K occurs at about 40°N. However, around the stratopause, the magnitude of the response is very close to 1 K for all latitudes. The simulated temperature response is significant at low latitudes down to the tropopause but at high latitudes only in the uppermost stratosphere.

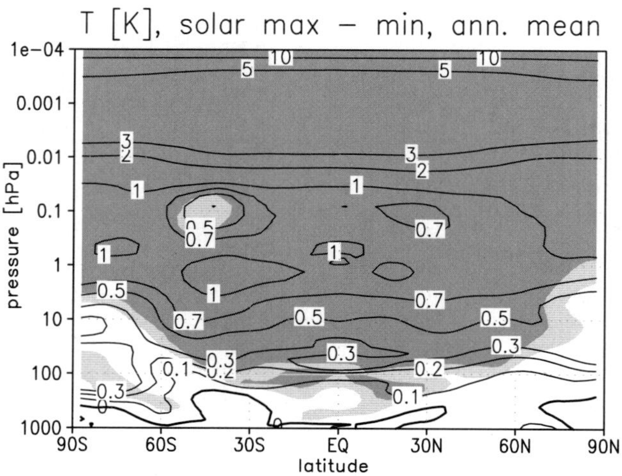


*Figure 2.* Difference of annual mean zonal mean temperatures (K) between solar max and min as simulated with HAMMONIA. Light and dark grey shading indicates statistical significance of 90 and 99%, respectively, as computed with Student's *T*-test.

Different analyses of satellite (Hood, 2004; Scaife *et al.*, 2000), rocket (Keckhut *et al.*, 2005), and reanalysis (Hood, 2004; Crooks and Gray, 2005) data show temperature response maxima of 0.5 to above 2 K in the equatorial upper stratosphere (~40 to 50 km) and most of them suggest a secondary maximum of smaller magnitude close to the equatorial tropopause. HAMMONIA shows a response of about 0.3 K slightly above 100 hPa (~16 km) which is in the range of observed values but appears only as a very weak local maximum.

Earlier modeling studies on the solar cycle effect on ozone or temperature (or both) are presented, e.g., by Haigh (1994), Shindell *et al.* (1999), Tourpali *et al.* (2003), Matthes *et al.* (2004), and Rozanov *et al.* (2004). In most cases, magnitude and height of the maximum ozone and temperature responses are comparable to our results. However, the exact pattern and longitudinal distribution of the response can differ significantly between models (in case of the temperature response see, e.g., the collection of results for five different models by Matthes *et al.*, 2003). A negative response of temperature or ozone in the equatorial middle stratosphere, as suggested by some observations, is in general not simulated by numerical models.

### 3.2. Northern Hemisphere Winter Zonal Wind

In studies based on NCEP (Kodera and Kuroda, 2002) and ERA40 (Crooks and Gray, 2005) reanalysis data, the response of stratospheric zonal mean zonal winds in the northern winter hemisphere is discussed. Kodera and Kuroda (2002) suggest that the strength of the subtropical jet in the upper stratosphere increases for solar max in early winter and that this increase propagates poleward and downward during the course of the winter. The response pattern simulated by HAMMONIA (Figure 3) for December is similar to the analysis of Kodera and Kuroda (2002) for January. The response simulated actually for January differs substantially from the reanalysis while the weakening of the stratospheric polar night jet in February occurs in the reanalysis and the simulation. A poleward and downward movement of the maximum of the stratospheric jet increase, as observed by Kodera and Kuroda (2002), is not simulated. In general, the statistical significance of the simulated wintertime wind response is small. It has to be noted that earlier simulations with a slightly different model version that yielded smaller absolute response values in ozone and temperatures produced a downward and poleward propagation of the wintertime jet very similar to that analyzed by Kodera and Kuroda (2002). However, also the response in this earlier simulation was of low statistical significance. We therefore conclude that the dynamical response in the stratosphere is not robust in our model. It seems that a prerequisite to simulate a strong early winter increase of the upper stratospheric jet is a strong latitudinal gradient of the temperature response in the same height region with strong positive values at low latitudes and negative values at high latitudes. The negative high latitude response is much stronger in the earlier simulations than in the simulations presented here. However, due to the

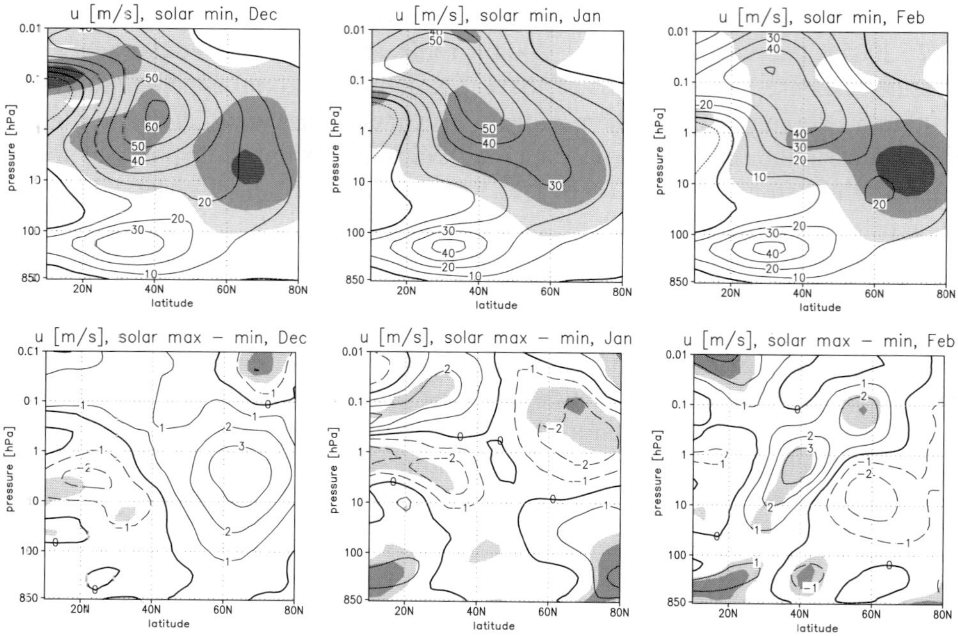

*Figure 3. Upper panels:* Zonal mean zonal winds (m/s) for solar min as simulated with HAMMONIA for the northern winter hemisphere from December to February. Light, medium and dark shading indicates the standard deviations larger than 4, 8, and 12 m/s, respectively. Compare to Figure 10 of Matthes *et al.* (2003). *Lower panels:* Difference of the respective zonal mean zonal winds (m/s) between solar max and min. Light and dark grey shading indicates statistical significance larger than 80 and 90%, respectively, as computed with Student's *T*-test. Compare to Figure 4 of Matthes *et al.* (2004).

strong interannual temperature variability, in both cases, this negative response is not significant.

Also other models seem to have difficulties to reproduce analyzed dynamical patterns as can be seen in the model intercomparisons by Matthes *et al.* (2003) and Shibata and Kodera (2005). Matthes *et al.* (2003) see the proper simulation of the background wind climatology (including the interannual variability) as a prerequisite to reproduce the solar cycle effect correctly. In our model (see Figure 3), we simulate a subtropical jet of realistic magnitude but a too weak polar night jet. During northern winter we observe a weak poleward and downward movement of the stratospheric jet maximum. The interannual variability of the winds is slightly weaker than observed but shows a double peak structure in the winter stratosphere as described by Matthes *et al.* (2003).

Matthes *et al.* (2004) have improved the dynamical model response in a GCM by prescribing a realistic QBO (which is not included in our simulation). Interactions between solar variability and QBO have been described by many authors (e.g., Labitzke and van Loon, 1988; Salby and Callaghan, 2000). Matthes *et al.* (2004)

point out the importance of the QBO in producing a realistic stratospheric wind variability also at higher latitudes. The inclusion of the QBO leads, e.g., to a more realistic semi-annual oscillation in the stratopause region (e.g., Garcia and Sassi, 1999), i.e., in a region where the direct solar signal is strong.

## 4. The Zonal Structure of the Temperature Response

In the previous section, the solar cycle effect on the atmosphere has been discussed only in terms of zonal mean signals. However, HAMMONIA includes comprehensive dynamics of the lower and middle atmosphere, should thereby provide a reasonable representation of upward propagating planetary waves and tides, and can be expected to be a useful tool to describe also the three-dimensional atmospheric response to solar variability. It should be noted that only the temperature response will be discussed in the following. However, also the simulated responses of winds and several chemical compounds exhibit a longitude dependence.

Figure 4 presents horizontal cross-sections for the January temperature difference between solar max and min at four different pressure levels (500, 30, 1, and 0.01 hPa). It is obvious that the temperature response in the northern (winter)

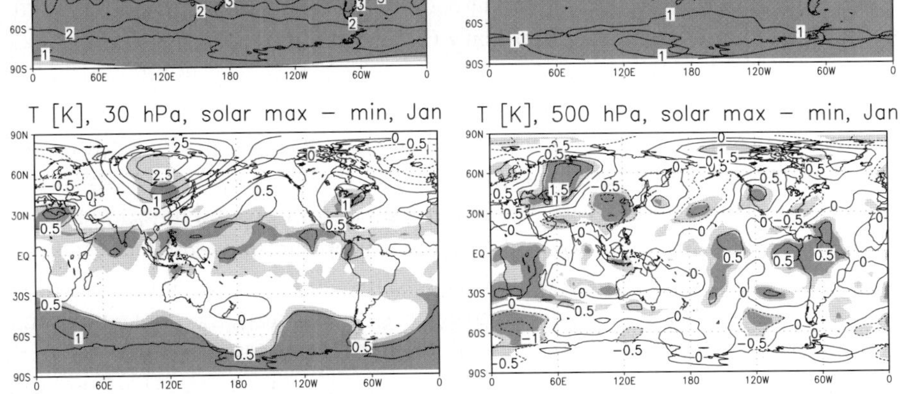


*Figure 4.* Horizontal cross sections of the differences of January mean temperatures (K) between solar max and min as simulated with HAMMONIA for the pressure levels of 500, 30, 1, and 0.01 hPa. Light and dark grey shading indicates statistical significance larger than 80 and 90%, respectively.

hemisphere strongly depends on longitude for all given altitudes. The pattern suggests that the standing planetary waves may be modified by the changing solar activity. Near the stratopause (1 hPa) at 60°N, the difference of the temperature signal for different longitudes lies between about +3 K and −2 K at 0°E and 120°E, respectively. The longitude dependence becomes smaller but is still clearly visible in the upper mesosphere (0.01 hPa). In the stratosphere (30 and 1 hPa), the temperature difference is dominated by a wave-1 pattern. The phases of the signals in the lower stratosphere (30 hPa) and at the stratopause (1 hPa) are shifted by almost 180°. In the southern (summer) hemisphere the temperature response is almost uniform in latitude at stratospheric heights. The same is true for the northern summer (not shown). The reason is that the wave activity itself is much stronger in winter than in summer. In the upper mesosphere, zonal structures can also be detected at equatorial and southern latitudes. Further analysis is needed to decide on the origin of this pattern. Changes in tides or in mesospheric planetary wave activity (of, e.g., the quasi-two day wave) may contribute to it. Figure 4 indicates also statistical significances of the temperature signals. In the mesosphere, the strong positive temperature signal is significant almost everywhere. In the stratosphere, the region of strong wave activity in high northern latitudes shows high statistical significance only at the maxima of the signal. This indicates that in this area of strong wave activity a longer simulation time is needed to obtain a significant mean wave pattern change. Southward of this region, significances are higher, except where the signal itself is very small as in some regions at 30 hPa. In the free troposphere, the temperature signal has a much less clear pattern than above. However, it is interesting to note that the statistical significance of several small scale maxima and minima is relatively large.

A longitudinal dependence of the solar cycle signal has been discovered earlier in other model studies. For the troposphere see for example the work from Haigh (1999). Recent studies with a GCM (Matthes *et al.*, 2004), with a CCM (Rozanov *et al.*, 2004), and with a mechanistic 3D-model using imposed planetary wave forcing at the lower model boundary (Hampson *et al.*, 2005; Chanin, 2006) show stratospheric wave patterns of comparable magnitude as our simulations. An evaluation of the simulated longitudinal dependence seems difficult. Satellite data are usually analyzed for zonal mean results because for any partition of longitude bands the number of available data is substantially smaller and the statistical significance of the results would be low. Wintertime data from ground based rocket and lidar observations for six northern mid-latitude stations are presented by Hampson *et al.* (2005). They seem to indicate that, depending on the location (and therefore maybe depending on longitude), the stratospheric temperature response may be of different magnitude. Most stations suggest a negative temperature response in the mid-latitude winter stratosphere. However, in general, the errors associated with the observations are large enough to allow either a positive or a negative response. As can be seen in Figure 4 the zonal mean model response in this region is close to zero and not significantly negative.

## 5. Discussion

In this paper, we have presented results from two 20-year simulations with the HAMMONIA CCM for solar min and max conditions. As in earlier studies that have used simpler 2D or 3D models (see, e.g., references in Schmidt *et al.*, 2006), the simulated magnitude of the stratospheric zonal mean ozone and temperature change between solar min and solar max is close to the observed values. However, a detailed evaluation of model results is difficult as also analyses of different data sources give different results. For example, the ozone response analyses for different satellite data (SAGE, Wang *et al.*, 1996; SBUV, McCormack and Hood, 1996; HALOE, Soukharev and Hood, 2005) do not agree completely in the position and magnitude of maxima and minima. One problem is certainly the relatively short time-series available from satellite instruments of in general not more than two solar cycles. Another problem in the comparison with numerical simulations becomes evident in the work of Lee and Smith (2003). The multilinear regression technique used in most data analysis studies may fail in properly separating signals from different sources (like the solar cycle and the QBO). The comparison of these results with numerical time-slice experiments like in the present study is therefore difficult and it is desirable to perform future numerical simulations with real transient solar forcing.

The stratospheric dynamical response to the solar cycle as simulated with HAMMONIA does not reproduce robustly and in a statistically significant way the analyses obtained by Kodera and Kuroda (2002). The same is to our knowledge true for most other earlier simulations with different numerical models. According to Matthes *et al.* (2003), an important reason for this behavior could be that most GCMs are still not able to reproduce exactly the stratospheric dynamical background conditions. On the other hand, analyses of observations rely mostly on only two solar cycles. So, it can not be excluded, that the solar cycle response of the real atmosphere is not robust. It may change between different solar cycles, in particular when the background atmospheric conditions undergo climate change. An option to improve the dynamical model response of the middle atmosphere may be the inclusion of the QBO (Matthes *et al.*, 2004) which is known to interact with the solar cycle signal (e.g., Labitzke, 2003). HAMMONIA is expected to be one of the few GCMs that are able to produce internally a realistic QBO (Giorgetta *et al.*, 2002). However, the vertical model resolution in the stratosphere would have to be increased substantially with respect to the current simulations which would make such experiments extremely costly in terms of computer time.

A problem of interpreting ground based observations with respect to the solar cycle signal becomes obvious when looking at the longitudinal variability of the HAMMONIA results. Provided our simulations are representative of the real atmosphere, it can be expected that up to the mesopause, the temperature response for example may be either positive or negative, depending on longitude. This is in particular true for middle to high latitudes in winter. Future numerical simulations will be analyzed in further detail with respect to the mechanisms that may cause changes in planetary wave activities. However, the future evaluation of modeling results for the solar cycle depends crucially on the availability of continuous observational (both ground based and satellite) time series over the next solar cycles. Only longer accurate time series will allow to obtain statistically significant patterns of the atmospheric response to solar variability.

## Acknowledgements

The authors are grateful to Judith Lean, Naval Research Laboratory, Washington, for providing the solar irradiance spectra. The work was partly financed by the German Ministry of Education and Research (BMBF) within the framework of the AFO2000 research program under contract 07ATF10. We also acknowledge the support of the German Climate Computing Center (DKRZ) where the numerical computations were performed.

## References

Banks, P. M., and Kockarts, G.: 1973, *Aeronomy, Part B*, New York, Academic Press.

Beres, J. H., Garcia, R. R., Boville, B. A., and Sassi, F.: 2005, 'Implementation of a gravity wave source spectrum parameterization dependent on the properties of convection in the whole atmosphere community climate model (WACCM)', *J. Geophys. Res.* **110**, doi:10.1029/2004JD005504.

Chanin, M.-L.: 2006, 'Signature of the 11-year cycle in the upper atmosphere', *Space Sci. Rev.*, this volume, doi: 10.1007/s11214-006-9062-5.

Crooks, S. A., and Gray, L. J.: 2005, 'Characterization of the 11-year solar signal using a multiple regression analysis of the ERA-40 dataset', *J. Climate* **18**, 996–1015.

Fomichev, V. I., and Blanchet, J.-P.: 1995, 'Development of the new CCC/GCM longwave radiation model for extension into the middle atmosphere', *Atmos. Ocean* **33**, 513–531.

Fomichev, V. I., Blanchet, J.-P., and Turner, D. S.: 1998, 'Matrix parameterization of the 15 $\mu$m $CO_2$ band cooling in the middle and upper atmosphere for variable $CO_2$ concentration', *J. Geophys. Res.* **103**, 11,505–11,528.

Fomichev, V. I., Ward, W. E., Beagley, S. R., McLandress, C., McConnell, J. C., McFarlane, N. A., Shepherd, T. G.: 2002, 'The extended Canadian middle atmosphere model: Zonal mean climatology and physical parameterizations', *J. Geophys. Res.* **107**, doi:10.1029/2001JD000479.

Froehlich, C.: 2004, 'Solar irradiance variability', in J. M. Pap and P. Fox (eds.), *Solar Variability and its Effects on Climate*, No. 141 in Geophys. Monogr. Ser., Washington, DC: AGU, pp. 97–110.

Garcia, R. R., and Sassi, F.: 1999, 'Modulation of the mesospheric semiannual oscillation by the quasi-biennial oscillation', *Earth Planets Space* **51**, 563–569.

Giorgetta, M. A., Manzini, E., and Roeckner, E.: 2002, 'Forcing of the quasi-biennial oscillation from a broad spectrum of atmospheric waves', *Geophys. Res. Lett.* **29**, doi:10.1029/2002GL014756.

Haigh, J. D.: 1994, 'The role of stratospheric ozone in modulating the solar radiative forcing of climate', *Nature* **370**, 544–546.

Haigh, J. D.: 1999, 'A GCM study of climate change in response to the 11-year solar cycle', *Quart. J. Roy. Meteor. Soc.* **125**, 871–892.

Hampson, J., Keckhut, P., Hauchecorne, A., and Chanin, M.-L.: 2005, 'The effect of the 11-year solar-cycle on the temperature in the upper-stratosphere and mesosphere: Part III investigations of zonal asymmetry', submitted to *J. Atm. Sol. Terr. Phys.*

Hines, C. O.: 1997a, 'Doppler-spread parameterization of gravity wave momentum deposition in the middle atmosphere. Part 1: Basic formulation', *J. Atm. Sol. Terr. Phys.* **59**, 371–386.

Hines, C. O.: 1997b, 'Doppler-spread parameterization of gravity wave momentum deposition in the middle atmosphere. Part 2: Broad and quasi monochromatic spectra, and implementation', *J. Atm. Sol. Terr. Phys.* **59**, 387–400.

Hong, S. S., and Lindzen, R. S.: 1976, 'Solar semidiurnal tide in the thermosphere'. *J. Atm. Sc.* **33**, 135–153.

Hood, L.: 2004, 'Effects of solar UV variability on the stratosphere', in J. M. Pap and P. Fox (eds.), *Solar Variability and its Effects on Climate*, No. 141 in Geophys. Monogr. Ser., Washington, DC: AGU, pp. 283–304.

Keckhut, P., Cagnazzo, C., Chanin, M.-L., Claud C., and Hauchecorne A.: 2005, 'Midlatitude long-term variability of the atmosphere: Trends and cyclic and episodic changes', *J. Atm. Sol. Terr. Phys.* **67**, 940–947.

Kinnison, D. E., Brasseur, G. P., Walters, S., Garcia, R. R., Sassi, F., Boville, B., Marsh, D., Harvey, L., Randall, C., Emmons, L., and Pan, R. W. L.: 2005, 'Sensitivity of chemical tracers to meteorological parameters in the MOZART-3 chemical transport model', submitted to *J. Geophys. Res.*

Kodera, K., and Kuroda, Y.: 2002, 'Dynamical response to the solar cycle', *J. Geophys. Res.* **107**, doi:10.1029/2002JD002224.

Labitzke, K.: 2003, 'The global signal of the 11-year sunspot cycle in the atmosphere: When do we need the QBO?', *Meteorol. Z.* **12**, 209–216.

Labitzke, K., and van Loon, H.: 1988, 'Associations between the 11-year solar cycle, the QBO and the atmosphere, part I: The troposphere and the stratosphere in the northern hemisphere winter', *J. Atmos. Terr. Phys.* **64**, 203–210.

Lean, J.: 2000, 'Evolution of the Sun's spectral irradiance since the Maunder minimum', *Geophys. Res. Lett.* **27**, 2425–2428.

Lean, J., Rottman, J., Kyle, G. J., Woods, H. L., Hickey, T. N., and Pugga, J. R.: 1997, 'Detection and parameterization of variations in solar mid and near-ultraviolet radiation (200–400 nm)', *J. Geophys. Res.* **102**, 29,939–29,956.

Lee, H., and Smith, A. K.: 2003, 'Simulation of the combined effects of solar cycle, quasi-biennial oscillation, and volcanic forcing on stratospheric ozone changes in recent decades', *J. Geophys. Res.* **108**, doi:10.1029/2001JD001503.

Manzini, E., McFarlane, N. A., and McLandress, C.: 1997, 'Impact of the Doppler spread parameterization on the simulation of the middle atmosphere circulation using the MA/ECHAM4 general circulation model', *J. Geophys. Res.* **102**, 25,751–25,762.

Matthes, K., Kodera, K., Haigh, J. D., Shindell, D. T., Shibata K., Langematz, U., and Rozanov, E.: 2003, 'GRIPS solar experiments intercomparison project: Initial results', *Papers in Meteorol. Geophys.* **54**, 71–90.

Matthes, K., Langematz, U., Gray, L. J., Kodera, K., and Labitzke, K.: 2004, 'Improved 11-year solar signal in the Freie Universität Berlin Climate Middle Atmosphere Model (FUB-CMAM)', *J. Geophys. Res.* **109**, doi:10.1029/2003JD004012.

McCormack, J. P., and Hood, L.: 1996, 'Apparent solar cycle variations of upper tratospheric ozone and temperature: Latitude and seasonal dependence', *J. Geophys. Res.* **101**, 20,933–20,944.

Richards, P. G., Fennelly, J. A., and Torr, D. G.: 1994, 'A solar EUV flux model for aeronomic calculations', *J. Geophys. Res.* **99**, 8981–8992. (Correction, JGR, 99, 13283, 1994).

Roeckner, E., Bäuml, G., Bonaventura, L., Brokopf, R., Esch, M., Giorgetta, M., Hagemann, S., Kirchner, I., Kornblueh, L., Manzini, E., Rhodin, A., Schlese, U., Schulzweida, U., and T. A.: 2003,

'The atmospheric general circulation model ECHAM 5. Part I: Model description', Technical Report 349, MPI for Meteorology, Hamburg, Germany.

Roeckner, E., Brokopf, R., Esch, M., Giorgetta, M., Hagemann, S., Kornblueh, L., Manzini, E., Schlese, U., and Schulzweida, U.: 2006, 'Sensitivity of simulated climate to horizontal and vertical resolution in the ECHAM5 atmosphere model', *J. Climate*, in press.

Rottman, G.: 2000, 'Variations of solar ultraviolet irradiance observed by the UARS SOLSTICE – 1991 to 1999', *Space Sc. Rev.* **94**, 83–91.

Rozanov, E. V., Schlesinger, M. E., Egorova, T. A., Li, B., Andronova, N., and Zubov, V. A.: 2004, 'Atmospheric response to the observed increase of solar UV radiation from solar minimum to solar maximum simulated by the University of Illinois at Urbana-Champaign climate-chemistry model', *J. Geophys. Res.* **109**, doi:10.1029/2003JD003796.

Salby, M., and Callaghan, P.: 2000, 'Connection between the solar cycle and the QBO: The missing link', *J. Climate* **13**, 2652–2662.

Scaife, A., Austin, J., Butchart, N., Pawson, S., Keil, M., Nash, J., and James, I. N.: 2000, 'Seasonal and interannual variability of the stratosphere diagnosed from UKMO TOVS analyses', *Quart. J. Roy. Meteor. Soc.* **126**, 2585–2604.

Schmidt, H., Brasseur, G. P., Charron, M., Manzini, E., Giorgetta, M. A., Fomichev, V. I., Kinnison, D., Marsh, D., and Walters, S.: 2006, 'The HAMMONIA chemistry climate model: Sensitivity of the mesopause region to the 11-year solar cycle and $CO_2$ doubling', *J. Climate* **19**, 3903–3931.

Shibata, K., and Kodera, K.: 2005, 'Simulation of radiative and dynamical responses of the middle atmosphere to the 11-year solar cycle', *J. Atm. Sol. Terr. Phys.* **67**, 125–143.

Shindell, D., Rind, D., Balachandran, N., Lean, J., and Lonergan, J.: 1999, 'Solar cycle variability, ozone, and climate', *Science* **284**, 305–308.

Simmons, A. J., *et al.*: 1989, 'The ECMWF medium-range prediction model: Development of the numerical formulations and the impact of increased resolution', *Meteorol. Atmos. Phys.* **40**, 28–60.

Siskind, D. E.: 2000, 'On the coupling between middle and upper atmospheric odd nitrogen', in D. E. Siskind, S. D. Eckermann, and M. E. Summers (eds.), *Atmospheric Science Across the Stratopause*, No. 123 in Geophys. Monogr. Ser., Washington, DC: AGU, pp. 101–116.

Soukharev, B., and Hood, L.: 2005, 'The 11-year solar cycle variation of stratospheric ozone as obtained from the SBUV and HALOE ozone profile measurements', Oral presentation IAGA2005-A-01264, at the IAGA assembly, Toulouse.

Tourpali, K., Schuurmans, C. J. E., van Dorland, R., Steil, B., and Brühl, C.: 2003, 'Stratospheric and tropospheric response to enhanced solar UV-radiation: A model study', *Geophys. Res. Lett.* **30**, doi:10.1029/2002GL016650.

Wang, H. J., Cunnold, D. M., and Bao, X.: 1996, 'A critical analysis of stratospheric aerosol and gas experiment ozone trends', *J. Geophys. Res.* **101**, 12,495–12,514.

# A POSSIBLE TRANSFER MECHANISM FOR THE 11-YEAR SOLAR CYCLE TO THE LOWER STRATOSPHERE

L. J. GRAY[1,*], S. A. CROOKS[2], M. A. PALMER[2], C. L. PASCOE[3] and S. SPARROW[2]
[1]*Centre for Global Atmospheric Modelling, Department of Meteorology, Reading University, Early Gate, PO Box 243, Reading RG6 6BB, UK*
[2]*Department of Atmospheric, Oceanic and Planetary Physics, Oxford University, UK*
[3]*Rutherford Appleton Laboratory, Chilton, Didcot, UK*
*(*Author for correspondence: E-mail: lesley@met.rdg.ac.uk)*

(Received 9 September 2005; Accepted in final form 7 March 2006)

**Abstract.** Observational evidence of the 11-year solar cycle (SC) modulation of stratosphere temperatures and winds from the ERA-40 dataset is reviewed, with emphasis on the Northern winter hemisphere. A frequency modulation of sudden warming events is noted, with warmings occurring earlier in solar minimum periods than in solar maximum periods. The observed interaction between the influence of the SC and the quasi biennial oscillation (QBO) on the frequency of sudden warmings is noted as a possible clue for understanding their mechanism of influence. A possible transfer route for the 11-year solar cycle from the equatorial stratopause region to the lowest part of the stratosphere is proposed, via an influence on sudden warming events and the associated induced meridional circulation. SC and QBO composites of zonal wind anomalies show anomalous wind distributions in the subtropical upper stratosphere in early winter. Mechanistic model experiments are reviewed that demonstrate a sensitivity of sudden warmings to small wind anomalies in this region. Various diagnostics from these experiments are shown, including EP fluxes and their divergence and also the synoptic evolution of the polar vortex, in order to understand the mechanism of the influence. Some recent GCM experiments to investigate the SC/QBO interaction are also described. They simulate reasonably well the observed SC/QBO interaction of sudden warming events and appear to support the hypothesis that tropical/subtropical upper stratospheric wind anomalies are an important influence on the timing of sudden warmings.

**Keywords:** stratosphere, quasi biennial oscillation, solar cycle, atmosphere, stratospheric sudden warming

## 1. Introduction

A prime challenge in studies of solar variability and its influence on the Earth's climate has been to find a plausible mechanism for the transfer and amplification of the extremely small direct solar variations at the top of the atmosphere down to the surface. It is reasonably well established that there is a significant 11-year signal in stratospheric temperatures of around 1–2 K in the region 40–50 km (Labitzke, 1987; Scaife *et al.*, 2000; Hood, 2004; Keckhut *et al.*, 2005; Crooks and Gray, 2005). It was proposed by Haigh (1994, 1996) that this is a response not only to heating due to the solar irradiance changes but also to heating due to changes in stratospheric ozone (Hood, 1997, 2004; Lee and Smith, 2003) that are caused by


DOI: 10.1007/s11214-006-9069-y

the 11-year solar cycle (SC) variations in the UV part of the solar spectrum. Both of these give rise to higher temperatures in solar maximum ($S_{max}$) than in solar minimum ($S_{min}$).

There is, additionally, an observed secondary SC signal in both temperature and ozone centred over equatorial latitudes in the lower stratosphere, at around 20–25 km. The origin of this feature is less clear, although most proposed mechanisms have been in terms of a solar influence on planetary wave propagation and changes to the accompanying large-scale mean meridional circulation (e.g., Balachandran and Rind, 1995; Kodera, 2006). Once the SC signal has reached the lower stratosphere, a number of mechanisms have been proposed for its transmission downwards into the troposphere and hence to the Earth's surface (see, e.g., Haigh and Blackburn, 2006; Kodera, 2006). In this paper, we focus on the lower stratosphere and investigate in more detail the solar cycle influence on planetary wave propagation and hence the dynamical transfer of the SC signal from the region of the equatorial stratopause to the lower stratosphere.

The occurrence of Northern Hemisphere stratospheric sudden warmings has shown some interesting dependencies on both the Quasi Biennial Oscillation (QBO) and the SC. Sudden warmings are characterised by substantial increases in polar winter temperatures in just a few days, associated with the disturbance (and sometimes destruction) of the polar vortex caused by vertically propagating planetary waves from the troposphere. The QBO relationships were first reported by Holton and Tan (1980, 1982) who found that sudden warmings occurred more often in the QBO/E phase (i.e., when the equatorial winds were from the east). The SC/QBO relationship was first reported by Labitzke (1987) and Labitzke and van Loon (1988), who found a relationship between sudden warmings and the SC that only emerged strongly when the years were separated into QBO-E and QBO-W years (see Labitzke, 2006). In fact, they showed that in $S_{max}$, warmings occurred more often when the QBO was in its westerly phase. This latter observation was rather surprising, and does not fit with the generally accepted Holton-Tan theory that the QBO influence on warmings is primarily through its control of the latitudinal boundary of the mid-latitude westerlies and hence the latitudinal extent of planetary wave propagation, which predicts that warmings occur preferentially in the QBO-E phase (Baldwin *et al.*, 2001). More recent data have, in general, upheld these unexpected observations of warmings in $S_{max}$/QBO-W years, so that the Holton-Tan relationship appears to hold well during $S_{min}$ years but not during $S_{max}$ years (Gray *et al.*, 2001a, 2004).

Kodera *et al.* (1990, 2000) and Kodera (1995) showed that when zonal wind anomalies in northern hemisphere winter were composited according to the SC and to the phase of the QBO, the monthly-averaged $S_{min}$ minus $S_{max}$ differences and the QBO/E minus QBO/W differences both display strikingly similar patterns of anomalous easterlies that move polewards and downwards with time during the winter. This suggested that the SC/QBO interaction might provide some clues for a possible mechanism for the signal transfer to the lower stratosphere (see also

Balachandran and Rind, 1995). However, as noted earlier, the main SC temperature signal is observed in the upper stratosphere, close to the stratopause level (40–50 km) whereas the QBO has its largest amplitude in the lower stratosphere (25–30 km), and the detailed mechanism for their interaction has been unclear.

Gray *et al.* (2001b) highlighted the fact that the QBO signal actually extends over a considerable depth of the stratosphere and although its amplitude diminishes with height, it is nevertheless still of the order of 5–10 $ms^{-1}$ at the stratopause (see also Baldwin and Gray, 2005; Pascoe *et al.*, 2005). There is a similarly small SC wind anomaly of around 5–10 $ms^{-1}$ in the subtropics at this level, associated with the latitudinal gradient in the SC temperature response (Crooks and Gray, 2005) and this has an impact on the variability of the mesospheric subtropical jet in early winter (Kodera and Kuroda, 2002; Kodera *et al.*, 2003). It is suggested that these QBO and SC wind anomalies at equatorial/subtropical latitudes in the upper stratosphere may influence the development of stratospheric warmings at polar latitudes, since mechanistic modelling studies of sudden warmings have shown strong sensitivity to small wind anomalies in these regions (Gray *et al.*, 2001b, 2004; Gray, 2003). In this paper, we summarise both the observational evidence and the modelling studies that have tested this hypothesis. In particular, the studies have suggested that the SC modulation of sudden warmings is a frequency modulation, rather than a more straightforward amplitude modulation. This may explain why the SC/QBO interaction has been difficult to characterise and disentangle in the observations, since simple techniques, such as linear regression analysis, are not capable of describing a non-linear interaction of this nature.

## 2. The Solar Cycle Signal in ERA-40 Analysis

In a recent linear regression analysis, Crooks and Gray (2005) used the ERA-40 reanalyses dataset for the period 1979–2001 to separate the 11-year SC signal from other sources of variability such as volcanic activity, the QBO, NAO, ENSO and long-term climate trend. The results (Figure 1) confirm the presence of the height-wise dipole structure in the solar component of the annual-mean zonal-mean temperature field, with a maximum amplitude of 1.75 K over the equator at 40–50 km and a secondary maximum in the lower stratosphere at 20–25 km. Interestingly, the equatorial lower stratospheric signal has a dipolar structure in latitude, with maxima of 0.5–0.75 K in the subtropics and a local minimum over the equator, which has a similar structure to the observed QBO (Baldwin *et al.*, 2001). It suggests a SC/QBO interaction in this region that the linear regression technique has been unable to separate.

There is also a relatively strong latitudinal temperature gradient at 40–50 km, especially in the Northern Hemisphere, which implies the presence of an associated zonal wind anomaly. This is confirmed in Figure 2. Analysis of the separate seasons (see Crooks and Gray, 2005) shows that the subtropical wind anomaly in each

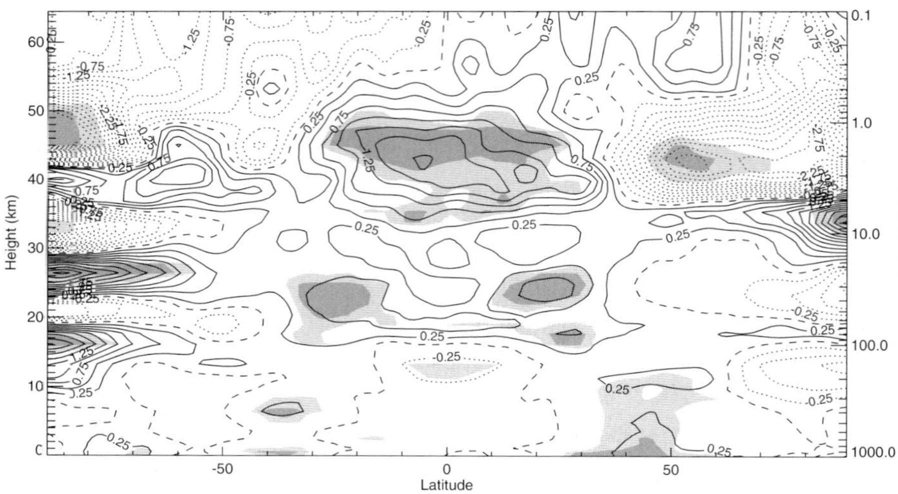

*Figure 1.* Annual-averaged, zonally-averaged solar signal in stratospheric temperature (Kelvin per solar maximum minus solar minimum) from a regression analysis of ERA-40 data. Shaded areas denote statistical significance at the 95% (light shading) and 99% (dark shading) levels. Figure adapted from Crooks and Gray (2005).

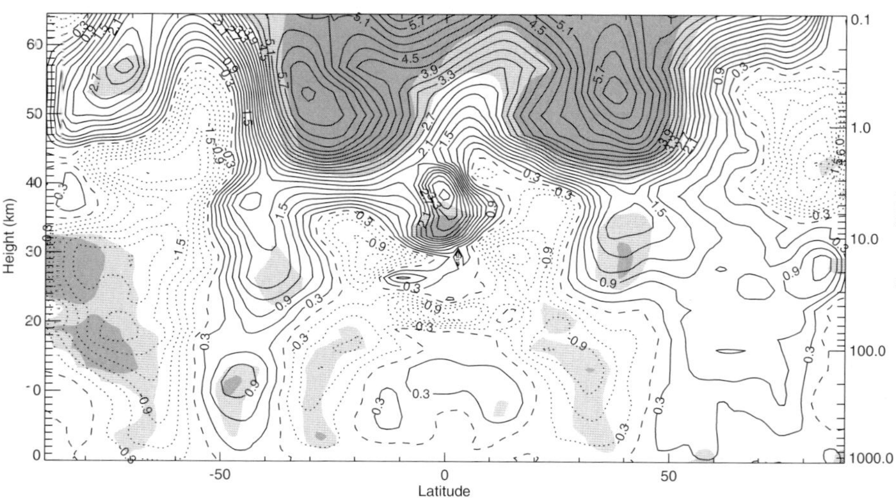

*Figure 2.* As for Figure 1 but for zonal wind ($ms^{-1}$ per solar maximum minus solar minimum).

hemisphere centred around 50 km is primarily a winter-time feature. This is further highlighted in Figure 3, which shows the monthly mean zonal wind anomalies for November through March composited into solar minimum and solar maximum years. The max minus min difference shown in the third column confirms the presence of a statistically significant SC wind anomaly that moves polewards and downwards throughout the winter, as first described by Kodera *et al.* (1990).

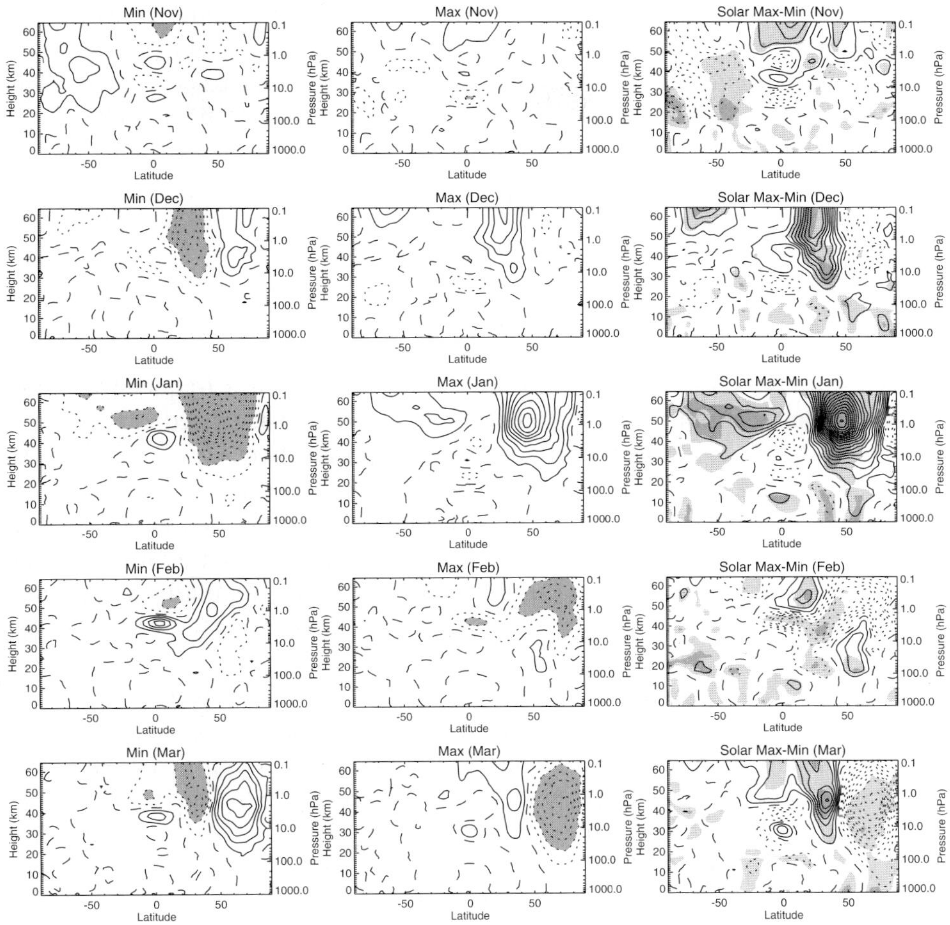

*Figure 3.* Solar cycle composites of monthly averaged latitude-height sections of zonally averaged zonal wind anomalies ($ms^{-1}$) for the period 1979–2001 in the months November to March, derived from the ERA-40 dataset. The left column shows the $S_{min}$ years, the middle column shows the $S_{max}$ years and the right column shows the $S_{max}$ minus $S_{min}$ differences. Contour interval is 2 $ms^{-1}$, dashed contour is zero, dotted contours are negative. In the left and middle columns, shaded values indicate values less than $-4$ $ms^{-1}$. In the right hand column, the light and dark shading indicate statistical significance at the 75% and 95% levels respectively, using a standard Student's *t* test. Figure adapted from Gray *et al.* (2004).

Examination of the individual SC composites provides additional insight into the nature of this poleward, downward propagating SC signal. During solar-min winters (left column) an easterly anomaly present at 50–60 km over the equator in November, grows in amplitude and moves polewards and extends downwards throughout December and January. This is a typical signature of the presence of sudden warming events in early-to-mid winter of these years (December, January), after which the vortex is re-established. Conversely, in the solar maximum composite (middle column), the vortex is unusually strong during this period. However, an easterly anomaly is nevertheless present in $S_{max}$ years, but is not present until February/March, indicating that warmings in these years tend to occur later than in solar-min years. This is therefore suggestive of a frequency modulation of the sudden warmings by the SC.

## 3. Mechanistic Modelling

### 3.1. Frequency Modulation of Sudden Warmings

The ERA-40 analysis in the previous section has suggested a frequency modulation of sudden warming events by the SC. In a series of idealised experiments using a mechanistic model of the stratosphere and mesosphere, Gray *et al.* (2003) showed that the strength of the tropospheric planetary wave forcing and the sign of the equatorial winds could each produce a similar frequency modulation. In one experiment, as the tropospheric wave forcing at the lower boundary of the model was increased, sudden warmings occurred earlier and earlier. Similarly, in a second experiment, as the imposed equatorial winds became more easterly (see Figure 4), the sudden warmings occurred earlier.

This second experiment to test the sensitivity to equatorial winds was, however, highly idealised, since the imposed equatorial winds were of the same sign at all heights in the stratosphere and therefore could not distinguish between the equatorial influence from different heights. In an extension to the study, Gray *et al.* (2004) examined the height dependence of this sensitivity, and showed that the warmings occurred significantly earlier when the equatorial winds were easterly in the lower stratosphere (as expected from Holton-Tan), and also when they were easterly in the upper stratosphere. In a comparative study of the influence of the upper versus lower equatorial stratosphere, Gray (2003) suggested that the lower stratospheric influence is greater during early winter, when the flow is more linear, but the upper stratosphere is more influential in mid-to-later winter, by which time the flow is highly non-linear.

### 3.2. Influence from the Subtropics

In a further extension to the study, Gray *et al.* (2004) examined the sensitivity of the timing of warmings to wind anomalies in the subtropics of the upper stratosphere. An easterly anomaly was imposed in the northern subtropical upper stratosphere (40–50 km; 0–30°N) for only the first 60 days of the run, in order to mimic the easterly anomaly in November/December in solar minimum years (left column, Figure 3). In the control run (with no imposed wind anomalies) the timing of the warmings was variable (see Figure 5, right hand column), on average reaching their peak at around day 120 of the run. In the subtropical anomaly experiment

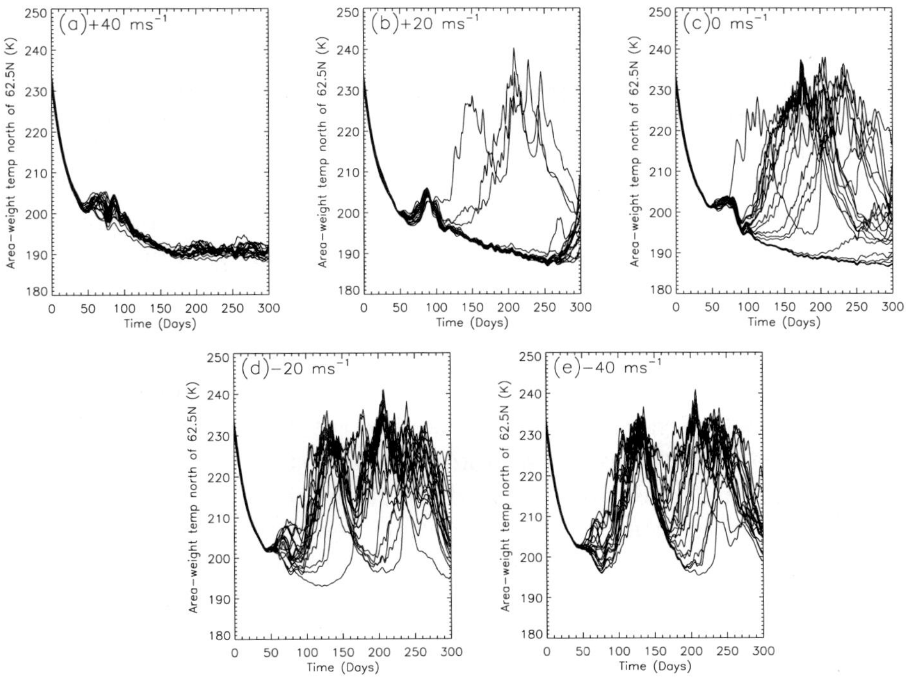


*Figure 4.* Time series of daily zonally averaged, area-weighted temperature (K) north of 62.5°N at 32 km for each of the 20 ensemble members from the mechanistic model experiments with imposed equatorial wind boundaries of (a) $+40\,\mathrm{ms}^{-1}$, (b) $+20\,\mathrm{ms}^{-1}$, (c) $0\,\mathrm{ms}^{-1}$, (d) $-20\,\mathrm{ms}^{-1}$, (e) $-40\,\mathrm{ms}^{-1}$. Figure adapted from Gray *et al.* (2003).

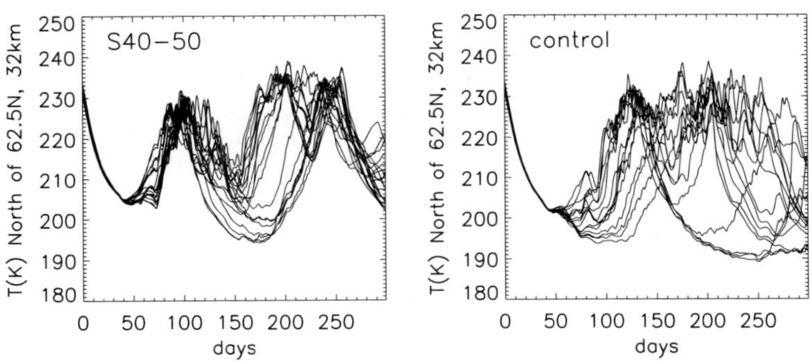


*Figure 5.* As Figure 4 but from (left) the experiment in which an easterly anomaly was imposed in the subtropics at 0–30°N and 40–50 km, and (right) the control experiment with no easterly forcing. Figure adapted from Gray *et al.* (2004).

(Figure 5, left hand column), however, the warmings were consistently earlier, peaking at around day 80–100. This supports the hypothesis that the timing of sudden warmings in this model is sensitive to relatively small wind anomalies in the equatorial/subtropical upper stratosphere in early winter.

## 3.3. E-P Flux Diagnostics

A detailed analysis of EP flux diagnostics suggests that wave forcing in the subtropical upper stratosphere in early winter is important to the later evolution of the warming. Figure 6 shows the composite E-P fluxes and divergence from the experiments described earlier (see Figure 3) in which easterlies and westerlies were imposed at the equator. The EP flux difference plots (right column, Figure 6) show stronger wave-mean-flow interaction in the upper stratosphere of the warm (easterly) composite, centred near 30–40°N, 40–50 km. Initially, it results from a convergence of flux vectors from the tropics and from the lower stratosphere (Figure 6c). The increased convergence from the tropics is due to the position of the zero wind line, which inhibits propagation across the tropics in the easterly composite. This is the mechanism behind the Holton-Tan hypothesis, operating primarily in the upper stratosphere. In the later period, as the warming develops, the strongest contribution to the enhanced flux convergence comes from enhanced fluxes through the lower boundary. This latter result has very interesting implications for the question of whether the stratosphere influences the troposphere below. The amplification of fluxes in the later stages suggests that the earlier upper stratospheric modifications

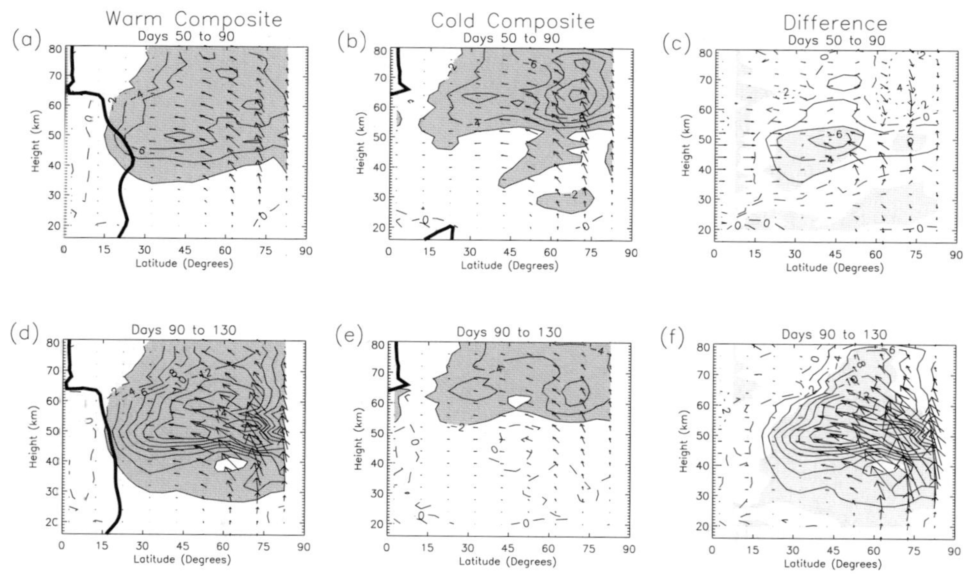


*Figure 6.* Eliassen-Palm flux diagnostics from left column: a composite of warm disturbed runs of the mechanistic model (primarily comprised of those runs shown in Figures 3a and 3b); middle column: a composite of cold undisturbed runs (primarily comprised of those runs shown in Figures 3d and 3e). The right hand column shows the warm minus cold difference. In the difference plot, the shading indicates statistical significance at the 99% level and the arrows have been magnified by a factor of two to improve legibility. For a definition of the EP flux diagnostics, see Gray *et al.* (2003). Figure adapted from Gray *et al.* (2003).

lead to enhanced wave fluxes of easterly momentum through the lower boundary of the model at 100 hPa.

## 3.4. Synoptic Developments

A diagnosis of the synoptic evolution of the polar vortex shows that it is the build-up of the Aleutian High in the subtropics of the upper stratosphere that is an important factor in the subsequent development of sudden warmings. Figure 7 shows the typical state of the vortex after just 50 days, from the subtropical easterly forcing experiment described above, compared with its control experiment. The easterly forcing experiment has a well established Aleutian High, centred at around 45°N in the upper left quadrant. There is a tongue of low PV subtropical air that extends and wraps around the edge of the vortex, apparently feeding the Aleutian High and building it up. In the control experiment, the same features are evident but are far less well developed. These differences can be readily understood, since the easterly forcing in the subtropics is in precisely the latitude band to reinforce the anticyclonicity of the Aleutian High and also the travelling anticyclones embedded within the tongue of subtropical air that are advected around the edge of the vortex and merge with the Aleutian High.

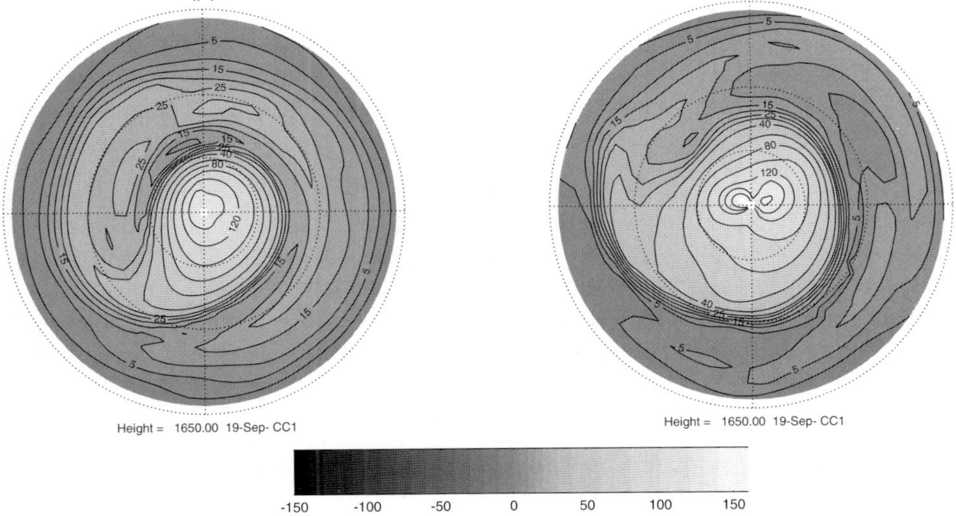


*Figure 7.* Typical distribution of the Northern Hemisphere Ertel potential vorticity on the 1650 K isentropic surface ($10^{-4}$ $Km^{-1}s^{-1}$) at day 50 from (left) the mechanistic model experiment (see Figure 5) in which an easterly forcing was applied in the subtropics 0–30°N, 40–50 km, and (right) the equivalent control run distribution. Figure adapted from Gray *et al.* (2004).

## 4. General Circulation Modelling

The sensitivity of the timing of stratospheric sudden warmings to imposed equatorial wind distributions has also been tested in a number of general circulation model (GCM) experiments (Matthes *et al.*, 2004; Palmer and Gray, 2005; Pascoe *et al.*, 2006). Pascoe *et al.* (2006) employed the UK Met. Office Unified Model (UM) and carried out a series of 28-year integrations in which the semi annual oscillation (SAO) was imposed (above 30 km) and then two slightly different QBO variations were superimposed, one that extended only to 35 km (the "shallow QBO" run) and one that extended throughout the whole depth of the stratosphere, as observed (the "deep QBO" run). Figure 8 summarises the results of these experiments and shows the North Pole temperature variance from each integration during winter. A high variance is a good indication of the timing of the sudden warmings. In the control run (solid line), which suffers from the typical "cold pole" problem, the maximum variance in North Pole temperatures occurs as late as February/March. When the SAO wind variation is imposed (dotted line), improving the representation of the equatorial upper stratosphere, the maximum variance occurs about a month earlier. When the shallow QBO is superimposed (dashed line), very little difference is made to the timing of the mid-winter warmings, although the early winter is noticeably more disturbed. However, when the deep QBO is superimposed (dash-dotted line), further improving the upper stratosphere representation, the peak in mid-winter variance moves earlier again, by another 2 weeks. Both the SAO and deep QBO results suggest that the timing of mid-winter warmings in this model is sensitive to the distribution of winds in the equatorial upper stratosphere.

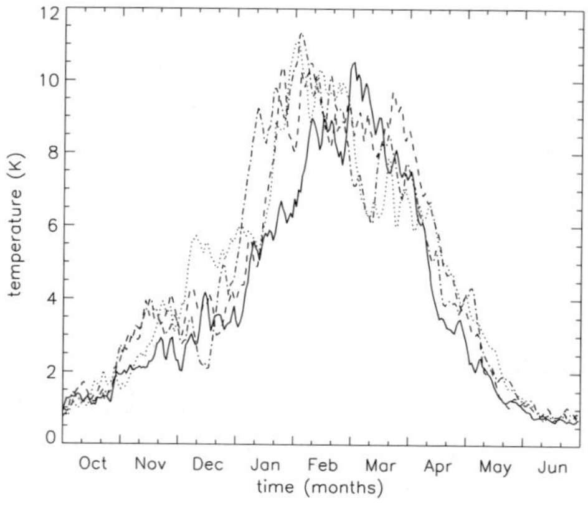


*Figure 8.* Standard deviation of daily North Pole temperature at 50 hPa for each of 4 simulations of the Met Office Unified Model, in which various equatorial wind distributions have been imposed (see text). Solid line: control experiment. Dotted line: SAO imposed only. Dashed line: SAO and shallow QBO (up to 35 km) imposed. Dash-dot line: SAO and deep QBO (to 70 km) imposed. Figure adapted from Pascoe *et al.* (2006).

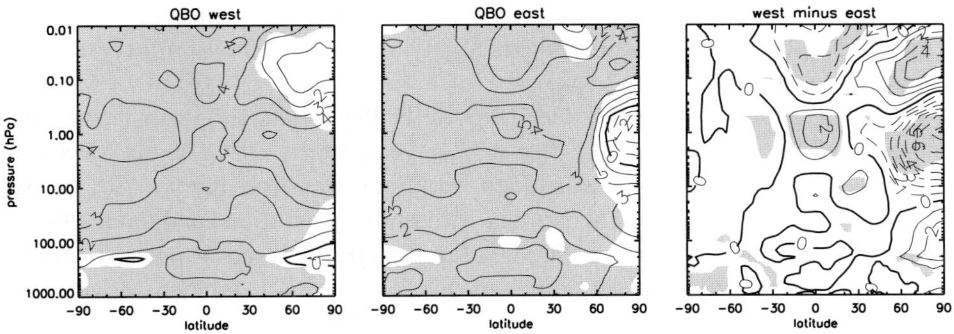


*Figure 9.* Solar maximum minus solar minimum zonal-mean temperature differences (K) for December–January–February from the Met Office Unified Model incorporating the Hines gravity wave parametrisation scheme to generate a QBO. The results have been divided into (a) QBO/W years, (b) QBO/E years. Panel (c) shows the QBO/E minus QBO/W differences. Contour interval is 1 K. Dashed contours denote negative values; bold contour is zero. Shading indicates statistical significance at the 95% level using the Student *t* test. Figure adapted from Palmer and Gray (2005).

In the experiments of Matthes *et al.* (2004) using the Free University of Berlin (FUB) GCM, several sets of perpetual $S_{min}/S_{max}$ experiments were carried out. In the first set, a typical distribution for QBO/E phase and QBO/W phase were imposed at the equator up to around 32 km, but this did not reproduce the observed SC/QBO polar temperature relationship. However, in a second set of runs, when the typical QBO wind profiles were imposed throughout the whole depth of the stratosphere, the observed relationship was reproduced, with warmer polar temperatures in $S_{min}$/E and $S_{max}$/W than in $S_{min}$/W and $S_{max}$/E. This result therefore supports the hypothesis of Gray (2003) that the equatorial winds in the upper stratosphere are influential and that the interaction of the SC and QBO signals may occur in this region.

In a similar experiment, Palmer and Gray (2005) carried out a set of 25-year perpetual $S_{min}/S_{max}$ experiments, but in this case the QBO was internally generated by the gravity-wave parametrisation scheme in the model, instead of being imposed as in Matthes *et al.* (2004). It thus allowed the possibility of a SC influence on the nature of the QBO (e.g., its period) although, unlike the Matthes *et al.* (2004) experiment, it does not allow a tropical influence on sudden warmings to be attributed to any particular height region of the tropics.

Figure 9 shows the difference in zonally-averaged DJF temperatures between the $S_{max}$ and $S_{min}$ experiments, divided into QBO/E and QBO/W years. As expected, the temperatures throughout the majority of the stratosphere are greater in $S_{max}$ than in $S_{min}$ conditions in both QBO/E and QBO/W, reflecting the radiative impact of the irradiance and ozone changes. (We note here that the forcings in this experiment were much larger than in reality, in order to test sensitivity, and hence the signal is much larger than the observations shown in Figure 1. The different patterns of distribution in Figure 9 compared with Figure 1 can be attributed to the differences between DJF and annual signals and to the fact that the model integrations were perpetual $S_{min}/S_{max}$ experiments while in reality the atmosphere is unlikely to reach such steady-state conditions).

However, at North polar latitudes (Figure 9), the results shows that while the temperatures are warmer in $S_{max}$ than in $S_{min}$ years in the QBO/W, the reverse is true in the QBO/E phase. Hence, sudden warmings are more likely in $S_{min}$/E and $S_{max}$/W years. This is in good agreement with both observations and the results of Matthes *et al.* (2004). Although the signal in the QBO/E years is not statistically significant, due to the large variability in this region, the difference between the QBO/E and QBO/W years (right panel) is statistically significant.

## 5. Discussion

Both mechanistic model studies and GCM studies appear to support the hypothesis that the development of sudden warmings is influenced by the presence of relatively small wind anomalies in the tropical/subtropical upper stratosphere. Both the SC and QBO have associated wind anomalies in this region. It is the timing of the sudden warmings that appears to be most affected, both in the observations and the model studies, suggesting a frequency modulation of the warmings by the SC/QBO.

Sudden warmings are generally initiated in the upper stratosphere and penetrate throughout the depth of the stratosphere as they evolve. Therefore, they provide a possible transfer route for SC/QBO influence from the equatorial/subtropical stratopause region to the polar lower stratosphere. In addition, angular momentum deposition by the planetary waves that give rise to sudden warmings induces a mean meridional circulation that extends to lower latitudes. This provides a possible mechanism for transfer of the SC influence to the equatorial lower stratosphere, via the strength of the equatorial upwelling.

A SC modulation of equatorial upwelling would give rise to both warmer temperatures and increased ozone during solar maximum, as observed (Hood, 1997). A SC modulation of equatorial upwelling is also likely to feed back onto the evolution of the QBO itself. There is some evidence for a SC modulation of the QBO frequency in the observations (Salby and Callaghan, 2000) although the more recent data have thrown some doubt on the exact nature of the modulation (Hamilton, 2002). 2D model studies (McCormack, 2003; Cordero and Nathan, 2005) and the GCM study described above (Palmer and Gray, 2005) have simulated a small SC modulation of the QBO period, although the changes are small and/or not statistically significant and longer runs are needed. It is not clear to what extent these modelling results arise from a direct influence of the SC on the QBO descent rate at equatorial latitudes and to what extent it arises due to the more indirect winter polar route described in this paper.

## References


Balachandran, N. K., and Rind, D.: 1995, 'Modeling the effects of UV variability and the QBO on the troposphere stratosphere system. I: The middle atmosphere', *J. Clim.* **8**, 2058–2079.

Baldwin, M. P., and Gray, L. J.: 2005, 'Tropical stratospheric zonal winds in ECMWF ERA-40 reanalysis, rocketsonde data, and rawinsonde data', *Geophys. Res. Letts.* **32**, doi:10.1029/2004GL022328.

Baldwin, M. P., *et al.*: 2001, 'The quasi biennial oscillation', *Rev. Geophys.* **39**, 179–229.

Cordero, E. C., and Nathan, T. R.: 2005, 'A new pathway for communicating the 11-year solar cycle signal to the QBO', *Geophys. Res. Letts.* **32**, doi:10.1029/2005GL023696.

Crooks, S., and Gray, L. J.: 2005, 'Characterisation of the 11-year solar signal using a multiple regression analysis of the ERA-40 dataset', *J. Clim.* **18,7**, 996–1015.

Gray, L. J.: 2003, 'The influence of the equatorial upper stratosphere on stratospheric sudden warmings', *Geophys. Res. Lett.* **30**, doi:10.1029/2002GL016430.

Gray, L. J., Phipps, S. J., Dunkerton, T. J., Baldwin, M. P., Drysdale, E. F., and Allen, M. R.: 2001a, 'A data study of the influence of the equatorial upper stratosphere on northern hemisphere stratospheric sudden warmings', *Q. J. R. Meteorol. Soc.* **127**, 1985–2003.

Gray, L. J., Drysdale, E. F., and Lawrence, B. N.: 2001b, 'Model studies of the interannual variability of the northern hemisphere stratospheric winter circulation: The role of the quasi biennial oscillation', *Q. J. R. Meteorol. Soc.* **127**, 1413–1432.

Gray, L. J., Sparrow, S., Juckes, M., O'Neill, A., and Andrews, D. G.: 2003, 'Flow regimes in the winter stratosphere of the northern hemisphere', *Q. J. R. Meteorol. Soc.* **129**, 925–945.

Gray, L. J., Crooks, S. A., Pascoe, C., and Sparrow, S.: 2004, 'Solar and QBO influences on the timings of stratospheric sudden warmings', *J. Atmos. Sci.* **61**, 2777–2796.

Haigh, J. D.: 1994, 'The role of stratospheric ozone in modulating the solar radiative forcing of climate', *Nature* **370**, 544–546.

Haigh, J. D.: 1996, 'The impact of solar variability on climate', *Science* **272**, 981–984.

Haigh, J. D., and Blackburn, M.: 2006, 'Solar influences on dynamical coupling between the stratosphere and troposphere', *Space Sci. Rev.*, this volume, doi: 10.1007/s11214-006-9067-0.

Hamilton, K.: 2002, 'On the quasi decadal modulation of the stratospheric QBO period', *J. Clim.* **15**, 2562–2565.

Holton, J. R., and Tan, H.-C.: 1980, 'The influence of the equatorial quasi-biennial oscillation on the global circulation at 50 mb', *J. Atmos. Sci.* **37**, 2200–2208.

Holton, J. R., and Tan, H.-C.: 1982, 'The quasi-biennial oscillation in the northern hemisphere lower stratosphere', *J. Meteor. Soc. Jpn.* **60**, 140–148.

Hood, L. L.: 1997, 'The solar cycle variation of total ozone: Dynamical forcing in the lower stratosphere', *J. Geophys. Res.* **102**, 1355–1370.

Hood, L. L.: 2004, 'Effects of solar UV variability on the stratosphere', in J. Pap, P. Fox, C. Frohlich, H. Hudson, J. Kuhn, J. McCormack, G. North, W. Sprigg, and S. T. Wu (eds.), *Solar variability and its effect on the Earth's atmosphere and climate system*, AGU Monograph Series, American Geophysical Union, Washington D.C.

Keckhut, P., Cagnazzo, C., Chanin, M.-L., Claud, C., and Hauchecorne, A.: 2005, 'The 11-year solar-cycle effects on temperature in the upper stratosphere and mesosphere. Part I: Assessment of observations', Submitted to *J. Atmos. Sol. Terr. Phys.*

Kodera, K.: 1995, 'On the origin and nature of the interannual variability of the winter stratospheric circulation in the northern hemisphere', *J. Geophys. Res.* **100**, 14,077–14,087.

Kodera, K.: 2006, 'The role of dynamics in solar forcing', *Space Sci. Rev.*, this volume, doi: 10.1007/s11214-006-9066-1.

Kodera, K., and Kuroda, Y.: 2002, 'Dynamical response to the solar cycle', *J. Geophys. Res.* **107**, art. No. 2224.

Kodera, K., Yamazaki, K., Chiba, M., and Shibata, K.: 1990, 'Downward propagation of upper stratospheric mean zonal wind perturbation to the troposphere', *Geophys. Res. Lett.* **17**, 1263–1266.

Kodera, K., Kuroda, Y., and Pawson, S.: 2000, 'Stratospheric sudden warmings and slowly propagating zonal-mean zonal wind anomalies', *J. Geophys. Res.* **105**, 12,351–12,359.

Kodera, K., Matthes, K., Shibata, K., Langematz, U., and Kuroda, Y.: 2003, 'Solar impact on the lower mesospheric subtropical jet: A comparative study with general circulation model simulations', *Geophys. Res. Lett.* **30**, doi:10.1029/2002GL016124.

Labitzke, K.: 1987, 'Sunspots, the QBO and the stratospheric temperature in the north polar region', *Geophys. Res. Lett.* **14**, 535–537.

Labitzke, K.: 2003, 'The global signal of the 11-year sunspot cycle in the atmosphere: When do we need the QBO?', *Meteorol. Zeitschrift* **12**, 209–216.

Labitzke, K.: 2006, 'Solar variation and stratospheric response', *Space Sci. Rev.*, this volume, doi: 10.1007/s11214-006-9061-6.

Labitzke, K., and van Loon, H.: 1988, 'Association between the 11-year solar cycle, the QBO and the atmosphere. Part I: The troposphere and stratosphere in northern hemisphere in winter', *J. Atmos. Terr. Phys.* **50**, 197–206.

Lee, H., and Smith, A. K.: 2003, 'Simulation of the combined effects of solar cycle, quasi-biennial oscillation and volcanic forcing on stratospheric ozone changes in recent decades', *J. Geophys. Res.* **108**, doi:10.1029/2001JD001503.

Matthes, K., Langematz, U., Gray, L. J., Kodera, K., and Labitzke, K.: 2004, 'Realistic solar signal in the FUB-CMAM', *J. Geophys. Res.* **109**, doi:10.1029/2003JD004012.

McCormack, J. P.: 2003, 'The influence of the 11-year solar cycle on the quasi biennial oscillation', *Geophys. Res. Lett.* **30**, doi:10.1029/2003GL018314.

Palmer, M. A., and Gray, L. J.: 2005, 'Modelling the atmospheric response to solar irradiance changes using a GCM with a realistic QBO', *Geophys. Res. Lett.* **32**, doi:10.1029/2005GL023809.

Pascoe, C. L., Gray, L. J., Crooks, S. A., Juckes, M. N., and Baldwin, M. P.: 2005, 'The quasi biennial oscillation: Analysis using ERA-40 data', *J. Geophys. Res.* **110** doi:10.1029/2004JD004941.

Pascoe, C. L., Gray, L. J., and Scaife, A. A.: 2006, 'A GCM study of the influence of equatorial winds on the timing of sudden stratospheric warmings', *Geophys. Res. Lett.* **33**, doi:10.1029/2005GL024715.

Salby, M., and Callaghan, P.: 2000, 'Connection between the solar cycle and the QBO: The missing link', *J. Clim.* **13**, 2652–2662.

Scaife, A. A., Austin, J., Butchart, N., Pawson, S., Keil, M., Nash, J., and James, I. N.: 2000, 'Seasonal and interannual variability of the stratosphere diagnosed from UKMO TOVS analyses', *Quart. J. Roy. Met. Soc.* **126**, 2585–2604.

# RECENT SPACE DATA
*Introductory Paper*

J. LANGEN
*ESA–ESTEC, Keplerlaan 1, 2200AG Noordwijk, The Netherlands*
*(E-mail: joerg.langen@esa.int)*

(Received 14 October 2005; Accepted 12 December 2005)

**Abstract.** This paper summarises the workshop session on recent space data. Most presentations addressed the intense solar storm in October–November 2003. Large perturbations of atmospheric trace gas concentrations, notably $NO_2$ and $HNO_3$, were found over extended areas around the magnetic poles in the mesosphere and stratosphere, extending over many weeks in the stratosphere. The impact on total ozone seems to be very limited although some more subtle investigations are still to be done. Several new space instruments with many innovative data products have been introduced. Very good coverage in vertically resolved observations of many chemical species is reached for stratospheric chemistry and dynamics research. Data have already been used to improve stratospheric models. Data continuity is an issue. However, the greatest concern is the lack of any suitable future space instrumentation for tropospheric research (air quality and climate forcing/carbon cycle) as well as UTLS problems (climate/chemistry interaction, stratosphere/troposphere exchange).

**Keywords:** atmospheric composition, space instrumentation

## 1. Introduction

As illustrated in the previous volume Chapters, solar variability exists at various time scales from five minutes to thousands of years. Space-borne data have only been available over the last three solar cycles, and not many parameters have a continuous record over that time period. Correlations between solar cycle and atmospheric temperature and ozone content have been looked at in the preceding chapters of the present volume. The purpose of the workshop session on "Recent Space Data" was to review the potential of new space missions to investigate the impact of solar variability on Earth's atmosphere, and in particular on its chemical composition. This decade is seeing an unprecedented set of new high quality space data on atmospheric chemistry which still need to be fully exploited. It is too early to use them for climate research, but process studies have already been performed. The papers that follow cover three aspects of these new space missions, namely

- the impact of solar storms, in particular the strong event in Fall 2003, on mesospheric and stratospheric trace gas abundances;

Space Science Reviews (2006) 125: 371–379
DOI: 10.1007/s11214-006-9070-5

– the introduction of a number of new space instruments for observation of atmospheric chemistry on American and European platforms, and examples of the data products generated from their measurements;
– some new insights into atmospheric chemistry and transport gained from the comparison of atmospheric models with these data.

## 2. Impact of Solar Proton Events on Atmospheric Composition

The current solar cycle 23 turns out to be extremely active in producing 7 of the 12 largest solar proton events (SPEs) during the last 40 years. The event observed at the end of October 2003 was ranked 4th in this time span. It was observed by all the new space instruments. Protons appear at geomagnetic latitudes $>60°$; they penetrate typically to 87 km altitude (at 1 MeV) down to 33 km (100 MeV) and deposit their energy by dissociation, ionisation and excitation of nitrogen and oxygen molecules. Resulting $HO_x$ and $NO_x$ radicals perturb the ozone concentration. $HO_x$ effects are of short duration (few hours) and limited to the mesosphere, $NO_x$ effects penetrate down into the stratosphere and can persist for weeks or even months, due to downward transport in the polar vortex and the long lifetime of $NO_x$ in the polar winter stratosphere. Eventually, the impact of the October 2003 event was stopped by a stratospheric warming, but then followed by an unprecedented strong polar vortex with very fast downward transport.

The temporal evolution of the $NO_x$ and $O_3$ anomalies in the Arctic and Antarctic following the SPE on 28 October 2003 can be seen in Figure 1. Huge $NO_x$ increases are observed; local ozone losses of more than 50% occurred in the mid mesosphere. A pronounced interhemispheric asymmetry is apparent, with the SPE perturbation much higher in the Arctic than the Antarctic. This is due to longer chemical lifetime and to downward transport of $NO_y$ into lower levels in the winter hemisphere. Enhanced concentrations of nitrogen reservoir species ($HNO_3$, $N_2O_5$, $ClONO_2$) were also observed following the SPE.

The impact of SPEs on ozone chemistry is reasonably well understood although interference with other variabilities requires more quantitative analysis. As a result of a model prediction, the loss of the total ozone column due to SPEs seldomly exceeds 1% at latitudes $<60°$, and nearly never exceeds 2% at any latitude. A caveat must be made with respect to the impact of moderately strong SPEs. Such an event was observed at the end of May 2003 and was followed by enhanced $NO_2$ concentrations and ozone loss in the Antarctic winter stratosphere. There are diverging views between the authors of this session as to whether the chemical anomaly might have been caused by the SPE or by energetic electron precipitation. Whether the sum of the larger number of moderate energy SPEs might lead to a significant increase of the overall impact on ozone will be subject to further investigation.

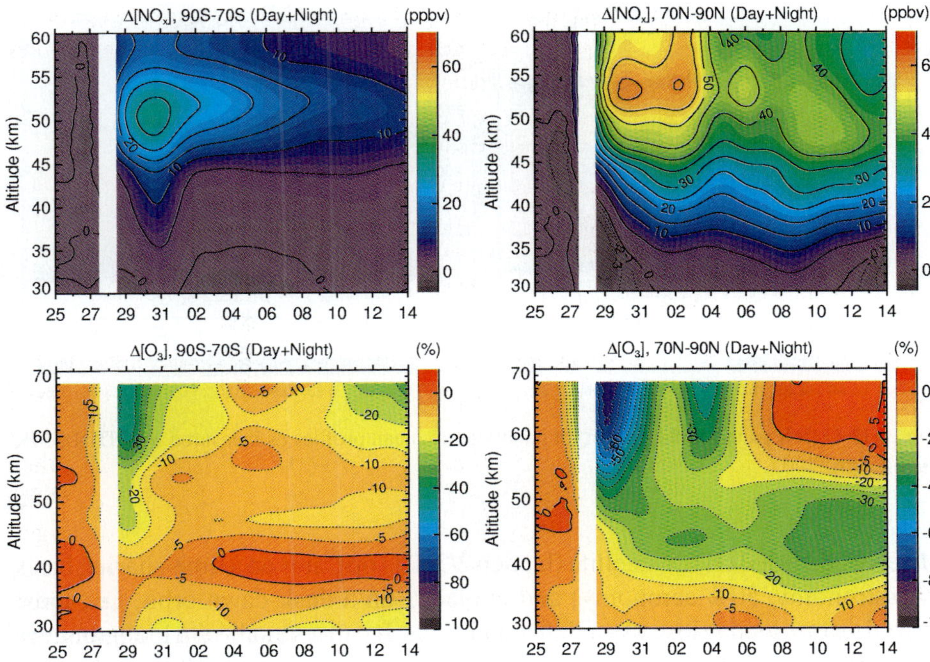


*Figure 1.* $NO_x$ (upper panels) and $O_3$ (lower panels) anomalies in the upper stratosphere and lower mesosphere following the SPE event of 28 October–7 November 2003, derived from Envisat-MIPAS data. Left panels show Antarctic data (70–90° S), right panels Arctic data (70–90° N). Horizontal axis: day of month October, November. Reproduced from López-Puertas *et al.* (2005) with permission of American Geophysical Union.

## 3. New Generation Atmospheric Chemistry Space Instruments

The need for 3D atmospheric composition data with global coverage is driven by four main objectives, namely, the better understanding and predictive capability of

1. the evolution of chemical depletion of stratospheric ozone which shields life on Earth from excessive UV-B radiation;
2. the various forms of climate-chemistry interaction, e.g., radiative forcing of greenhouse gases and aerosols affecting temperature and wind (including the atmospheric compartment of the carbon cycle); the impact of anthropogenic aerosols on precipitation patterns;
3. the air quality: pollution sources, chemical processing, transport pathways;
4. the oxidising capacity of the atmosphere, closely related to the OH chemistry in the troposphere.

Several new satellites with major atmospheric chemistry payload have been deployed during the first half of this decade, to address these issues. Some of them have been introduced during the workshop session.

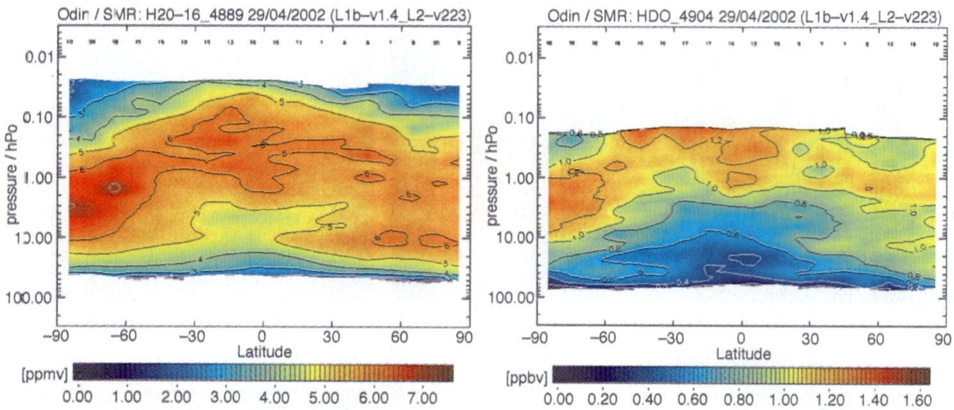


*Figure 2.* $H_2O$ (left panel) and HDO (right panel) zonal mean profiles as measured by ODIN-SMR on 29–30 April 2002.

The Swedish-led ODIN satellite (launch 2/2001) operates on a time-sharing basis (50% astronomy, 50% aeronomy) and is placed on a dusk-dawn orbit, i.e., most observations are made close to twilight conditions. It carries a sub-mm limb-sounder "SMR" and a UV-VIS-NIR limb backscatter spectrometer "OSIRIS" including a 3-band NIR imager; both observing stratospheric and mesospheric trace gases. Spectral bands can partly be switched to optimise for specific observation priorities. Innovative data products include: ClO with good signal to noise ratio, CO without non-LTE contamination, and the complete set of $H_2O$ isotopes (see Figure 2 for an example).

ESA's Envisat platform (launched 3/2002 on a sun-synchronous orbit at 10:00 local time) has a payload complement of 10 instruments, three of them are devoted to atmospheric chemistry:

GOMOS is a UV-VIS-NIR spectrometer including two fast photometers in stellar occultation, best performing at night. After a break due to mechanical problems, GOMOS now resumed operations. The main purpose of the instrument is high vertical resolution and accuracy monitoring of stratospheric and mesospheric ozone, but several other trace gases such as $NO_2$, $NO_3$ and water vapour as well as aerosol are also observed. GOMOS is the first to see $NO_3$ with a reasonable coverage from space, see Figure 3. High correlation with temperature is found in the upper stratosphere, as expected from chemistry. GOMOS also measures the dispersion of the atmospheric refractive index between red and blue. This quantity is converted to a time delay and finally to a high resolution mid-stratospheric temperature profile.

MIPAS, a Fourier-Transform spectrometer in the infrared range, observes emissions of stratospheric and some mesospheric trace gases (also in the upper troposphere in cloud-free conditions) at day and night from pole to pole. A long list of data products includes the complete nitrogen family and many tracers, but also,

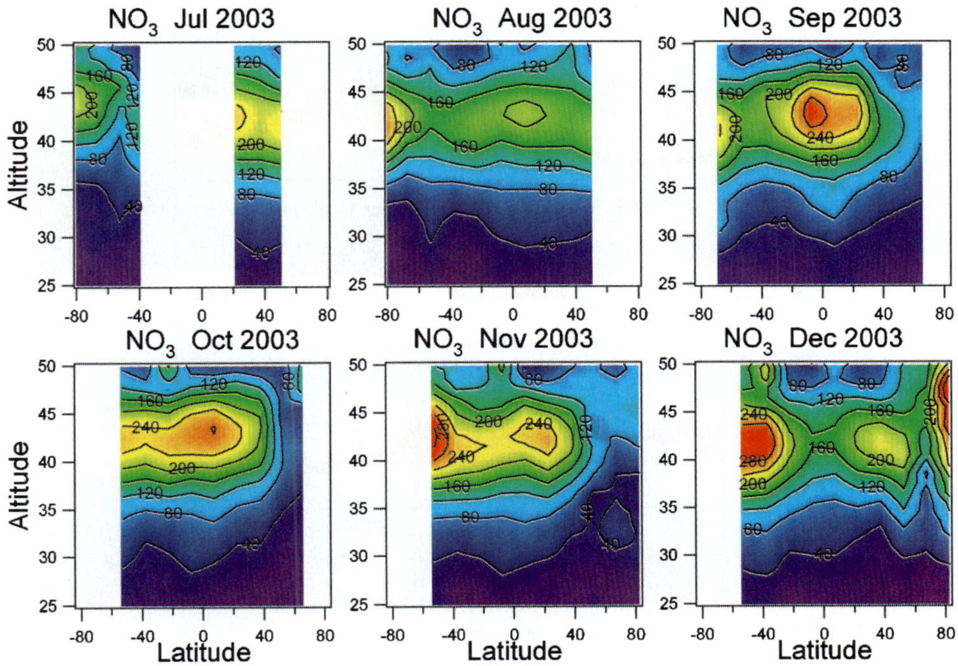


*Figure 3.* Stratospheric $NO_3$ during the second half of the year 2003, as seen by Envisat-GOMOS. Units: ppt.

e.g., $H_2O_2$ (zonal averages) and ClO (when enhanced). Due to mechanical problems with the interferometer drive, MIPAS now operates with reduced duty cycle and spectral resolution.

Huge variabilities of mesospheric and stratospheric $NO_2$ have been observed by MIPAS. This has two primary reasons: firstly the dynamical behaviour of the polar winter atmosphere, and secondly the sources such as auroral electron precipitations and sporadic solar proton events. $NO_2$ variabililty propagates down to other chemically related species, such as $HNO_3$ and $N_2O_5$. SCIAMACHY is an 8-band spectrometer (240–2400 nm) observing backscattered radiation alternating in limb and nadir directions, with across-track scan, while on the dayside. It performs also solar and lunar occultations whenever possible. The longest wavelength channels are affected by ice condensation on the detector. The instrument targets many species in the stratosphere and troposphere, and some in the mesosphere. For the nadir UV-VIS observations, it provides continuity to GOME measurements which started in 1995. A combined GOME/SCIAMACHY Mg-II index is shown in Figure 4 and shows very good agreement with the NOAA Mg-II index. Through its improved horizontal resolution (in comparison to GOME), SCIAMACHY provides detailed maps of tropospheric $NO_2$. SCIAMACHY is the first instrument to provide maps of $CH_4$ over land with boundary layer sensitivity (Figure 5). Many other data have

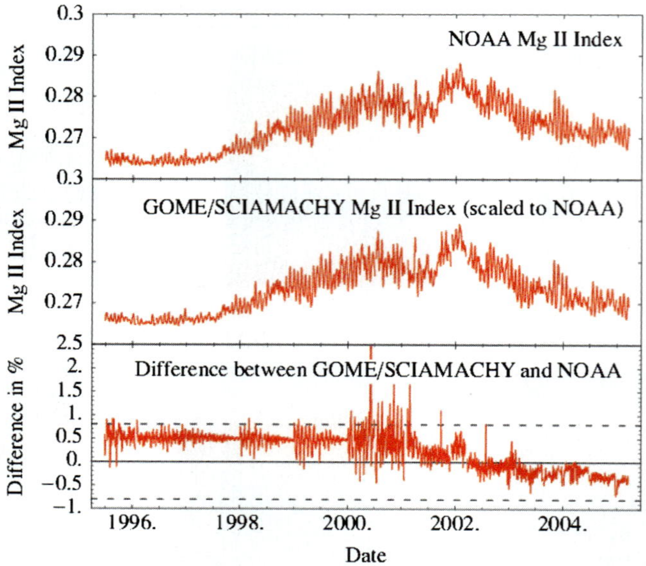


*Figure 4.* NOAA (upper panel) vs combined GOME/SCIAMACHY (middle panel) Mg-II index over GOME/SCIAMACHY lifetime, and relative difference (lower panel).

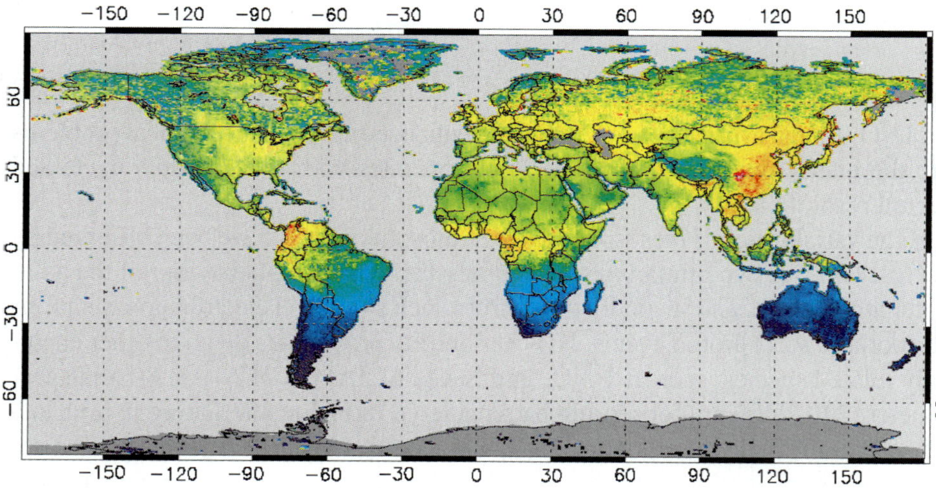

*Figure 5.* SCIAMACHY annual mean $CH_4$ dry air mixing ratio (ppb).

been produced although the generation of efficient and accurate retrieval algorithms is a challenge.

The EOS-Aura satellite was launched in July 2004 and is NASA's second large satellite dedicated to atmospheric composition research after UARS.

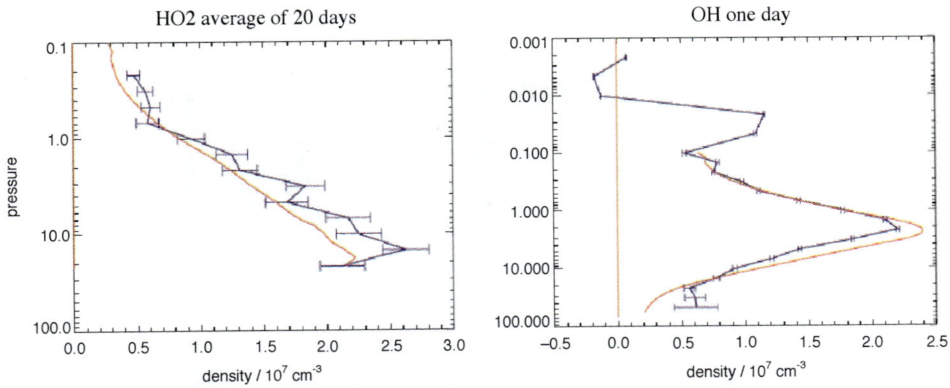


*Figure 6.* Zonal mean (20°N–20°S) day-night differences of $HO_2$ (left panel) and OH (right panel) as measured by Aura-MLS (blue) and predicted by the SLIMCAT model (red lines).

Aura carries four instruments:

- MLS, a five-band mm, sub-mm and far infrared limb-sounder, extending the mission of its predecessor on UARS with increased sensitivity and coverage of stratospheric chlorine species, the first long-term space measurements of OH and $HO_2$ (see Figure 6), and improved CO data quality;
- TES, a mid-IR Fourier Transform spectrometer, is very well suited for vertically resolved nadir observations of the troposphere, given its small ground pixel size and high spectral resolution (Figure 7). It is designed for limb geometry as well (similar to MIPAS) but, due to instrumental issues, most of its data are taken in nadir view;
- HIRDLS, a multi-channel infrared limb-sounding radiometer with 6 across-track view angles. This instrument was designed to resolve fine horizontal and vertical structures in the upper troposphere and the stratosphere; however unfortunately all views except part of one are obstructed;
- OMI, a nadir-viewing push-broom UV-VIS spectrometer with high horizontal resolution and full daily global coverage. This instrument marks the transition from predecessors (GOME, GOME-2) which focussed on stratospheric ozone to future instruments which are expected to be optimised for tropospheric observations. Significant synergism is expected from the combination of OMI and TES data, e.g., tropospheric ozone for which OMI provides an integrated column value with boundary layer sensitivity and TES a vertical profile with highest sensitivity to mid and upper troposphere (Figure 7).

Several other atmospheric chemistry space instruments have been launched in the last years, such as IMG and ILAS on ADEOS-1 and ILAS-2 on ADEOS-2 (Japan), ACE and MAESTRO on the Canadian SCISAT platform, and various aerosol instruments. All these data provide very good coverage for the stratospheric ozone issue. In the near future the NASA OCO mission, and probably the JAXA

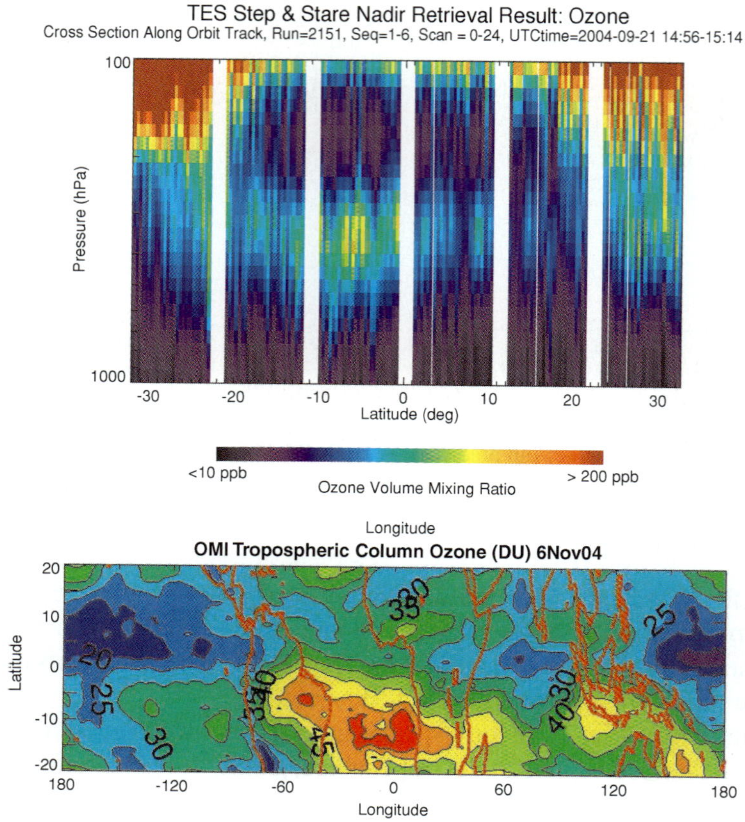


*Figure 7. Upper panel:* TES tropospheric ozone cross section over the tropical Atlantic featuring a high ozone event in the mid troposphere. *Lower panel:* OMI tropospheric ozone column map for a similar event.

GOSAT mission, both dedicated to $CO_2$ observations for carbon cycle research, will enter operations. However, beyond 2010 there is no provision for atmospheric composition data from space beyond the moderate contributions of NOAA and EUMETSAT meteorological programmes. Besides the need for continuity of existing missions, the main data gaps will be for

- high temporal/spatial resolution space-based measurements of tropospheric (boundary layer) composition for application to air quality;
- high vertical resolution measurements in the UT/LS region for ozone and climate applications (climate-chemistry interactions, stratosphere-troposphere exchange, stratospheric ozone);
- climate gases ($CO_2$, $CH_4$ and CO) and aerosol monitoring with sensitivity to the boundary layer.

## 4. New Developments in Stratospheric Chemistry and Dynamics

The use of the new Envisat data in modelling of stratospheric chemistry and dynamics was illustrated with several examples.

An intercomparison was performed between 4 stratospheric chemistry transport models (CTMs) and Envisat MIPAS data. For $O_3$ and $CH_4$, good agreement was obtained. $HNO_3$ varied between the models, and in particular was much higher in MIPAS data.

Runs of the SLIMCAT model, once with an equilibrium scheme for PSCs, and once with a full PSC microphysics scheme included, showed that observed denitrification could be much better reproduced using the full microphysics. This points to a deficiency of many CTMs using simplified schemes.

Preliminary tests of $HO_x$ chemistry with MLS data show good agreement for $HO_2$ and OH day-night differences. MIPAS $H_2O_2$ fields show a similar pattern but mostly smaller values than SLIMCAT, and a strong diurnal variation that is not reproduced by the model.

Chemical data assimilation turns out to be a powerful tool. The prediction of abundances of unmeasured constituents (in this case $Cl_y$) was demonstrated by assimilating HALOE data of a range of other species.

## Acknowledgements

All material in this summary paper has been drawn from the presentations by J. Burrows, M. Chipperfield, A. Hauchecorne, E. Hilsenrath, C. Jackman, M. López-Puertas, and D. Murtagh in the workshop session on "Recent Space Data".

## Reference

López-Puertas, M., Funke, B., Gil-López, S., von Clarmann, T., Stiller, G. P., Höpfner, M., *et al.*: 2005, 'Observation of $NO_x$ enhancement and ozone depletion in the Northern and Southern Hemispheres after the October–November 2003 solar proton events', *J. Geophys. Res.* **110**, doi:10.1029/2005JA011050.

# SATELLITE MEASUREMENTS OF MIDDLE ATMOSPHERIC IMPACTS BY SOLAR PROTON EVENTS IN SOLAR CYCLE 23

C. H. JACKMAN[1,*], M. T. DELAND[1,2], G. J. LABOW[1,2], E. L. FLEMING[1,2] and M. LÓPEZ-PUERTAS[3]

[1]*NASA Goddard Space Flight Center, Code 613.3, Greenbelt, MD 20771, USA*

[2]*Science Systems and Applications, Inc., Lanham, MD, USA*

[3]*Instituto de Astrofisica de Andalucia, CSIC, Granada, SPAIN*

*(*Author for correspondence: E-mail: charles.h.jackman@nasa.gov)*

(Received 17 August 2005; Accepted in final form 26 October 2005)

**Abstract.** Solar cycle 23 was extremely active with seven of the largest twelve solar proton events (SPEs) in the past forty years recorded. These events caused significant polar middle atmospheric changes that were observed by a number of satellites. The highly energetic protons produced ionizations, excitations, dissociations, and dissociative ionizations of the background constituents in the polar cap regions (> 60 degrees geomagnetic latitude), which led to the production of $HO_x$ (H, OH, $HO_2$) and $NO_y$ (N, NO, $NO_2$, $NO_3$, $N_2O_5$, $HNO_3$, $HO_2NO_2$, $BrONO_2$, $ClONO_2$). The $HO_x$ increases led to short-lived ozone decreases in the polar mesosphere and upper stratosphere due to the short lifetimes of the $HO_x$ constituents. Polar middle mesospheric ozone decreases greater than 50% were observed and computed to last for hours to days due to the enhanced $HO_x$. The $NO_y$ increases led to long-lived polar stratospheric ozone changes because of the long lifetime of the $NO_y$ family in this region. Upper stratospheric ozone decreases of >10% were computed to last for several months past the solar events in the winter polar regions because of the enhanced $NO_y$.

**Keywords:** ozone, $NO_y$, solar proton event, middle atmosphere, stratosphere, mesosphere

## 1. Introduction

Solar eruptive events sometimes result in large fluxes of high-energy solar protons at the Earth. This period of time, wherein the solar proton flux is generally elevated for a few days, is known as a solar proton event (SPE) and primarily occurs near solar maximum. Solar cycle 23 experienced several very large SPEs, which occurred in July and November 2000, September and November 2001, October/November 2003, and January 2005. The twelve largest SPEs in the past 40 years are given in Table I and over half of them (seven) occurred in the past solar maximum period.

The Earth's magnetic field guides the solar protons into the northern and southern polar cap regions (>60° geomagnetic latitude), e.g., see Jackman and McPeters (2004). The protons interact with the neutral middle atmosphere (stratosphere and mesosphere) and produce ionizations, dissociations, dissociative ionizations, and excitations.

Both $HO_x$ (H, OH, $HO_2$) and $NO_y$ (N, NO, $NO_2$, $NO_3$, $N_2O_5$, $HNO_3$, $HO_2NO_2$, $BrONO_2$, $ClONO_2$) can be enhanced either directly by the protons and their

Space Science Reviews (2006) 125: 381–391
DOI: 10.1007/s11214-006-9071-4

TABLE I

Largest twelve solar proton events in the past forty years.

| Date of SPEs | Rank in size | $NO_y$ production in the middle atmosphere (# of molecules) |
|---|---|---|
| October 19–27, 1989 | 1 | $6.7 \times 10^{33}$ |
| August 2–10, 1972 | 2 | $3.6 \times 10^{33}$ |
| July 14–16, 2000 | 3 | $3.5 \times 10^{33}$ |
| October 28–31, 2003 | 4 | $3.4 \times 10^{33}$ |
| November 5–7, 2001 | 5 | $3.2 \times 10^{33}$ |
| November 9–11, 2000 | 6 | $2.3 \times 10^{33}$ |
| September 24–30, 2001 | 7 | $2.0 \times 10^{33}$ |
| August 13–26, 1989 | 8 | $1.8 \times 10^{33}$ |
| November 23–25, 2001 | 9 | $1.7 \times 10^{33}$ |
| September 2–7, 1966 | 10 | $1.2 \times 10^{33}$ |
| January 15–23, 2005 | 11 | $1.1 \times 10^{33}$ |
| Sep. 29–Oct. 3, 1989 | 12 | $1.0 \times 10^{33}$ |

associated secondary electrons or through a photochemical sequence initiated by the protons impacting the atmosphere (e.g., Warneck, 1972; Swider and Keneshea, 1973; Crutzen *et al.*, 1975; Porter *et al.*, 1976; Frederick, 1976; Jackman *et al.*, 1980, 1995, 2000, 2001; Solomon *et al.*, 1981; Rusch *et al.*, 1981; McPeters, 1986; Zadorozhny *et al.*, 1992; Randall *et al.*, 2001). Ozone can be influenced by the solar protons through photochemical depletion processes caused by the enhanced $HO_x$ and $NO_y$ (e.g., Weeks *et al.*, 1972; Heath *et al.*, 1977; McPeters *et al.*, 1981; Solomon and Crutzen, 1981; Thomas *et al.*, 1983; Solomon *et al.*, 1983; McPeters and Jackman, 1985; Jackman and McPeters, 1985, 1987; Jackman *et al.*, 1995, 2000, 2001; Reid *et al.*, 1991; Seppala *et al.*, 2004; Degenstein *et al.*, 2005).

The first middle atmospheric impact of solar protons was measured with a rocket by Weeks *et al.* (1972). These observations showed a very dramatic reduction in mesospheric ozone caused by the November 1969 SPE, which was explained in Swider and Keneshea (1973) as caused by $HO_x$ enhancements.

Crutzen *et al.* (1975) postulated that stratospheric ozone should be impacted by NO generated through interaction of the protons with the atmosphere. Two years later Heath *et al.* (1977) confirmed with satellite observations that stratospheric ozone was dramatically reduced during the August 1972 SPE.

In this paper we discuss satellite observations of middle atmospheric impacts by protons ejected by the Sun in solar cycle 23. There have been several publications in the past few years on this subject including Jackman *et al.* (2001, 2005a,b), Randall *et al.* (2001), Krivolutsky *et al.* (2003, 2005), Seppala *et al.* (2004), Degenstein *et al.* (2005), Semeniuk *et al.* (2005), Verronen *et al.* (2005), López-Puertas *et al.* (2005a,b, 2006), and Rohen *et al.* (2005). López-Puertas *et al.* (2006) as well as other papers (e.g., Natarajan *et al.*, 2004; Randall *et al.*, 2005; Orsolini *et al.*, 2005; Rinsland *et al.*, 2005) showed certain solar influences on the polar upper stratosphere from effects that may not have been caused by solar protons, but rather by energetic electrons and/or X-rays associated with the solar storms of October–November 2003, and will not be discussed here.

This paper contains seven primary sections, including the Introduction. The solar proton flux and production of $HO_x$ and $NO_y$ are discussed in Section 2. The observations of $NO_y$ constituents, ClO and HOCl constituents, and ozone change as a result of solar protons in solar cycle 23 are given in Sections 3, 4, and 5, respectively. Global model results during certain disturbed time periods are presented in Section 6. Finally, conclusions are given in Section 7.

## 2. Solar Proton Flux and Production of $HO_x$ and $NO_y$

The National Oceanic and Atmospheric Administration (NOAA) Space Environment Center (SEC) provides a proton flux dataset for the NOAA Geostationary Operational Environmental Satellites (GOES) on their website (see `http://sec.noaa.gov/Data/goes.html`). GOES proton fluxes are provided at this site in several energy intervals (>1 MeV, >5 MeV, >10 MeV, >30 MeV, >50 MeV, and >100 MeV), which are updated every five minutes.

As an example of a very large solar proton flux time period that occurred in solar cycle 23, we present the proton flux measured by GOES-11 for the October 26 through November 7, 2003 time period in Figure 1. The proton fluxes are given for 1 and 100 MeV, which reach to altitudes of about 87 and 33 km, respectively.

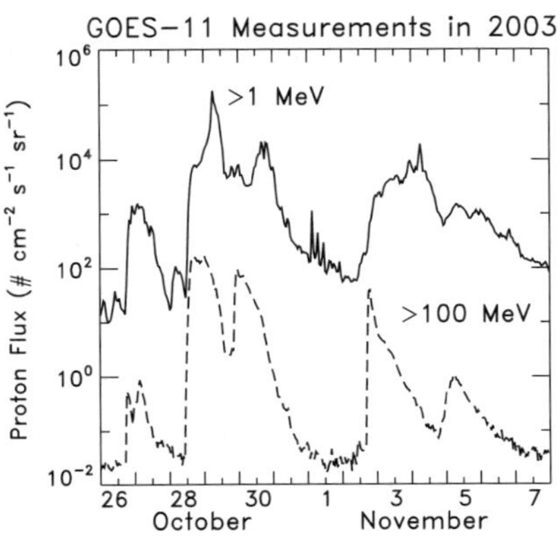


*Figure 1.* GOES-11 proton flux measurements in 2003 for energies >1 MeV (solid line) and >100 MeV (dashed line). These data are provided by the NOAA SEC at their website (`http://sec.noaa.gov/Data/goes.html`). Taken from Figure 1 of Jackman *et al.* (2005a).

The very fast moving protons with energies >100 MeV arrive at Earth early in the solar event, whereas the slower moving protons with energies >1 MeV mostly arrive later. The most intense time period of proton fluxes occurred during October 28–30.

Using the energy deposition methodology discussed in Vitt and Jackman (1996), the daily average ion pair production can be computed. $HO_x$ is created through ion chemistry (e.g., Solomon *et al.*, 1981) wherein: (1) two ($HO_x$) constituents are produced per ion pair up to about 70 km; and (2) less than two ($HO_x$) constituents are produced per ion pair above 70 km.

The solar protons and associated secondary electrons (produced in ionization events) create atomic nitrogen through dissociations, predissociations, or dissociative ionization collisions with $N_2$. Porter *et al.* (1976) showed that approximately 1.25 N atoms are produced per ion pair. About 45% or 0.55 per ion pair of the N atoms are in the ground state and about 55% or 0.7 per ion pair are in excited states. Although excited state N atoms are more likely to produce other $NO_y$ constituents, both types of N atoms can have an impact on the middle atmosphere.

## 3. Satellite Measurements of $NO_y$ Constituent Changes from SPEs

### 3.1. $NO_x$ ($NO + NO_2$) Observations

Substantial changes in $NO_y$ constituents as a result of SPEs have been measured by several satellite instruments during solar cycle 23. Very large fluxes of solar protons in July 2000 produced huge increases (>50 ppbv in the mesosphere) in polar Northern Hemisphere $NO_x$ measured by the Upper Atmosphere Research Satellite (UARS) Halogen Occultation Experiment (HALOE) (Jackman *et al.*, 2001). Randall *et al.* (2001) demonstrated that the polar Southern Hemisphere $NO_x$ was also influenced by this very large SPE. Using UARS HALOE and Polar Ozone and Aerosol Measurement (POAM) III data, Randall *et al.* (2001) showed evidence of large $NO_x$ enhancements in September 2000 in the middle stratosphere that were almost certainly caused by the July 2000 SPE.

The large solar storms in late October and early November 2003 also caused very large proton fluxes (see Figure 1) that created $NO_x$ and was measured by UARS HALOE (Jackman *et al.*, 2005a) and Envisat Michelson Interferometer for Passive Atmospheric Sounding (MIPAS) (López-Puertas *et al.*, 2005a). Jackman *et al.* (2005a) showed polar Southern Hemisphere $NO_x$ enhancements during the very large SPEs that were nearly as large as those that were measured during the July 2000 SPE. López-Puertas *et al.* (2005a) showed huge $NO_x$ increases in both the polar Northern and Southern Hemispheres, which lasted for at least two weeks after the major SPEs in the Northern polar regions. These measurements by Envisat MIPAS showed polar SPE-driven changes in both the winter (dark) and summer (daylight) hemispheres (López-Puertas *et al.*, 2005a). A plot of this interhemispheric difference is shown in this issue of *Space Science Reviews* in

Figure 1 of Langen (2006), which is similar to that given in Figure 4 of López-Puertas *et al.* (2005a). Seppala *et al.* (2004) showed $NO_2$ enhancements over several hundred per cent in the middle to upper stratosphere using Envisat Global Ozone Monitoring by Occultation of Stars (GOMOS) data.

### 3.2. $HNO_3$, $N_2O_5$, AND $ClONO_2$ OBSERVATIONS

Other $NO_y$ constituents were also elevated as a result of the huge SPEs that occurred in Oct–Nov 2003. López-Puertas *et al.* (2005b) showed polar $HNO_3$, $N_2O_5$, and $ClONO_2$ changes using Envisat MIPAS. Nitric acid ($HNO_3$) enhancements of 1–2 ppbv (100%) in the middle to upper stratosphere occurred in conjunction with the SPEs in both Hemispheres. Given the sudden nature of this $HNO_3$ enhancement, it was suggested to be most likely caused by gas phase chemistry (López-Puertas *et al.*, 2005b), in particular, $NO_2 + OH + M \rightarrow HNO_3 + M$. There were measured increases in $NO_2$ (e.g., Jackman *et al.*, 2005a; López-Puertas *et al.*, 2005b) and predicted increases in OH (Jackman *et al.*, 2005a) as a result of the Oct–Nov 2003 solar events, thus the reaction would likely be accelerated. Orsolini *et al.* (2005) examined the anomalous $HNO_3$ layer observed by MIPAS over a longer period of time (November 4, 2003 to February 20, 2004). They found that there were two periods of enhanced upper stratospheric $HNO_3$ in November 2003, one observed November 4–5 (early) and the other observed November 20–30 (late). Only the early enhanced $HNO_3$ period was thought to be caused directly by solar protons. The late November enhanced $HNO_3$ period was likely caused by production of $NO_x$ by energetic particle precipitation in the mesosphere followed by descent to the stratosphere (Orsolini *et al.*, 2005).

Dinitrogen pentoxide ($N_2O_5$) was increased around 40 km by about 0.5 to 1.2 ppbv (20–60%) in the polar Northern middle to upper stratosphere several days after the very large SPEs (López-Puertas *et al.*, 2005b). These enhancements were delayed because the most likely production mechanism $NO_2 + NO_3 + M \rightarrow N_2O_5 + M$ is relatively slow.

Chlorine nitrate ($ClONO_2$) was enhanced because of the previously discussed $NO_2$ increases speeding up the $ClO + NO_2 + M \rightarrow ClONO_2 + M$ reaction. López-Puertas *et al.* (2005b) showed polar Northern $ClONO_2$ enhancements of a maximum of about 0.4 ppbv (40%) near 32 km.

## 4. Satellite Measurements of ClO and HOCl Constituents Change from SPEs

Envisat MIPAS measurements showed significant increases of ClO and HOCl in Oct–Nov 2003 (von Clarmann *et al.*, 2005). These observations gave indirect proof of enhanced $HO_x$ abundances as a result of the large SPEs in late October 2003. Northern polar ClO increased by 0.2 to 0.4 ppbv and HOCl increased by 0.3 ppbv at altitudes above 32 km (von Clarmann *et al.*, 2005). These measurements, along with the observed $ClONO_2$ enhancements discussed in Section 3.2, suggest that HCl was destroyed either through reaction with OH or via ion cluster chemistry (von Clarmann *et al.*, 2005).

## 5. Satellite Measurements of Ozone Changes from SPEs

Several satellite instruments measured ozone decreases during solar cycle 23 as a result of SPEs including the UARS HALOE; NOAA 14 and 16 Solar Backscatter Ultraviolet 2 (SBUV/2) instruments; POAM III; Envisat GOMOS, MIPAS and Scanning Imaging Absorption spectrometer for Atmospheric ChartograpHY (SCIAMACHY); and Odin OSIRIS. Substantial mesospheric and upper stratospheric ozone decreases during and after the July 2000 SPE were measured by UARS HALOE and NOAA 14 SBUV/2 (Jackman *et al.*, 2001). Randall *et al.* (2001) used POAM III observations to show middle stratospheric ozone decreases in September 2000 as a result of the July 2000 SPE.

The most-studied period (to date) of ozone decreases as a result of SPEs occurred in October 26–November 7, 2003. Seppala *et al.* (2004) and Verronen *et al.* (2005) showed long lasting ozone depletions of 20 to 60% in the Northern Hemisphere polar lower mesosphere and upper stratosphere as a result of the events using Envisat GOMOS. Jackman *et al.* (2005a) found short-term ozone depletions of 40% in the Southern Hemisphere polar lower mesosphere with ozone depletions of 5–8% lasting days beyond the events in the upper stratosphere using NOAA 16 SBUV/2. The ozone decreases from a quiescent baseline day (October 25) over the October 28–November 1, 2003 disturbed period are given in Figure 2a.

López-Puertas *et al.* (2005a) showed significant polar lower mesosphere and upper stratosphere ozone decreases (10–70%) during the events using Envisat MIPAS. They also showed large differences between the two Hemispheres, with substantially more polar ozone depletion in the Northern (>20%) than in the Southern Hemisphere (5–10%). This interhemispheric difference is shown in this issue of *Space Science Review* in Figure 1 of Langen (2006), which is similar to that given in Figure 4 of López-Puertas *et al.* (2005a). Envisat SCIAMACHY measurements generally agreed with these Envisat MIPAS observations of ozone depletion (Rohen *et al.*, 2005).

## 6. Model Predictions of SPE Influences

Models have also been used to help interpret the influence of SPEs on the atmosphere during solar cycle 23. Jackman *et al.* (2001) model predictions showed reasonable agreement with measured $NO_x$ enhancements and ozone depletions caused by the July 2000 SPE. Verronen *et al.* (2005) studied the diurnal variation of ozone de-

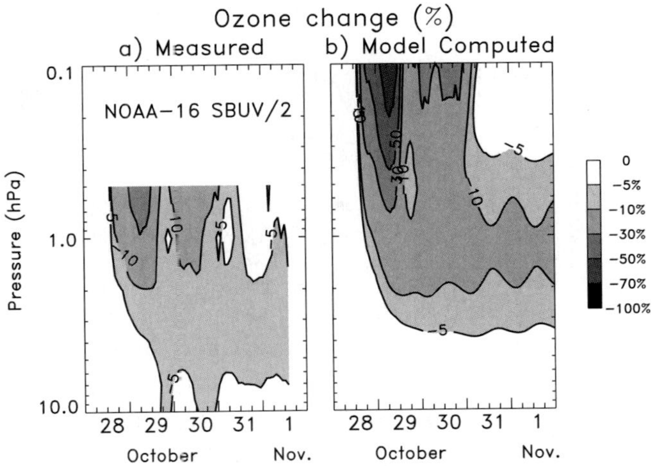


*Figure 2.* Percentage ozone change from October 25, 2003 values for the polar Southern Hemisphere from (a) NOAA 16 SBUV/2 measurements (top level of 0.5 hPa) and (b) the GSFC 2D model over the October 28–November 1, 2003 period. These plots show the influence of the Oct–Nov 2003 SPEs. Contour levels plotted are −5%, −10%, −30%, −50%, and −70%.

pletion during the Oct–Nov 2003 SPEs and found that maximum ozone depletions were predicted at the maximum solar zenith angles during sunrise and sunset, similar to the results that Solomon *et al.* (1983) obtained in predicting the July 1982 SPE atmospheric impact. Krivolutsky *et al.* (2003) calculated significant ozone depletion due to SPEs in July 2000, November 2000, September 2001, and November 2001. Krivolutsky *et al.* (2005) showed computations of ozone depletion for the three very large SPE periods in July 2000, November 2001, and Oct–Nov 2003.

The Jackman *et al.* (2005a) simulations using the Goddard Space Flight Center (GSFC) two-dimensional (2D) model for ozone decreases during and after the Oct–Nov 2003 very large SPEs are presented in Figure 2b. Huge ozone depletions (>70%) are calculated near 0.1 hPa on October 29 during the period of maximum proton flux intensity (see Figure 1). The short-lived depletion is caused by the enhanced $HO_x$ constituents, which last only during and for a few hours after the events. The longer-lived ozone depletion between about 0.3 and 3 hPa is caused by the $NO_y$ family. The modeled ozone change is in reasonable agreement during the SPEs. After the SPEs, the computed ozone change is slightly higher (lower) than that measured for the 0.3 to 2 hPa (2 to 7 hPa) altitude region. Ozone is also undergoing other changes at this time of year not connected with the SPEs, thus the apparent model/measurement disagreement does require further analysis.

It is important to note that the UARS HALOE measured and GSFC 2D model computed $NO_x$ changes are in fairly good agreement in the lower mesosphere and upper mesosphere over the October 30 to November 7 period (Jackman *et al.*, 2005a). The long-lived $NO_x$ enhancements drive the ozone changes over the months after the very large SPEs.

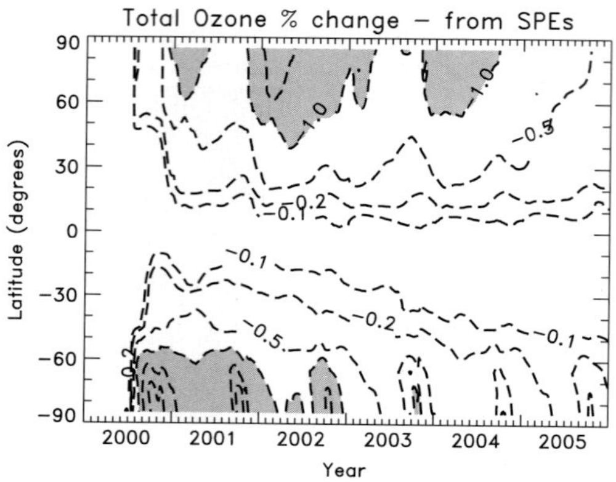


*Figure 3.* Model computed percentage total ozone changes from 2000–2005 resulting from SPEs in 2000–2003. Contour intervals are −3%, −2%, −1%, −0.5%, −0.2%, and −0.1%. The gray highlighted areas indicate total ozone decreases greater than 1%. Taken from Figure 4 of Jackman *et al.* (2005b).

Jackman *et al.* (2005a) and Rohen *et al.* (2005) both show model results confirming the very large interhemispheric differences observed in the Oct–Nov 2003 SPEs impacts on the middle atmosphere. These studies showed a much larger impact from these SPEs in the Northern than in the Southern Hemisphere. The $NO_y$ family has a much longer lifetime in the late Fall/Winter than in the late Spring/Summer for two reasons:

1. the $NO_y$ is lost through the mechanism $NO + h\nu$ (<191 nm) $\rightarrow N + O$ followed by $N - NO \rightarrow N_2 + O$, thus when the sunlight is intense (e.g., Summer and late Spring) the loss is greater;
2. the $NO_y$ is transported to lower levels in the atmosphere where it is more effectively shielded from loss by sunlight.

Jackman *et al.* (2005b) simulated all the SPEs in the 2000–2003 period with the GSFC 2D model. The computed impact on total ozone in this simulation is shown in Figure 3 for the 2000–2005 time period. Total ozone is reduced by a maximum of about 3% in the polar Southern latitudes in late 2001 and 2002 that is primarily driven by the July 2000 SPE, a Winter SPE in this hemisphere. Both polar regions had extended periods of depleted ozone greater than 1% as a result of SPEs. Although the SPE-driven total ozone change is significant at polar latitudes, the computed annually-averaged global ozone change is fairly small (<0.5%).

## 7. Conclusions

Solar cycle 23 was extremely active with seven of the largest twelve SPEs in the past forty years recorded, especially in the years 2000, 2001, 2003, and 2005.

These events caused significant polar middle atmospheric changes that were observed by a number of satellites. The atmospheric impacts from the July 2000 and Oct–Nov 2003 SPEs were discussed in several papers published between 2001 and 2005. These SPEs produced both $HO_x$ and $NO_y$ constituents. Most of the substantial polar ozone decreases caused by the SPEs were short-lived and driven by the $HO_x$ enhancements, however, the longer-lived ozone depletions caused by $NO_y$ increases were observed to last for months past the events in the late Fall and Winter seasons. These longer term impacts from SPEs should be considered when quantifying other natural as well as anthropogenic influences on the middle atmosphere, especially near solar maximum in the polar regions.

## Acknowledgements

We acknowledge all the satellite teams involved in collecting observations of solar proton event influences on the atmosphere and the NOAA SEC GOES team for providing solar proton flux data over the internet. This work was supported by the NASA Living with a Star Targeted Research and Technology Program and the NASA Atmospheric Composition Data and Analysis Program.

## References

Crutzen, P. J., Isaksen, I. S. A., and Reid, G. C.: 1975, 'Solar proton events: Stratospheric sources of nitric oxide', *Science* **189**, 457–458.

Degenstein, D. A., Lloyd, N. D., Bourassa, A. E., Gattinger, R. L., and Llewellyn, E. J.: 2005, 'Observations of mesospheric ozone depletion during the October 28, 2003 Solar proton event by OSIRIS', *Geophys. Res. Lett.* **32**, doi:10.1029/2004GL021521.

Frederick, J. E.: 1976, 'Solar corpuscular emission and neutral chemistry in the Earth's middle atmosphere', *J. Geophys. Res.* **81**, 3179–3186.

Heath, D. F., Krueger, A. J., and Crutzen, P. J.: 1977, 'Solar proton event influences on stratospheric ozone', *Science* **197**, 886–889.

Jackman, C. H., and McPeters, R. D.: 1985, 'The response of ozone to solar proton events during solar cycle 21: A theoretical interpretation', *J. Geophys. Res.* **90**, 7955–7966.

Jackman, C. H., and McPeters, R. D.: 1987, 'Solar proton events as tests for the fidelity of middle atmosphere models', *Physica Scripta* **T18**, 309–316.

Jackman, C. H., and McPeters, R. D.: 2004, 'The effect of solar proton events on ozone and other constituents', in: J. M. Pap, and P. Fox (eds.), *Solar Variability and Its Effects on Climate*, Washington, DC, *Geophys. Monograph* **141**, 305–319.

Jackman, C. H., Frederick, J. E., and Stolarski, R. S.: 1980, 'Production of odd nitrogen in the stratosphere and mesosphere: An intercomparison of source strengths', *J. Geophys. Res.* **85**, 7495–7505.

Jackman, C. H., Cerniglia, M. C., Nielsen, J. E., Allen, D. J., Zawodny, J. M., McPeters R. D., Douglass, A. R., Rosenfield, J. E., and Rood, R. B.: 1995, 'Two-dimensional and three-dimensional model simulations, measurements, and interpretation of the influence of the October 1989 solar proton events on the middle atmosphere', *J. Geophys. Res.* **100**, 11,641–11,660.

Jackman, C. H., Fleming, E. L., and Vitt, F. M.: 2000, 'Influence of extremely large solar proton events in a changing stratosphere', *J. Geophys. Res.* **105**, 11,659–11,670.

Jackman, C. H., McPeters R. D., Labow, G. J., Fleming, E. L., Praderas, C. J., and Russell, J. M.: 2001, 'Northern hemisphere atmospheric effects due to the July 2000 solar proton event', *Geophys. Res. Lett.* **28**, 2883–2886.

Jackman, C. H., DeLand, M. T., Labow, G. J., Fleming, E. L., Weisenstein, D. K., Ko, M. K. W., Sinnhuber, M., and Russell, J. M.: 2005a, 'Neutral atmospheric influences of the solar proton events in October–November 2003', *J. Geophys. Res.* **110**, doi:10.1029/2004JA010888.

Jackman, C. H., DeLand M. T., Labow, G. J., Fleming, E. L., Weisenstein, D. K., Ko, M. K. W., Sinnhuber, M., Anderson, J., and Russell, J. M.: 2005b, 'The influence of the several very large solar proton events in years 2000–2003 on the neutral middle atmosphere', *Adv. Space Res.* **35**, 445–450.

Krivolutsky, A., Kuminov, A., Vyushkova, T., Pereyaslova, N., and Nazarova, M.: 2003, 'Proton activity of the sun during 23rd solar maximum and its response in ozonosphere of the earth', *Adv. Space Res.* **31**, 2151–2156.

Krivolutsky, A., Kuminov, A., and Vyushkova, T.: 2005, 'Ionization of the atmosphere caused by solar protons and its influence on ozonosphere of the earth during 1994–2003', *J. Atmos. Sol.-Terr. Phys.* **67**, 105–117.

Langen, J.: 2006, 'Recent space data – introductory paper', *Space Sci. Rev.*, this volume, doi: 10.1007/s11214-006-9070-5.

López-Puertas, M., Funke, B., Gil-Lopez, S., von Clarmann, T., Stiller, G. P., Hopfner, M., Kellmann, S., Fischer, H., and Jackman, C. H.: 2005a, 'Observation of $NO_x$ enhancement and ozone depletion in the northern and southern hemispheres after the October–November 2003 solar proton events', *J. Geophys. Res.* **110**, doi:10.1029/2005JA011050.

López-Puertas, M., Funke, B., Gil-Lopez, S., Mengistu Tsidu, G., Fischer, H., and Jackman, C. H.: 2005b, '$HNO_3$, $N_2O_5$ and $ClONO_2$ Enhancements after the October–November 2003 solar proton events', *J. Geophys. Res.* **110**, doi:10.1029/2005JA011051.

López-Puertas, M., Funke, B., Clarmann, T. V., Fischer, H., and Stiller, G. P.: 2006, 'The stratospheric and mesospheric $NO_y$ in the 2002–2004 polar winters as measured by MIPAS/Envisat', *Space Sci. Rev.*, this volume, doi: 10.1007/s11214-006-9073-2.

McPeters, R. D.: 1986, 'A nitric oxide increase observed following the July 1982 solar proton event', *Geophys. Res. Lett.* **13**, 667–670.

McPeters, R. D., and Jackman, C. H.: 1985, 'The response of ozone to solar proton events during solar cycle 21: The observations', *J. Geophys. Res.* **90**, 7945–7954.

McPeters, R. D., Jackman, C. H., and Stassinopoulos, E. G.: 1981, 'Observations of ozone depletion associated with solar proton events', *J. Geophys. Res.* **86**, 12,071–12,081.

Natarajan, M., Remsberg, E. E., and Deaver, L. E.: 2004, 'Anomalously high levels of $NO_x$ in the polar upper stratosphere during April, 2004: Photochemical consistency of HALOE observations', *Geophys. Res. Lett.* **31**, doi:10.1029/2004GL020566.

Orsolini, Y. J., Manney, G. L., Santee, M. L., and Randall, C. E.: 2005, 'An upper stratospheric layer of enhanced $HNO_3$ following exceptional solar storms', *Geophys. Res. Lett.* **32**, doi:10.1029/2004GL021588.

Porter, H. S., Jackman, C. H., and Green, A. E. S.: 1976, 'Efficiencies for production of atomic nitrogen and oxygen by relativistic proton impact in air', *J. Chem. Phys.* **65**, 154–167.

Randall, C. E., Siskind, D. E., and Bevilacqua, R. M.: 2001, 'Stratospheric $NO_x$ Enhancements in the Southern Hemisphere Polar Vortex in Winter and Spring of 2000', *Geophys. Res. Lett.* **28**, 2385–2388.

Randall, C. E., Harvey, V. L., Manney, G. L., Orsolini, Y., Codrescu, M., Sioris, C., Brohede, S., Haley, C. S., Gordley, L. L., Zawodny, J. M., and Russell, J. M.: 2005, 'Stratospheric effects of energetic particle precipitation in 2003–2004', *Geophys. Res. Lett.* **32**, doi:10.1029/2004GL022003.

Reid, G. C., Solomon, S., and Garcia, R. R.: 1991, 'Response of the middle atmosphere to the solar proton events of August–December 1989', *Geophys. Res. Lett.* **18**, 1019–1022.

Rinsland, C. P., Boone, C., Nassar, R., Walker, K., Bernath, P., McConnell, J. C., and Chiou, L.: 2005, 'Atmospheric Chemistry Experiment (ACE) arctic stratospheric measurements of $NO_x$ during February and March 2004: Impact of intense Solar Flares', *Geophys. Res. Lett.* **32**, doi:10.1029/2005GL022425.

Rohen, G., von Savigny, C., Sinnhuber, M., Llewellyn, E. J., Kaiser, J. W., Jackman, C. H., Kallenrode, M.-B., Schroter, J., Eichmann, K. -U., Bovensmann, H., and Burrows, J. P.: 2005, 'Ozone depletion during the solar proton events of Oct./Nov. 2003 as Seen by SCIAMACHY', *J. Geophys. Res.* **110**, doi: 10.1029/2004JA010984.

Rusch, D. W., Gerard, J.-C., Solomon, S., Crutzen, P. J., and Reid, G. C.: 1981, 'The effect of particle precipitation events on the neutral and ion chemistry of the middle atmosphere, 1, odd nitrogen', *Planet Space Sci.* **29**, 767–774.

Semeniuk, K., McConnell, J. C., and Jackman, C. H.: 2005, 'Simulation of the October–November 2003 solar proton events in the CMAM GCM: Comparison with observations', *Geophys. Res. Lett.* **32**, doi:10.1029/2004GL022392.

Seppala, A., Verronen, P. T., Kyrola, E., Hassinen, S., Backman, L., Hauchecorne, A., Bertaux, J. L., and Fussen, D.: 2004, 'Solar proton events of October–November 2003: Ozone depletion in the northern hemisphere polar winter as seen by GOMOS/Envisat', *Geophys. Res. Lett.* **31**, doi:10.1029/2004GL021042.

Solomon, S., and Crutzen, P. J.: 1981, 'Analysis of the August 1972 solar proton event including chlorine chemistry', *J. Geophys. Res.* **86**, 1140–1146.

Solomon, S., Rusch, D. W., Gerard, J.-C., Reid, G. C., and Crutzen, P. J.: 1981, 'The effect of particle precipitation events on the neutral and ion chemistry of the middle atmosphere, 2, odd hydrogen', *Planet Space Sci.* **29**, 885–893.

Solomon, S., Reid, G. C., Rusch, D. W., and Thomas, R. J.: 1983, 'Mesospheric ozone depletion during the solar proton event of July 13, 1982, 2, Comparison between theory and measurements', *Geophys. Res. Lett.* **10**, 257–260.

Swider, W., and Keneshea, T. J.: 1973, 'Decrease of ozone and atomic oxygen in the lower Mesosphere during a PCA event', *Planet Space Sci.* **21**, 1969–1973.

Thomas, R. J., Barth, C. A., Rottman, G. J., Rusch, D. W., Mount, G. H., Lawrence, G. M., Sanders, R. W., Thomas, G. E., and Clemens, L. E.: 1983, 'Mesospheric ozone depletion during the solar proton event of July 13, 1982, 1, measurements', *Geophys. Res. Lett.* **10**, 253–255.

Verronen, P. T., Seppala, A., Clilverd, M. A., Rodger, C. J., Kyrola, E., C.-F. Enell, Ulich, T., and Turunen, E.: 2005, 'Diurnal variation of ozone depletion during the October–November 2003 solar proton events', *J. Geophys. Res.* **110**, doi:10.1029/2004JA010932.

Vitt, F. M., and Jackman, C. H.: 1996, 'A comparison of sources of odd nitrogen production from 1974 through 1993 in the earth's middle atmosphere as calculated using a two-dimensional model', *J. Geophys. Res.* **101**, 6729–6739.

von Clarmann, T., Glatthor, N., Hopfner, M., Kellmann, S., Ruhnke, R., Stiller, G. P., and Fischer, H.: 2005, 'Experimental evidence of perturbed odd hydrogen and chlorine chemistry after the October 2003 solar proton events', *J. Geophys. Res.* **110**, doi:10.1029/2005JA011053.

Warneck, P.: 1972, 'Cosmic radiation as a source of odd nitrogen in the stratosphere', *J. Geophys. Res.* **77**, 6589–6591.

Weeks, L. H., CuiKay R. S., and Corbin, J. R.: 1972, 'Ozone measurements in the mesosphere during the solar proton event of 2 November 1969', *J. Atmos. Sci.* **29**, 1138–1142.

Zadorozhny, A. M., Tuchkov, G. A., Kikhtenko, V. N., Lastovicka, J., Boska, J., and Novak, A.: 1992, 'Nitric oxide and lower ionosphere quantities during solar particle events of October 1989 after rocket and ground-based measurements', *J. Atmos. Terr. Phys.* **54**, 183–192.

# IMPACT OF SOLAR ACTIVITY ON STRATOSPHERIC OZONE AND $NO_2$ OBSERVED BY GOMOS/ENVISAT

A. HAUCHECORNE*, J.-L. BERTAUX and R. LALLEMENT
*Service d'Aéronomie/IPSL, CNRS, Université de Versailles-Saint-Quentin, Université Pierre et Marie Curie, BP3, 91371 Verrières-le-Buisson, France*
*(*Author for correspondence: E-mail: alain.hauchecorne@aerov.jussieu.fr)*

(Received 2 September 2005; Accepted 25 November 2005)

**Abstract.** Energetic proton precipitation occurring during solar events can increase the production of odd nitrogen in the upper stratosphere and mesosphere. A very intense solar proton event (SPE) occurred on 28 October 2003. Its impact on the composition of the middle atmosphere was observed in details due to the availability of several space instruments. Here we present GOMOS observations of a strong $NO_2$increase and a related ozone decrease in the upper stratosphere at north polar latitude. The perturbation of the chemical composition of the stratosphere was observed until the middle of December 2003. A strong $NO_2$ increase was also observed in the south polar vortex in June–July 2003. It is tentatively attributed to the effect of an SPE with protons of moderate energy occurring on 29 May 2003. If this hypothesis is confirmed, it will imply that the global effect of SPEs on the composition of the stratosphere is underestimated when only strong energy SPEs are considered.

**Keywords:** stratosphere, ozone, solar activity

## 1. Introduction

It is well recognized that energetic proton precipitation occurring during solar events can increase the production of odd nitrogen in the upper stratosphere and mesosphere (Crutzen, 1975). Odd nitrogen is important to the ozone balance due to catalytic reactions playing a role in ozone destruction. Effect of large solar proton events on ozone and odd nitrogen have been modelled by several authors (Rusch *et al.*, 1981; Reid *et al.*, 1991; Jackman *et al.*, 2000; Verronen *et al.*, 2002). Until very recent years, very few simultaneous ozone and odd nitrogen measurements from space were available. During the large July 2000 SPE space observations from HALOE and NOAA14 SBUV/2 showed a simultaneous $NO_x$ ($NO + NO_2$) increase and $O_3$ depletion in the upper stratosphere at north polar latitudes (Jackman *et al.*, 2001). The large solar storm happened end October 2003 caused the fourth largest SPE in the last 40 years. The effect of this event on $NO_x$ and $O_3$ in the upper stratosphere has been reported by several authors due to the availability of measurements from various space sounders. Seppälä *et al.* (2004) and Verronen *et al.* (2005) from ENVISAT GOMOS observed a strong enhancement of $NO_2$ and depletion of $O_3$ in the north polar vortex until mid-December when the vortex

Space Science Reviews (2006) 125: 393–402
DOI: 10.1007/s11214-006-9072-3

was destroyed by a stratospheric warming. Other observations were published by Rohen *et al.* (2005) from ENVISAT SCIAMACHY, Jackman *et al.* (2005) from HALOE and SBUV/2 and Semeniuk *et al.* (2005) from SciSat-I ACE. Degenstein *et al.* (2005) reported ODIN OSIRIS observations showing a strong mesospheric ozone depletion a few hours after the maximum proton flux. Orsolini *et al.* (2005) showed, from ENVISAT MIPAS data, a layer with a high $HNO_3$ mixing ratio in the upper stratosphere during early winter, descending to 30 km in January 2004.

An enhancement of stratospheric NO and $NO_2$ was observed in spring 2004 (March to May) in the north polar vortex by several space instruments (HALOE, MIPAS, SAGE II, POAM III) (Natarajan *et al.*, 2004; Randall *et al.*, 2005). The interpretation of these observations is still in discussion. Natarajan *et al.* (2004) attribute this enhancement to the large production of NO in the mesosphere/thermosphere during the October–November 2003 solar storm and its diabatic descent to the stratosphere during the following months when Randall *et al.* (2005) consider this hypothesis unlikely due to the strong mixing of stratospheric polar air after the December warming and propose a production of $NO_x$ by energetic particle precipitation in early 2004.

Here, using data from GOMOS star occultation spectrometer, on board the European satellite ENVISAT launched 1st March 2002, we give further observational evidences of the impact of the large October 2003 SPE on ozone depletion and $NO_2$ production in the North polar vortex. We present also observations of a strong upper stratospheric $NO_2$ enhancement in the south polar vortex in June–July 2004, tentatively attributed to an SPE with protons of moderate energy occurring in late May 2004.

## 2. Instrument

GOMOS (Global Ozone Monitoring by Occultation of Stars) is a stellar occultation spectrometer aimed to build a climatology of ozone and related species in the middle atmosphere (15 to 100 km) with a very high accuracy using the technique of stellar occultation (Bertaux *et al.*, 2004). When a star sets behind the atmosphere, its light crosses quasi-horizontally the atmosphere in a limb geometry and travels a very long distance in layers just above the tangent point defined as the location of lowest altitude. During its travel, the star light is absorbed by the atmospheric constituents. Each constituent can be identified by its absorption spectrum. The atmospheric transmission spectrum is equal to the ratio between the star spectrum absorbed by the atmosphere and the reference star spectrum measured outside the atmosphere. As the reference spectrum is measured at the beginning of each occultation, we can consider that GOMOS is a self-calibrated instrument, independent of any radiometric calibration. Furthermore the stellar occultation technique allows a perfect knowledge of the tangent altitude, depending only on the geometry of the light path between the star and the satellite. The 250–680 nm spectral domain is used for the determination of $O_3$, $NO_2$, $NO_3$, aerosols and atmospheric density from Rayleigh extinction from the upper troposphere to the mesosphere. In addition, two high spectral resolution channels centred at 760 and 940 nm allow measuring $O_2$ and $H_2O$ and two fast photometers (1 kHz sampling rate) are used to correct star scintillation perturbations and to determine high vertical resolution temperature profiles. Global latitude coverage is obtained with up to 40 stellar occultations per orbit from South Pole to North Pole. Data acquired on dark limb (night-time) are of better quality than on bright limb (day-time) because of a smaller perturbation by background light.

## 3. Data Processing

Results presented here are obtained using version V6.0 of the algorithms in the prototype processor. This version has been implemented in April 2004 for the full reprocessing of 2003 data. Main characteristics of the data inversion in V6.0 are summarized below. More details are given in Hauchecorne *et al.* (2005). The inversion of spectrometer data from Level 1b (calibrated transmission spectra) to Level 2 (vertical profiles of constituents) is made in 2 steps, the spectral inversion and the vertical inversion.

The goal of the spectral inversion is to determine the number of molecules along the horizontally integrated line of sight densities (slant densities) of each absorber from the transmission spectrum. The spectral inversion is made globally on all retrieved species using a Levenberg-Marquardt method except for $NO_2$ and $NO_3$ for which a DOAS (differential optical absorption spectroscopy) method is applied. The temperature dependence of $O_3$, $NO_2$ and $NO_3$ cross-sections is taken into account using ECMWF temperature along the line of sight.

The goal of the vertical inversion is to compute local density profiles as a function of altitude from slant density profiles as a function of tangent altitude. The method used in V6.0 is a Tikhonov regularization with an imposed vertical resolution depending on the species. It is fixed to a constant value equal to 4 km for $NO_2$ and $NO_3$ and it increases from 2 km below 20 km to 4 km above 30 km for $O_3$.

The uncertainty on local densities includes a random part due to photon counting and detector noise and a systematic part due to uncertainties in the data processing parameters (cross-sections, instrument spectral resolution). For an individual profile, the random error is in the order of 5% for $O_3$ between 15 and 60 km and 20% for $NO_2$ and 30% for $NO_3$ between 25 and 40 km and increases above and below these altitude ranges. The systematic error, dominated by uncertainties in absorption cross-sections and on their dependence on

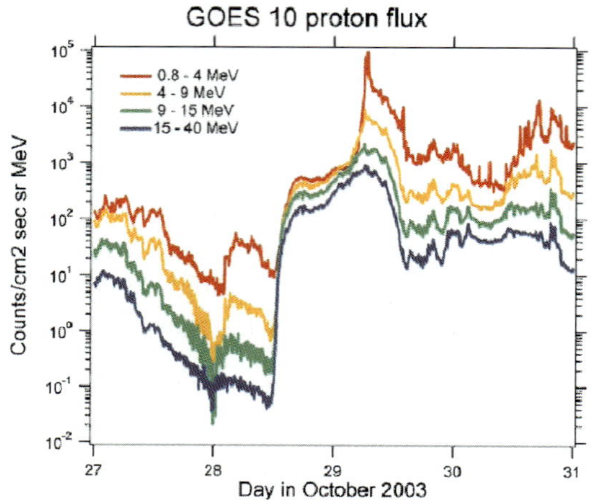


*Figure 1.* Proton flux measured by GOES 10 in 4 energy bands between 27 October and 30 October 2003 (from `http://www.sec.noaa.gov/Data/goes.html`).

temperature, can be roughly estimated to 3% for $O_3$, 10% for $NO_2$ and 15% for $NO_3$.

## 4. October 2003 Solar Proton Event

On 28–30 October 2003 a very large SPE occurred. The solar proton >100 MeV flux reached values >100 $cm^{-2}s^{-1}sr^{-1}$ and the time integrated flux was estimated to $1.2 \times 10^7$ $cm^{-2}sr^{-1}$ (Figure 1). This is the fourth largest event since 1962. The Thule neutron monitor indicates a 20% forbush decrease (not shown).

During the period from 16 October to 09 November 2003 GOMOS performed a series of occultation of star β Cet (visual magnitude 2.04, brightness temperature 4500 K) at a high and almost constant latitude (69°N–71°N). This star was chosen because it was the only one occulted in the auroral zone on dark limb around the SPE period. On 28th October GOMOS instrument was off due to an orbit correction of ENVISAT platform.

The evolution of $O_3$ concentration at 48 km during the few days before and after the SPE shows a decrease of about 50% (Figure 2). During the same time, a spectacular increase of $NO_2$ is observed from values near the noise level to about $5 \times 10^8$ mol $cm^{-3}$ (∼20 ppb). There is no indication of a return to previous values during the following 6 days (29 October to 3 November). The comparison of the mean profile for the period before and after the 28th October (Figures 3 and 4) shows that $O_3$ is depleted down to at least 35 km. On the same time, the $NO_2$ increase is observed at all altitudes above 37 km. The vertical integration of the difference between the $O_3$ profile before and after the SPE gives a decrease of 7 Dobson Units (DU), more than 2% of the total column estimated to be about 300 DU at 70°N

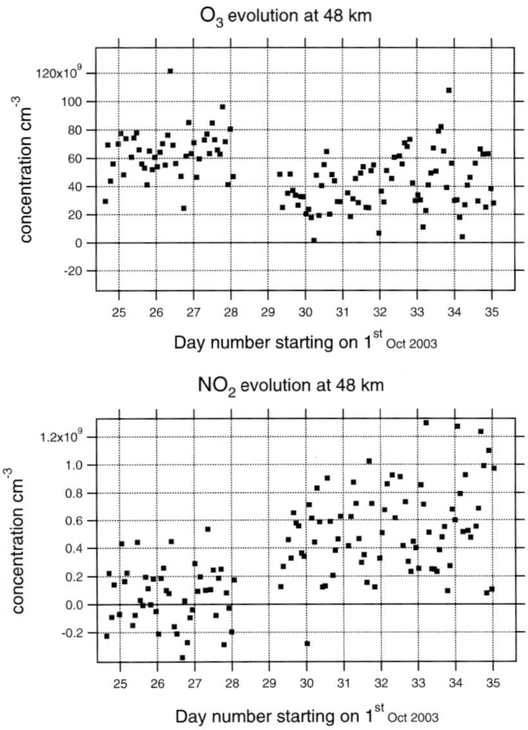


*Figure 2.* Evolution of $O_3$ and $NO_2$ at 48 km around 70°N between 25 October and 4 November 2003.

in October–November. This has to be compared with model estimations. Jackman *et al.* (2000) estimated a 1% total ozone depletion at high latitudes after the very strong August 1972 and October 1989 SPEs.

## 5. Austral Winter 2003

During austral winter 2003, a strong enhancement of $NO_2$ has been observed by GOMOS in the high latitude upper stratosphere (Hauchecorne *et al.*, 2005). The development of this enhanced $NO_2$ can be followed on the monthly latitude-height distribution maps (Figure 5). In April, the $NO_2$ distribution is near the climatology with highest mixing ratios at low latitudes. From 2 to 30 May GOMOS was stopped due to instrumental problems and it is not possible to compute monthly mean profiles in May. A maximum appears above 40 km at southern high latitude in June and develops and extends downwards in July. In August, the maximum decreases and peaks at 40 km. In September, the $NO_2$ distribution is no more perturbed. There is no major SPE reported during this period to explain the strong enhancement of upper

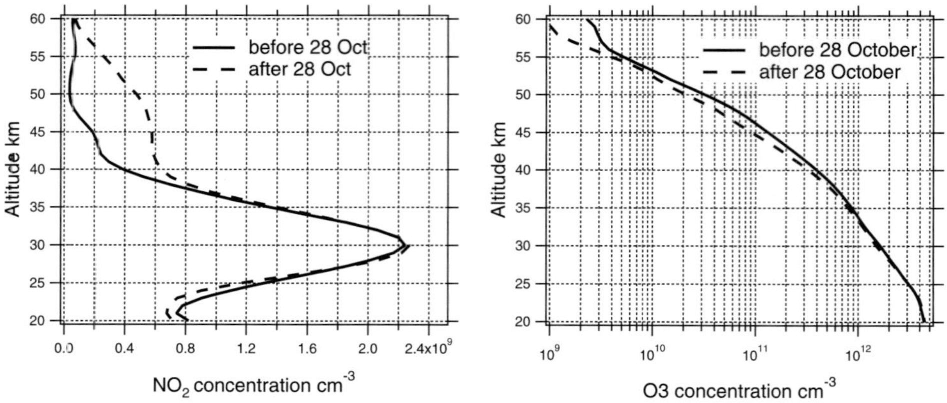


*Figure 3.* Mean $NO_2$ (left) and $O_3$ (right) vertical profile around 70°N before and after the 28 October 2003 SPE.

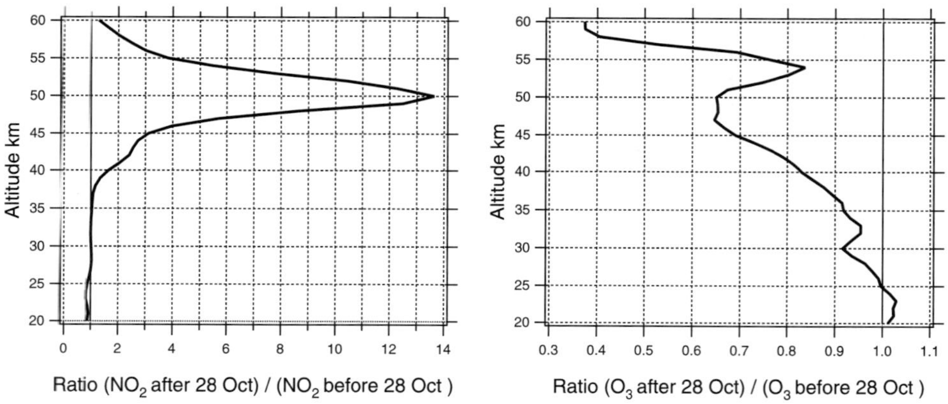


*Figure 4.* Ratio of $NO_2$ (left) and $O_3$ (right) vertical profiles around 70°N before and after the 28 October SPE.

stratospheric $NO_2$. However, there is one SPE reported on 29 May 2003 with a weak flux of high energy protons but with a relatively strong flux of low energy protons. The proton flux measured by GOES 10 in the band 0.8–4 MeV (Figure 6) peaks to $3 \times 10^4$ cm$^{-2}$s$^{-1}$sr$^{-1}$ against $8 \times 10^4$ cm$^{-2}$s$^{-1}$sr$^{-1}$ for the 28 October 2003 SPE. These low energy protons deposit most of their energy in the mesosphere (Jackman *et al.*, 2005) and can produce an important amount of $NO_x$ there. In June–July there is a strong diabatic descent in the south polar vortex and $NO_x$ produced in the mesosphere will descend in the upper stratosphere. This is a possible explanation of the strong $NO_2$ enhancement observed in June–July at high southern latitudes. This hypothesis is supported by the evolution of $NO_2$ concentration profile between May and August (Figure 7). In May, the $NO_2$ concentration is near the level of detection above 45 km. On 13 June, $NO_2$ concentration increases above 45 km with of a secondary maximum at 51 km. This maximum descends to 42 km on

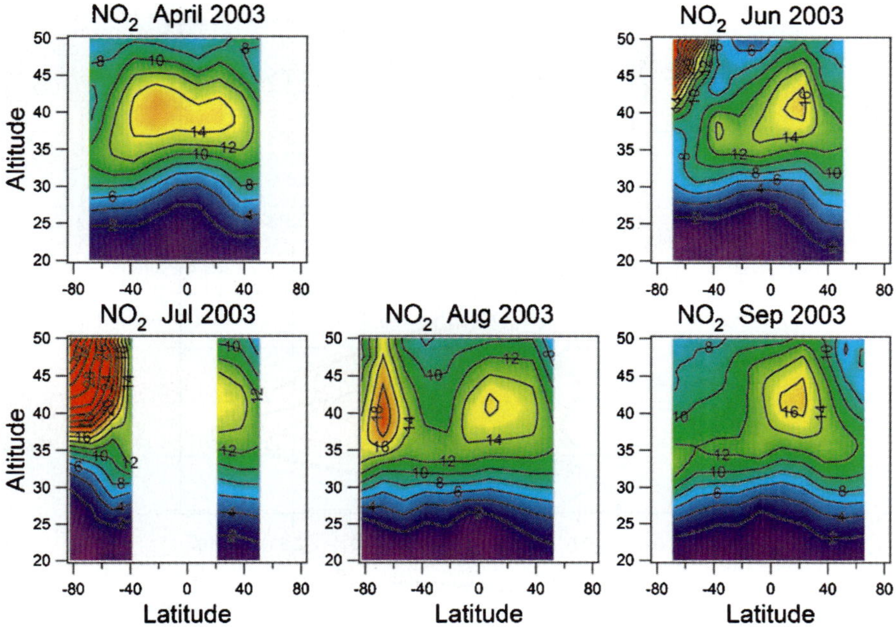


*Figure 5.* Latitude-height section of monthly mean $NO_2$ mixing ratio (in ppbv) from April to September 2003.

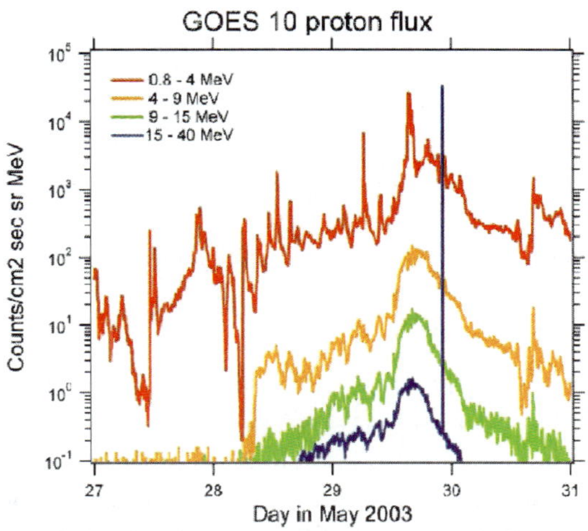


*Figure 6.* Proton flux measured by GOES 10 in 4 energy bands between 27 May and 30 May 2003 (from `http://www.sec.noaa.gov/Data/goes.html`).

27 June, 38 km on 17 July, 36 km on 1st August, 34 km on 15 August and 32 km on 26 August. Variations in $NO_2$ concentration observed below 30 km are due to the seasonal evolution of $NO_x$ in the high latitude middle stratosphere and are not discussed here. The evolution of $O_3$ profile from 15 April to 31 May (Figure 8) indicates a possible depletion of up to 40% around 55 km just after the SPE.

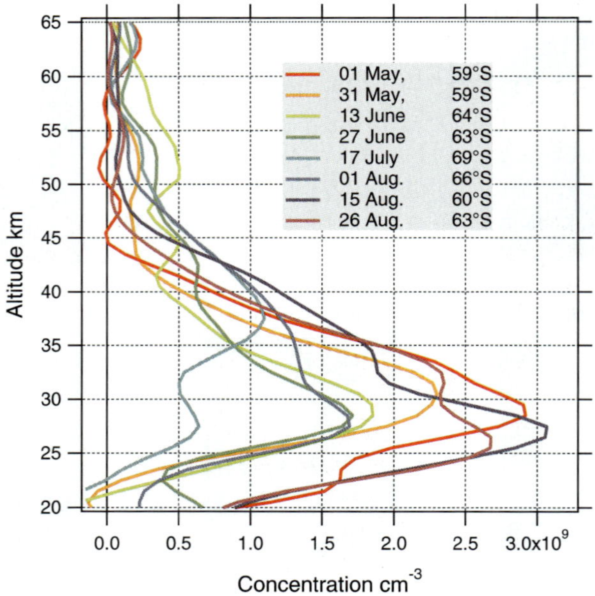


*Figure 7.* Evolution of $NO_2$ profile at southern high latitude from 1 May to 26 August 2003.

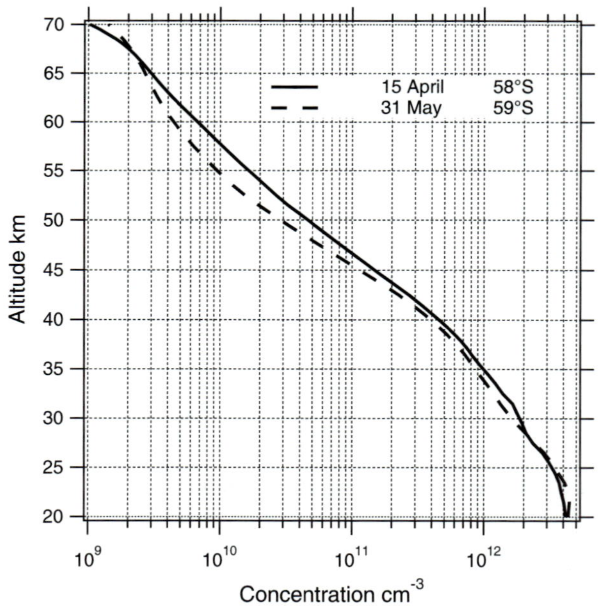


*Figure 8.* Evolution of $O_3$ profile at southern high latitude from 15 April to 31 May.

The possibility of a large mesosphere/thermospere production of $NO_x$ by energetic particle precipitation (Randall *et al.*, 2005) and subsequent diabatic descent in the polar vortex should be also considered to explain these observations.

## 6. Conclusion

A large $NO_2$ increase and $O_3$ depletion was observed by GOMOS in the Northern polar stratosphere after the strong 28 October 2003 SPE. During such an event, $NO_x$ is formed in the upper stratosphere/mesosphere in the polar cap which coincides approximately to the polar vortex during the winter period. The total ozone column in the polar vortex was decreased by about 2% north of 70°N, about twice model estimations (Jackman *et al.*, 2000). The $NO_2$ perturbation was observable until mid-December when the polar vortex was destroyed by an upper stratospheric warming. The impact of an SPE on the stratosphere is very dependent on the season and on the meteorology of the polar stratosphere. The impact is important if $NO_x$ is transported in the polar night downwards into the middle stratosphere where it can stays during several months without being destroyed. A large $NO_2$ increase was also observed in June–July 2003 in the south polar stratosphere with some indications of a simultaneous ozone decrease. The origin of this $NO_2$ increase is still not clear. It may be related to the production of $NO_x$ by energetic particle precipitation as proposed by Randall *et al.* (2005). We propose here another mechanism. A SPE with protons of moderate energy occurred on 29 May 2003. Most of the proton energy was confined in the band 0.8–4 MeV which means that $NO_x$ production is limited to the mesosphere. However, it occurred during a period where the south polar vortex is very stable and in the polar night and the diabatic descent inside the vortex is very strong. In these conditions, one can expect $NO_x$ to be rapidly transported into the upper stratosphere without being photo-chemically destroyed. Further studies are needed to evaluate the efficiency of this mechanism. If it is confirmed, it means that the impact of SPEs on the stratospheric chemistry may be strongly underestimated by the models taking into account only the strong energy events.

## Acknowledgements

We thank European Space Agency and Centre National d'Etudes Spatiales for their support. We are very grateful to the members of the GOMOS Team for their contribution to the success of the GOMOS mission.

## References

Bertaux, J.-L., Hauchecorne, A., Dalaudier, F., Cot, C., Kyrölä, E., Fussen, D., Tamminen, J., Leppelmeier, G.-W., Sofieva, V., Hassinen, S., Fanton O., d'Andon, Barrot, G., Mangin, A., Théodore, B., Guirlet, M., Korablev, O., Snoeij, P., Koopman, R., and Fraisse, R.: 2004, 'First results on GOMOS/ENVISAT', *Adv. Space Res.* **33**, 1029–1035.

Crutzen, P. J.: 1975., 'Solar proton events: Stratospheric sources of nitric oxide', *Science* **189**, 457–458.

Degenstein, D. A., Lloyd, N. D., Bourassa, A. E., Gattinger, R. L., and Llewellyn, E. J.: 2005, 'Observations of mesospheric ozone depletion during the October 28, 2003 solar proton event by OSIRIS', *Geophys. Res. Lett.* **32**, doi:10.1029/2004GL021521.

Hauchecorne, A., Bertaux, J.-L., Dalaudier F., Cot, C., Lebrun, J.-C., Bekki, S., Marchand, M., Kyrölä, E., Tamminen J., Sofieva, V., Fussen, D., Vanhellemont, F., Fanton d'Andon, O., Barrot G., Mangin, A., Théodore, B. Guirlet, M., Snoeij, P., Koopman, R., Saavedra de Miguel, L., Fraisse, R., and Renard, J.-B.: 2005, 'First simultaneous global climatologies of night-time stratospheric $NO_2$ and $NO_3$ observed by GOMOS/ENVISAT in 2003' , *J. Geophys. Res.* **110**, doi:10.1029/2004JD005711.

Jackman, C. H., Fleming, E. L., and Vitt, F. M.: 2000, 'Influence of extremely large solar proton events in a changing stratosphere', *J. Geophys. Res.* **105**, 11,659–11,670.

Jackman, C. H., McPeters, R. D., Labow, G. J., Fleming, E. L., Praderas, C. J., and Russell, J. M.: 2001, 'Northern hemisphere atmospheric effects due to the July 2000 solar proton event', *Geophys. Res. Lett.* **28**, 2883–2886.

Jackman, C. H., DeLand M. T., Labow, G. J., Fleming, E. L., Weisenstein, D. K., Ko, M. K. W., Sinnhuber, M., and Russell, J. M.: 2005, 'Neutral atmospheric influences of the solar proton events in October–November 2003', *J. Geophys. Res.* **110**, doi:10.1029/2004JA010888.

Natarajan, M., Remsberg, E. E., Deaver, L. E., and Russell III, J. M.: 2004, 'Anomalously high levels of $NO_x$ in the polar upper stratosphere during April, 2004: Photochemical consistency of HALOE observations', *Geophys. Res. Lett.* **31**, doi:10.1029/2004GL020566.

Orsolini, Y. J., Manney, G. L., Santee, M. L., and Randall, C. E.: 2005, 'An upper stratospheric layer of enhanced $HNO_3$ following exceptional solar storms', *Geophys. Res. Lett.* **32**, doi:10.1029/2004GL021588.

Randall, C. E., *et al.*: 2005, 'Stratospheric effects of energetic particle precipitation in 2003–2004', *Geophys. Res. Lett.* **32**, doi:10.1029/2004GL022003.

Reid, G. C., Solomon, S., and Garcia, R. R.: 1991, 'Response of the middle atmosphere to the solar proton events of August–December, 1989', *Geophys. Res. Lett.* **18**, 1019–1022.

Rohen, G., *et al.*: 2005, 'Ozone depletion during the solar proton events of October/November 2003 as seen by SCIAMACHY', *J. Geophys. Res.* **110**, doi:10.1029/2004JA010984.

Rusch, D. W., Gérard, J.-C., Solomon, S., Crutzen, P. J., and Reid, G. C.: 1981, 'The effect of particle precipitation events on the neutral and ion chemistry of the middle atmosphere – I. Odd nitrogen', *Plan. Space Sc.* **29**, 767–774.

Semeniuk, K., McConnell, J. C., and Jackman, C.nH.: 2005, 'Simulation of the October–November 2003 solar proton events in the CMAM GCM: Comparison with observations', *Geophys. Res. Lett.* **32**, doi:10.1029/2005GL022392.

Seppälä, A., Verronen, P. T., Kyrölä E., Hassinen, S., Backman, L., Hauchecorne, A., Bertaux, J. L., and Fussen, D.: 2004, 'Solar proton events of October–November 2003: Ozone depletion in the Northern Hemisphere polar winter as seen by GOMOS/Envisat', *Geophys. Res. Lett.* **31**, doi:10.1029/2004GL021042.

Verronen, P. T., Turunen, E., Ulich Th., and Kyrölä, E.: 2002, 'Modelling the effects of the October 1989 solar proton event on mesospheric odd nitrogen using a detailed ion and neutral chemistry model', *Annales Geohysicae* **20**, 1967–1976.

Verronen, P. T., Seppälä, A., Clilverd, M. A., Rodger, C. J., Kyrölä, E., Enell, C., Ulich, T., and Turunen, E.: 2005, 'Diurnal variation of ozone depletion during the October–November 2003 solar proton events', *J. Geophys. Res.* **110**, doi:10.1029/2004JA010932.

# THE STRATOSPHERIC AND MESOSPHERIC $NO_y$ IN THE 2002–2004 POLAR WINTERS AS MEASURED BY MIPAS/ENVISAT

M. LÓPEZ-PUERTAS[1,*], B. FUNKE[1], T. VON CLARMANN[2], H. FISCHER[2] and G. P. STILLER[2]

[1]*Instituto de Astrofísica de Andalucía (CSIC), Apdo. Postal 3004, 18008, Granada, Spain*

[2]*Institut für Meteorologie und Klimaforschung, Forschungszentrum Karlsruhe und Universität Karlsruhe, Germany*

*(*Author for correspondence: E-mail: puertas@iaa.es)*

(Received 19 September 2005; Accepted in final form 26 October 2005)

**Abstract.** The Michelson Interferometer for Passive Atmospheric Sounding (MIPAS) on board ENVISAT, provided global (pole-to-pole, the polar night winter regions) measurements of nearly all constituents of the $NO_y$ family (including NO, $NO_2$, $HNO_3$ and $N_2O_5$) from July 2002 to the end of March 2004 from the upper stratosphere up to the middle mesosphere. The inter-annual variability of the $NO_2$ and $HNO_3$ abundances in the Arctic and Antarctic winters from September 2002 through March 2004 was enormous with tremendous hemispheric asymmetry and extraordinary values in two winters. The origin of these variations and of the extreme measured values has been analyzed on the basis of the changing atmospheric dynamics (using the $CH_4$ tracer) and solar activity, including the extraordinary solar protons events of Oct–Nov 2003.

**Keywords:** $NO_y$, stratosphere, mesosphere, MIPAS, ENVISAT

## 1. Introduction

During polar winter, large amounts of upper atmospheric $NO_x$ ($=NO + NO_2$) can be transported down to the stratosphere by the meridional circulation without being photochemically destroyed. Once transported into the stratosphere, $NO_x$ is photo-chemically stable and thus may contribute to the stratospheric $NO_y$ budget, which largely controls the $O_3$ abundance. This solar-terrestial coupling mechanism was already proposed in the early 1980s (Crutzen, 1979; Solomon *et al.*, 1982; Frederick and Orsini, 1982), and its first experimental evidence was given by the Limb Infrared Monitor of the Stratosphere (LIMS) which measured enhanced upper stratospheric and mesospheric $NO_2$ in the 1978–1979 polar night Northern hemisphere (Russell *et al.*, 1988). Later, several satellite occultation measurements have corroborated this mechanism and helped in understanding it. Thus, the Halogen Occultation Experiment (HALOE) showed the existence of upper atmospheric $NO_x$-rich air in the Southern Hemisphere (SH) polar winter stratosphere during 1992–1996 (Siskind *et al.*, 2000). Similar observations were made by the Atmospheric Trace Molecule Spectroscopy (ATMOS) experiment (Rinsland *et al.*, 1999), the Polar Ozone and Aerosol Measurement instrument (POAM) (Randall

Space Science Reviews (2006) 125: 403–416
DOI: 10.1007/s11214-006-9073-2

*et al.*, 1998), and by the Stratospheric Aerosol and Gas Experiment (SAGE) (Callis *et al.*, 1998, 2001) for different Northern Hemisphere (NH) and SH winters. These measurements suggest that: (1) the $NO_x$ descent is well confined to the polar vortex; (2) there is a pronounced hemispheric asymmetry, with larger $NO_x$ values in the Antarctic winters than in the Arctic winters; and (3) they have large inter-annual variations closely correlated to the solar and geomagnetic activity (see, e.g., Randall *et al.*, 1998; Siskind *et al.*, 2000; Callis *et al.*, 2001).

$NO_x$ is produced in the polar regions by different processes at different altitudes. In the lower thermosphere, the main production mechanism is electron precipitation (with auroral origin or produced in the outer trapping region of the magnetosphere) dissociating $N_2$, and then N reacting with $O_2$ to form NO. This is a continuous source although highly variable since it is modulated by the electron fluxes at different energies in short (days) an long (solar cycle) scales (Callis *et al.*, 1998, 2001). In addition, NO is produced at lower altitudes, from the upper mesosphere down to the stratosphere, by solar protons. The protons reach these atmospheric regions during large solar eruptions when large fluxes of high-energy solar protons reach the Earth (see, e.g., Jackman *et al.*, 2006). This is a rather sporadic source, although also affected by the long-term solar cycle.

The major loss of NO in the mesosphere and lower thermosphere is given by its photo-dissociation, $NO + h\nu \rightarrow N + O$, followed by its recombination with atomic nitrogen, $NO+N \rightarrow N_2 + O$. The photo-dissociation rate of NO decreases drastically below around 50 km and hence leads $NO_x$ to have a long (longer than months) chemical lifetime below that altitude (Brasseur and Solomon, 1986).

The $NO_x$ produced in the upper atmosphere is transported downwards during polar winter by the descending branch of the meridional circulation. Hence, the major interest in the past was to determine the amount of $NO_x$ that is produced in the upper atmosphere and injected in the polar stratosphere. This is mainly demanded by our needs to fully understand the sources and sinks of stratospheric ozone, strongly dominated by the catalytic $NO_x$ destruction cycle in the stratosphere. The major factors controlling the $NO_x$ stratospheric deposition are twofold: (1) the production of NO, largely modulated by the solar activity through both upper atmospheric electron production, and through the sporadic solar proton events which extent from the upper mesosphere down to the mid-stratosphere; and (2) the factors controlling the dynamics of this region. To make the $NO_x$ downward transport efficient, dark conditions are required. However, the descending air within the polar vortex is not always confined to the polar night region, but moves away from it by planetary wave activity exposing $NO_x$ to sunlight. The descent is also quite often decelerated by the appearance of stratospheric warmings, which also bring polar air into illuminated mid-latitude regions where it is photodissociated, and interrupt the descent of $NO_x$ within the polar night region.

Despite the numerous measurements confirming the downward transport of $NO_x$ in polar winter, the quantitative assessment of the injection of $NO_x$ in the polar winter stratosphere still remained to be made; partially because of the lack of wintertime $NO_x$ data polewards of 50° (Siskind, 2000).

The Michelson Interferometer for Passive Atmospheric Sounding (MIPAS) measures most of the $NO_y$ constituents, including NO, $NO_2$, $HNO_3$ and $N_2O_5$, as well as dynamical tracers as $CH_4$ and CO, with global coverage and independent of illumination conditions. It is thus perfectly suited for studying polar winter $NO_x$ enhancements in the mesosphere and upper stratosphere, its downward transport, and its effect on the stratospheric $NO_y$ budget. Several studies on $NO_y$ enhancements using MIPAS measurements have been undertaken (López-Puertas *et al.*, 2005a,b; Funke *et al.*, 2005; Stiller *et al.*, 2005). Here, we summarize these results and extend the analysis to the measurements of $NO_2$ and $HNO_3$ for the four consecutive polar winters from September 2002 to March 2004 in order to assess the interannual variability of this solar-terrestrial coupling mechanism. $CH_4$ abundances are also used to help understanding the dynamical effects on the $NO_2$ and $HNO_3$ distributions.

## 2. MIPAS Data

MIPAS is a limb emission Fourier transform spectrometer, operating in the mid infrared spectral region, designed for measurement of atmospheric trace species from space (Fischer and Oelhaf, 1996). It is on board the Environmental Satellite (ENVISAT) which was launched into its sun-synchronous polar orbit of 98.55° inclination at about 800 km altitude on 1 March 2002. MIPAS operated from July 2002 to March 2004 at spectral resolution of 0.035 $cm^{-1}$. It observes the atmosphere during day and night with global coverage (pole-to-pole), including the polar night. In its nominal measurement mode, MIPAS covered tangent heights of 6–68 km with steps of 3 km from 6 to 42 km and further tangent heights at 47, 52, 60, and 68 km. MIPAS orbits the Earth at 14.3 times a day, taking about 72 profiles per orbit. The retrieval of gas abundances from L1b MIPAS spectra are being carried out by different groups. In this paper we analyze the L2 operational data (versions 4.61/4.62) processed under ESA responsibility. The operational retrieval algorithm and the characterization of L2 products are described by Ridolfi *et al.* (2000) and Raspollini *et al.* (2006). In this paper we analyze the $NO_2$ and $HNO_3$ distributions from September 2002 to March 2004. Concentrations of the $CH_4$ tracer are also used to show the effects of dynamics on the $NO_2$ and $HNO_3$ fields. The L2 profiles were interpolated to fixed pressure levels and then averaged for each day for the 65–90° polar cap. Typically between 50 and 100 profiles were averaged for each day. The errors for these species are reported by Raspollini *et al.* (2006). Since the number of averaged profiles is rather large, errors are dominated by systematic sources. These are similar for $NO_2$, $HNO_3$ and $CH_4$ with values about 10% or smaller below 40 km (30 km for $HNO_3$) and around 20% above this altitude.

## 3. Variability in $NO_2$

We describe in this section the nighttime distribution of $NO_2$ measured by MIPAS in the polar cap (65–90°) for the four successive winters in the Arctic and Antarctic from September 2002 until March 2004 (see Figure 1; measurements during the Antarctic 2002 winter are limited to a few days and hence are not shown). Due to the rapid conversion of NO into $NO_2$ in the absence of sunlight, night-time $NO_2$ can be considered as a reasonable proxy for the total $NO_x$.

The $NO_2$ in the Arctic 2002–2003 started to enhance in the middle mesosphere (~0.1 hPa, ~65 km) in late October and continued until the arrival of the sunlight by the end of February. The mean polar cap values at ~0.1 hPa were around 40 ppbv (80 ppbv near the North pole) which, although high, are much smaller than in other winters (see below). The figure shows clearly the descent of the $NO_x$ layer in the polar cap through the winter with a profound interruption in late December/early January when the $NO_x$ abundance was abruptly and significantly reduced, particularly above ~50 km (~1 hPa). The $CH_4$ field shows a similar behavior (see Figure 2) with descent of upper atmosphere $CH_4$-poor air in the early winter and a sudden increase in late December-early January. This change was caused by a minor stratospheric warming occurring in early January followed by a major one in late January (Manney *et al.*, 2005b). During such a warming, the upward propagation and breaking of planetary waves produces a large mixing and exchange of polar and mid-latitude air masses. In the upper stratosphere and mesosphere, mid-latitude $NO_x$-poor air enters the polar cap, significantly reducing the $NO_x$ concentration, while the opposite occurs in the mid-stratosphere. Thus, the $NO_2$ increase in late Dec/early Jan around ~2 hPa was due to mixing with mid-latitude $NO_x$-rich air rather than downward transport. The $NO_x$ that moves away from the polar night region into mid-latitudes may persist below ~50 km but it is rapidly photodissociated above, thus reducing the downward transport of $NO_x$. In addition, the stratospheric warmings also reduce the descent rate of the meridional circulation in the mesosphere and thus further weaken the downward transport of $NO_x$ (Liu and Roble, 2002). Their effects in the stratosphere are, however, opposite, inducing a rapid descent. As a result, the $NO_x$ abundance at the end of this winter in the stratospheric polar cap was only slightly larger than in autumn, just above 8 ppbv. Funke *et al.* (2005) have estimated that the amount of $NO_x$ deposited in the stratosphere in this winter was negligible.

The middle panel of Figure 1 shows the time series of the nighttime $NO_2$ in the Antarctic 2003 winter. Two major features are clearly distinguishable in comparison with the previous Arctic winter: (1) the higher $NO_2$ in the middle mesosphere (0.1 hPa, ~65 km), which reach mean values of up to 150 ppbv in the polar cap and of 300 ppbv near the South pole; and (2) the absence of any break in the continuous descent of $NO_x$ throughout the whole winter. The latter is corroborated by the $CH_4$ field (Figure 2, middle panel) which shows the $CH_4$ abundance continuously decreasing until mid-October, i.e., well into the southern spring. Note also the

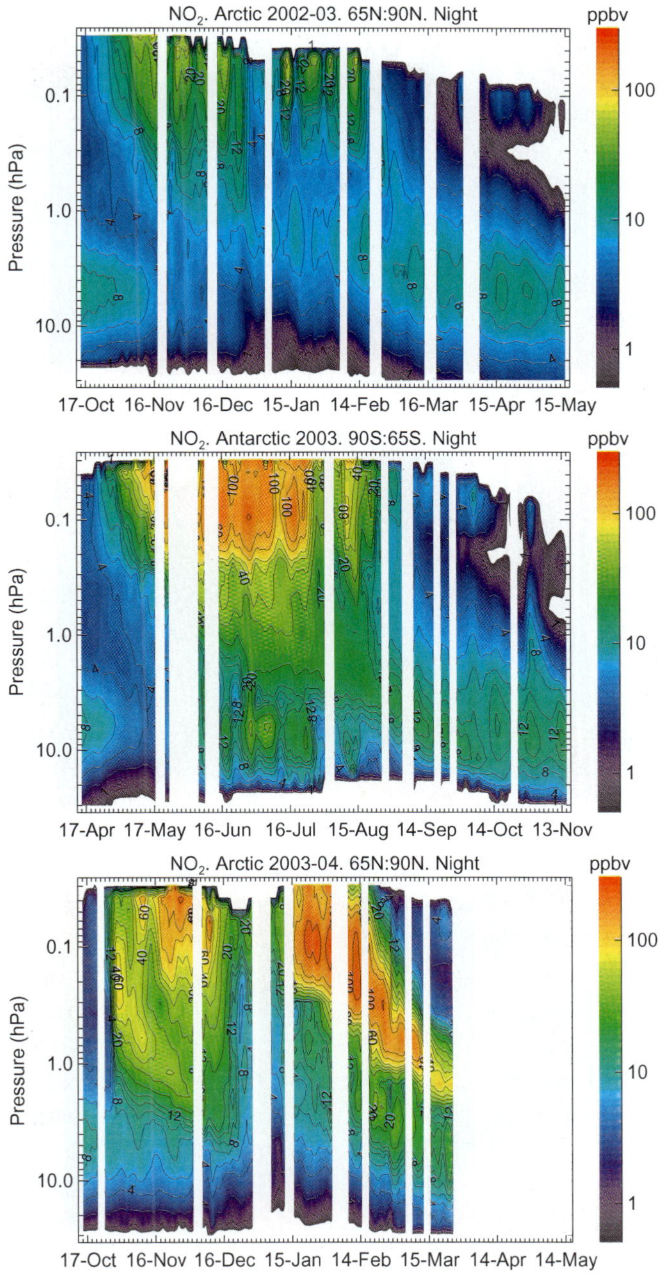


*Figure 1.* Temporal evolution of the $NO_2$ nighttime abundance in the winter polar cap (65°–90°). *Upper panel:* Arctic winter 2002–2003; *Middle panel:* Antarctic winter 2003; *Lower panel:* Arctic winter 2003–2004. White bands represents lack of measurements. Other white regions are values outside the scale. Note the logarithmic scale used for $NO_2$.

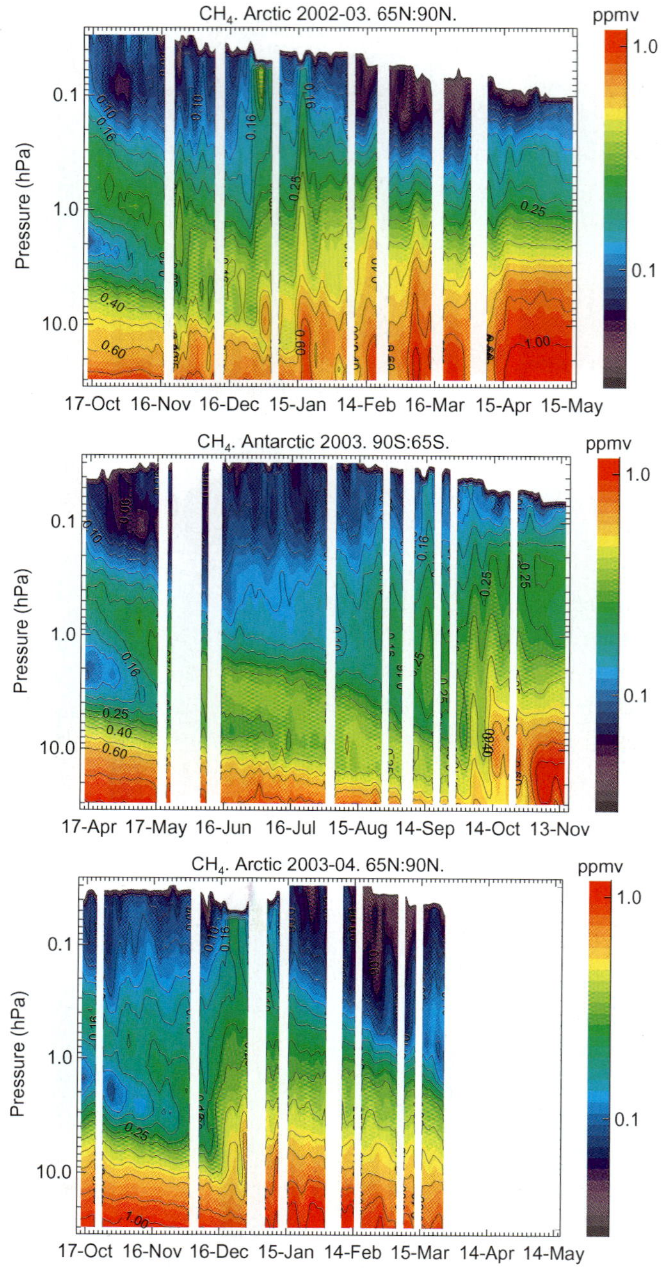


*Figure 2.* As Figure 1 but for the $CH_4$ mean day/night distribution.

significant enhancement of $NO_2$ in the upper stratosphere and lower mesosphere (~2–0.07 hPa) in late October which was produced by the solar proton events (SPEs) of October 2003 (the effects of these SPEs in the Northern hemisphere are discussed below). The rather high $NO_2$ values at 4–20 hPa from mid-June to mid-July seems to be a retrieval artifact. A test performed show that the chi-squared values for these data are enhanced (Raspollini *et al.*, 2006). Also, independent retrievals performed using the IMK-IAA retrieval code do not show these high $NO_2$ values in this region (Funke *et al.*, 2005). In consequence the $NO_2$ abundances in this data version should not be trusted.

The production mechanisms and downward transport of the upper mesospheric $NO_x$ during this winter have been analyzed in detail by Funke *et al.* (2005) using the ECMWF data and the CO, $CH_4$, NO and $NO_2$ fields measured by MIPAS. They have attributed the observed $NO_x$ enhancements to production by electron precipitation in the upper mesosphere and lower thermosphere (~80–110 km), and subsequent descent with the meridional circulation. The $A_p$ geomagnetic index, a good proxy of the auroral activity and electron precipitation and hence of NO production in the upper atmosphere, was considerably larger compared to the Arctic 2002–2003 winter and the 12-year seasonal mean (23 versus 15 and 12, respectively). Funke *et al.* (2005) have also shown that the downward transport of $NO_x$ took place mainly inside the polar vortex. A very extended mesospheric vortex was found with high values of $NO_x$ extending to latitudes as far as 35° equatorwards. As suggested by Figure 2 (middle panel), the vortex was found to be very stable in and above the upper stratosphere until August with the absence of stratospheric warmings (as it is usual in the Antarctic winter). As a result of both these favorable conditions, large amounts of $NO_x$ were injected into the stratosphere leading to high $NO_2$ abundances of up to 12 ppbv in Sep–Oct compared to 8 ppbv in May. Funke *et al.* (2005) have estimated that about 2.4 Gigamoles of $NO_x$ were deposited in the stratosphere during this winter. This represents a 9% of the amount produced by $N_2O$ oxidation in the SH, and should therefore be accounted for in chemistry-transport models.

By comparing the stratospheric $NO_x$ profile in the late winter (early October) with HALOE occultation measurements for the 1991–1996 winters (Siskind *et al.*, 2000) we found that $NO_x$ in the 2003 winter is larger than in other Antarctic winters but of comparable magnitude. In particular, in 1991, when the mean seasonal $A_p$ was largely increased, the profiles agree rather well. The electron precipitation activity is expected to have a solar cycle variability with larger values in the declining phase (Rangarajan and Barreto, 2000), e.g., in 1991 and 2003. Thus, the $NO_x$ values measured by MIPAS in the 2003 Antarctic winter were rather high within a decade time frame, but seem typical in the context of the solar cycle progression.

The $NO_2$ distribution in the Arctic winter 2003–2004 shows very different features than in the previous year (see Figure 1, lower panel). First, early in the winter (late October), the $NO_2$ was largely enhanced in the stratosphere and middle mesosphere. This sudden increase was due to the production of $NO_x$ by the very large SPEs that occurred at this time (see, e.g., Jackman *et al.*, 2006). The $NO_x$ distribution in this period, and its effects on ozone and other $NO_y$ compounds, have been analyzed in detail by López-Puertas *et al.* (2005a,b). The $NO_x$-enhanced layer was transported downwards into the mid-low stratosphere, favored by a large dark area in the polar cap, and thus producing an unusual deposition of $NO_x$ in the stratosphere in the first half of the winter.

A large $NO_2$ enhancement appeared at 60–70 km from around mid-November until mid-December, at the same time of the year but larger than in the previous Arctic winter. In situ production is discarded since no significant proton or high energy electron (>300 keV) event impacted the atmosphere at that time. The $NO_2$ enhancement seems to be due to a more abundant $NO_x$ above ~70 km, produced by enhanced electron fluxes and highly energetic protons originated by the SPEs, which was then transported downwards and reached the middle mesosphere a few weeks later. An additional contribution due to the high auroral activity during the two weeks preceding the SPEs may be possible, since the high energy electron flux (>100 keV) in this period was a factor of 3 larger than in 2002.

The $NO_x$ downward transport was also perturbed in this winter by a major stratospheric warming that started in late December (see lower panel of Figure 2 and Manney *et al.*, 2005b). The $NO_2$ abundance in the polar cap was significantly reduced in this period at practically all measured altitudes for nearly one month.

$NO_2$ abundances increased enormously from mid-January on in the mid- and lower-mesosphere, reaching mean values of 180 ppbv in the polar cap and 350 ppbv near the North pole; even larger than in the Antarctic 2003 winter. Similarly huge $NO_2$ concentrations have also been measured by other instruments from February until May (Natarajan *et al.*, 2004; Randall *et al.*, 2005; Rinsland *et al.*, 2005; Jackman *et al.*, 2005).

An in situ production of this $NO_2$ enhancement is excluded since no major SPEs or high energy (>300 keV) electron precipitation occurred. These exceptionally high $NO_2$ abundances seem to be produced by both dynamical and chemical processes. A very strong and compact polar vortex started to develop by early January which, in February and March became the strongest since regular observations began in 1979 (see Manney *et al.*, 2005b and the $CH_4$ distribution in Figure 2). This produces a fast downward transport of $NO_x$, confined to the polar night region (i.e., shielded from sunlight destruction), which led to the enormous concentrations of $NO_2$ measured in the mid- low-mesosphere and upper stratosphere. In addition, a larger production of $NO_x$ in the upper mesosphere and lower thermosphere seems to have contributed to the high $NO_x$ abundances. It is not clear, however, which is the source of this $NO_x$. On one side, the strong solar storms in late October/early November, with associated solar protons and electron precipitation, generated large amounts of $NO_x$ from the upper stratosphere through the lower thermosphere during and shortly after the storms (see Figure 1 and Jackman *et al.*, 2005). Semeniuk *et al.* (2005) have found, however, that the contribution of the solar protons is clearly insufficient to explain such high $NO_x$ values. Another source could be the continuous production of NO along the winter by auroral electrons. This winter had an unusually high geomagnetic activity. The auroral electron fluxes integrated for the post-storm period (5 Nov–31 Dec) are larger than those for the storm period (30 Oct–4 Nov) by factors of 3.2 and 1.6 for electrons with energies of >30 keV and

>100 keV, respectively; and are comparable to the electron fluxes measured in the high-$NO_x$ Antarctic winter of 2003. It is then expected that the NO produced during the post-storm period is the major contributor to the $NO_x$ enhancement observed after mid-January.

The enormous $NO_2$ in the middle atmosphere descended rapidly and reached the upper stratosphere by mid-March, producing a large injection of $NO_2$ in the upper stratosphere. At this time, however, a large fraction of the polar cap was already illuminated and hence limited the stratospheric injection. In summary, in this unusual Arctic winter, the amount of $NO_x$ transported through the mesosphere was very large, probably even larger than in the Antarctic 2003 winter. An accurate estimation of the $NO_x$ deposited in the stratosphere has not been performed yet.

MIPAS also measured during a few days in the Antarctic winter of 2002 (not shown). The $NO_2$ abundances for July and August 2002 in the upper stratosphere and low/mid-mesosphere reached maximum values of 20 ppbv, one order of magnitude smaller than in the Antarctic winter of 2003. This Antarctic 2002 winter was, however, very unusual since, for the first time, a stratospheric warming was observed in the SH producing the split of the vortex (Manney *et al.*, 2005a). The high planetary wave activity during most of the winter, precursor of the vortex split, mixed up polar and mid-latitude air masses and thus reduced the $NO_2$ abundances in the polar region. This Antarctic winter was, however, very unique and, at least so far, the Antarctic winters are more similar to that of 2003 than to this one.

## 4. Variability in $HNO_3$

Given the large variability in $NO_2$ measured by MIPAS in the different winters, one would expect changes also in other $NO_y$ species, namely $HNO_3$. Enhancements of $HNO_3$ in the polar winter upper stratosphere have been observed by several instruments (see Stiller *et al.*, 2005 and references therein). MIPAS offers, however, a pole-to-pole spatial coverage, high altitude range and temporal coverage, together with simultaneous measurements of other $NO_y$ species. Hence, we discuss here the $HNO_3$ distributions in the upper stratosphere for the three major winters observed by MIPAS (see Figure 3).

The $HNO_3$ abundance in the Arctic 2002–2003 winter does not show a clear enhancement in the upper stratosphere. We note two minor enhancements (~2 ppbv) in the mid-stratosphere (2–6 hPa) but these are associated with the stratospheric warmings that occurred in early and late January (see Section 3). The absence of an $HNO_3$ enhancement is in consonance with the low $NO_2$ abundances measured during this winter (Figure 1, upper panel).

In contrast, the Antarctic 2003 winter shows an enormous enhancement in upper stratospheric $HNO_3$, with mean polar cap values up to 8 ppbv (single profiles up to 14 ppbv) covering an extended region and lasting over two months. This enhancement have been studied in detail by Stiller *et al.* (2005) and it was

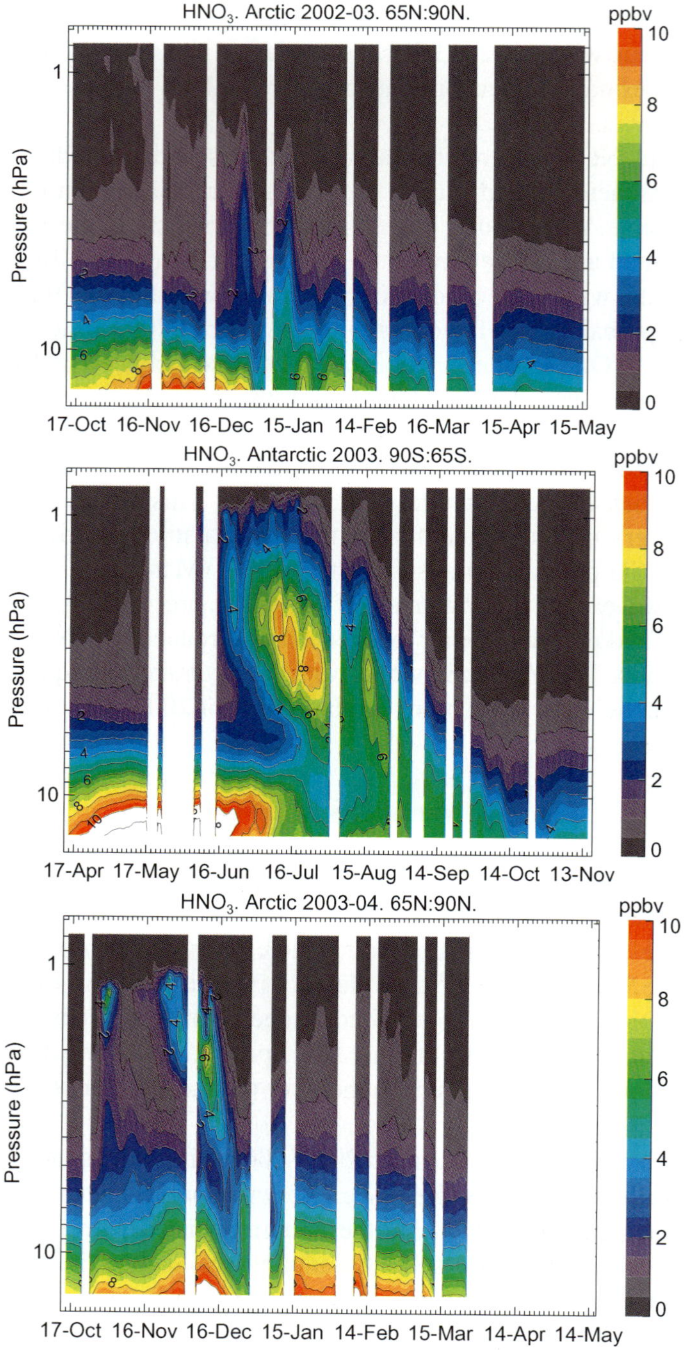


*Figure 3.* As Figure 1 but for the $HNO_3$ mean day/night distribution. Note the pressure range is different than in Figures 1 and 2. The figure in the middle panel is taken from Stiller *et al.* (2005). Reproduced by permission of American Geophysical Union.

found to be produced by chemical reactions involving water ion clusters and $N_2O_5$ formed from $NO_x$ (de Zafra *et al.*, 1997): $N_2O_5 + H_2O$ (ion clusters) $\rightarrow 2\,HNO_3$. Stiller *et al.* (2005) have shown the consistency between the measured $NO_x$, $N_2O_5$, and $HNO_3$ fields, and have demonstrated that the large $HNO_3$ enhancement is caused by the strong downward transport of $NO_x$ in this winter (see Figure 1, middle panel). The descent of $HNO_3$ in the measured field is also apparent, although the chemical losses of $HNO_3$ also plays an important role in its evolution (Stiller *et al.*, 2005).

The $HNO_3$ distribution in the Arctic 2003–2004 winter is very different from the other two winters. Normally, the upper stratosphere $HNO_3$ enhancement in the polar region is very small in the Arctic winters. However, this winter was very anomalous with large $NO_x$ abundances in the upper stratosphere and mesosphere (see Section 3). $HNO_3$ was enhanced in the upper stratosphere (~1.5 hPa) in late October. Its origin is not related to the cluster ion chemistry but to the gas phase chemistry due to the enhancements of $NO_2$ and OH produced by the Oct/Nov 2003 SPEs (López-Puertas *et al.*, 2005b): $NO_2 + OH + M \rightarrow HNO_3 + M$. A second $HNO_3$ enhancement is observed starting around 1.5 hPa in late November. Another SPE occurred on 20 Nov but it was much weaker than that of October while the $HNO_3$ enhancement is larger. Hence, it is unlikely that it was produced by local gas phase chemistry. Note that this $HNO_3$ increase is well timed with the arrival of the $NO_2$ enhancement at the altitude of the $HNO_3$ maximum (see Figure 1, lower panel). Instead, it seems more likely that it was produced by the cluster ion chemistry in an $NO_x$-rich environment. The arrival of the stratospheric warming on mid-December mixed up the $HNO_3$-rich polar air with the low-$HNO_3$ mid-latitude air and hence drastically reduced the polar cap $HNO_3$ enhancement. In the second half of the winter, although the amount of mesospheric $NO_x$ was very large, it did not reach the upper stratosphere until the end of February/early March. At this time, the significant illumination prevented a build up of $HNO_3$ since both, its precursor $N_2O_5$ and $HNO_3$ itself, were photodissociated.

## 5. Discussion and Conclusions

In this paper we have discussed the MIPAS measurements of $NO_2$ and $HNO_3$ in the stratosphere and mesosphere during two Arctic and two Antarctic polar winters between 2002 and 2004. The most outstanding feature in the data is the large variability of these species both on inter-annual and inter-hemispheric scales. Two winters, Antarctic 2002 and Arctic 2002–2003, show rather low $NO_x$ abundances in those regions; but the other two, Antarctic 2003 and Arctic 2003–2004, exhibited very enhanced $NO_x$ concentrations. Two major factors seem to control the $NO_x$ abundances: (1) the dynamical behavior of the polar winter stratosphere and mesosphere; and (2) the source of $NO_x$, its production, mostly in the mid-upper atmosphere and lower thermosphere by auroral electron precipitations and, sporadically, its production by solar protons in the stratosphere and mesosphere after solar flares events.

Thus, the combination of high auroral activity during the Antarctic 2003 winter together with its rather usual very stable vortex and steady descent of the meridional circulation gave rise to very enhanced $NO_x$ abundances in the polar stratosphere and mesosphere. These values, although higher than usual because of the high auroral activity expected at the declining phase of the solar cycle, can be considered as typical, as, for example, HALOE measurements have shown. In the previous Antarctic winter, 2002, the auroral activity was not so high but an unusual high planetary waves activity took place, producing the first observation of a vortex split in the SH. This weakened the downward transport of $NO_x$ and moved the polar night air to illuminated latitudes where $NO_x$ was destroyed. As a result, this winter showed very low $NO_x$ values in the upper stratosphere and low-mid mesosphere, but should be considered as an atypical Antarctic winter.

In the Arctic 2002–2003 winter, where auroral activity was moderate, the usual stratospheric warming in the middle of the winter led to low-to-moderate $NO_x$ abundances. This can be considered as typical for an Arctic winter where the presence of stratospheric warming is rather usual. On the other hand, in 2003–2004, two unusual phenomena occurred. First, besides the high auroral activity during this winter, several SPEs occurred which, on one hand, enhanced the $NO_x$ production by the direct impact of solar protons on the atmosphere and, on the other, favored the electron fluxes associated to geomagnetic perturbations. In addition, after an usual stratospheric warming, a strong (the record so far measured) polar vortex developed in late winter. Both conditions finally led to very large $NO_x$ abundances in the upper stratosphere at the end of the winter.

This variability in $NO_x$ has also triggered upper stratospheric $HNO_3$ changes which, in addition, shows its own peculiarities. E.g., $HNO_3$ abundances are further modulated by illumination-dependent $N_2O_5$ chemistry (Stiller *et al.*, 2005). The deposition of upper atmospheric $NO_x$ into the stratosphere for favorable conditions, e.g., the Antarctic 2003 winter, has been estimated at values as large as 9% of the $NO_y$ production by the oxidation of $N_2O$ averaged over the SH (Funke *et al.*, 2005). This then suggests that the variability observed in $NO_x$ should have an impact on the polar upper stratospheric $O_3$, and maybe on the overall stratospheric ozone. This analysis, however, still remains to be done.

## Acknowledgements

We wish to thank the ESA L2 quality working group and in particular Anu Dudhia for providing and helping us with the use of ESA MIPAS L2 offline data. The IAA team has been supported by Spanish MEC under projects REN2001–3249/CLI and ESP2004–01556, and EC FEDER funds. IMK was partially supported by German projects SACADA (07ATF53) and KODYACS (07ATF43).

## References

Brasseur, G., and Solomon, S.: 1986, *Aeronomy of the Middle Atmosphere*, 2nd edition, D. Reidel Publishing Co., Norwell, Mass., 1986.

Callis, L. B., Natarajan, M., Lambeth J. D., and Baker D. N.: 1998, 'Solar-atmospheric coupling by electrons (SOLACE), 2, Calculated stratospheric effects of precipitating electrons, 1979–1988', *J. Geophys. Res.* **103**, 28,421–28,438.

Callis, L. B., Natarajan, M., and Lambeth, J. D.: 2001, 'Solar-atmospheric coupling by electrons (SOLACE) 3. Comparisons of simulations and observations, 1979–1997, issues and implications', *J. Geophys. Res.* **106**, 7523–7539.

Crutzen, P. J.: 1979, 'The role of NO and $NO_2$ in the chemistry of the troposphere and stratosphere', *Ann. Rev. Earth Planet. Sci.* **7**, 443.

de Zafra, R. L., Chan, V., Crewell, S., Trimble, C., and Reeves, J. M.: 1997, 'Millimeter wave spectroscopic measurements over the South Pole – 3. The behavior of stratospheric nitric acid through polar fall, winter, and spring', *J. Geophys. Res.* **102**, 1399–1410.

Fischer, H., and Oelhaf, H.: 1996, 'Remote sensing of vertical profiles of atmospheric trace constituents with MIPAS limb-emission spectrometers', *Appl. Opt.* **35**, 2787–2796.

Frederick, J. E., and Orsini, N.: 1982, 'The distribution and variability of mesospheric odd nitrogen: A theoretical investigation', *J. Atm. Terr. Phys.* **44**, 479–488.

Funke, B., López-Puertas, M., Gil-López, S., Clarmann, T. v., Stiller, G. P., *et al.*: 2005, 'Downward transport of upper atmospheric $NO_x$ into the polar stratosphere and lower mesosphere during the Antarctic 2003 and Arctic 2002–03 winters', *J. Geophys. Res.*, in press.

Jackman, C. H., DeLand M. T., Labow, G. J., Fleming, E. L., Weisenstein, D. K., *et al.*: 2005, 'Neutral atmospheric influences of the solar proton events in October–November 2003', *J. Geophys. Res.* **110**, doi: 10.1029/2004JA010888.

Jackman, C. H., Deland, M. T., Labow, G. J., Fleming, E. L., and López-Puertas, M.: 2006, 'Satellite measurements of middle atmospheric impacts by solar proton events in solar cycle 23', *Space Sci. Rev.*, this volume, doi: 10.1007/s11214-006-9071-4.

Liu, H.-L., and Roble, R. G.: 2002, 'A study of a self-generated stratospheric sudden warming and its mesospheric-lower thermospheric impacts using the coupled TIME-GCM/CCM3', *J. Geophys. Res.* **107**, doi: 10.1029/2001JD001533.

López-Puertas, M., Funke, B., Gil-López, S., Clarmann, T. v., Stiller, G. P., *et al.*: 2005a, 'Observation of $NO_x$ enhancements and ozone depletion in the northern and southern hemispheres after the October–November 2003 solar proton events', *J. Geophys. Res.* **110**, doi: 10.1029/2005JA011050.

López-Puertas, M., Funke, B., Gil-López, S., Clarmann, T. v., Stiller, G. P. *et al.*: 2005b, '$HNO_3$, $N_2O_5$ and $ClONO_2$ enhancements after the October–November 2003 solar proton events', *J. Geophys. Res.* **110**, doi: 10.1029/2005JA011051.

Manney, G. L., Sabutis, J. L., Allen, D. R., Lahoz, W. A., Scaife, A. A. *et al.*: 2005a, 'Simulations of dynamics and transport during the September 2002 Antarctic major warming', *J. Atmos. Sci.* **62**, 690–707.

Manney, G. L., Krüger, K., Sabutis J. L., Sena, S. A., and Pawson, S.: 2005b, 'The remarkable 2003–2004 winter and other recent warm winters in the Arctic stratosphere since the late 1990s', *J. Geophys. Res.* **110**, doi: 10.1029/2004JD005367.

Natarajan, M., Remsberg, E. E., Deaver, L. E., and Russell III, J. M.: 2004, 'Anomalously high levels of $NO_x$ en the polar upper stratosphere during April 2004: Photochemical consistency of HALOE observations', *Geophys. Res. Lett.* **31**, doi: 10.1029/2004GL020566.

Randall, C. E., Rusch, D. W., Bevilacqua, R. M., Hoppel, K. W., and Lumpe, J. D.: 1998, 'Polar ozone and aerosol measurement (POAM) II stratospheric $NO_2$, 1993–1996', *J. Geophys. Res.* **103**, 28,361–28,371.

Randall, C. E., Harvey, V. L., Manney, G. L., Orsolini, Y., Codrescu, M., *et al.*: 2005, 'Stratospheric effects of energetic particle precipitation in 2003–2004', *Geophys. Res. Lett.* **32**, doi: 10.1029/2004GL022003.

Rangarajan, G., and Barreto, L. M.: 2000, 'Long term variability in solar wind velocity and IMF intensity and the relationship between solar wind parameters and geomagnetic activity', *Earth Planets Space* **52**, 121–132.

Raspollini, P., Belotti, C., Burgess, A., Carli, B., Carlotti, M., *et al.*: 2006, 'MIPAS Level 2 operational analysis with ORM code', *Atmos. Chem. Phys.* **6**, 6525–6585.

Ridolfi, M., Carli, B., Carlotti, M., Clarmann, T. v., Dinelli, B. M., *et al.*: 2000, 'Optimized forward model and retrieval scheme for MIPAS near-real-time data processing', *Appl. Opt.* **39**, 1323–1340.

Rinsland, C. P., Salawitch, R. J., Gunson, M. R., Solomon, S., Zander, R., *et al.*: 1999, 'Polar stratospheric descent of $NO_y$ and CO and Arctic denitrification during winter 1992–1993', *J. Geophys. Res.* **104**, 1847–1861.

Rinsland, C. P., Boone, C., Nassar, R., Walker, K., Bernath, P., *et al.*: 2005, 'Atmospheric chemistry experiment (ACE) Arctic stratospheric measurements of $NO_x$ during February and March 2004: Impact of intense solar flares', *Geophys. Res. Lett.* **32**, doi: 10.1029/2005GL022425.

Russell III, J. M., Farmer, C., Rinsland, C., Zander, R., Froidevaux, L., *et al.*: 1988, 'Measurements of odd nitrogen compounds in the stratosphere by the ATMOS experiment on spacelab 3', *J. Geophys. Res.* **93**, 1718–1736.

Semeniuk, K., McConnell J. C., and Jackman C. H.: 2005, 'Simulation of the October–November 2003 solar proton events in the CMAM GCM: Comparison with observations', *Geophys. Res. Lett.* **32**, doi: 10.1029/2005GL022392.

Siskind, D. E.: 2000, 'On the coupling between middle and upper atmospheric odd nitrogen', In *Atmospheric Science Across the Stratopause*, Geophysical Monograph, American Geophysical Union.

Siskind, E., Nedoluha, G. E., Randall, C. E., Fromm, M., and Russell, M.: 2000, 'An assessment of Southern Hemisphere stratospheric $NO_x$ enhancements due to transport from the upper atmosphere', *Geophys. Res. Lett.* **27**, 329–332.

Solomon, S., Crutzen, P. J., and Roble, R. G.: 1982, 'Photochemical coupling of the thermosphere and the lower atmosphere 1. Odd nitrogen from 50 to 120 km', *J. Geophys. Res.* **87**, 7206–7220.

Stiller, G. P., Mengistu Tsidu, G., Clarmann T. v., Glatthor, N., Höpfner, M., *et al.*: 2005, 'An enhanced $HNO_3$ second maximum in the Antarctic mid-winter upper stratosphere 2003', *J. Geophys. Res.* **110**, doi: 10.1029/2005JD006011.

# EARLY DATA FROM AURA AND CONTINUITY FROM UARS AND TOMS

E. HILSENRATH[1,*], M. R. SCHOEBERL[1], A. R. DOUGLASS[1], P. K. BHARTIA[1], J. BARNETT[2], R. BEER[3], J. WATERS[3], M. GUNSON[3], L. FROIDEVAUX[3], J. GILLE[4] and P. F. LEVELT[5]

[1]*NASA Goddard Space Flight Center, Greenbelt Road, 20771 Greenbelt, MD, USA*
[2]*Oxford University, Oxford, UK*
[3]*Jet Propulsion Laboratory, Pasadena, CA, USA*
[4]*University of Colorado, Boulder, CO, USA*
[5]*Royal Netherlands Meteorological Institute, De Bilt, The Netherlands*
*(*Author for correspondence: E-mail: ernest.hilsenrath@nasa.gov)*

(Received 26 September 2005; Accepted in final form 30 November 2005)

**Abstract.** Aura, the last of the large EOS observatories, was launched on July 15, 2004. Aura is designed to make comprehensive stratospheric and tropospheric composition measurements from its four instruments, HIRDLS, MLS, OMI and TES. These four instruments work in synergy to provide data on ozone trends, air quality and climate change. The instruments observe in the nadir and limb and provide the best horizontal and vertical resolution ever achieved from space. After over one year in orbit the instruments are nearly operational and providing data to the scientific community. We summarize the mission, instruments, and initial results and give examples of how Aura will provide continuity to earlier chemistry missions.

**Keywords:** satellite observations, atmospheric composition

## 1. Introduction

NASA's Earth Observing System Program began in the late 1980s with the selection of a large number of Earth science instruments and interdisciplinary science teams. The present EOS compliment of satellites consists of three core platforms, Terra, Aqua and AURA and several smaller satellites such as SORCE and ICESAT. Terra, launched in late 1999, focuses on land processes. Aqua, launched in 2002, focuses on the hydrological cycle. Aura, whose focus is atmospheric composition, was launched July 15, 2004 into an ascending node 705 km sun-synchronous polar orbit with a 98° inclination with an equator-crossing time of 13:45. Aura flies in formation about 15 minutes behind Aqua. The Cloud-Aerosol Lidar and Infrared Pathfinder Satellite Observation (CALIPSO) and Cloudsat, that were launched together in the Summer of 2006, fly a few minutes behind Aqua. This group of satellites, including the CNES PARASOL satellite, launched in December 2004, and the ESSP Orbiting Carbon Observatory (OCO), scheduled for launch in 2008, are referred to as the "A-Train" The measurements from Aura will be within 30 minutes of these other


DOI: 10.1007/s11214-006-9074-1

TABLE I
Summary of Aura's four instruments.

| Acronym | Name | Instrument PI | Constituent | Instrument description |
|---|---|---|---|---|
| HIRDLS | High Resolution Dynamics Limb Sounder | John Gille, NCAR & U. of Colorado, Boulder, USA; John Barnett, Oxford University | Profiles of T, $O_3$, $H_2O$, $CH_4$, $N_2O$, $NO_2$, $HNO_3$, $N_2O_5$, $CF_3Cl$, $CF_2Cl_2$, $ClONO_2$, Aerosols | Limb IR filter radiometer from 6.2 $\mu$m to 17.76 $\mu$m. 1.2 km vertical resolution up to 50 km. |
| MLS | Microwave Limb Sounder | Joe Waters, JPL, Pasadena, USA | Profiles of T, $H_2O$, $O_3$, ClO, BrO, HCl, OH, $HO_2$, $HNO_3$, HCN, $N_2O$, CO, Cloud ice | Microwave limb sounder, 118 GHz–2.5 THz. 1.5–3 km vertical resolution. |
| OMI | Ozone Monitoring Instrument | Pieternel Levelt, KNMI, The Netherlands | Column $O_3$, $SO_2$, aerosols, $NO_2$, BrO, OClO, HCHO, UVB, cloud top pressure, $O_3$ profiles | Hyperspectral nadir imager, 114° FOV, 270–500 nm. 13 × 24 km footprint for $O_3$ and aerosols. |
| TES | Tropospheric Emission Spectrometer | Reinhard Beer, Mike Gunson, JPL, Pasadena, USA | Profiles of T, $O_3$, $NO_2$, CO, $HNO_3$, $CH_4$, $H_2O$ | Limb (to 34 km) and nadir Fourier Transform IR spectrometer, 3.2–15.4 $\mu$m. Nadir footprint 5.3 × 8.5 km, limb 2.3 km. |

platforms. The A-Train can be thought of as a extended instrument package focusing on climate change. Although NASA's TOMS, SBUV, UARS and NOAA's SBUV/2 series of instruments are not formally part of EOS, they have played a key role in NASA's EOS program by collecting a highly accurate and comprehensive long term data set of ozone and atmospheric species active in its chemistry.

The Aura spacecraft carries four chemistry instruments, the High Resolution Dynamics Limb Sounder (HIRDLS), the Microwave Limb Sounder (MLS), the Ozone Monitoring Instrument (OMI) and the Tropospheric Emission Spectrometer (TES), that operate in the wavelength range from the ultraviolet to the millimeter wave region. The instruments and their data are summarized in Table I. These instruments were selected because of (1) their complementary measurements; (2) their technological heritage; and (3) the new capabilities they bring to measuring the

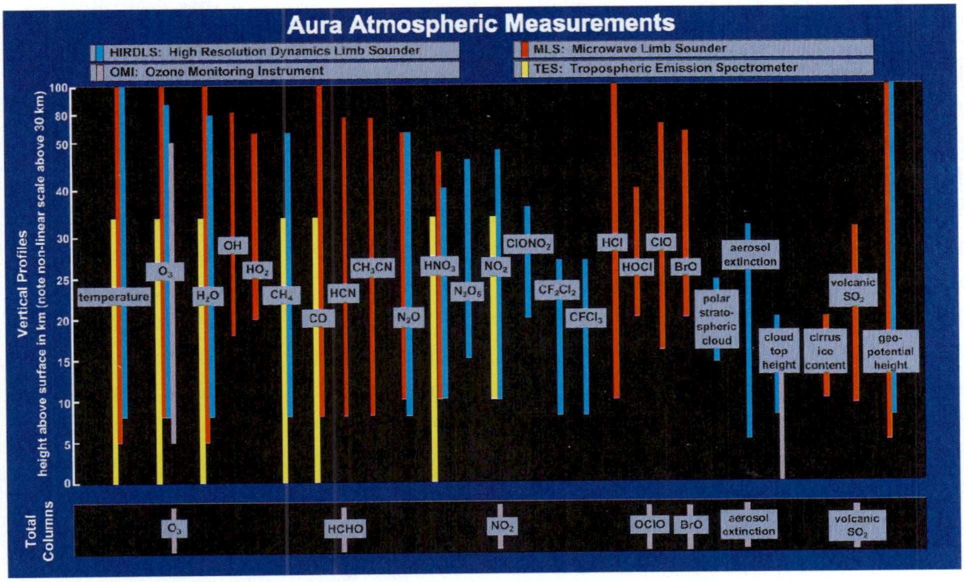


*Figure 1.* Aura data products and their height range depicted here are based on pre-launch design. For HIRDLS, full capability is not yet demonstrated, but 50 km is the upper limit of the measurements and measures in one azimuth direction only because of a field of view obscuration. TES limb measurements are limited because of mirror drive problems. Column measurements by OMI are indicated in the lower part of the figure (Figure courtesy of J. Waters).

Earth's atmosphere. Below we describe the objectives and science strategy of the Aura mission. Figure 1 graphically shows the vertical range of the Aura instrument data products.

## 2. Science Objectives of the Aura Mission

The objective of the Aura mission is to attack three principal science questions: Is the ozone layer changing as expected and is there a response to the Montreal Protocol? Where are the sources and what are the processes that control tropospheric pollutants? What are the roles of upper tropospheric aerosols, water vapor, and ozone in climate change? These questions are being answered with the data sets compiled as shown in Table I and illustrated in Figure 1. Most importantly these data are being collected with high vertical and horizontal resolution throughout the atmosphere. Also, when combined with measurements from ground, balloon, aircraft missions, and other measurements from the A-Train, Envisat, ODIN, and ACE, they will provide unprecedented insights into the chemical and dynamical processes associated with our atmosphere. The Aura science questions and the impact the Aura instruments will have on these questions are reviewed below.

## 2.1. Is the Ozone Layer Changing as Expected?

Total Ozone Mapping Spectrometer (TOMS) observations from 1978 to the present show decreasing trends in ozone both at midlatitudes and in polar regions. Although the Antarctic ozone hole appears to be no longer growing and chlorine levels in the stratosphere are beginning to decline, very significant ozone depletions have occurred in the Arctic (Newman *et al.*, 1997; WMO, 2002). As a result of international agreements, tropospheric chlorofluorocarbons concentrations have begun to decrease. HALOE (UARS) data show a flattening in the stratospheric chlorine reservoir concentrations (Anderson *et al.*, 2000) although an unambiguous decrease has not been detected. A decrease in chlorine should lead to recovery of the ozone layer, but the recovery in the polar regions may be delayed by increases in greenhouse gases which can cool the stratosphere and with possible increases in stratospheric water giving rise to more frequent and persistent polar stratospheric clouds (Solomon, 1999). Stratospheric water vapor does appear to be increasing faster than can be accounted for by the secular trend in methane. As a result of the uncertainty in stratospheric trace gas measurements, current models used to assess the ozone layer do not agree on the timing of the recovery of ozone (WMO, 2002).

Determining the recovery of ozone and the effectiveness of the CFC regulating protocols is a major science objective for HIRDLS, MLS and OMI. The stratospheric measurements made by these instruments will permit a very complete assessment of the chemical processes controlling ozone. All three of the major radicals that destroy ozone (ClO, OH and NO) are being made by HIRDLS and MLS along with the main reservoir gases, HCl, $ClONO_2$ and $HNO_3$. In addition, the Aura instruments payload make measurements of chlorine and nitrogen sources gases as well as long lived tracers of motion, e.g., $N_2O$, $H_2O$, and $CH_4$. OMI continues the column and profile ozone trends from TOMS/SBUV but with higher horizontal resolution.

## 2.2. Where Are Sources of and What Are the Processes that Control Tropospheric Pollutants?

Observations have shown that human activities have likely increased surface-level ozone concentrations (Guicherit and Roemer, 2000; Logan, 1985). Decades of research and regulations now reveal that a global approach is required to understand sources and sinks of pollutants such as ozone and its precursors. Tropospheric ozone production occurs when volatile organic compounds (VOCs) and nitrogen oxides ($NO_x$) are exposed to sunlight in the presence of water vapor. Since the emissions of these ozone precursors are directly linked to today's urban and industrial lifestyle, reductions are both socially and economically costly. For this reason, policy makers have sought guidance from the scientific community in determining effective ways to meet health-based ozone standards or goals. The response of ozone to changes in VOCs and $NO_x$ emissions can be quite complex and variable. Moreover, winds can transport both ozone and its precursors over large distances; and, as a result, exposure to elevated ozone can arise from both local and distant sources. The Aura mission is designed to produce a global assessment of tropospheric ozone, its precursors and controlling gases. The measurements from TES (CO and $CH_4$) as well as measurements from OMI ($NO_2$, ozone, and aerosols) combined are providing important data on these processes. Both OMI and TES make complimentary measurements of tropospheric ozone.

### 2.3. How Do Upper Tropospheric Aerosols, Water Vapor and Ozone Affect Climate Change?

The importance of the upper troposphere/lower stratosphere (UT/LS) in chemistry and climate problems is well stated in a number of international assessment documents (i.e., Houghton *et al.*, 2001). Within the context of chemical problems one of the largest uncertainties is how much ozone and odd nitrogen is supplied to the troposphere from the stratosphere (Hauglustaine *et al.*, 1998). Even estimates of the cross-tropopause flux of ozone from the stratosphere are uncertain to within 15% or greater (Gettelman *et al.*, 1997; WMO, 2002). There are fundamental climate change questions related to moistening or drying of the upper troposphere as convective activity changes. For example, it is now understood that the IR cooling of the atmosphere is strongly influenced by upper tropospheric/lower stratospheric water vapor, aerosols and ozone (Houghton *et al.*, 2001). On decadal time scales, signals from greenhouse gas changes and signals from the changes in reactive constituents are intertwined and difficult to untangle. The tropopause is a complex internal boundary within the atmosphere related to both radiative and dynamical mechanisms. Because the concentrations of many trace gases vary greatly across the tropopause, changes in the UT/LS impact not only the radiative attributes of the atmosphere, but also the chemical environment.

The location and intensity of both tropical and mid-latitude stratosphere-troposphere exchange is key to our ability to quantify the properties of the UT/LS. One of the major science questions Aura measurements addresses is why stratospheric water vapor is increasing. Stratospheric water vapor concentrations are increasing faster than can be accounted for by increases in methane, which is the major in situ source of water vapor in the stratosphere. A number of hypotheses suggest that changes in the freeze-dry mechanism at the tropical tropopause are producing an increase in water transport into the stratosphere. Untangling the complex freeze-dry mechanism of the tropical tropopause is one of the goals of the Aura science team. Relative to the upper troposphere / lower stratosphere dynamics and chemistry, Aura instruments will make key measurements of ozone, water vapor, ice particles and long lived trace gases that will give lead to a better understanding of the dynamics and chemistry of that region. HIRDLS high vertical resolution measurements in the UT/LS are especially important.

*Figure 2.* Aura instrument fields of view are shown as colored beams looking from behind the spacecraft. MLS performs forward limb sounding (green). OMI nadir measurement is the blue swath. TES limb and nadir measurements are pink. HIRDLS will only be able to make measurements in the brighter yellow scan position rather than the planned five azimuth positions, because of the blockage.

## 3. Spacecraft and Instrument Descriptions

The Aura spacecraft orbits at 705 km in a sun-synchronous orbit (98° inclination) with a 13:45 equator crossing time. Aura limb instruments were all designed to observe roughly along the orbit plane; however, with the HIRDLS anomaly discussed below, HIRDLS observations are limited to East (dayside) of the Aura ground track. MLS measures in the forward velocity direction. HIRDLS and TES make limb soundings in the anti-velocity direction. OMI and TES make nadir soundings as shown in Figure 2. The advantage of this instrument configuration is that each of the instruments observes the same air mass within minutes.

### 3.1. HIRDLS

HIRDLS is a 21-channel infrared limb-scanning filter radiometer designed to make the measurements listed in Table I (Gille *et al.*, 2003). HIRDLS can also determine the altitude of polar stratospheric clouds and detect sub-visible cirrus clouds. The HIRDLS instrument has a long heritage extending back to Nimbus-6 (1975), and was designed to obtain profiles over the entire globe, including the poles, both day and night. Complete Earth coverage (including polar night) could be obtained in 12 hours. HIRDLS was designed to achieve high horizontal resolution using commandable azimuth scans which, in conjunction with a rapid elevation scan, would provide profiles up to 3000 km apart in an across-track swath. The primary advantage of HIRDLS over all other previous infrared limb instruments is its high vertical and horizontal resolution which extends from the upper troposphere throughout the stratosphere.

*Current status:* After launch engineering studies concluded that a piece of thermal blanketing material ruptured from the back of the instrument during the explosive decompression of launch and covered most of the scan mirror. Even with this blockage, high vertical resolution measurements are still made at one scan angle. As of this writing, the HIRDLS team has demonstrated temperature and ozone retrievals with the instrument and they believe that they can retrieve most of the other constituents as planned. HIRDLS will no longer have its designed horizontal coverage and the partially blocked scan mirror will not allow measurements over the Antarctic.

### 3.2. MLS

MLS uses microwave emission to measure stratospheric and upper tropospheric temperature and constituents (Table I) (Waters *et al.*, 1999, 2005). MLS also has unique capability to measure upper tropospheric water vapor in the presence of tropical cirrus, and also the cirrus ice content. Aura MLS continues the successful effort of UARS MLS (Waters *et al.*, 1993) using advanced technology to provide new measurements. These measurements are especially valuable for diagnosing the potential for severe loss of Arctic ozone during the critical period when abundances of stratospheric chlorine will still be high and slight cooling of the stratosphere could exacerbate ozone loss due to chlorine chemistry. MLS is making the first global measurements of OH, $HO_2$ and BrO, constituents that play an important role in stratospheric chemistry.

MLS made comprehensive measurements of the evolution of the Antarctic ozone hole. Figure 3 illustrates MLS observations during the 2004 polar vortex breakup at the 520 K level (20 km). The vortex broke into several fragments between 5 and 11 December, and the evolution and erosion of those fragments is shown through late December, when only one weak fragment remained. After chlorine deactivation by early October, high vortex HCl and very strong HCl gradients across the vortex edge made it an excellent tracer of vortex evolution and morphology.

The MLS instrument aboard UARS has demonstrated the capability of measuring upper tropospheric water vapor profiles (Read *et al.*, 1995; Sandor *et al.*, 1998). Aura MLS measurements are being made the presence of tropical cirrus, where important processes affecting climate variability occur. MLS also provides unique measurements of cirrus ice content. The simultaneous MLS measurements of upper tropospheric water vapor, ice content, and temperature, under all conditions and with good vertical resolution, will make a major contribution to improving our understanding of large scale meteorological systems (such as

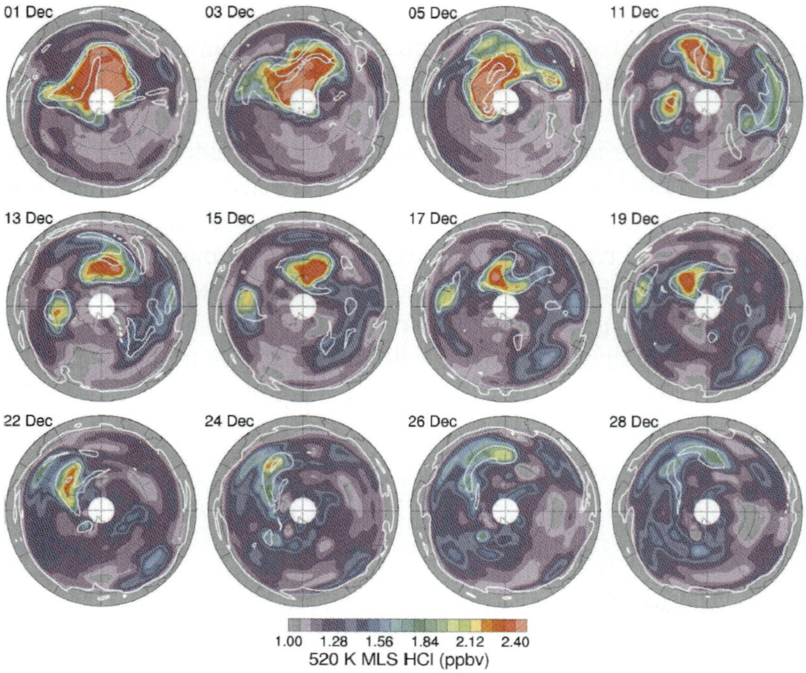


*Figure 3.* MLS HCl in the lower stratosphere (520 K, 20 km) detailing the springtime breakup of the 2004 Antarctic vortex.

El Niño) affecting the distribution of atmospheric water, climate variability, and tropospheric-stratospheric exchange. The simultaneous measurements of dynamical tracers CO and $N_2O$ enhance the value of this data set by helping identify stratospheric or tropospheric source regions of the observed air masses.

MLS' ability to measure water vapor in the UT/LS is illustrated in Figure 4, which shows the tropical water vapor "tape recorder" (Mote *et al.*, 1996) as seen by MLS. HALOE observed this phenomenon over the lifetime of UARS and illustrated a quasibiennial effect. Continuation of this record will provide important data in understanding the stratospheric water vapor budget. MLS will also continue the HCl record produced by HALOE on UARS (Anderson *et al.*, 2000). There will be at a least one year overlap between Aura and UARS which will insure that possible biases between instruments will be reconciled. HCl is the major reservoir for active chlorine in the stratosphere.

## 3.3. OMI

The OMI instrument is a contribution of the Netherlands's Agency for Aerospace Programs (NIVR) in collaboration with the Finnish Meteorological Institute (FMI) to the EOS Aura mission. OMI will continue the TOMS record for total ozone

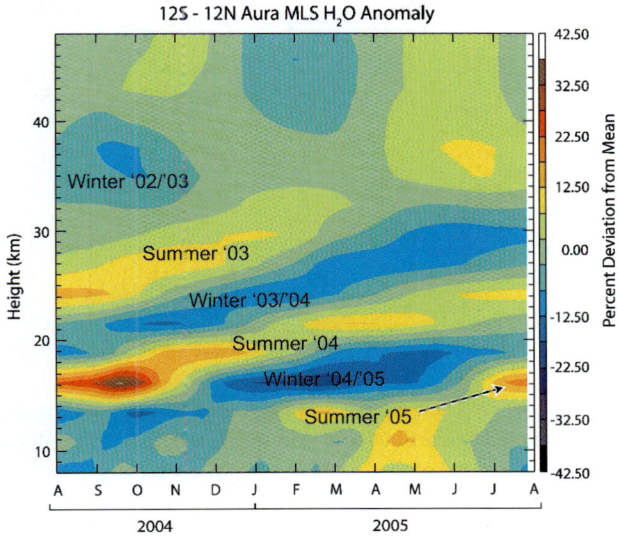


*Figure 4.* Zonal mean tropical water vapor anomalies recorded by MLS showing the upward propagating dry and wet regions associated with the annual modulation of the tropical tropopause temperature. Date labels show when the bands formed. Average mixing ratios are shown in red.

and other atmospheric parameters related to ozone chemistry and climate (Levelt *et al.*, 2005). The OMI instrument employs hyperspectral imaging in a push-broom mode to observe solar backscatter radiation in the visible and ultraviolet. The Earth is viewed in 740 wavelength bands along the satellite track with a swath large enough to provide global coverage in one day (14 orbits) with a nominal $13 \times 24$ km spatial resolution in the nadir. The hyperspectral capabilities will improve the accuracy and precision of the total ozone amounts and will also allow for accurate radiometric and wavelength self calibration over the long term. Aside from the measurements listed in Table I, the OMI instrument can distinguish between aerosol types, such as smoke, dust, and sulfates, and can measure cloud pressure and coverage, which provide data to derive tropospheric ozone. A combination of algorithms including TOMS version 8, differential optical Absorption Spectroscopy (DOAS), hyperspectral BUV retrievals and forward modeling will be used together to extract the various OMI data products. NASA will no longer fly TOMS instruments, therefore OMI will continue the long term ozone record produced by TOMS and in particular will provide daily maps of total ozone including the polar regions where the evolution and extend of the Antarctic ozone hole and Arctic depletion are being mapped. The careful calibration techniques developed for TOMS has also been applied to OMI data. The extent of the depletion has been quantified and added to the TOMS long term record, which is illustrated in Figure 5 and reported to the Intergovernmental Panel on Climate Change (IPCC, 2005).

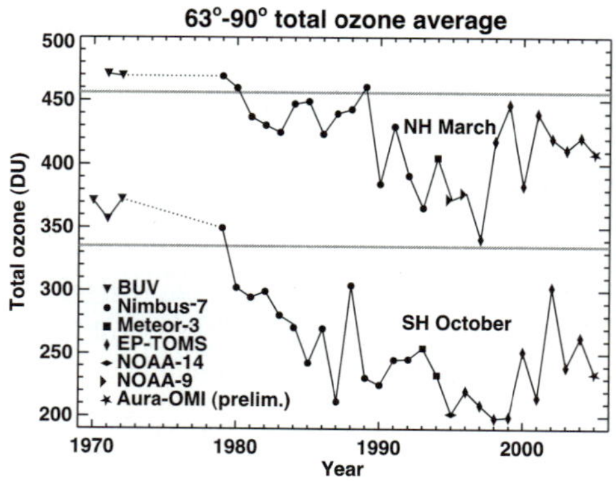


*Figure 5.* Average column ozone poleward of 63°N latitude in the Spring time of each hemisphere (March for the North and October for the South), in Dobson units, based on various indicated satellites. OMI data is the point (star) at the end of the Northern hemisphere record.

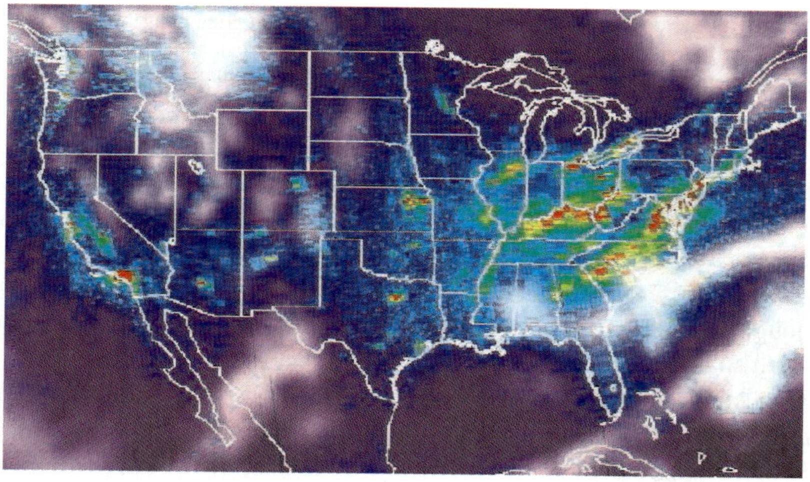

*Figure 6.* $NO_2$ measured by OMI over the United States on April 14, 2005. Red is high and blue are low values. White areas are clouds.

Another objective of OMI is to measure tropospheric ozone and one of its precursors, $NO_2$. OMI measures this constituent with a spatial resolution of $13 \times 24$ km at nadir on a daily global basis. This product is under intensive validation, however a sample is shown in Figure 6. OMI's high resolution clearly shows excess amounts over major cities and industrial areas as unique sources.

OMI is producing all of the other data products listed in Table I for which there are similar examples. However the size limitation for this paper prevents describing those products.

in the stratosphere and in some cases the troposphere. A unique opportunity exists to discriminate diurnal variations since the crossing times of these satellites differ for a given location. Very careful calibration and validation are needed to make these discriminations.

## 4. Summary

The EOS Aura mission was successfully launched on July 15, 2004. With the exception of HIRDLS, all of the instruments are functioning as designed, although to preserve instrument life, TES is now operating only in the nadir mode. Aura is providing the next level of measurements needed by the stratospheric and tropospheric community to advance the science and to answer the crucial broad questions: Is the stratospheric ozone layer recovering? How is the chemistry of the troposphere changing? What are the roles of upper tropospheric aerosols, water vapor and ozone in climate change? The four instruments on Aura provide the needed sets of measurements to answer these broad questions. Further synergies are expected with the A-Train and other international missions. The Aura measurements will continue important atmospheric composition measurements started with earlier satellites such as TOMS and UARS. Good overlap will occur thus providing valuable data on ozone and climate assessments.

For more information on the Aura platform and instruments, please refer to the web site `http://aura.gsfc.nasa.gov`. A pre-launch version of this paper was published in EOS (Schoeberl *et al.*, 2004). A special issue of the IEEE journal of *Transactions on Geoscience and Remote Sensing* will appear in 2006 which will highlight Aura, its instruments, retrieval algorithms, and data examples. Many results will soon be appearing in the atmospheric science literature.

## References

Anderson, J., Russell III, J. M., Solomon, S., and Deaver, L. E.: 2000, 'Halogen occultation experiment confirmation of stratospheric chlorine decreases in accordance with the montreal protocol', *J. Geophys. Res.* **105**, 4483–4490.

Beer, R., Glavich, T. A., and Rider, D. M.: 2001, 'Tropospheric emission spectrometer for the Earth observing systems Aura satellite', *Appl. Opt.* **40**, 2356–2367.

Gettelman, A., Holton, J., and Rosenlof, K.: 1997, 'Mass fluxes of $O_3$, $CH_4$, $N_2O$ and $CF_2Cl_2$ in the lower stratosphere calculated from observational data', *J. Geophys. Res.* **102**, 19,149–19,159.

Gille, J., Barnett, J., Whitney, J., Dials, M., Woodard, D., Lambert, A., and Mankin, W.: 2003, 'The high resolution dynamics limb sounder (HIRDLS) experiment on Aura', *Proc. SPIE* **5152**, 162–171.

Glavich, T. and Beer, R.: 1991, 'Tropospheric emission spectrometer for the Earth observing system', in: *Infrared Technology XVII*, Proc. SPIE **1540**, 148–159.

Guicherit, R. and Roemer, M.: 2000, 'Tropospheric ozone trends', *Chemosphere Global Change Sci.* **2**, 167–183.

Hauglustaine, D., *et al.*: 1998, 'MOZART, a global chemical transport model for ozone and related chemical tracers 2. Model results and evaluation', *J. Geophys. Res.* **103**, 28,291–28,335.

Houghton, J., *et al.*: 2001, 'Climate Change 2001: The Scientific Basis, IPCC, Intergovermental Panel on Climate Change', Cambridge University Press, Cambridge, England, 944 pp.

IPCC: 2005, 'Safeguarding the ozone layer and the global climate system: Issues related to the hydrofluorocarbons and perfluorocarbons, Summary for Policy Makers', IPCC/TEAP Special Report, WMO/UNEP.

Levelt, P. F., Hilsenrath, E., Leppelmeier, G. W., van den Oord, G. H. J., Bhartia, P. K., Tamminen, J., de Haan, J. F., Veefkind, J. P., and Leppelmeier, G. W.: 2006, 'Science objectives of the ozone monitoring instrument', *IEEE Transactions on Geoscience and Remote Sensing*, 44, pp. 1199–1208, May 2006.

Logan, J. A.: 1985, 'Tropospheric ozone – seasonal behavior, trends, and anthropogenic influence', *J. Geophys. Res.* **90**, 10,463–10,482.

Mote, P., *et al.*: 1996, 'An atmospheric tape recorder: The imprint of tropical tropopause temperatures on stratospheric water vapor', *J. Geophys. Res.* **101**, 3989–4006.

Newman, P. A., Gleason, J. F., McPeters, R. D., and Stolarski, R. S.: 1997, 'Anomalously low ozone over the Arctic', *Geophys. Res. Lett.* **24**, 2689–2692.

Read, W., *et al.*: 1995, 'Upper tropospheric water vapor from UARS MLS', *Bull. Amer. Met. Soc.* **76**, 2381–2389.

Sandor, B. J., *et al.*: 1998, 'Seasonal behavior of tropical to mid-latitude upper tropospheric water vapor from UARS MLS', *J. Geophys. Res.* **103**, 25,935–25,947.

Schoeberl, M. R., *et al.*: 2004, 'Earth observing systems benefit atmospheric research', *AGU EOS* **85**, 177–178.

Solomon, S.: 1999, 'Stratospheric ozone depletion: A review of concepts and history', *Rev. Geophys.* **37**, 275–316.

Waters, J. W., *et al.*: 1993, 'Stratospheric ClO and ozone from the microwave limb sounder on the upper atmosphere research satellite', *Nature* **362**, 597–602.

Waters, J. W., *et al.*: 1999, 'The UARS and EOS microwave limb sounder (MLS) experiments', *J. Atmos. Sci.* **56**, 194–218.

Waters, J. W., *et al.*: 2005, 'The Earth observing system microwave limb sounder (EOS MLS) on the Aura Satellite', IEEE, submitted to *Transactions on Geoscience and Remote Sensing*.

WMO (World Meteorological Organization): 2002, *Scientific Assessment of Ozone Depletion: 2002*, Global Ozone Research and Monitoring Project–Report No. 47, WMO/UNEP, Geneva, 498 pp.

# WHAT DO WE KNOW ABOUT THE CLIMATE OF TERRESTRIAL PLANETS?

*Introductory Paper*

R.-M. BONNET
*International Space Science Institute, Hallerstrasse 6, CH-3012 Bern, Switzerland*
*(E-mail: rmbonnet@issibern.ch)*

(Received 6 June 2006; Accepted in final form 12 June 2006)

Confronted to the often heard criticisms pronounced by Earth scientists, that "we know Mars better than the Earth" – a way to complain that valuable research money is deviated from useful science (that of the Earth) to less useful science (that of the Solar System) – planetary scientists react and position their research, stating that "studying terrestrial planets can help better understanding the problems of the Earth". Indeed, at a time when numerous planetary systems are found around other stars, comparative planetology offers a potentially powerful tool for solving some of the problems facing planetary scientists.

In preparing the program of this workshop, the conveners had in mind to assess whether there were any similarities or peculiarities between the Earth, Venus, Mars and Titan that would help placing the problem of the Earth's climate on some more general grounds. This is the reason why this Section has been included in the present book. The outcome is amazing!

Very little was said about the effect of solar variability on the climate of Venus, Mars and Titan, mostly because we know little about the climate itself of these bodies and because each of them is differently influenced by the Sun, its variability being the least of problems. At 10 astronomical units from the Sun, Titan receives 100 times less solar energy than the Earth, resulting in a surface temperature of $-180°$C, and solar radiation variability of a few tenths of percents plays very little role there. On Venus, at a distance of only 0.7 astronomical units, the effect is opposite, resulting in a dramatic greenhouse effect which chokes Venus atmosphere with a surface temperature of some 450°C, and, again, a few tenths of percents of variation of the solar TSI has little influence on the climate. Mars is the body that is clearly most comparable to the Earth, which probably explains why it has received so much attention from the scientific community and by the main space agencies.

It seems therefore natural that instead of addressing the effect of solar variability, contributing authors in this Section discuss primarily the main causes of variability of the three bodies' respective climates. The situation, as it appears through this set of papers, is that we know very little about each of them and that we are a long way from using them as examples to better understand the climate of the Earth, which, in spite of its complexity, seems to be much better analyzed and understood than that

Space Science Reviews (2006) 125: 431–434
DOI: 10.1007/s11214-006-9075-0

of Venus, Mars or Titan. Furthermore, some topics appear as highly controversial, with some fairly radical conclusions being drawn that seem difficult to substantiate based on the relatively small set of observations coming yet from ground based telescopes or from space missions.

Two papers are missing that were originally planned for that Section: an overview of present and future space missions to Mars, and a discussion of thermospheric responses to solar variability, their authors having not found the time to send their manuscript within the assigned deadline. In this context, the group of Editors could have decided to drop the entire Section. This option was rejected, for two reasons: first, it would have been unfair to those who have considered their role in this workshop as essential to the topic and have consequently sent their manuscripts in due time. Second, the sketchy and controversial status of the situation is by itself interesting, sending the clear message that much more data are needed to properly understand the properties, and in particular the influences on the climate, of these three bodies.

Of all three, Mars is clearly the best known object and this can be judged through the excellent and well detailed article of F. Montmessin, which offers an uncontroversial description of the effect of orbital forcing or of the Milankovitch cycle on the climate of Mars. On Earth, obliquity changes have played a critical role in pacing glacial and interglacial eras. For Mars, such orbital changes have been far greater and have generated extreme variations in solar forcing, which seem to properly explain some of the surface structures of Mars. This is the case in particular of the signatures associated with the presence of water ice reservoirs identified at various positions across the surface, during periods of different orbital configurations. It is proposed that Mars is currently evolving between ice ages. Amazingly, it is the tools developed for the description of the Earth's climate that have yielded the most spectacular progress on Mars. The advent of climate tools has indeed given a theoretical frame to the study of orbitally-induced climate changes on Mars. These models have provided an explanation to the many puzzling observations which, when put together, have permitted reconstruction of almost the entire history of Mars in the last 10 million years.

As recalled by J.-L. Bertaux in his summary paper, the most dramatic climatic event on Mars has been the disappearance of liquid water, most probably due to the loss of the $CO_2$ atmosphere and consequently also of its strong greenhouse effect, sometime around 3.5 to 3.8 billion years before present. Bertaux discusses two possible causes of atmospheric erosion: impacting bodies, and the effect of the strong solar wind of the young Sun. Both would have had a dramatic effect on the Martian climate. Bertaux favours the first option because of its simplicity. The subject is obviously not closed and deserves further attention, in particular through more modelling.

The case of Venus seems to be much more controversial. In his paper, F. Taylor discusses the problem of the planet's strong greenhouse effect, that results mainly from IR radiation absorption by atmospheric $CO_2$, $SO_2$ and water vapour, implying that climate changes on Venus could result from variations in the concentration of these constituents. Taylor supports the hypothesis of active volcanism on Venus as a source of atmospheric $SO_2$. The author indicates that "the present concentration of $SO_2$ in the lower atmosphere is 10 to 100 times the value expected for equilibrium with the surface", which leads him to conclude, based on the results of an analysis published more than 20 years ago, that "the volcanic activity is current or recent (last 20 Myr), while changes in the $SO_2$ abundance near the clouds of a factor 10 over a 10-year period indicates contemporary volcanism". This argument is discussed also by J.-L. Bertaux, one of the Principal Investigators of the French-Soviet Vega-1 and Vega-2 ISAV experiments that measured and studied the vertical profile of $SO_2$ between the Venusian clouds and almost down to the surface. Unfortunately, the latter work is not mentioned in Taylor's paper. Furthermore, observations with the more recent NASA Magellan radar orbiter indeed clearly evidenced the existence of volcanoes on the surface of Venus, but not of any form of active volcanism during the 4 years of its operations, casting doubts as to the validity of the above discussed hypothesis.

Another important problem related to Venus' greenhouse effect is regarding $CO_2$, and more precisely its chemical equilibrium, which is governed by the balance between calcium carbonate, silica, and calcium silicate at the surface of the planet through the Bullock and Grinspoon process. If this were to be the main process controlling the abundance of $CO_2$ in the planets' atmosphere, a large abundance of carbonates should exist on the surface of Venus. Verification of this hypothesis might come in the future from a Venus lander or an even more hypothetical Venus Sample Return mission, of the type studied some 10 years ago by ESA upon the request of the French Research minister – who happened to be a geologist! Similarly, if this process is operating on Venus, it would be interesting to understand why it does not operate on the surface of Mars where no carbonates were found neither by the different NASA landers and rovers, nor by the OMEGA experiment onboard Mars Express that would have detected them through their infrared band at 4.2 $\mu$m. These relatively simple questions await an answer in the future, coming from new analysis and from new data.

Finally, Taylor discusses the case of Titan, the moon of Saturn. With an atmosphere composed essentially of nitrogen with a few percents of methane, it is apparently very similar to the Earth's atmosphere some 4.5 billions of years ago, while the cloud cover is reminiscent of the case of Venus. Methane is continuously removed from the atmosphere by photolysis. Without methane, whose greenhouse effect contributes to raise the temperature, the atmosphere of Titan would collapse. The source of methane on Titan is therefore of crucial importance and is presently widely discussed especially after the successful landing of the European Huygens probe and the continuous arrival of data collected by NASA's Cassini orbiter payload. Taylor discusses the effect of the variation of a few percents in the solar TSI – which have not been observed yet in the recent history of the Sun – that could raise the temperature of Titan substantially. It foresees Earth-like temperatures if the TSI were to increase ten times above the present level, which may well occur during the red dwarf phase of the Sun. We are here entering nearly fiction, but clearly, Titan is an obvious target for future space missions.

In conclusion, anthropogenic pollution for the Earth, volcanism for Venus – if it is still active – , the re-supply of methane for Titan, and, for Mars, atmospheric erosion as well as the Milankovitch cycle, seem to be the main factors of climate forcing for these bodies. The problem is that apart from the latter, we know very little of the Venusian volcanism and of the source of methane on Titan. Much more work both experimental and of modeling, as well as more frequent and open confrontation of controversial ideas, might in the future lead to some progress in this delicate field of planetary climatology.

# SOLAR VARIABILITY AND CLIMATE IMPACT ON TERRESTRIAL PLANETS

J.-L. BERTAUX
*Service d'Aéronomie du CNRS/IPSL, BP3, 91371 Verrières-le-Buisson, France*
*(E-mail: bertaux@aerov.jussieu.fr)*

(Received 20 February 2006; Accepted in final form 13 June 2006)

**Abstract.** Some possible factors of climate changes and of long term climate evolution are discussed with regard of the three terrestrial planets, Earth, Venus and Mars. Two positive feedback mechanisms involving liquid water, i.e., the albedo mechanism and the greenhouse effect of water vapour, are described. These feedback mechanisms respond to small external forcings, such as resulting from solar or astronomical constants variability, which might thus result in large influences on climatic changes on Earth. On Venus, reactions of the atmosphere with surface minerals play an important role in the climate system, but the involved time scales are much larger. On Mars, climate is changing through variations of the polar axis inclination over time scales of $\sim 10^5$–$10^6$ years. Growing evidence also exists that a major climatic change happened on Mars some 3.5 to 3.8 Gigayears ago, leading to the disappearance of liquid water on the planet surface by eliminating most of the $CO_2$ atmosphere greenhouse power. This change might be due to a large surge of the solar wind, or to atmospheric erosion by large bodies impacts. Indeed, except for their thermospheric temperature response, there is currently little evidence for an effect of long-term solar variability on the climate of Venus and Mars. This fact is possibly due to the absence of liquid water on these terrestrial planets.

**Keywords:** climate evolution, Earth, Mars, Venus, atmospheres

## 1. Introduction

The climate of a planet at a given time may be primarily characterized by its surface temperature, and by the variation of the latter with latitude and season. A second parameter that can be taken into account is rain and snow precipitation, which is at least of relevance for Earth and Mars, and possibly also for Titan. In an exercise of comparative planetology, we might wonder about lessons to be learned about the Earth climate and its changes from the study of other planets.

As we shall see, the time scales of interest in the context of climate studies on Earth and on the other terrestrial planets are considerably different. This difference can be attributed to the presence of a large quantity of liquid water on the Earth's surface (about 2.8 km equivalent thickness). Water is playing a very important role in Earth's climate system, with ocean currents and atmospheric circulation cells redistributing heat deposited near the equator to higher latitudes. But water on Earth is also a major factor of climate instability. Fast redistribution of heat also implies the possibility for the occurrence of fast climatic changes, in response to changes in


DOI: 10.1007/s11214-006-9076-z

global oceanic and atmospheric circulation. The situation is very different for the other planets, due to the lack of liquid water on their surfaces. However, the climate on these planets might respond to other chemical or thermodynamical feedback effects. The following discussion will focus more particularly at the climate of the terrestrial planets, Earth, Venus, and Mars, while Titan is addressed in more details by Taylor (2006). Titan is not a planet, but, as Donald Hunten said once, "*Titan is more like a planet than some planet*". Clearly he had in mind that Titan is more like a planet than Mercury, which has no decent atmosphere and is not discussed here.

## 2. The Earth

Geological records show that climate on Earth has changed in the past on various time scales. The recent climate of Earth has been characterized by an alternation of glacial and interglacial periods, for the last ~ three million years. But before, the climate was much warmer than now, with no glacial periods for several tens of millions of years. It is now widely admitted that this recent alternation has been dictated by the Milankovitch mechanism: the variation of polar axis inclination, the precession of equinoxes, and the variations of orbital eccentricity combine to modulate the solar insolation at high latitudes. This dictates more snow precipitations during winter in some phases, which are not completely melted the following summer. The existence of a large quantity of water (about 2.8 km equivalent thickness; had it been ~4 times larger, then there would not be any continent emerged) is playing on Earth a very important role to determine the climate. Ocean currents are redistributing in latitude the heat which is mostly deposited near the equator, as well as latent heat of water vapor transported by winds at other latitudes. Rain fall rates are indeed used to the definition of climate on Earth: a region is declared as desertic if the yearly rate is less than 200 mm per year. But water on Earth is also a major factor of climate instability, because of two positive feedback mechanisms.

### 2.1. The Albedo Effect

Water can exist under three forms: liquid, solid, and vapour, with liquid phase requiring a temperature $T > 273$ K. However, the albedo $A$ of liquid water is extremely low ($A = 0.04$), while the albedo of ice or snow is extremely high ($A = 0.95$) (Figure 1).

Therefore, for the same solar input $F_s$, the black-body equilibrium temperature is drastically different in the two cases:

$$\sigma T_s^4 = F_s(1 - A) \tag{1}$$

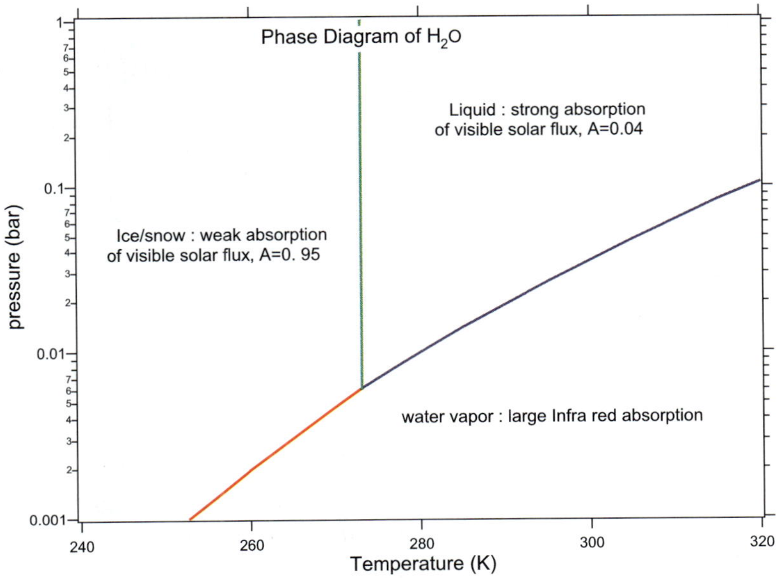


*Figure 1.* The domain of existence of the three phases of water, as a function of pressure and temperature. The main radiative properties are indicated in each domain. Changing from ice to liquid, the albedo is decreasing, the solar flux is more absorbed, which increases the heating. This is a strong positive feedback mechanism, which can explain the easiness to go from a glacial to an interglacial state with a small external perturbation. Note the exponential variation of saturating vapour pressure (red and blue lines) with temperature, another positive feedback mechanism: the higher the surface temperature, the more water vapour in the atmosphere, the larger the greenhouse effect.

The equilibrium temperature as a function of latitude on Earth can be computed for the case of ice cover, or for the case of liquid water cover. For ice, $T_s$ is $<273$ K at all latitudes, and the ice is stable everywhere, including at the equator (Figure 2 is a result from Pierrehumbert, 2004, more elaborated than can be derived from Equation (1)). Such a state of global glaciation is possible, and has been described in the literature as the "hard snow ball" state, which could have happened as recently as 543 to 900 million years ago (Hoffman *et al.*, 1998). For liquid water, because of the low albedo, most of the solar flux is absorbed, and the equilibrium temperature is $>273$ K over a large range of latitudes. This is a most important positive feedback of water, when changing phase. Therefore, because of the water albedo positive feedback, it is likely that, during a period of time with oscillations between glacial and interglacial periods, a transition from one state to the other may be triggered by a small perturbation: the astronomical influence (change of obliquity, orbital excentricity, etc...), which is small, is such a trigger. But this does not happen permanently: three million years ago, the average temperature was higher, and there were no glaciation. In the near future, the extrapolation of the modelled astronomical influence indicates that there should not be a new glaciation (M. Crucifix, private communication).

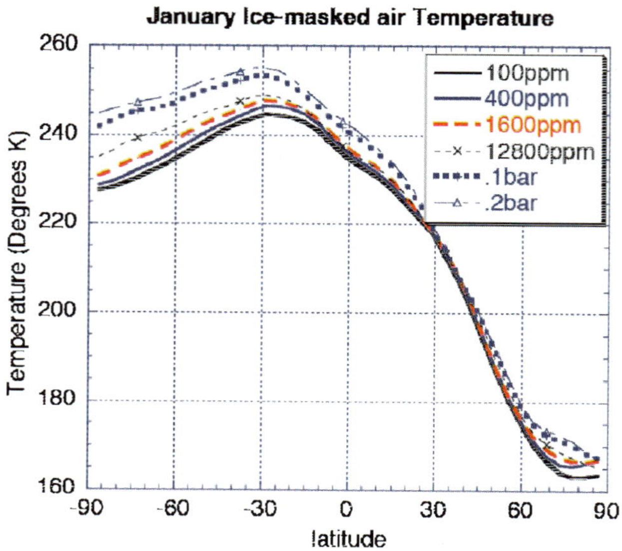


*Figure 2.* January air temperature as a function of latitude, for a "snowball Earth" entirely covered with ice, and with various quantities of $CO_2$. The greenhouse effect of $CO_2$ is small enough, preventing to reach $H_2O$ ice melting at 273 K, even with 0.2 bar of $CO_2$. From Pierrehumbert (2004).

## 2.2. The Greenhouse Effect of Water Vapor

There is another important water positive feedback mechanism, connected to the infrared absorption properties of water vapour, acting as a powerful greenhouse gas. As shown in Figure 1, a small change of temperature induces a large change of water vapour pressure. But since the greenhouse power is proportional to the abundance of $H_2O$ vapour in the atmosphere, a change of 3 K at 260 K will not produce a significant $H_2O$ greenhouse effect, while the same change of 3 K at 290 K will produce a large increase of the $H_2O$ greenhouse effect. Water vapour is an extraordinary powerful amplifier of the small changes in greenhouse effect resulting from changes in $CO_2$ abundance, but only if the temperature is already >280 K (at present, $H_2O$ accounts for 80 to 95% of the greenhouse effect). A contrario, a recent model calculation (Pierrehumbert, 2004) was used to find how much $CO_2$ would be needed in the atmosphere to terminate a global glaciation by its greenhouse effect. Even with a pressure of 0.2 bar of $CO_2$ (550 times the present content), it would be insufficient to melt the ice anywhere (Figure 2), because at these low temperatures, the change in water vapour abundance associated to the warming by the $CO_2$ is negligible (Figure 1).

## 2.3. Solar Influence on Shorter and Longer Time Scales

On the shorter time scales, there are some evidence from lacustrine deposits and tree rings of a connection with solar cycles, rather on a regional scale than on a global scale. Both signs are linked to variations of precipitations. Yet, there is no mechanism widely admitted which would relate solar activity with increased precipitations. Atmospheric electricity may play a role there. The recent discovery of sprites (lightnings above the tropopause, possibly up to the ionosphere) is certainly something of interest for study in this context: the space project Taranis (presently studied by CNES) is dedicated to the study of sprites and blue jets, which might turn out to be the missing link between solar activity and precipitations. There is a very clear signature of the thermosphere temperature response to solar variations, both in the 27 days and the 11-year cycle, with about a 600 K heating at solar maximum (G. Keating, private communication). Also, some heating occurs in association with charged particles precipitations, in particular in the polar regions, well documented. However, it is not clear if these changes at high altitude ($>$150 km) are influencing the atmosphere below, and the surface climate. On the other hand, there is a newly discovered cooling trend of the lower thermosphere apparently due to the anthropogenic increase of $CO_2$ (radiative cooling, G. Keating, private communication).

Considering the whole Earth's history, it seems that liquid water has existed permanently on Earth (except for possible episodic total glaciations mentioned above). This requires a certain range of ground temperature, which is the result of solar absorption and greenhouse warming. The early atmosphere of Earth was containing 100 bar of $CO_2$, producing a large greenhouse effect; however the Sun is believed to have been less bright 4.6 billion years ago (70% of present output), yielding a lower temperature and allowing liquid water instead of steam as on Venus. Then $CO_2$ may dissolve in water and precipitate under the form of $CaCO_3$, reducing the greenhouse effect. Also, $CO_2$ (and $H_2O$) might have been exhausted from the interior through volcanism after the final accretion. The permanent conditions favourable for liquid water might have been the result of the fortuitous combination of two totally independent factors: the history of the solar constant, and the history of $CO_2$ outgassing. What a piece of luck!

## 3. Climate on Venus

The variability of climate on Venus is reviewed by Taylor (2006), and its recent evolution was modeled by Bullock and Grinspoon (2001). Venus is an extreme case of greenhouse effect, produced by the combination of $CO_2$, $H_2O$, and cloud particles made mainly of sulphuric acid droplets, produced by $SO_2$. What is the origin of this $SO_2$? Thermo-chemical reactions with the surface minerals (Fegley *et al.*, 1997) indicates that the $SO_2$ mixing ratio in the atmosphere should be of the order of 5 ppmv. The fact that several descent probes found around 40–50 km a mixing ratio of 180 ppmv has been used to suggest that $SO_2$ is of volcanic origin, and may be fairly recent. This view is adopted by Taylor (2006), even though the radar mapping of Venus by NASA's Magellan orbiter did not find any evidence of active volcanism. Therefore, he suggests that, when volcanoes are stopping their activity, $SO_2$ might disappear from the atmosphere by recombining in the ground. The atmosphere would be cleared from clouds, and hotter than now because of decreased albedo. In support of this view was the measured decrease of $SO_2$ above the clouds over several years, as reported by Esposito (1984) from Pioneer Venus UV spectra, interpreted as a sign of present decline of $SO_2$ after an eruption some years (or tens of years) ago. However, two things have to be considered in the discussion. First, the French-Russian UV spectroscopy ISAV experiment on board the descent probes of Vega-1 and Vega-2 have actually measured the vertical profile of $SO_2$ in the clouds, below the clouds and almost down to the surface (Bertaux *et al.*, 1996). The $SO_2$ mixing ratio was found at 140–200 ppmv at cloud level (different between Vega 1 and Vega2), decreasing below, down to ~10–20 ppmv at 5–10 km altitude (Figure 3). Therefore, it indicates that indeed $SO_2$ could be in equilibrium with the surface. Second, $SO_2$ is destroyed very rapidly above the cloud by solar UV, and its density decreases very fast. Therefore, the $SO_2$ downward trend reported by Esposito (1984) could be rather a subtle change in the vertical structure of the clouds and haze above the clouds. The forthcoming Venus Express mission offers a means of confirming or not the presence of active volcanoes on Venus now. On the other hand, if no volcanic activity is detected, it is hoped that, by measuring

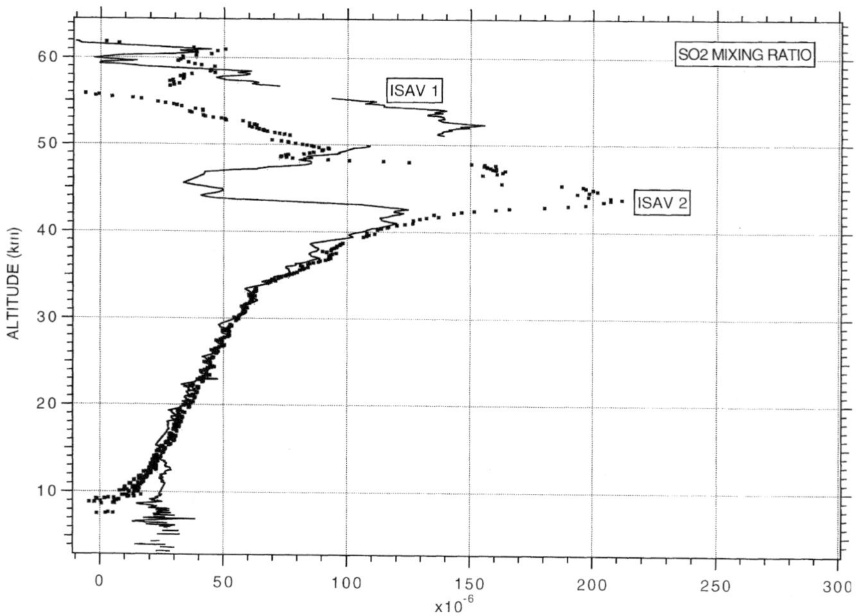


*Figure 3.* Measurements of $SO_2$ mixing ratio vertical profile obtained with ISAV instruments in Vega 1 and Vega 2 descent probes. Above 40 km, the vertical structure is different for the two probes, while below 40 km, they measure essentially identical profiles, decreasing when approaching the ground, and compatible with a value of 5 ppmv, the equilibrium value with surface minerals at 750 K (Bertaux *et al.*, 1996).

$SO_2$ vertical profiles just above the clouds, and monitoring some possible changes, we may be able to understand the origin of the variations observed by Esposito.

Venus is extremely dry. $H_2O$ is only present in the atmosphere. If condensed in water, it would make a 3-cm thick layer only. This translates into 4600 km$^3$, which could have been brought by one single comet impact, with a diameter of 30 km: smaller than the estimated size of the recent Hale-Bopp comet! And the whole deuterated water of Venus (about 0.3 mm) could have been brought by one single comet (or Kuiper object) with a size of 90 km only. G. Keating at the workshop reviewed what was learned on the upper atmosphere of the three planets from atmospheric drag effects on satellites, or from accelerometers measurements. It was found that the temperature of the thermosphere of Venus and Mars is at ~300 K, much smaller than the Earth's (1000 K), and much smaller than early theoretical predictions. This is because there is a newly understood cooling mechanism by $CO_2$ emission, involving energy exchange with O atoms (Dodd *et al.*, 2005). There is also a clear response of the thermospheric temperature to solar changes, at 27 days (rotating Sun and solar active regions) and 11-year time scales. The Venus night side of the thermosphere is at ~100 K only, while on the day side, there is a temperature minimum at 170 K around 95 km of altitude (Schofield and Taylor, 1983). This low temperature is a bottleneck for the transport of $H_2O$ at higher altitudes, where it is photodissociated into H and O atoms.

Let us imagine the early Venus, with the same $CO_2$ content, but in addition with 300 bar of steam, corresponding to the Earth's present content of water. The size of the atmosphere would be only modestly increased by less than two scale heights (about 5 km). The $CO_2$ cooling properties would still dictate a region of minimum temperature, providing an efficient barrier to the transport of $H_2O$ in the upper region of photodissociation, and not much of atmospheric escape of H or O through thermal or non-thermal exospheric mechanisms. The present escape rate for the Earth, at $2 \times 10^8$ H atoms/cm$^2$s, considered as high, would eliminate only 4 m of water over the age of the solar system (Bertaux, 1975). This is why I believe that the atmosphere of Venus, if ever heavily charged in $H_2O$, evolved to its dry state rather through so-called hydrodynamic escape: a mechanical type of escape with no differentiation, all atoms and molecules escaping at the same rate, and the isotopic/chemical composition of the atmosphere being not changed. It is my contention, however, that this hydrodynamic escape exists only during catastrophic events, like the collision of the planet with big planetesimals, or planetary embryos the size of Mars. This is the process of atmospheric erosion, studied with some details by Genda and Abe (2004, 2005) for instance.

## 4. Mars

Most of our knowledge on Mars relies on space missions, as was reviewed by O. Korablev during the workshop. The orbit of Mars is eccentric, inducing a large change of TSI (Total Solar Irradiance) of 40% on a yearly basis. Like on the Earth, we can expect an 11 years cycle of ozone depletion (−2 to −4%) due to increased UV at solar maximum. On the longer term, the application of GCM modelling shows an exchange of water ice between poles and equator when the inclination of the polar axis is changing as discussed extensively by F. Montmessin in this volume Section. There is a Milankovitch type of climate variation, as on the Earth. One difference though: the amplitude of variation is much larger on Mars (because of the stabilizing influence of the moon for Earth). As a result, the Milankovitch cycle is effective for Earth only for minor parts of the life of the Earth (we just happen to be at the right time, it is finishing), while it is always important for Mars. It should be emphasized that the reconstruction for Mars is totally accurate for at least the last 10 million years (Levrard *et al.*, 2004) showing a major change of the average inclination 5 million years ago, from 35 to 25°. This explains a number of geological observations, in particular around Olympus Mons: at 35° inclination, GCMs predict the formation of glaciers on the flanks of the volcano, which disappear when the inclination drops to 25°. There is mounting geological evidence that liquid water existed on Mars, but only during the Noachian epoch, the earliest epoch in Mars history ending some time between 3.8 and 3.5 billion years ago. The recent identification of phyllosilicates (clays) by the OMEGA experiment on board Mars Express, found to be existing only in the oldest terrains (Poulet *et al.*, 2005) is the most recent of a series of findings pointing to a major climatic change on Mars 3.5 to 3.8 billion years ago, when the $CO_2$ atmospheric pressure went down, and its greenhouse effect became insufficient to allow stable liquid water on the ground of Mars. How did the 1–2 bar of $CO_2$ necessary for such a greenhouse effect disappear? We describe below two different hypotheses.

### 4.1. Erosion by the Solar Wind?

Lundin (2005) is reporting an atmospheric loss of 100 tons per day at present (derived from ASPERA measurements on the soviet Phobos 2 mission), due to ionization and stripping by the solar wind. He is also arguing that the solar wind of the young Sun was much higher, and could have disposed of the 1–2 bar of $CO_2$ early on. He refers to the work of Wood *et al.* (2005), who estimated the present stellar wind of a few young stars with different ages analog to the Sun. Wood *et al.* (2005) derived the amount of stellar wind from the interaction of the astrosphere (the region of space around the star containing the stellar wind) with the surrounding interstellar medium, which creates a population of decelerated H atoms at the astrosphere interface. When measuring the star's Lyman-$\alpha$ line at very high spectral resolution with the Hubble Space Telescope, those atoms manifest themselves in absorption, displaced with respect to the huge interstellar line, due to their different velocity. This effect was first identified for the heliospheric interface, and then observed on a few stars. Wood *et al.* (2005), in a revision of their earlier work, found that the stellar mass loss (and hence the solar wind) was moderate at the beginning, for 0.7 Gyr, then increased abruptly to 100 times the present value at 0.7 Gyr, then declined slowly to its present value. Therefore, in this scheme, the Sun would have had a strong influence on the evolution of the martian climate. This age of 0.7 Gyr coincides nicely with the end of the Noachian, and one may speculate that the $CO_2$ atmosphere was blown away at this time by a strong solar wind in a few hundred million years.

### 4.2. Erosion by Impacting Bodies?

There is another way to lose an atmosphere, and this is through erosion from impacting bodies. Indeed, the ancient terrains of Mars are heavily cratered, and the end of the Noachian epoch is marked by a considerable decrease of the impact rate.

Which process was more important to get rid of the atmosphere of Mars? I would dare to say that we have a very clear evidence (with craters) of the heavy bombardment, while the line of reasoning invoking a strong early solar wind is very elegant, but intricated and still fragile. Better modelling of the atmospheric escape with impact erosion would still be worthwhile in the future.

## 5. Conclusions

The climatic changes on the three terrestrial planets are drastically different. This is due to the presence of liquid water on Earth which is a major factor of instability, because of two positive feedback mechanisms: the enormously different albedo of ice and liquid water, and the water vapor pressure exponential variation with temperature coupled with its green house effect.

*Note added in proofs:*
New measurements of non-thermal escape of Mars made by Aspera-3 instrument on board Mars Express have been recently analyzed, yielding an escape rate of about 1 ton per day, (Stas Barabash, private communications, 2006) instead of 100 tons per day as inferred from Phobos 2 analysis.

## Acknowledgements

I wish to thank Raymond Pierrehumbert for his permission to reproduce his Figure on the temperature of a snowball Earth.

## References

Bertaux, J. L.: 1975, 'Observed variations of the exospheric hydrogen density with exospheric temperature', *J. Geophys. Res.* **80**, 639–642.

Bertaux, J. L., Widemann, T., Hauchecorne, A., Moroz, V. I., and Ekonomov, A. P.: 1996, 'Vega-1 and Vega-2 entry probes: an investigation of local UV absorption (220–400 nm) in the atmosphere of Venus (SO2, aerosols, Cloud Structure)', *J. Geophys. Res.* **101**, 12,709–12,745.

Bullock, M. A. and Grinspoon, D. H.: 2001,'The recent evolution of climate of venus', *Icarus* **150**, 19–37.

Dodd, J. A., Castle, K. J., Rhinehart, J. M., and Hwang, E. S.: 2005, 'Quenching of $CO^2$(nu2) by O: New Results and Analysis', AGU, Fall Meeting 2005, abstract no. SA11A–0222.

Esposito, L. W.: 1984, 'Sulfur dioxide: Episodic injection shows evidence for active venus volcanism', *Science* **223**, 1072–1074.

Fegley, B., Zolotov, M. Y., and Lodders, K.: 1997, 'The oxidation state of the lower atmosphere and surface of venus', *Icarus* **125**, 416–439.

Genda, H., and Abe, Y.: 2004, 'Survival of a proto-atmosphere through the stage of giant impacts: The mechanical aspects', *Icarus* **164**, 149–162.

Genda, H., and Abe, Y.: 2005, 'Enhanced atmospheric loss on protoplanets at the giant impact phase in the presence of oceans', *Nature* **433**, 842–844.

Hoffman, P. F., Kaufman, A. J., Halverson, G. P., and Schrag D. P.: 1998, 'A neoproterozoic snowball earth', *Science* **281**, 1342–1346.

Levrard, B., Forget, F., Montmessin, F., and Laskar, J.: 2004, 'Recent ice-rich deposits formed at high latitudes on Mars by sublimation of unstable equatorial ice during low obliquity', *Nature* **431**, 1072–1075.

Lundin, R.: 2005, 'Planetary magnetic fields and solar forcing-critixal aspects for the evolution of the Earth-like planets', ISSI Workshop on "Geology and Habitability of Terrestrial Planets", *Space Sci. Rev.*, volume in preparation.

Montmessin, F.: 2006, 'The orbital forcing of climate changes on Mars', *Space Sci. Rev.*, this volume, doi: 10.1007/s11214-006-9078-x.

Pierrehumbert, R. T.: 2004, 'High Levels of atmospheric carbon dioxide necessary for the termination of the global glaciation', *Science* **429**, 646–648.

Poulet, F., Bibring, J.-P., Mustard, J. F., Gendrin, A., Mangold, N., Langevin, Y., Arvidson, R. E., Gondet, B., and Gomez, C.: 2005, 'Phyllosilicates on Mars and implications for early martian climate', *Nature* **438**, 638.

Schofield, J. T., and Taylor, F. W.: 1983, 'Measurements of the mean, solar-fixed temperature and cloud structure of the middle atmosphere of Venus', *Quart. J. Roy. Meteorol. Soc.* **109**, 57–80.

Taylor, F.: 2006, 'Climate variability on Venus and Titan', *Space Sci. Rev.*, this volume, doi: 10.1007/s11214-006-9077-y.

Wood, B. E., Mller, H.-R., Zank, G. P. J. L., Linsky and Redfield, S.: 2005, 'New mass-loss measurements from astrospheric ly $\alpha$ absorption', *Astrophys. J.* **628**, L143–L146.

# CLIMATE VARIABILITY ON VENUS AND TITAN

F. W. TAYLOR
*Atmospheric, Oceanic and Planetary Physics, University of Oxford,*
*Clarendon Laboratory, Oxford, OX1 3PU, England*
*(E-mail: fwt@atm.ox.ac.uk)*

(Received 27 July 2005; Accepted in final form 22 February 2006)

**Abstract.** Venus and Titan are both slowly-rotating, approximately Earth-sized bodies with cloudy, dynamic atmospheres. Each has a complex climate system, even less well understood than the terrestrial equivalent, and the processes that appear to maintain the climate near the surface on both bodies have interesting similarities and differences with each other and with the Earth. By considering these factors and their possible evolution with the aid of elementary climate models, some interesting, albeit tentative, conclusions can be reached concerning the stability of climate on Earth-like planets, and the likely nature of past and future climate change.

**Keywords:** Venus, Titan, climate, atmosphere, planetary

## 1. Introduction

Enough is now known about the atmospheres of the Earth's nearest planetary neighbour, Venus, and the large, Earth-like moon of Saturn, Titan, to support theoretical studies of the climate on these bodies and to consider whether these climates are subject to change in response to any likely stimulus. A possible reason for considering the two together is their similarity to each other, and to the Earth, in several respects. Some of the fixed parameters affecting climate such as size, rotation rate, and distance from the Sun are compared in Table I. Also compared are the key variable factors representing the response of the climate system in each case, such as surface temperature and pressure, cloud cover, atmospheric composition (including greenhouse gases), and the equilibrium temperature of the climate system, defined as the equivalent blackbody temperature at which the planet radiates heat to space.

A suitable definition of climate for present purposes is the global mean temperature profile for the surface and lower atmosphere. On all three planets, the atmosphere can be separated into an optically thick region, nearest to the surface (the troposphere), and an optically thin region above (the stratosphere). Here we intend optically thick or thin in a wavelength-averaged sense, which means that in the troposphere heat is transported vertically mainly by convection, while in the stratosphere radiation is more efficient and vertical motions are suppressed.

Simple model atmospheres can be constructed for each planet by assuming convective equilibrium in the troposphere and radiative equilibrium in the stratosphere (Taylor, 2005, Chapter 3). The principal assumptions are that the temperature

Space Science Reviews (2006) 125: 445–455
DOI: 10.1007/s11214-006-9077-y

TABLE I

Approximate values, for use in models, of data relevant to the climate on Venus, Earth and Titan.

| | Venus | | Earth | | Titan | |
|---|---|---|---|---|---|---|
| Distance from Sun (AU) | 0.7 | | 1 | | 10 | |
| Radius (km) | 6052 | | 6376 | | 2575 | |
| Surface pressure (bar) | 92 | | 1 | | 1.5 | |
| Surface temperature (K) | 730 | | 288 | | 94 | |
| Rotation period (Earth days) | 243 | | 1 | | 16 | |
| Main atmospheric Constituents (*key greenhouse gases) | *$CO_2$ | 97% | $N_2$ | 79% | $N_2$ | 95% |
| | $N_2$ | 3% | $O_2$ | 18% | *$CH_4$ | 5% |
| | | | Ar | 1% | | |
| | | | *$H_2O$ | 2% | | |
| | | | *$CO_2$ | 0.035% | | |
| Bond Albedo | 0.76 | | 0.3 | | 0.2 | |
| Radiative Equilibrium temperature (K) | 230 | | 255 | | 85 | |

gradient is constant, and equal to the adiabatic lapse rate, in the troposphere, and constant, and equal to zero, in the stratosphere. Then, three parameters (surface temperature, tropospheric lapse rate and stratospheric temperature) provide a complete specification of the model climate as defined above. Despite their approximate nature, such models provide remarkably good approximation to the mean conditions on all of the terrestrial planets (Taylor, 2005, Chapter 12).

To formulate the model, the perfect gas law, the first law of thermodynamics and the hydrostatic equation are used to obtain the vertical gradient of temperature $T$ with height $z$ in the troposphere. For a "dry" atmosphere, in which latent heat effects can be ignored, this is just $dT/dz = -g/c_p$, where $c_p$ is the specific heat at constant pressure and $g$ is the acceleration due to gravity. A more complicated but still straightforward expression can be obtained by introducing the Clausius-Clapeyron equation to allow for the latent heat effects of condensable species, primarily water on Earth and Venus and methane on Titan. Under the assumption of optically thin layers in radiative balance, the temperature $T_s$ in the stratosphere is constant with height and given by $T_s = T_e/2^{1/4}$, where $T_e$ is the equivalent blackbody temperature at which the planet cools to space. This, and the height of the tropopause at which the radiative and convective regimes change over, can be calculated either precisely by the detailed application of radiative transfer theory, or estimated by any of a range of approximate methods (Goody and Yung, 1989).

The simplest approach is to assume energy balance for the planet as a whole and then to calculate the temperature at which the emission of infrared radiation to space matches the solar irradiance. In a similar way, the mean surface temperature can be found by calculating the emission from the surface that matches the downward

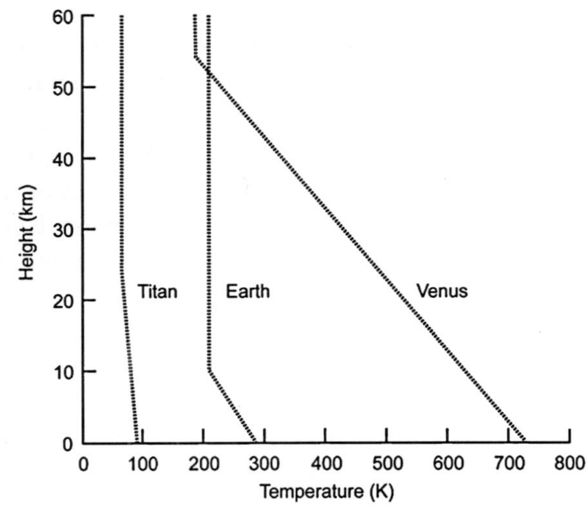


*Figure 1.* Mean temperature profiles for Venus, Titan, and Earth, on a common altitude scale, represented by simple radiative-convective equilibrium models. These fail completely, especially for Earth and Titan, at higher levels because they take no account of the heating due to ozone and aerosol respectively, which greatly modifies the profile on those planets. However, they compare quite well for all three bodies in the middle to lower atmosphere, and at the surface.

flux. The latter has two components – directly transmitted or scattered radiation from the Sun, plus thermal emission from the atmosphere (the "greenhouse" warming), while the former is defined to be the blackbody emission from a surface at temperature $T_0$, which can be compared to measurements (see Table I) to verify the model atmospheres. The resulting profiles for Venus, Titan and Earth are shown and compared in Figure 1.

The model parameters for Venus and Titan are sensitive to the same factors that enter any discussion of climate change on the Earth, viz (i) changes in the output of energy from the Sun; (ii) changes in the concentration of greenhouse gases; or (iii) changes in the cloud or aerosol amount, through their effect on the albedo of the planet. Examples of the application to the Earth in each case to estimate the change in surface temperatures resulting from solar variations and pollution induced global warming and global dimming may be found in Taylor (2005, Chapter 7). The solutions obtained there are sufficiently consistent with predictions obtained by studies using more advanced models as to lend credibility to the use of the same approach when speculating about climate change in the less well-known atmospheres of Venus and Titan.

## 2. The Climate on Venus

The remarkably high surface temperature and pressure on Venus are maintained by a case of the greenhouse effect that is extreme compared to the Earth, Mars or Titan, raising the surface temperature by around 500 K compared to an airless globe with the same albedo and separation from the Sun. On Venus the principal greenhouse gases are (i) the main atmospheric constituent $CO_2$; (ii) sulphurous gases of volcanic origin, especially $SO_2$; and (iii) small but non-negligible amounts of water vapour, around 30 ppm in the troposphere. The planet-wide cloud cover also plays an important role in the energy balance, due both to its infrared opacity and its contribution to the very high albedo of the planet, although as with the Earth this is difficult to evaluate reliably in detail, because of its variability and complex structure and microphysics.

The Venusian clouds are revealed by measurements made in-situ and remotely to be highly irregular in both vertical and horizontal distribution (Knollenberg and Hunten, 1980; Carlson *et al.*, 1993). In the upper layers at least, the composition is known from spectropolarimetry to be 75% $H_2SO_4$ and 25% $H_2O$, both components probably originating in the flux from volcanoes on the surface. The present concentration of $SO_2$ in the lower atmosphere is 10 to 100 times the value expected for equilibrium with the surface, indicating that the volcanic activity is current or recent (last 20 Myr), while changes in the $SO_2$ abundance near the clouds of a factor 10 over a 10-year period indicates contemporary volcanism (Esposito, 1984). Secular changes in cloud cover would be expected if the level of volcanic activity on the planet declines and eventually ceases.

The high surface temperature on Venus, and the accompanying pressure of nearly 100 bar, were first detected by radio astronomers, and verified by the first spacecraft to visit the planet, beginning with Mariner 2 in 1964 and continuing with a series of spectacular landings by the Venera series produced in the Soviet Union. Although modern greenhouse models, with some assumptions about the cloud properties in particular, are able to show that an atmosphere this deep can produce such high temperatures, this does not explain why Venus is in such a condition in the first place.

As long ago as 1952, when it was known that the atmosphere was predominantly $CO_2$, although not that it was so hot or so dense, Harold Urey suggested that the reaction of atmospheric carbon dioxide with common surface minerals might have a role in determining the surface pressure. On theoretical and abundance grounds, the mostly likely process to dominate (Bullock and Grinspoon, 1996) is the equilibrium balance between calcium carbonate, silica and calcium silicate:

$$CaCO_3 + SiO_2 \leftrightarrow CaSiO_3 + CO_2$$

and indeed we now know that the temperature and pressure on the surface of Venus at the present time fall on the equilibrium phase curve for this reaction. Not only that, but if we also plot the surface temperature as a function of surface pressure in the simple radiative-convective equilibrium model described in outline above, it can be seen that the two intersect at a point which is strikingly close to the surface conditions measured by the Venera landers (Figure 2).

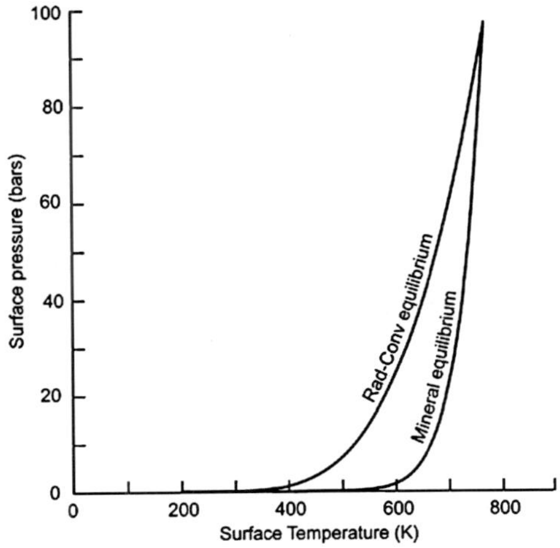


*Figure 2.* Phase curves for the surface temperature and pressure on Venus, corresponding to radiative-convective equilibrium in the atmosphere (calculated using the simple model described above), and to chemical equilibrium between $CO_2$ in the atmosphere and in surface minerals (calculated using the data provided by Adamcik and Draper, 1963).

This agreement strongly suggests that the high pressure, high temperature state of the climate on Venus can be attributed to the balance for a pure $CO_2$ atmosphere in equilibrium with the common minerals likely to be abundant on Venus. Although possibly no more than a coincidence, the existence of a straightforward account for the extreme state of Venus' climate, which is otherwise very difficult to explain, makes it hard to set it aside, at least without a further examination using more complex models and more data from new missions to the planet.

However, Bullock and Grinspoon (1996) and Hashimoto *et al.* (1997), who carried out similar calculations, noted that the temperature dependence of the mineral equilibrium is such that the present climate would be unstable if it was produced in this way. The strong dependence of atmospheric $CO_2$ amount on surface temperature represented by the phase curve shown in Figure 2 means that any small perturbation in the surface temperature, either positive or negative, leads to positive feedback and dramatic changes in atmospheric $CO_2$, resulting in runaway heating or cooling respectively. The simultaneous requirement that the atmosphere seeks to remain in radiative-convective equilibrium, which changes the dynamical cooling of the surface in response to a temperature perturbation, acts in the opposite sense, and much more quickly, to slow down the change, but on its own cannot altogether prevent it from happening. For instance, in the hypothetical case of a large perturbation where the surface temperature is somehow forced to increase rapidly from the present 735 to 1000 K, the mineral equilibrium surface pressure would then be over 3000 bar, while the atmospheric equilibrium value is only about 300 bar. The tropospheric temperature gradient corresponding to either of these would be massively super-adiabatic, convection would increase rapidly and cooling of the surface would result. Assuming the source of the perturbation was removed, and all other factors such as atmospheric composition and cloud opacity remained the same, the surface temperature would relax to a value slightly higher than before the perturbation was applied, because the atmospheric $CO_2$ content would be slightly higher. This stable atmospheric state would not correspond to mineral equilibrium, and the emission of carbon dioxide from the surface would continue, resulting in further warming. This cycle would not cease until the surface supply of $CO_2$ became exhausted. A similar line of argument can be followed to show that the result of a cool perturbation would be that all of the atmospheric $CO_2$ would end up in the surface.

This is still only the behaviour of a very simple model. Even if the regolith on Venus has the necessary supply of carbonate and silicate minerals, it is not clear that they have ever been in sufficiently intimate contact with the atmosphere. Also, the surface of Venus is not at a single uniform temperature; as Hashimoto *et al.* (1997) point out, topography alone leads to temperature contrasts of more than 100 K, corresponding to differences in $CO_2$ pressure for Urey equilibrium that would be far out of hydrostatic balance. Other surface-atmosphere reactions are undoubtedly involved, and may be important. For instance, the question of how the atmosphere and the surface "mix" may find an answer during those phases, some thought to be relatively recent (Nimmo and McKenzie, 1998) when there was extensive flooding of the surface by volcanic magma. Some aspect of these and other complexities not yet understood may act to stabilise the climate, at least against moderate perturbations, somewhat as the complexity of Earth's climate achieves an overall quasi-stable regime despite the disequilibrium state of most of its components. The strong suggestion that heterogeneous reactions involving the atmosphere and the regolith are involved indicates that missions to the surface of Venus, possibly involving deep core drilling and including the return of samples to the Earth, will be required before the question is resolved. It will be interesting to see from new data what feature of climate physics on Venus is required to explain in detail the remarkable success of the simple model in predicting the observed climate, which seems too good to be simply a coincidence.

While there is much still to be understood about the present climate, we can begin to investigate how Venus might respond to changes in external forcing, for example any past or future variation in total solar irradiance. For a change of 0.1% in solar output, similar to those measured currently by spacecraft orbiting the Earth, the simple model predicts a change in mean surface temperature of only 0.05 K. The spread in surface temperatures expected due to topographical variations is around three orders of magnitude greater than this. Volcanism seems much more likely to be an important agent for climate change in the relatively short term, particularly since there is evidence that Venus may be much more volcanically active than Earth. This evidence includes the observed disequilibrium in the atmosphere of the sulphurous gases that play a large role in the production and maintenance of the cloud layer,

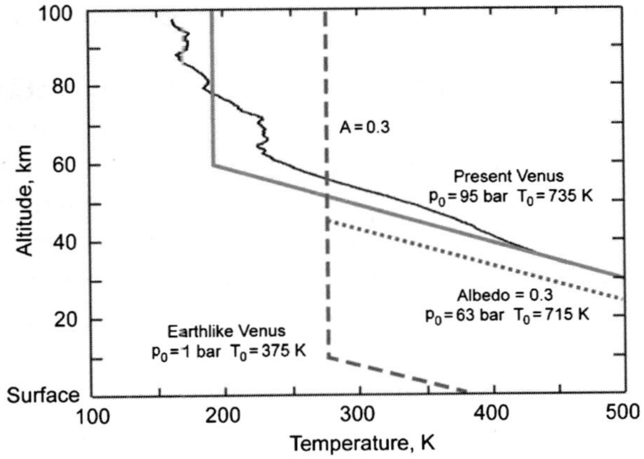


*Figure 3.* Simple climate models for Venus, corresponding to (a) the present day (solid gray line), (b) a decrease in albedo to 0.3 (dotted line), and (c) atmospheric loss of $CO_2$ such that the surface pressure fell to 1 bar (dashed line). A measured profile, obtained by the radio-occultation experiment on the Magellan orbiter on 5 October 1991 at latitude 67°N, is shown for comparison (solid black curve). While individual profiles do not necessarily coincide with the global mean represented by the model, the main difference seems to be due to heating by the absorption of solar radiation in the cloud layer in the 40–70 km height range.

and consequently the high planetary albedo. If this is the case, and if at some point in the future volcanism ceases on Venus and the cloud regime relaxes to one much more like the Earth, we can imagine that the albedo falls from the present 0.76 to an Earth-like 0.3. Neglecting for the moment all other consequences, such as atmospheric composition changes (except those in $CO_2$ due to adjustment to the Urey equilibrium at the surface), the model stratospheric temperature increases to around 273 K, while the surface temperature and pressure fall to about 715 K and 63 bar respectively (Figure 3).

A further consequence of a complete cessation in volcanic activity would be that the greenhouse effect would become less efficient through the removal from the atmosphere of some of the more chemically active minor constituents, such as $SO_2$, $H_2S$ and COS, as they move closer to their equilibrium values with the surface. The loss of atmospheric water due to dissociation and exospheric escape of hydrogen might no longer be compensated by emission from the interior, as presumably it is at present. Any loss of atmospheric infrared opacity, due to weaker absorption in the bands of these species, moves the tropopause downwards to a higher pressure level, resulting in falling surface temperatures and pressures as radiative-convective equilibrium is maintained. This large cooling perturbation leads to the steady take-up of atmospheric $CO_2$ by the surface, as described above, although it remains to be shown that this is physically possible in the absence of liquid water to facilitate the exchange. Hashimoto and Abe (2005) note that the carbon dioxide content of the current atmosphere is sufficient to produce a layer of carbonate ~1 km thick if entirely converted to surface rock.

Venus without active volcanism might thus become more Earth-like. Venus has about the same amount of $N_2$ in its present-day atmosphere as the Earth; the selective removal of $CO_2$ could at some point leave a composition of 99% $N_2$, 0.5% Ar, 0.5% $H_2O$, and 0.035% $CO_2$, with a surface pressure in the region of 1 bar. In this case, the mean surface temperature would be a relatively balmy 70°C. This is the Venus that was imagined by our forebears, by extrapolation from the Earth, before the importance of the greenhouse effect and the true state of Venus' climate had been realised.

## 3. Climate Change on Titan

Titan's atmosphere has the same principal constituent as the Earth, molecular nitrogen, but with a few percent of methane as the next most abundant molecule. The cloud cover, with a global, high-altitude haze of organic material produced by photosynthesis and patchy, lower-level methane clouds, is reminiscent in structure and total $N_2$ content (if not overall composition), of that of Venus. Methane is the principal greenhouse gas and, allowing also for contributions by the pressure-induced bands of nitrogen and hydrogen, and by the clouds, its presence raises the surface temperature by about 12 K above the radiative equilibrium temperature of 82 K (McKay *et al.*, 1989). The Huygens probe recorded a mean lapse rate of about 1.2 K km$^{-1}$ during its descent (the dry lapse rate $g/c_p$ is about 1.3 K km$^{-1}$) and a temperature of 93.65 ± 0.25 K at a pressure of 1467 ± 15 hPa at its landing site (Fulchignoni *et al.*, 2005).

The profile labelled "present" in Figure 4 shows a radiative-convective model like that developed above for Venus but now calculated with the parameters for Titan taken from Table I. Over the vertical range shown, up to about 50 km altitude, this simple model is in reasonable agreement with the profiles measured by Voyager 2 in 1981 and by Huygens in 2005. As with Venus, we can use the model in a preliminary investigation of climate change scenarios for Titan.

Since methane is continuously removed from Titan's atmosphere by photolysis, there must be a reservoir on the surface or in the interior with access to the surface. The precise nature of this reservoir is being actively researched using the new data from the current Cassini-Huygens mission, but meanwhile Lorenz *et al.* (1997a) have noted that, should the reservoir become depleted, methane would be removed in a few million years and the surface temperature would drop below the condensation point for nitrogen, so Titan's atmosphere would collapse. The albedo at present is rather low at ~ 0.2, but this could increase to 0.8 as aerosol and Raleigh scattering become less important and fresh ice forms on the surface, further accelerating the cooling. Lorenz *et al.* (1997a) calculate that the residual opacity of nitrogen itself

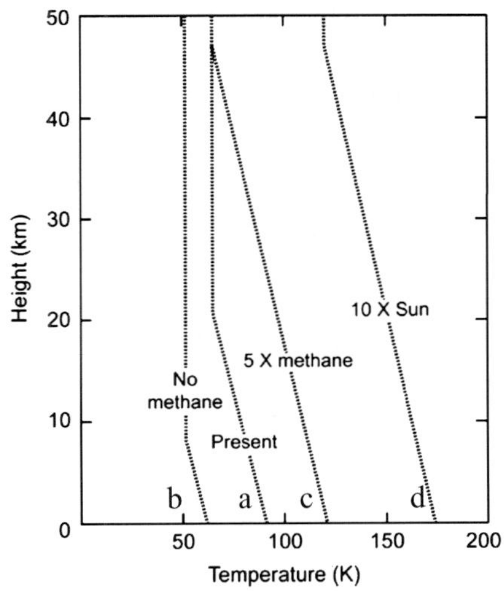


*Figure 4.* The vertical temperature profile on Titan according to radiative-convective models, derived as described in the text, for four scenarios: (a) the present day; (b) after the methane reservoir is completely depleted by photolysis; (c) with abundant methane in a "runaway greenhouse" scenario; and (d) 6 Gyr hence when the Sun is a red dwarf.

would allow a surface pressure of perhaps one tenth the present value, and a temperature of 65 K, to remain (Figure 4b). If the absorptivity of the surface subsequently increases with time due to the photochemical conversion of material in the ice or accumulation of dark material from space, perhaps organics similar to those seen on the surfaces of some of the other Saturnian moons, the surface temperature will increase, perhaps to the point where nitrogen and methane could re-evaporate and build up the atmosphere again. Such cycles may have occurred in the past, and it is interesting to speculate what record of this could remain in Titan's icy strata to be exhumed by sophisticated human or robotic explorers a century or so from now.

Conversely, if the methane supply is abundant, a perturbation such as a small increase in insolation over the present value could produce a dramatic warming of Titan's climate. For example, if the methane reservoir referred to above was a liquid methane ocean, any increase in heating would raise the atmospheric mixing ratio of methane rapidly and produce an enhanced greenhouse effect that would further increase the temperature of the surface, leading to more evaporation of methane and its photochemical products, principally ethane. Lorenz *et al.* (1999) modelled this behaviour and found that surface temperatures and pressures of 130 K and 6 bar were reached for a 10% increase in solar luminosity over the present value, assuming a large reservoir of methane was available in the crust, with access to the atmosphere (Figure 4c). If there are no volatiles near the surface so this sort of positive feedback is absent, the sensitivity to changes in solar output is much less; Lorenz *et al.* (1997b) find a surface temperature increase of only about 6 K for a doubling in solar irradiance. In either case, changes in the solar spectrum can be expected to change the haze amount and its radiative properties, producing climate feedback that is difficult to quantify reliably given our present ignorance of the cloud composition and microphysics.

Lorenz *et al.* (1997b) went on to consider changes in solar irradiance all the way on to the red dwarf phase of the Sun, expected in another ~6 Gyr. Then, the irradiance might be about 10 times the present level, which in the simple radiative-convective model, with no other changes, predicts a surface temperature of around 177 K (Figure 4d). In this temperature range, other volatiles such as ammonia, which is likely to be abundant on Titan but currently frozen, could start to contribute to the greenhouse opacity leading to further warming. In some scenarios, near Earth-like temperatures could eventually prevail on Titan.

## 4. Summary and Conclusions

Simple radiative-convective models provide a reasonable approximation to the mean climate on all three planetary bodies Venus, Titan and Earth. With such a model, we can estimate the effect on mean surface temperature of such climate forcing factors as change in total insolation, cloud properties expressed as Bond albedo, and atmospheric composition. The last of these is conjectured to vary due to changes in active volcanism on Venus, by anthropogenic pollution on Earth, and by depletion of the methane inventory on Titan.

The sensitivity of the model surface temperature to small changes in insolation is proportional to the third power of the radiative equilibrium temperature $T_e$, which in turn is proportional to the solar irradiance $S$ and the albedo $A$. The calculated effect of a 0.1% change in either $S$ or $A$ is <1 K on all three planets.

On Venus, the surface temperature and pressure seem to be held at their present high values by the buffering effect of the temperature dependence of the Urey reaction, which regulates the amount of atmospheric $CO_2$ in equilibrium with carbonates and silicates on the surface. The factor most likely to lead to significant shifts in the equilibrium point is changes in atmospheric composition due to the eventual decline in what appears currently to be a very active volcanic regime. It seems possible to speculate that, when this happens, the initial cooling will trigger the removal of most of the atmospheric carbon dioxide by carbonate rock formation, eventually reaching a more Earth-like climate on Venus. For example, should the Venusian atmosphere become identical to the Earth's in every respect except for the oxygen content, the surface would be around 50 K warmer than Earth due to the larger solar irradiance.

On Titan, the main amplifying factor for climate change is the evaporation and condensation of volatiles, most notably methane. For larger negative temperature excursions the main atmospheric constituent, nitrogen, can condense and, in the other, warmer, direction, significantly increased greenhouse gas contributions from higher hydrocarbons, ammonia, and even water vapour are possible. The most likely fate for the planet-sized moon is an eventual decline in the emission of methane into the atmosphere from surface lakes or from the interior. Then the greenhouse will collapse in a few million years and most of the atmosphere will freeze on the surface, as already happened on Neptune's large satellite Triton. A very long time into the future, major changes in the Sun could regenerate and enhance the atmosphere, possibly even achieving Earth-like temperatures at the surface.

## References


Adamcik, J. A., and Draper, A. L.: 1963, 'The temperature dependence of the Urey Equilibrium and the problem of the carbon dioxide content of the atmosphere of Venus', *Plan. Sp. Sc.* **11**, 1303–1307.

Bullock, M. A., and Grinspoon, D. G.: 1996, 'The stability of climate on Venus', *J. Geophys. Res.* **101**, 7521–7529.

Carlson, R. W., Kamp, L., Baines, K., Pollack, J. B., Grinspoon D., Encrenaz Th., Drossart, P., and Taylor, F. W.: 1993, 'Variations in Venus cloud-particle properties: A new view of Venus's cloud morphology as observed by the Galileo near infrared mapping spectrometer', *Plan. Sp. Sc.* **41**, 477–486.

Esposito, L. W.: 1984, 'Sulfur dioxide: Episodic injection shows evidence for active Venus volcanism', *Science* **223**, 1072–1074.

Fulchignoni, M., Ferri, F. *et al.*: 2005, 'In situ measurements of the physical characteristics of Titan's environment', *Nature Online*, 30 November 2005.

Goody, R. M., and Yung, Y. L.: 1989, *Atmospheric Radiation*, Oxford University Press. G

Hashimoto, G. L., and Abe, Y.: 2005, 'Climate control on Venus: Comparison of the carbonate and pyrite models', *Plan. Sp. Sc.* **53**, 839–848.

Hashimoto, G. L., Abe, Y., and Sasaki, S.: 1997, '$CO_2$ amount on Venus constrained by a criterion of topographic-greenhouse instability', *Geophys. Res. Lett.* **24**, 289–292.

Knollenberg, R. G., and Hunten, D. M.: 1980, 'The microphysics of the clouds of Venus: Results of the Pioneer Venus particle size spectrometer experiments', *J. Geophys. Res.* **85**, 8039–8058.

Lorenz, R. D., McKay C. P., and Lunine J. I.: 1997a, 'Photochemically-driven collapse of Titan's atmosphere', *Science* **275**, 642–644.

Lorenz, R. D., Lunine, J. I., and McKay, C. P.: 1997b, 'Titan under a red giant sun: A new kind of "Habitable" world', *Geophys. Res. Lett.* **24**, 2905–2908.

Lorenz, R. D., McKay C. P., and Lunine J. I.: 1999, 'Analytic investigation of climate stability on Titan: Sensitivity to volatile inventory', *Plan. Sp. Sc.* **47**, 1503–1515.

McKay, C. P., Pollack, J. B., and Courtin, R.: 1989, 'The thermal structure of Titan's atmosphere', *Icarus* **80**, 23–53.

Nimmo, F., and McKenzie, D.: 1998, 'Volcanism and tectonics on Venus', *Ann. Rev. Earth Plan. Sc.* **26**, 23–51.

Taylor, F. W.: 2005, *Elementary Climate Physics*, Oxford University Press.

Urey, H. C.: 1952, *The Planets*, Yale University Press, New Haven.

# THE ORBITAL FORCING OF CLIMATE CHANGES ON MARS

F. MONTMESSIN
*Service d'Aéronomie du CNRS, Réduit de Verrières, Route des Gatines, 91371 Verrières le Buisson, France*
*(E-mail: montmes@aero.jussieu.fr)*

(Received 24 January 2006; Accepted in final form 25 April 2006)

**Abstract.** Solar variability influences the climate of a planet by radiatively forcing changes over a certain timescale; orbital variations of a planet, which yield similar solar forcing modulations, can be studied within the same scientific context. It is known for Earth that obliquity changes have played a critical role in pacing glacial and interglacial eras. For Mars, such orbital changes have been far greater and have generated extreme variations in insolation. Signatures associated with the presence of water ice reservoirs at various positions across the surface of Mars during periods of different orbital configurations have been identified. For this reason, it has been proposed that Mars is currently evolving between ice ages. The advent of climate tools has given a theoretical frame to the study of orbitally-induced climate changes on Mars. These models have provided an explanation to many puzzling observations, which when put together have permitted reconstruction of almost the entire history of Mars in the last 10 million years. This paper proposes to give an overview of the scientific work dedicated to this topic.

**Keywords:** planetary sciences, mars, climate, water cycle, milankovitch cycle

## 1. Introduction

After the analysis of the data returned by the Mariner 4 probe sent to Mars in the early 1960s, there was not much room left for the belief that Mars could currently host life. Besides the discovery of a tenuous $CO_2$ atmosphere (surface pressure $\sim$6 mbar), the surface appeared covered by billion year old craters which had remained almost pristine since the last heavy bombardment. At first glance, water had left landscapes almost intact, and Mars looked little different from our moon. This view has changed substantially over the decades that followed these first discoveries. Thanks to a succession of successful planetary missions, it eventually became evident that running water had carved a myriad of channels and river valleys. These water-related landforms were revealed by increasingly higher resolution pictures of the martian terrains. Recently, dendritic valleys presumably associated with precipitation episodes have been discovered in the equatorial regions (Mangold *et al.*, 2004), strongly supporting the idea that Mars once had an active hydrological cycle, with liquid water exchanging between the atmosphere, the surface and the subsurface. However, these features refer to the early youth of Mars when most of the atmosphere (several bars of pressure) was still present, and various greenhouse gases probably helped maintain the surface temperature above the melting point of


DOI: 10.1007/s11214-006-9078-x

water (Pollack *et al.*, 1987; Kasting, 1997; Sagan and Chyba, 1997). Aeons have passed since then, during which the atmosphere has progressively lost most of its original mass, making thermodynamical conditions inappropriate for liquid water. And while remnants of localized sporadic discharges have been localized (Malin and Edgett, 2000), large flows of liquid water have certainly not run on the surface in the recent past. It is water ice instead which has left numerous marks of its work on the geology. The picture is now growing that recent Mars ages (the last hundreds of millions of years) have seen multiple events of ice mobilization across the planet. Glaciers, debris aprons, and other ice-flow features are observed at several locations. With the aid of crater counting, many of these geomorphological features have been dated to around several millions of years, which incidentally corresponds to the last excursion of Mars' rotation axis into a regime of high obliquity. This is not a coincidence. Many authors have hypothesized that the planet has encountered multiple climatic changes related to the cyclic variations of its orbital parameters and that this has probably shaped a lot of the Mars we now know (Mustard *et al.*, 2001; Head *et al.*, 2003).

With the accumulation of data, models of various kinds have been developed to interpret observations. Our understanding of Mars history has made considerable progress with the advent of climate simulation tools. Originally developed for Earth, they were later modified to accommodate the specific martian conditions. Compared to Earth however, Mars models have to represent far fewer climatic components and thus suffer less complexity. While Earth climate modelers must deal with the impact of oceans, vegetation and human activity, simulating Mars climate essentially amounts to simulating conditions of a gigantic desert. In turn, Mars offers a good opportunity to exercise models by testing some of their basic foundations. These numerical tools have improved to the point that they now incorporate a fully coupled and self-consistent description of atmospheric circulation, radiative transfer, tracer-related processes and surface interactions. First used to investigate current climate, it has been realized that simply changing the model parameters controlling the orbit would provide an easy and natural spin-off to study Mars recent past climates (those which prevailed during the last hundreds of millions of years). This has been initiated near the end of the 1990s (Haberle *et al.*, 2000), and the interest of the scientific community has never stopped growing since.

In this paper, I attempt to describe the current state of knowledge about recent climatic changes on Mars and their effects on the spatial distribution of water at the surface. This is particularly timely as many pieces of the puzzle have been recently put together thanks to the use of climate simulation tools. As we shall see, Mars is far from being the dead planet some used to think it was.

## 2. Today's Mars Water Cycle

The current martian atmosphere is very cold (200 K) and dry. Only a thin layer of a few microns would form at the surface if all the water vapor precipitated out. The only known reservoir that appears able to exchange with the atmosphere is located at the North Pole and is commonly referred to as the Northern Permanent Cap (NPC). 3 km thick, 1000 km in diameter (Figure 1, left panel), it would create a global ocean of a few meters deep if melted. When exposed to sunlight in spring/summer ($L_S$ between 60°and 150°), the cap surface warms up to 240 K, and water vapor is forced to sublimate. Despite a sluggish polar atmospheric circulation, water vapor spreads out equatorward, ultimately reaching southern hemisphere latitudes (see Figure 2 for the seasonal water vapor cycle). This transfer to the south is made possible by an

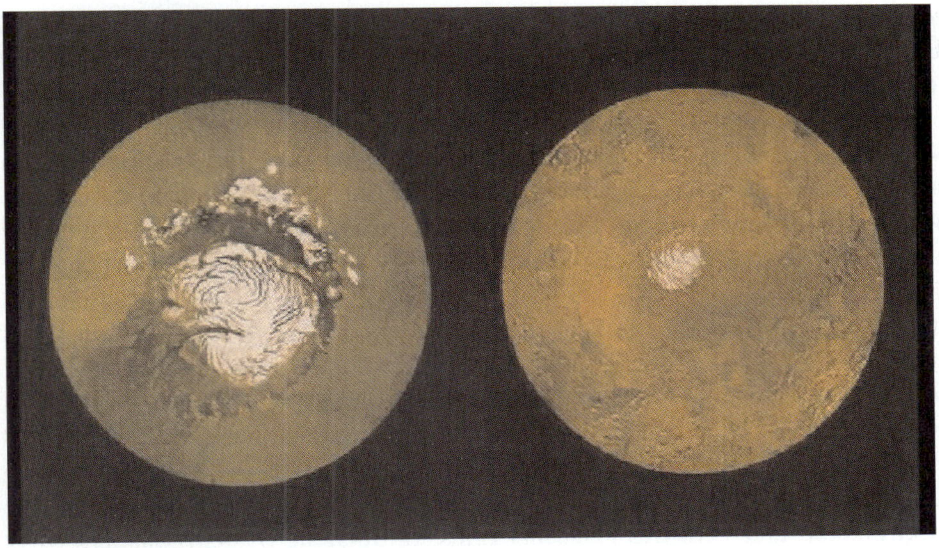

*Figure 1.* The northern polar cap (left) is composed of water and dust, and the surrounding terrain is relatively smooth. The southern cap (right) is smaller, and is thought to contain both water and carbon dioxide. These pictures are mosaics of images taken by the NASA Viking Orbiter spacecraft, and show each hemisphere from 65° latitude to the poles. Figure reprinted from Gierasch (Nature "News and Views", 2002).

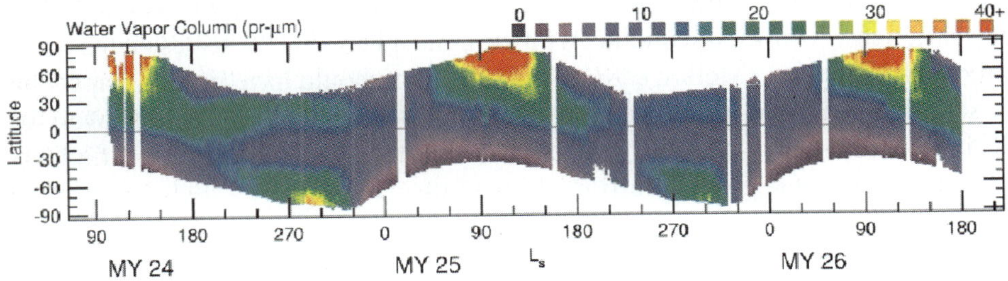


*Figure 2.* Multi-annual variations of water vapor abundances (in precipitable microns) in the martian atmosphere as observed by the Thermal Emission Spectrometer onboard Mars Global Surveyor. The $x$-axis gives the evolution of time expressed in aerocentric solar longitude, $L_S$. In this time reference, a Martian year corresponds to 360°, and northern summer solstice occurs at $L_S = 90°$.

overturning atmospheric cell extending between the tropics. A Hadley cell indeed fully develops at both solstices and reverses its orientation during a short season around each equinox. The seasonal cycle of water is characterized by a factor of 2 change in total humidity (Smith, 2002), with the largest fluctuations found at high latitudes where seasonal contrasts are strongest. Interestingly, summertime water vapor extraction from the cap is almost exactly balanced by a seasonal return flow occurring shortly before sublimation season. This has been first predicted by Global Climate/Circulation Model (GCM) computations (Houben *et al.*, 1997; Richardson and Wilson, 2002a) and is now confirmed by several observations showing the recurrent presence of an annulus of water ice slowly creeping to the poles during the spring retreat of the $CO_2$ seasonal frost (Titus and Kieffer, 2003; Bibring *et al.*, 2005). This return of water closes the annual cycle which thus seems to evolve in a quasi-stationary state.

While it was rapidly discovered that the NPC was made of water ice an dust (Kieffer *et al.*, 1976), the nature of the south residual cap (Figure 1, right panel) has long remained a subject of controversy. Thermal emissions from the surface indicate a temperature close to $CO_2$ saturation all year long, the result of an annually residing $CO_2$ ice cover slightly offset from the pole. The debate concerned the actual thickness of the residual $CO_2$ cap – was it thin? And if so, could it buffer the $CO_2$ atmospheric reservoir? Theoretical work (Byrne and Ingersoll, 2003) has suggested the $CO_2$ ice is only a few meters deep and stacked upon what can be considered the "true" component of the south residual cap: water ice. Recent mapping in the near-infrared (where $H_2O$ and $CO_2$ ices can be identified and separated) by OMEGA (Observatoire pour la Minéralogie, l'Eau, les Glaces et l'Activité) onboard Mars Express and earlier mapping in the far infrared by TES (Thermal Emission Spectrometer) and THEMIS (Thermal Emission Imaging System) onboard Mars Global Surveyor (Titus *et al.*, 2003) confirm this: the $CO_2$ residual layer indeed forms a thin veneer above a much thicker layer of water ice which actually extends slightly beyond the limits of $CO_2$. Water ice is even seen at the surface away from the cap as detached and extended bright patches.

Besides the presence of the caps at the poles, water also resides in the shallow subsurface of the high latitude regions in both hemispheres (Feldman *et al.*, 2004). This buried reservoir should interact with the atmosphere on timescales on the order of thousands of years (Mellon *et al.*, 2004) and thus should have limited impact on the seasonal evolution of the water cycle (Böttger *et al.*, 2005). The discovery of subsurface water ice in such large quantities constitutes one of the greatest findings in recent years, yet begs the question of its origin.

## 3. Mars and Its Shaking Orbit...

If compared to the few watts change in the Earth energetic balance caused by human release of greenhouse gases, orbital cycles on Mars set a much higher standard of

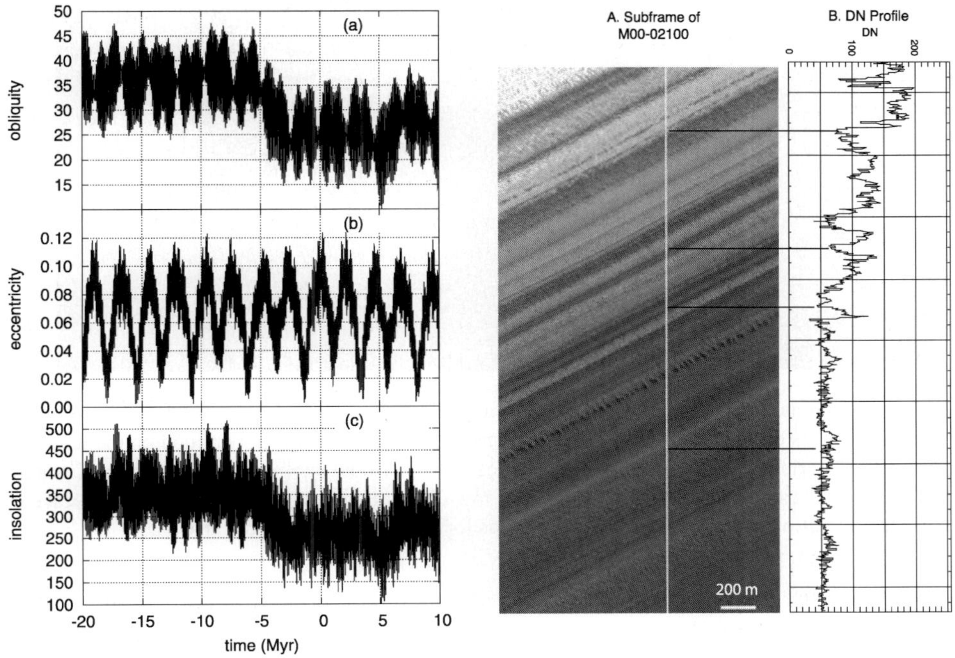


*Figure 3. Left panel:* (a) Obliquity, (b) eccentricity, and (c) insolation at the north pole surface at the summer equinox over the last 20 Myr and for the next 10 Myr. Figure reprinted from Laskar *et al.* (2004). *Right panel:* (a) Subframe of a MGS/MOC image showing an example of the succesion of bright and dark layers down the trough walls of the north polar layered deposits. (b) DN profile extracted from the same MOC image along the white line traced in the image. Reprinted from Milkovich and Head (2005).

climate perturbations. As computed by Laskar and Robutel (1993) from numerical integration of the motion of the solar system bodies, planetary secular perturbations cause the orbit of Mars to experience large variations, largely because no moon can stabilize it like it does for Earth. Over the last 100 million years (Myr henceforth), Mars' obliquity has evolved in a chaotic zone ranging from 0° to 60°. Obliquity oscillations can nevertheless be reconstructed for the last 20 Myr (Laskar *et al.*, 2004; Figure 3, left panel). Changing inclination of the rotation axis has caused insolation at the poles to vary by more than a factor of three during this period. Five million years ago, when obliquity was oscillating around a higher value (35°) than today (25°), pole insolation was about 50% greater than today.

Mars orbit changes are generally represented by their impact on the summer insolation at the poles since the only reservoirs of water directly in contact with the atmosphere are located there. The sequence of bright and dark layers found within the scarps and troughs of the NPC indicate variation in erosional and depositional rates of water ice, likely reflecting cyclic insolation conditions (Laskar *et al.*, 2002; Milkovich and Head, 2005; Figure 3, right panel). In the polar regions, the solar zenith angle in summer decreases when obliquity increases and surface

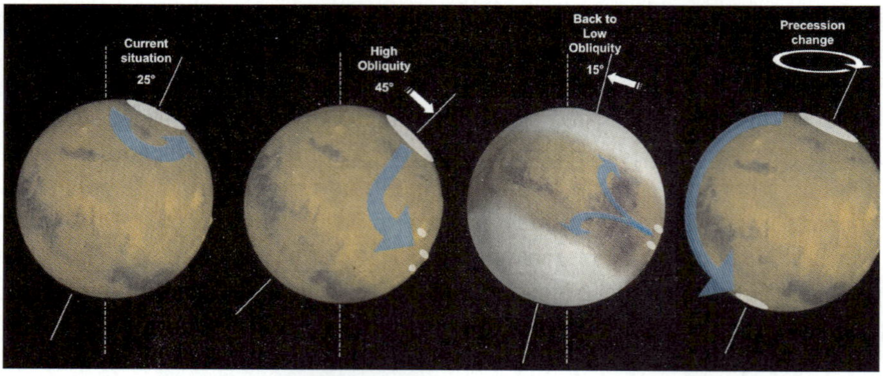


*Figure 4.* A summary of the recent (last millions of year) changes in the Mars orbital parameters and their consequences on the mobilization of water across the planet. In the first cartoon on the left, the blue arrow is reoriented to the north pole to indicate the seasonal sublimation and return of water to the permanent cap. All other cases with high, low obliquity and precession changes generate a permanent extraction of water from the cap towards locations indicated by the arrows.

temperatures are raised accordingly. It is estimated that the temperature at the North pole in summer (currently 240 K) would reach or exceed 270 K at obliquities greater than 45°. By virtue of the Clausius-Clapeyron law, this 30 K difference is exponentially translated into one or two orders of magnitude larger vapor pressure of water and thus into a sublimation process orders of magnitude stronger for the ice at the poles (Jakosky and Carr, 1985; Haberle *et al.*, 2000; Richardson and Wilson, 2002a; Mischna *et al.*, 2003; Levrard *et al.*, 2004; Forget *et al.*, 2006).

Obliquity alone has not paced Mars climate in the recent ages; eccentricity and precession changes have yielded substantial, though milder, solar forcing variations. These two parameters are intimately related as the effect of precession on climate depends on eccentricity. Precession refers to the circular motion of the rotation axis (Figure 4) which cycles on a 50 kyr timescale. While eccentricity makes the climate seasonally asymmetric, precession determines the timing of closest approach to the Sun and thus decides which summer hemisphere receives more sunlight than the other (currently the South). Many studies (Laskar *et al.*, 2002; Hecht, 2003; Milkovich and Head, 2005) favor the precession factor, which has dominated the last 0.5 Myr cycle of insolation, as being one of the main controls on deposition of at least the first hundreds meters of the north polar terrains. This implies that even the smooth climatic changes caused by the precessing rotation axis of Mars have been sufficient to produce observable consequences on the geology. Presumably, changes due to obliquity may have been even more dramatic.

## 4. Precession Changes

One of the most puzzling aspects of martian geology is the great disparity in resurfacing ages between the polar layered deposits of the Northern and Southern hemispheres (Herkenhoff and Plaut, 2000). While the North polar layered deposits surface appears young (10–100 kyr or less), devoid of large impact craters, the South polar layered terrains has recorded impact events millions of years old (Plaut *et al.*, 1988). Astronomical forcing has no preferred pole, hence the existence of an additional mechanism of likely endogenic origin is required to produce such hemispheric asymmetry. Richardson and Wilson (2002b) have discovered part of the answer in their analysis of the annually averaged circulation pattern on Mars.

Volatile transfer between hemispheres is achieved by Hadley cells which are primarily driven by differential heating across latitudes and which tend to reduce temperature contrast by moving hot air masses to cold regions. On Mars, two components combine to make this process highly asymmetrical about the equator: topography and orbit eccentricity. The latter forces southern spring/summer to be currently 30% more exposed to the Sun than its northern counterpart as southern spring/summer seasons coincide with perihelion passage.

The argument involving topography is based on the fact that the southern hemisphere constitutes a plateau which is 2 or 3 kilometers higher than the northern plains. This has consequences for the meridional transport of volatiles. For reasons still not fully understood, circulation models indicate that the southern summer Hadley cell is many times stronger than its northern summer counterpart (Richardson and Wilson, 2002b), regardless of the eccentricity argument. This bias is in fact independent of any orbital factor and has acted since the formation of the North-to-South topographic dichotomy. Alone, the topographic forcing on circulation favors the accumulation of volatiles in the northern hemisphere and thus the storage of water and dust at the surface of the North polar region. If enhanced accumulation leads to a more effective resurfacing process, then the (Richardson and Wilson, 2002b) mechanism may have participated in making the north pole look younger than the South. This topographic bias led (Richardson and Wilson, 2002b) to speculate that water accumulation at the South pole would not be permitted on timescales shorter than 1 Myr. However, water transport does not uniquely depend on circulation strength; it also depends directly on atmospheric thermal conditions.

Martian Hadley cells are created by air masses converging in the summer tropics where they rise. At higher altitudes, they move to the opposite hemisphere where they sink into colder regions. The loop is completed as air masses cross the equator again, following a near-surface return flow motion. A similar circulation pattern is responsible on Earth for the well-known trade winds. With water, complications arise as it condenses during its ascent within the tropical upwelling zone. At aphelion (northern spring/summer), upward motion and subsequent adiabatic cooling is so efficient that the atmosphere becomes water-saturated before water vapor can reach the upper horizontal branch of the Hadley cell and be carried to the South. This large-scale condensation process has observable consequences since it produces the "Aphelion Cloud Belt" (Clancy *et al.*, 1996); i.e., a ring of clouds enshrouding the northern low latitudes. The same does not currently occur at perihelion since enhanced sunlight pushes the water condensation to sufficiently high altitudes so as to suppress cloud formation in the southern summer tropics. Clancy *et al.* (1996) proposed that when water sublimates from the pole whose summer coincides with the colder conditions of aphelion, it is forced to pool in the same hemisphere as a result of cloud precipitation and sequestration of humidity in the aphelion convergence zone. For the current orbital configuration, this process helps the northern hemisphere to retain water. This cloud-induced mechanism is called the "Clancy Effect" and has been supported and even quantified by several climate models (Richardson and Wilson, 2002a; Montmessin *et al.*, 2004).

Today, both topography and eccentricity conspire to make the North pole the preferred destination for water. However, the $CO_2$ residual cap imposes year-round freezing conditions which force water to condense and be permanently trapped when it is transported near the surface of the South pole. Stability of the residual $CO_2$ ice cap is another long-standing issue of the Mars climate, but it has been suggested that it may not survive more than a few hundreds or thousands of years (Byrne and Ingersoll, 2003). With such a $CO_2$ cold trap, a net North-to-South flux of water is unavoidable. But when the trap is absent, topography and eccentricity should have ample time to work against the presence of water ice at the South pole. If water ice instead of $CO_2$ was covering the South pole today, it would rapidly sublimate and transfer to the North (Richardson and Wilson, 2002b).

Half a precession cycle ago (25 kyr), perihelion passage was synchronized with northern summer (hereafter called "reversed perihelion"). The eccentricity effect was then favourable to volatile accumulation in the southern hemisphere and thus was opposed to the topography effect. With climate models gaining in maturity, it has become possible to confront the theory of Clancy *et al.* (1996) with that of Richardson and Wilson (2002a) to ascertain whether the cloud-induced mechanism could be strong enough in case of "reversed perihelion" to overcome the topographic forcing on circulation and let water accumulate in the South. This study was conducted with a Mars GCM developed at the Laboratoire de Météorologie Dynamique (LMD-France), where the only changes from the original version of the model were the removal of the current $CO_2$ cold trap at the south pole and a perihelion timing phased with northern summer (Montmessin *et al.*, 2004). In such a martian world, where the northern summer is warmer than the south, sublimation from the NPC is reinforced, and clouds no longer restrain the transport of water to the South. As expected by Clancy *et al.* (1996), water accumulates at the South pole where thermodynamic conditions are more favorable (Figure 5).

However, the model also predicts that the thermal structure of the atmosphere, which is highly sensitive to the abundance of airborne dust particles, would not vary seasonally with the same contrast as today. The potential for dust lifting should considerably diminish in a "reversed perihelion" climate. While Montmessin *et al.* (2004) can not yet propose a complete explanation for this behavior, it appears intuitively linked to the absence in the North of regional sources of dust as productive as in the South, like the Hellas basin where dust storms are regularly observed. Ultimately, the dust cycle, which currently has a pronounced maximum during

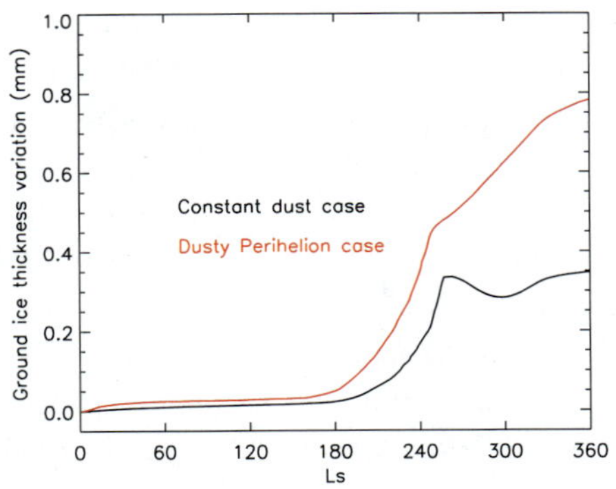


*Figure 5.* Results obtained by the model in a situation of reversed perihelion. The two curves show the thickness variation of water ice at the surface of the south pole during a martian year. In both cases, water ice accumulates at rates slightly less than 1 mm/yr. The red curve shows what would happen if the reversed perihelion climate had the same dust cycle and thus the same seasonal contrast as today. The black curve shows what would happen if the dust loading was low and had no seasonal variation, as might be expected from dust lifting predictions in a reversed perihelion climate.

southern summer, would remain constant in a reversed perihelion climate, substantially reducing the thermal contrast between the two seasons. This effect would not keep water from migrating to the south pole, rather transfer would be two or three times less efficient than in a reversed perihelion climate having the same seasonal contrast as today (see the comparison of the two curves in Figure 5).

Provided the obliquity is not too high, precession changes can be responsible for a pole-to-pole migration of water. Very recent ($\sim$10 kyr) orbital configurations may have allowed water to accumulate in the South, in contrast with the million years timescale hypothesized by Richardson and Wilson (2002a). Is the exposed water ice in the South a remnant of such precession induced water transfer? Possibly, but it requires additional understanding of the layering process in the South polar region. Does it explain the discrepant ages of the two poles? Partially, since the proposed mechanism is made asymmetric by its relation to the dust cycle. Still, the existence of a precession-induced water transport mechanism is consistent with the thickest portion of the NPC which shows strong correlation with the recent precession-dominated insolation signal. As modelled by Mischna *et al.* (2003), at times of high obliquity, precession changes could also have led to a redistribution of water between the two hemispheres, but at low latitudes.

## 5. Mars Water Cycle at High Obliquity

The effect of obliquity has long been suspected to have had strong consequences for the martian water cycle. Based on calculated insolation variations at the poles, Jakosky and Carr (1985) have suggested that episodes of high obliquity could have

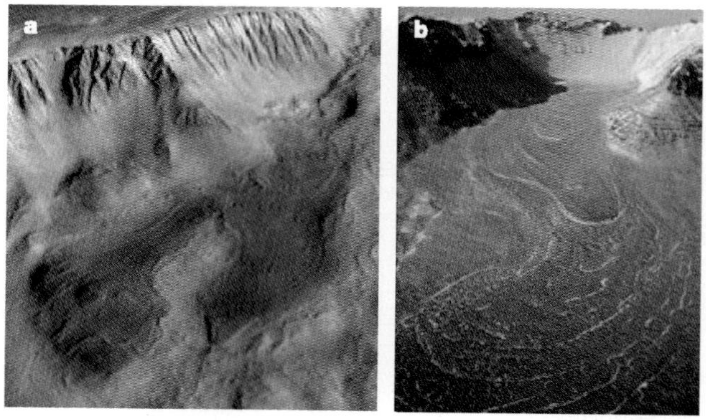


*Figure 6.* Deposits from a young lobate rock glacier at the base of the Olympus Mons scarp (138°W, 18°N). (a) Perspective view looking southwest towards the 6 km-high scarp. Note lobate deposits extending about 20–25 km from the base of the alcoves (from right centre towards lower left) darkened for emphasis. HRSC data from Mars Express. (b) Perspective view of the upper ~5 m of a debris-covered rock glacier emerging from a cirque in Mullins Valley, Antarctic Dry Valleys on Earth. Note the morphologic similarity to features at the base of Olympus Mons. Figure and caption reprinted from Head *et al.* (2005).

caused water to precipitate at low latitudes very early in the planet's history, when the Tharsis topographic bulge was not even formed.

Jakosky and Carr (1985) have probably inspired our modern view of recent climate changes on Mars, although their conclusion may have been somewhat misleading. Their study was based on simple theoretical arguments, assuming thermodynamical conditions at the surface determine the zones where water ice can be stable. At high obliquity, these zones should be confined to near the equator where Sun exposure in summer is weaker than at the poles. However, the first simulations based on a GCM set with a high obliquity configuration (Haberle *et al.*, 2000) have showed that if water indeed accumulates at low latitudes, the pattern is not uniform in longitude, but instead concentrates almost uniquely near elevated places like the Tharsis region. Mischna *et al.* (2003), who also pioneered simulation of the water cycle at high obliquity with a GCM, made similar conclusions as Haberle *et al.* (2000). However, they found a correlation with thermal inertia in addition to that with topography, a result somewhat contrasting with other climate models.

Surface properties are indeed not uniform, but albedo and thermal inertia alone are not sufficient to explain the preferred ice accumulation zones predicted by the model. In fact, another factor not accounted for by Jakosky and Carr (1985) controls these emplacements, and circulation is the key.

As illustrated by Figure 6, some Tharsis volcanoes exhibit the presence of features on their flanks that bear a striking resemblance to features created by precipitation-fed glaciers on Earth (Head and Marchant, 2003). Remarkably, these glaciers are about 4 Myr old (Neukum *et al.*, 2004), contemporaneous with the most recent regime of high obliquity on Mars (later, obliquity transitioned to lower

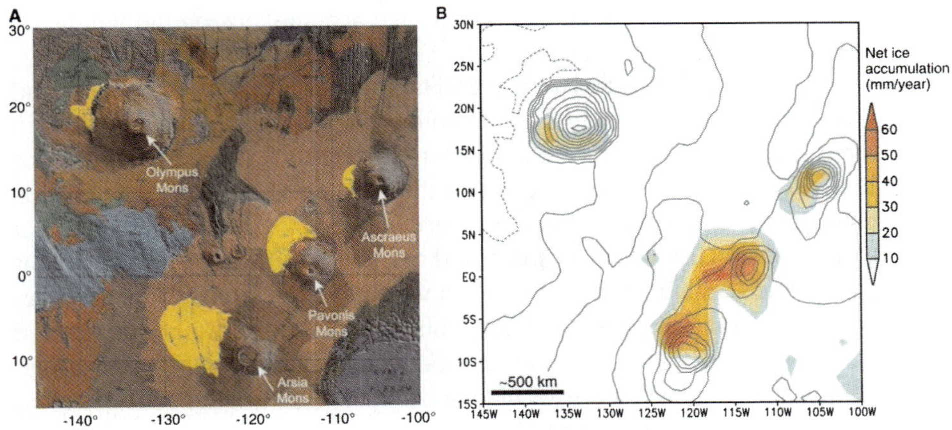


*Figure 7.* (a) Geologic map of the Tharsis region showing the location of fan-shaped deposits of Amazonian age (yellow) located on the northwest slopes of the Tharsis Montes and Olympus Mons. (b) Net surface water ice accumulation in the Tharsis region simulated with a 45° obliquity and assuming that surface water ice is present but sublimating at the North pole. Superimposed topography contours. Reprinted from Forget *et al.* (2006).

values, as at present). To account for the formation of these putative glaciers, Forget *et al.* (2006) have added to previous work by running the LMD model at much higher spatial resolution (with a grid about 8 times more refined) and by using a more sophisticated cloud model wherein cloud particle size is predicted, rather than set constant and left unchanged from its present value as in previous works (Mischna *et al.*, 2003; in the work of Haberle *et al.*, 2000, ice clouds are not modelled, the atmospheric holding capacity is limited by water saturation). If precipitation is indeed a key aspect of glacier formation, then it demands that models properly represent cloud particle fall, which depends almost uniquely on particle size. The higher spatial resolution adds a significant benefit as it allows one to resolve the dimensions of many observed glacial features, not to mention the fact that it also simulates wind patterns more accurately.

Like other GCMs, the Forget *et al.* (2006) LMD model produces a much more intense water cycle at high obliquity than it does at the present obliquity, with 30 times more water sublimating from the NPC. However, the locations of accumulation are restricted to a few small areas while other models predict much more extended and sometimes totally different zones. These areas are found on the flanks of volcanoes in the Tharsis region, but also on Olympus (Figure 7) and Elysium Mons, precisely where glaciers are thought to have developed. Provided the predicted accumulation rates existed for thousands of years, the Forget *et al.* (2006) model suggests that precipitation could have created water ice edifices hundreds of meters thick. Water only precipitates in these regions because a monsoon-like circulation pattern conveys it to the volcanoes. As this wet air blows over the flanks, it cools and condenses, forming icy particles ten times larger than in today's martian clouds. Particles rapidly precipitate, and water ice accumulates on the windward side of the volcanoes.

The Forget *et al.* (2006) study also predicts precipitation in the southern mid latitudes, explaining new observations of geologically recent ice flow features in a small area located east of the Hellas crater rim. The working hypothesis for the explanation of these features is based on the results presented in the previous section which indicate that precession changes could lead to water accumulation at the south pole. Eventually, such transfer could create a significant reservoir similar to the NPC, but this time in the South. If such was the case 5 Myr ago, when Mars was in the high obliquity regime, huge amounts of water may have been extracted from this southern ice cap, intensifying the water cycle just as did the NPC at high obliquity. This time however, water would not only precipitate in the Tharsis region but would also be deposited east of Hellas. Again, this is precisely where many geomorphological features (tongue-shaped lobes, hourglass-shaped craters filled by debris-covered glacier) have been detected, and, again, this is due to a particular circulation pattern that is imposed by the huge topographic depression of Hellas.

## 6. From High to Low Obliquity

In the previous part, we have seen how the high obliquity climate that prevailed several Myr ago has created several localized ice reservoirs outside the polar regions. Even though these sources are now extinct, they were probably active during the rapid transition from the high to low mean obliquity regimes that occured 4 Myr ago. To understand to where water could have been mobilized in such circumstances, the LMD model was used again (Levrard *et al.*, 2004). Figure 8 shows the predicted map of water accumulation when the model assumes only one source of water located in the Tharsis region and an obliquity decreased to 15°, a value reached several

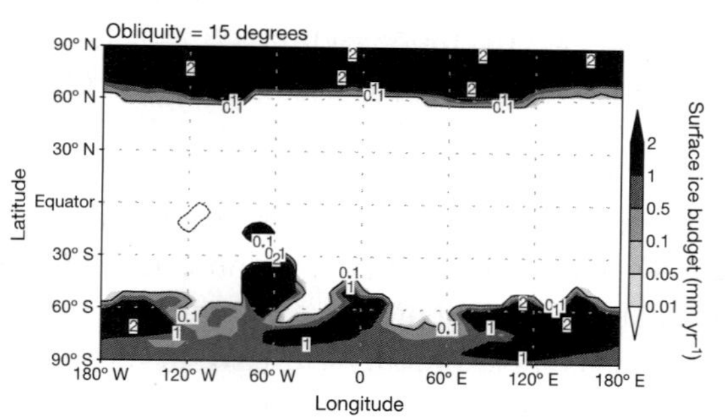


*Figure 8.* Surface water ice budget in mm per martian year after ten years of simulation for an obliquity of 15°and with an equatorial ice reservoir situated in the Tharsis region whose boundaries are indicated by a thick solid line. This reservoir is set to be the only active water source. Reprinted from Levrard *et al.* (2004).

times since the obliquity transition. The model suggests that high latitudes would become the preferred location for water at the surface. Even when the equatorial water source is exhausted, the stability pattern remains almost unchanged. Except for the outer latitudinal edges which are slightly eroded, water resides poleward of 60° in both hemispheres. Nearly 10 meters could have accumulated during an obliquity cycle of 15°, lasting 60 kyr.

In that case, atmospheric circulation does not solely determine where water is stable on the surface, but rather a simple energy balance argument can explain the result. With decreasing tilt of the rotation axis, the equatorial reservoir becomes highly exposed, and water sublimates massively in spring and summer. Released water accumulates in the fall/winter high latitude regions where temperature conditions (imposed by the presence of a seasonal $CO_2$ frost cover) favor ice stability at the surface. Deposition rates are so large compared to sublimation rates that ice is able to survive through spring and summer. Ice accumulation does not exhibit significant zonal structure even though it is a little less uniformly distributed in longitude and latitude in the southern than in the northern hemisphere. The more pronounced structure in the South is probably the result of a strong poleward stationary flow heavily loaded in humidity, escaping from the Hellas basin in spring and leading to a preferred deposition of water in the eastern hemisphere.

Again, these results find an echo with the recent observations of the Gamma Ray Spectrometer (GRS) (Mars Odyssey Mission) which have revealed the presence of subsurface water ice in the first meters of the martian regolith. Data indicate that closer to the equator, water ice is buried deeper. There is a sharp latitudinal cut-off around 60°, equatorward of which water ice is either not detected or in very low quantity. Levrard *et al.* (2004) propose that the return to present obliquity, with the equatorial source completely exhausted, desiccated the upper portion of the high latitude ice that is predicted by the model to have been deposited under conditions of low obliquity. Because water ice crystals that precipitate on the surface are formed on airborne dust particles, sublimation of such ice would leave dust particles behind and progressively create a protective lag which would act as a thermal insulator, preventing further sublimation. In this explanation, what has been seen by GRS would simply be the remainder of the water ice sublimated from equatorial glaciers at low obliquity and deposited at higher latitudes then hidden below a dust layer which thickened during the desiccation process.

Another convincing explanation has been proposed to explain the GRS discovery. Mellon *et al.* (2004) and, more recently, Mellon and Feldman (2005) showed that the GRS mapping of subsurface water ice coincides almost exactly with the stability zones of ground ice which have been deduced from a 1D regolith model. This would suggest water was emplaced by diffusion from the atmosphere through the regolith pores, instead of having been deposited at the surface during the obliquity transition. In the Mellon and Feldman (2005) explanation, climate change is not invoked since ground-ice distribution reflects the mean climatic conditions that have prevailed during the last hundred to thousand years.

## 7. Conclusion

The story of recent martian climates is slowly emerging (a summary of it is given in Figure 4). We now see Mars as a planet whose water has cycled on the surface among several specific areas during the last hundreds of millions of years. In fact, computations show that the most probable obliquity over the last 4 billions of years has been 41.8° (Laskar *et al.*, 2004), which implies that the planet has spent most of its time with the poles facing the Sun much more than today. As a consequence, current Mars probably renders a false image of its usual appearance. Until recently, active glaciers at low latitudes and no ice at the poles may have been closer to its common look.

Now that most orbital configurations have been explored, it will be hard for climate models, if they remain in their current state, to produce again such original findings. They need to evolve. Important improvements concern the implementation of several feedbacks still ignored. All the water cycle studies presented in this paper have been conducted without the effects of wind on dust lifting, without the radiative effects of water ice clouds, without the effects of latent heat... Other efforts will have to be made regarding the representation of surface modifications resulting from evaporation and deposition processes. Such modifications subsequently affect the exchange of water between the soil and the atmosphere and they also determine the aspect of layered deposits and thus have implication for our interpretation of Mars climate history.

Although models still need significant improvements, they have already provided believable, although still debated, answers to many questions posed by the Mars geology. It is fascinating to realize these tools can be used to study a planet other than that for which they were originally designed. Terrestrial models find in Mars an alternative application that suggests an underlying level of consistency. Is it now the time for Mars to give back to Earth research? Possibly. While more and more scientific projects are now turned to the characterization of climates in desert environments, it would be interesting to see what can be learned from Mars conditions that could help with the understanding of those encountered in terrestrial places. Among the possible by-products of the study of Mars is the refinement of dust/sand storm predictions, and the understanding of how mineral and condensate particles interact. This would provide more perspectives into the so-called indirect effect of aerosols on the Earth climate; as on Mars, clouds and dust are perfectly isolated from other kinds of particles. Mars, like other telluric objects, must be used as a full-scale experimentation chamber that nature has given to humanity to understand its own habitat.

## Acknowledgements

I wish to thank the organizer of this workshop for having been able to gather climate scientists from various fields around such an interesting subject. The work presented here is the result of years of effort, and I wish to thank all of the people involved in the theoretical study of Mars' climate for having inspired the essence of this manuscript. The author is indebted to K. Fishbaugh and to an anonymous reviewer whose corrections greatly improved this manuscript.

## References

Bibring, J. P., Langevin, Y., Poulet, F., Gendrin, A., Gondet, B., Berthé M., Soufflot A., Drossart, P., Combes, M., Bellucci, G., Moroz, V., Mangold, N., Schmitt, B., and the OMEGA team: 2004, 'Perennial water ice identied in the south polar cap of Mars', *Nature* **428**, 627–630.

Bibring J. P., Langevin, Y., Gendrin, A., Gondet, B., Poulet, F., Berthé, M., Soufflot A., Arvidson, R., Mangold, N., Mustard, J., Drossart, P., and the OMEGA team: 2005, 'Mars surface diversity as revealed by the OMEGA/Mars express observations', *Science* **307**, 1576–1581.

Böttger, H. M., Lewis, S. R., Read, P. L., and Forget, F.: 2005, 'The effects of the Martian regolith on GCM water cycle simulations', *Icarus* **177**, 174–189.

Byrne S., and Ingersoll, A. P.: 2003, 'A sublimation model for Martian south polar ice features', *Science* **299**, 1051–1053.

Clancy, R. T., Grossman, A. W., Wolff, M. J., James, P. B., Rudy D. J., Billawala, Y. N., Sandor, B. J., Lee, S. W., and Muhleman D. O.: 1996, 'Water vapor saturation at low latitudes around aphelion: A key to Mars climate?', *Icarus* **122**, 36–62.

Feldman, W. C., *et al.*: 2004, 'The global distribution of near-surface hydrogen on Mars', *J. Geophys. Res.* **109**, doi:10.1029/2003JE00216.

Forget, F., Haberle, R. M., Montmessin, F., Levrard, B., and Head, J. W.: 2006, 'Formation of glaciers on Mars by atmospheric precipitation at high obliquity', *Science* **311**, 368–371.

Haberle, R. M., McKay C. P., Schaeffer, J., Joshi, M., Cabrol, N. A., and Grin, E. A.: 2000, 'Meteorological control on the formation of Martian Paleolakes', In *Lunar and Planetary Institute Abstracts*, 31st Annual Lunar and Planetary Science Conference, Houston, Texas, abstract no. 1509.

Head, J. W., and Marchant, D. R.: 2003, 'Cold-based mountain glaciers on Mars: Western Arsia mons', *Geology* **31**, 641–644.

Head, J. W., Mustard, J. F., Kreslavsky, M. A., Milliken, R. E., and Marchant, D. R.: 2003, 'Recent ice age son Mars', *Nature* **426**, 797–802.

Head, J. W., Neukum, G., Jaumann, R., Hiesinger, H., Hauber, E., Carr, M., Masson, P., Foing, B., Hoffmann, H., Kreslavsky, M., Werner, S., Milkovich, S., van Gasselt, S., and the HRSC team: 2005, 'Tropical to mid-latitude snow and ice accumulation, flow and glaciation on Mars', *Nature* **434**, 346–351.

Hecht, M. H.: 2003, 'What is the time scale for orbital forcing of the Martian water cycle?', In *International Conference on Mars Abstracts*, Sixth International Conference on Mars, Pasadena, California, abstract no. 3285.

Herkenhoff, K. E., and Plaut, J. J.: 2000, 'Surface ages and resurfacing rates of the polar layered deposits on Mars', *Icarus* **144**, 243–253.

Houben, H., Haberle, R. M., Young, R. E., and Zent, A. P.: 1997, 'Modeling the Martian seasonal water cycle', *J. Geophys. Res.* **102**, 9069–9083.

Jakosky, B. M., and Carr, M. H.: 1985, 'Possible precipitation of ice at low latitudes of Mars during periods of high obliquity', *Nature* **315**, 559.

Kasting, J. F.: 1997 'Warming early Earth and Mars', *Science* **276**, 1213.

Kieffer, H. H., Martin, T. Z., Chase, S. C., Miner, E. D., and Palluconi, F. D.: 1976, 'Martian north pole summer temperatures – dirty water ice', *Science* **194**, 1341–1344.

Laskar, J., and Robutel, P.: 1993, 'The chaotic obliquity of the planets', *Nature* **361**, 608–612.

Laskar, J., Levrard, B., and Mustard, J. F.: 2002, 'Orbital forcing of the martian polar layered deposits', *Nature* **419**, 375–377.

Laskar, J., Correia, A. C. M., Gastineau, M., Joutel, F., Levrard, B., and Robutel, P.: 2004, 'Long term evolution and chaotic diffusion of the insolation quantities of Mars', *Icarus* **170**, 343–364.

Levrard, B., Forget, F., Montmessin, F., and Laskar, J.: 2004. 'Recent ice-rich deposits formed at high latitudes on Mars by sublimation of unstable equatorial ice during low obliquity', *Nature*, **431**, 1072–1075.

Malin, M. C., and Edgett, K. S.: 2000, 'Evidence for recent groundwater seepage and surface runoff on Mars', *Science* **288**, 2330–2335.

Mangold, N., Quantin, C., Ansan, V., Delacourt, C., and Allemand, P.: 2004, 'Evidence for precipitation on Mars from dendritic valleys in the valles Marineris area', *Science* **305**, 78–81.

Mellon, M. T., and Feldman, W. C.: 2005, 'Martian ground ice and equilibrium with atmospheric diffusion', *AGU Fall Meeting Abstract.*

Mellon, M. T., Feldman, W. C., and Prettyman, T. H.: 2004, 'The presence and stability of ground ice in the southern hemisphere of Mars', *Icarus* **169**, 324–340.

Milkovich, S. M., and Head, J. W.: 2005, 'North polar cap of Mars: Polar layered deposit characterization and identification of a fundamental climate signal', *J. Geophys. Res.* **110**, E01005, doi:10.1029/2004JE002349.

Mischna, M. A., Richardson, M. I., Wilson, R. J., and McCleese, D. J.: 2003, 'On the orbital forcing of Martian water and CO2 cycles: A general circulation model study with simplified volatile schemes', *J. Geophys. Res.* **108**, 5062, doi:10.1029/2003JE002051.

Montmessin, F., Haberle, R. M., and Forget, F.: 2004, 'Making water ice permanent at the South pole 25,000 years ago', In *Lunar Planet. Sci. Conference Abstracts*, 35th Lunar and Planetary Science Conference, League City, Texas, abstract no. 1312.

Mustard, J. F., *et al.*: 2001, 'Evidence for recent climate change on Mars from the identification of youthful near-surface ground ice', *Nature* **412**, 411–414.

Neukum, G., *et al.*: 2004, 'Recent and episodic volcanic and glacial activity on Mars revealed by the high resolution stereo camera', *Nature* **432**, 971–979.

Plaut, J. J., Kahn, R., Guinness, E. A., and Arvidson, R. E.: 1988, 'Accumulation of sedimentary debris in the south polar region of Mars and implications for climate history', *Icarus* **76**, 357–377.

Pollack, J. B., Kasting, J. F., Richardson, S. M., and Poliakoff, K.: 1987, 'The case for a wet, warm climate on early Mars', *Icarus* **71**, 203–224.

Richardson, M. I., and Wilson, R. J.: 2002a, 'Investigation of the nature and stability of the Martian seasonal water cycle with a general circulation model', *J. Geophys. Res.* **107**, 5031, doi:10.1029/2001JE001536.

Richardson, M. I., and Wilson, R. J.: 2002b, 'A topographically forced asymmetry in the Martian circulation and climate', *Nature* **416**, 298–301.

Sagan, C., and Chyba, C.: 1997, 'The early faint sun paradox: Organic shielding of ultraviolet-labile greenhouse gases', *Science* **276**, 1217–1221.

Smith, M. D.: 2002, 'The annual cycle of water vapor on Mars as observed by the thermal emission spectrometer', *J. Geophys. Res.* **107**, doi:10.1029/2001JE001522.

Smith, M. D.: 2004, 'Annual variability in TES atmospheric observations of Mars during 1999–2003', *Icarus* **167**, 148–165.

Titus, T. N., and Kieffer, H. H.: 2003, 'Temporal and spatial distribution of $CO_2$ snow and ice', *International Conference on Mars Abstracts*, Sixth International Conference on Mars, Pasadena, California, abstract no. 3273

Titus, T. N., Kieffer, H. H., and Christensen, P. R.: 2003, 'Exposed water ice discovered near the south pole of Mars', *Science* **299**, 1048–1051.

# Space Science Series of ISSI

1. R. von Steiger, R. Lallement and M.A. Lee (eds.): *The Heliosphere in the Local Insterstellar Medium.* 1996 ISBN 0-7923-4320-4
2. B. Hultqvist and M. Øieroset (eds.): *Transport Across the Boundaries of the Magnetosphere.* 1997 ISBN 0-7923-4788-9
3. L.A. Fisk, J.R. Jokipii, G.M. Simnett, R. von Steiger and K.-P. Wenzel (eds.): *Cosmic Rays in the Heliospehre.* 1998 ISBN 0-7923-5069-3
4. N. Prantzos, M. Tosi and R. von Steiger (eds.): *Primordial Nuclei and Their Galactic Evolution.* 1998 ISBN 0-7923-5114-2
5. C. Fröhlich, M.C.E. Huber, S.K. Solanki and R. von Steiger (eds.): *Solar Composition and its Evolution – From Core to Corona.* 1998 ISBN 0-7923-5496-6
6. B. Hultqvist, M. Øieroset, Goetz Paschmann and R. Treumann (eds.): *Magnetospheric Plasma Sources and Losses.* 1999 ISBN 0-7923-5846-5
7. A. Balogh, J.T. Gosling, J.R. Jokipii, R. Kallenbach and H. Kunow (eds.): *Co-rotating Interaction Regions.* 1999 ISBN 0-7923-6080-X
8. K. Altwegg, P. Ehrenfreund, J. Geiss and W. Huebner (eds.): *Composition and Origin of Cometary Materials.* 1999 ISBN 0-7923-6154-7
9. W. Benz, R. Kallenbach and G.W. Lugmair (eds.): *From Dust to Terrestrial Planets.* 2000 ISBN 0-7923-6467-8
10. J.W. Bieber, E. Eroshenko, P. Evenson, E.O. Flückiger and R. Kallenbach (eds.): *Cosmic Rays and Earth.* 2000 ISBN 0-7923-6712-X
11. E. Friis-Christensen, C. Fröhlich, J.D. Haigh, M. Schüssler and R. von Steiger (eds.): *Solar Variability and Climate.* 2000 ISBN 0-7923-6741-3
12. R. Kallenbach, J. Geiss and W.K. Hartmann (eds.): *Chronology and Evolution of Mars.* 2001 ISBN 0-7923-7051-1
13. R. Diehl, E. Parizot, R. Kallenbach and R. von Steiger (eds.): *The Astrophysics of Galactic Cosmic Rays.* 2001 ISBN 0-7923-7051-1
14. Ph. Jetzer, K. Pretzl and R. von Steiger (eds.): *Matter in the Universe.* 2001 ISBN 1-4020-0666-7
15. G. Paschmann, S. Haaland and R. Treumann (eds.): *Auroral Plasma Physics.* 2002 ISBN 1-4020-0963-1
16. R. Kallenbach, T. Encrenaz, J. Geiss, K. Mauersberger, T.C. Owen and F. Robert (eds.): *Solar System History from Isotopic Signatures of Volatile Elements.* 2003 ISBN 1-4020-1177-6
17. G. Beutler, M.R. Drinkwater, R. Rummel and R. Von Steiger (eds.): *Earth Gravity Field from Space – from Sensors to Earth Sciences.* 2003 ISBN 1-4020-1408-2
18. D. Winterhalter, M. Acuña and A. Zakharov (eds.): *"Mars" Magnetism and its Interaction with the Solar Wind.* 2004 ISBN 1-4020-2048-1
19. T. Encrenaz, R. Kallenbach, T.C. Owen and C. Sotin: *The Outer Planets and their Moons.* ISBN 1-4020-3362-1
20. G. Paschmann, S.J. Schwartz, C.P. Escoubet and S. Haaland (eds.): *Outer Magnetospheric Boundaries: Cluster Results* ISBN 1-4020-3488-1
21. H. Kunow, N.U. Crooker, J.A. Linker, R. Schwenn and R. von Steiger (eds.): *Coronal Mass Ejections* ISBN 978-0-387-45086-5

22. D.N. Baker, B. Klecker, S.J. Schwartz, R. Schwenn and R. von Steiger (eds.): *Solar Dynamics and its Effects on the Heliosphere and Earth* ISBN 978-0-387-69531-0
23. Y. Calisesi, R.-M. Bonnet, L. Gray, J. Langen and M. Lockwood (eds.): *Solar Variability and Planetary Climates* ISBN 978-0-387-48339-9

---

Springer – Dordrecht / Boston / London